FLORA ZAMBESIACA

Flora terrarum Zambesii aquis conjunctarum

VOLUME SIX: PART FIVE

FLORA ZAMBESIACA

MOZAMBIQUE, MALAWI, ZAMBIA, ZIMBABWE, BOTSWANA

VOLUME SIX: PART FIVE

Author

D.J. NICHOLAS HIND *(Royal Botanic Gardens, Kew, U.K.)*

Edited by

B.F.P. LOEUILLE & M.A. GARCÍA
Royal Botanic Gardens, Kew
Real Jardin Botánico (CSIC), Madrid, Spain

on behalf of the Editorial Board:

FRANCES CHASE
National Botanical Research Institute, Windhoek, Namibia

MARTIN CHEEK
Royal Botanic Gardens, Kew, U.K.

DAVID CHUBA
School of Natural Sciences, UNZA, Lusaka, Zambia

IAIN DARBYSHIRE
Royal Botanic Gardens, Kew, U.K.

DAVID GOYDER
Royal Botanic Gardens, Kew, U.K.

SHAKKIE KATIVU
University of Zimbabwe, Harare, Zimbabwe

JAMESON SEYANI
Herbarium and National Botanical Gardens, Malawi

Published by the Royal Botanic Gardens, Kew
for the Flora Zambesiaca Managing Committee
2025

© The Flora Zambesiaca Managing Committee, Kew 2025
Illustrations copyright © contributing artists

Citation: Hind, D.J.N. (2025). Compositae (part 5).
In: B.F.P. Loeuille & M.A. García (eds), *Flora Zambesiaca*, Vol. 6(5). Royal Botanic Gardens, Kew.

First published in 2025 by
Royal Botanic Gardens, Kew,
Richmond, Surrey, TW9 3AE, UK
www.kew.org

Distributed on behalf of the Royal Botanic Gardens, Kew in North America by the University of Chicago Press, 1427 East 60th Street, Chicago, IL 60637, USA

ISBN 978-1-84246-808-1
eISBN 978-1-84246-809-8

British Library Cataloguing in Publication Data
A catalogue record for this book is available from the British Library

Design and page layout: Nicola Thompson, Culver Design
Production management: Georgina Hills

Printed in the UK by Halston

EU Authorised Representative: Easy Access System Europe Oü, 16879218.
Mustamäe tee 50, 10621, Tallinn, Estonia (email: gpsr.requests@easproject.com).

For information or to purchase all Kew titles please visit
shop.kew.org/kewbooksonline or email publishing@kew.org

Kew's mission is to understand and protect plants and fungi, for the wellbeing of people and the future of all life on Earth.

Kew receives approximately one third of its running costs from Government through the Department for Environment, Food and Rural Affairs (Defra). All other funding needed to support Kew's vital work comes from members, foundations, donors and commercial activities including book sales.

CONTENTS

LIST OF TRIBES INCLUDED IN VOLUME 6, PART 5   vi

GENERAL SYSTEMATIC TEXT   6,5: **1**

INDEX TO BOTANICAL NAMES   6,5: **213**

FAMILY 97. **COMPOSITAE**

LIST OF TRIBES INCLUDED IN VOLUME 6, PART 5

12. Calenduleae
13. Heliantheae
14. Eupatorieae

No new names or combinations published in this part.

97. COMPOSITAE

by D.J. Nicholas Hind

Preface to Flora Zambesiaca Compositae parts 6(2–5)[1]

Since the publication of Flora Zambesiaca **6**(1) in 1992 (where Pope provided a general conspectus to the arrangement of tribes in the then remaining two parts, and Jeffrey & Pope provided a key to the tribes found in the Flora area), much has happened in Compositae research.

Bremer, Asterac. Cladist. Classif. (1994), provided the first of the modern treatments of the family, recognising 17 tribes in three subfamilies, with an estimate of 1535 genera and 23,000 known species ('excluding microspecies'). His treatment was an alphabetical coverage of genera within each of the infratribal groupings (each genus provided with a synoptic description highlighting 'unusual' or 'diagnostic' characters in italics), but lacked keys. With partially systematic partially alphabetical, tribal treatments, the Compositae volume in Kubitzki's Families and genera of vascular plants (FGVP) (Kadereit & Jeffrey, Fam. Gen. Vasc. Pl. **8**, [2006] 2007) provided the keys, but the tribal, infratribal, and infrafamilial taxa had dramatically changed since both Bremer's and Pope's volumes. In Jeffrey's opening summary the number of genera had increased to 'over 1,600' (1620 were treated), and subfamilies to four; treatments in the volume had increased the number of tribes to 30. Funk et al.'s 2009 'Compositae' volume (Funk *et al.*, Syst. Evol. Biogeogr. Compositae, 2009), following on from the conference on 'Systematics & Evolution of the Compositae' in Barcelona in 2006, summarized subsequent changes. Five subfamilies and 40+ tribes were recognized – too much to be able to communicate easily to students attempting to come to grips with such a large family: Jeffrey & Pope's tribal key in Flora Zambesiaca 6(1) was just under two pages (Royal Octavo); Jeffrey's global key to the family in the FGVP volume ran to eight pages, of two columns per page (of A4).

We are essentially following Pope's conspectus for tribal order (and ignoring subfamilial arrangements), but it is now convenient to recognize two additional tribes – the Platycarpheae and Athroismeae. The Platycarpheae is only represented by *Platycarphella*. The genera of the Athroismeae (*Anisopappus*, *Artemisiopsis*, *Athroisma*, *Blepharispermum*) are now placed after the Inuleae s.s. where part (*Anisopappus* and *Artemisiopsis*) would key out in Jeffrey & Pope's tribal key. Purely to maintain a manageable production line the coverage of tribes in remaining part are proposed as follows:

Volume 6 part 2	*Volume 6 part 3*	*Volume 6 part 4*	*Volume 6 part 5*
6. Platycarpheae	8. Athroismeae	11. Senecioneae	12. Calenduleae
7. Inuleae	9. Astereae		13. Heliantheae
	10. Anthemideae		14. Eupatorieae

Publication of these parts will be taking place in reverse order as they are completed.

The Inuleae are treated sensu lato (including former tribal splits of the Plucheeae and the Gnaphalieae). The Heliantheae are treated sensu lato (including the Helenieae) and not split as under following the proposals of Panero (see his accounts in the FGVP volume)

---

[1] By D.J.N. Hind and B. Loeuille

where 20 tribes were recognized, instead of one. The other two genera of the Athroismeae (*Athroisma* and *Blepharispermum*) key out next to the Heliantheae because of the presence of black, carbonized achenes. We were tempted to opt for Jeffrey's proposal to go to the level of recognising 'supersubtribes' because of the conceived problem with the Eupatorieae being nested within the Perityleae-Millerieae-Madieae clade in the so-called 'Heliantheae alliance'! However, this was a step too far: the concept of the Eupatorieae is one that is easy to communicate to students; how many know of supersubtribes?

As originally acknowledged by Pope, the 'foundations [of several generic treatments] were laid by Professor H. Wild', with the significant exception of the Senecioneae, which has been worked on 'from scratch'. Hiram Wild's 'foundations' are also gratefully acknowledged in the remaining treatments where published, or manuscript, accounts were available. There are, needless to say, many generic and tribal revisions that have been published since Wild's accounts appeared in Kirkia. Full advantage has been taken of these, but, in addition, taxonomic problems unearthed in writing several accounts has resulted in several synopses being produced in advance of the Flora accounts. One or two modified concepts will be presented where necessary.

1. Capitula ligulate; corollas all ligulate and ligule with 5 apical teeth . . . . . . **4. Lactuceae**
- Corollas pseudoligulate or radiate, ray limb with 4 or less apical teeth, and disc corollas actinomorphic, or all florets actinomorphic and capitula discoid, or all corollas actinomorphic and outer radiant, or capitula disciform and outer florets filiform and female and disc florets cylindrical and hermaphrodite or functionally male. 2
2. Style arms usually long, well exserted from corolla and with conspicuous papillose appendages; capitula homogamous and discoid; florets all hermaphrodite; mature achenes black . . . . **14. Eupatorieae**
- Style arms short or long, but lacking papillose appendages; capitula heterogamous and radiate, radiant, or disciform, or homogamous and discoid; mature achenes of various colours, sometimes black 3
3. Phyllaries uniseriate, coherent by overlapping margins, or partially or wholly connate, calyculate or ecalyculate; pappus present 4
- Phyllaries imbricate, 2- or more- seriate, free or connate, if uniseriate then free or pappus absent, or capitula unisexual or achenes densely villous and long-hairy from base 5
4. Phyllaries with evident elongated oil glands; achenes black when mature . . . .
  . . . . **13. Heliantheae** (*Tagetes*, cult. *Dyssodia*)
- Phyllaries lacking elongated oil glands; achenes brownish when mature . . . .
  . . . . **11. Senecioneae**
5. Style arms long, gradually attenuate-acute, short-hairy abaxially; style shaft similarly hair on upper part; capitula homogamous, and florets hermaphrodite . . . . . . **3. Vernonieae**
- Style and capitula not showing above combination of character states, or capitula unisexual 6
6. Capitula with all or only outer floret corollas bilabiate, corollas with 3-toothed outer lip and 2-lobed inner lip . . . . **1. Mutisieae**
- Capitula with florets lacking any bilabiate corollas 7
7. Capitula homogamous and discoid; phyllaries lacking scarious unlobed appendages or apices, and lacking scarious markings, or capitula heterogamous and radiant with outer florets enlarged and sterile, or capitula unisexual and plants dioecious; phyllary appendages, if present, spiniform and/or pinnately divided 8
- Capitula heterogamous and radiate or disciform but not radiant, or capitula homogamous and phyllaries with unlobed scarious often white or coloured appendages or apices, or with scarious usually brownish margins, but not with spiniform and/or pinnately divided appendages; or capitula unisexual and plants monoecious. 13

8. Leaves spiny or bristly-spiny at least towards base. 9
 – Leaves spineless or not bristly-spiny 10
9. Anthers tailed; receptacle densely setose. . . . . **2. Cardueae**
 – Anthers tail-less; receptacle alveolate with fringed alveolae . . . . **5. Arctotideae**
10. Leaves rosettiform; inflorescences of a dense glomerule of small capitula on crown of plant . . . . **6. Platycarpheae**
 – Leaves alternate or opposite; inflorescences of solitary capitula or capitula variously aggregated into lax to relatively dense, variously branched, inflorescences, or if glomerulate then not sessile on crown of plant 11
11. Corolla lobes of tubular florets much longer than wide; leaves alternate; capitula of more than one floret, not aggregated into glomerules 12
 – Corolla lobes about as long as wide, or if much longer than wide then leaves opposite; capitula sometimes of only one floret, or capitula aggregated into glomerules 14
12. Receptacle densely setose; inner florets hermaphrodite, outer sterile and usually corollas radiant . . . . **2. Cardueae**
 – Receptacle naked or scaly, not setose; all florets hermaphrodite or unisexual, corollas never radiant . . . . **1. Mutisieae**
13. Style arms of disc florets connate in lower part, connate part thicker than style shaft and abruptly marked off from it by a ring of short hairs at base; capitula always hermaphrodite. . . . . **5. Arctotideae**
 – Style arms of disc florets long to short, or absent, but connate lower part of style arms if present thicker than style shaft, or capitula unisexual and plants monoecious 14
14. Receptacle scaly and leaves opposite, or leaves alternate and mature achenes black and/or capitula glomerulate, or receptacle not scaly and pappus of scales and/or leaves opposite and/or mature achenes black and/or capitula unisexual and/or capitula glomerulate; pappus of scales or often barbellate coarse setae or bristles but not of slender setae, or if of slender setae then setae plumose. 15
 – Receptacle not scaly (although sometimes hairy or fimbrillate), or scaly and leaves alternate (although sometimes opposite on upper stem) and achenes variously brown, sometimes dark brown, or pale-coloured when mature, or capitula not unisexual; capitula glomerulate or inflorescences variously branched and capitula lax or dense on branches; if lower leaves opposite then pappus of slender barbellate setae, at least in disc florets, or pappus absent, receptacle naked and capitula not glomerulate 16
15. Anthers tailed. . . . . **8. Athroismeae** (*Athroisma, Blepharispermum*)
 – Anthers not tailed . . . . **13. Heliantheae**
16. Style arms of hermaphrodite or functionally male florets each with a subulate to triangular papillose appendage. 17
 – Style arms of hermaphrodite or functionally male florets acute to rounded, or truncate and fringed with short hairs or papillae, or short-conical at apex with subdistal fringe of hairs, unappendaged; or style undivided. 18
17. Receptacle alveolate; pappus a deeply and unequally laciniate corona or cupule; capitula disciform . . . . **13. Heliantheae**
 – Receptacle usually alveolate or foveolate, usually glabrous, sometimes fimbriate or paleaceous; pappus usually of barbellate capillary setae, sometimes reduced to scales or awns, or pappus absent; capitula radiate, cryptically radiate, discoid (and then capitula unisexual), or disciform (but then pappus never a corona) . . . . **9. Astereae**
18. Phyllaries with scarious usually brown often erose margins, not appendaged; leaves often pinnatipartite; style arms apically truncate and fringed, or style undivided and truncate; pappus absent, or if present then a short lacerate crown or lobed auricle, or leaves pinnatipartite and outer pappus of 5 white petaloid scales . . . . **10. Anthemideae**
 – Phyllaries green and herbaceous (though often with hyaline margins), or with scarious membranous often white or coloured appendages or apices, or rarely appendages foliose;

leaves never pinnatipartite; pappus of hairs or scales present in at least some florets, or if absent then style arms acute, obtuse, rounded or truncate, sometimes with external sweeping hairs, but not truncate and fringed. 19

19. Pappus absent; fruits large, curved or angular, or winged achenes, or smooth drupes; capitula radiate . . . .**12. Calenduleae**

– Pappus present or absent; fruits small achenes, fusiform, rarely distinctly beaked, terete, ribbed or angled, very rarely flattened; capitula radiate or disciform 20

20. Capitula radiate; ray florets neuter, or if fertile then style arms short-conical at apex with a subdistal fringe of hairs, never apically papillate . . . . **1. Mutisieae**

– Capitula radiate or disciform; ray florets if present female; style arm apices acute, obtuse or truncate, and with sweeping stigmatic hairs, sometimes apically papillate. 21

21. Carpopodium glabrous on upper margin; capitula radiate or disciform; pappus setae usually of capillary barbellate or plumose setae, or of rigid awns or scales, sometimes absent. 22

– Carpopodium setuliferous on upper margin; capitula disciform; pappus setae biseriate, outer series of scales, inner series of few, free, rapidly caducous setae . . . .**8. Athroismeae** (*Artemisiopsis*)

22. Receptacle epaleaceous (if paleaceous the achenes dimorphic and stems winged or wingless {*Callilepis, Neojeffreya*}, or achenes monomorphic and stems winged and receptacle covered in numerous bristles {*Geigeria*}, or plants spiny and leaves decussate or alternate on main stems or on brachyblasts {*Rosenia*}); phyllaries papery or sometimes herbaceous; pappus of capillary barbellate or plumose setae, or of rigid awns or scales, connate at base or free, sometimes absent. . . . . **7. Inuleae sensu lato**

– Receptacle usually paleaceous, sometimes epaleaceous; phyllaries herbaceous; stems never winged; achenes monomorphic, dark brown or blackish; pappus of short scales or absent. . . . . **8. Athroismeae** (*Anisopappus*)

Tribe 12. **Calenduleae** Cass.

Calenduleae Cass. in J. Phys. Chim. Hist. Nat. Arts **88**: 161 (1819). —Norlindh in Stud. Calenduleae **1**: 1–432 (1943). —Lisowski, (Asterac. Fl. Afr. Cent. 2) Fragm. Flor. Geobot. **36** Suppl. 1: 496–500 (1991). —Beentje in F.T.E.A., Compositae **2**: 531–537 (2002). — Nordenstam in Bremer, Asterac. Cladist. Classif.: 365–376 (1994); in Kubitzki, Fam. Gen. Vasc. Pl. **8**: 241–245 [2006](2007). —Nordenstam & Källersjö in Funk *et al.*, Syst. Evol. Biogeogr. Compositae: 527–538 (2009).

Herbs, subshrubs, shrubs or small trees, unarmed or sometimes spinescent. Leaves alternate or opposite, sessile or petiolate, glabrous, glabrescent or variously pubescent, sometimes glandular, margins entire or variously lobed or dissected. Inflorescences of solitary capitula or corymbose, rarely racemose. Capitula pedicellate or rarely sessile, heterogamous, radiate; phyllaries 1–3-seriate, imbricate; receptacle naked. Ray florets female, fertile or sterile, rarely neuter, corollas yellow to orange or white, pink to purple or blue, frequently discolorous above and beneath. Disc florets hermaphrodite, perfect or functionally male, corollas 5-lobed, yellow to orange or reddish; anthers caudate; apical appendage triangular-ovate, flat; endothecial tissue polarized; style fertile or sterile, entire and undivided or shortly bilobed to bifurcate; stigmatic areas divided at least basally, sweeping hairs in a subapical collar or rarely extending down the style-branches. Achenes homomorphic to polymorphic, terete, triquetrous or flattened, winged (and sometimes conspicuously fenestrate) or wingless, straight or curved, sometimes rostrate, glabrous, in some taxa with a fleshy coloured or whitish exocarp; carpopodium usually inconspicuous if present, and when present a narrow annulus concolorous with base of achene; pappus absent.

A small tribe of 12 genera and c. 120 spp. The tribe is well understood at specific level although generic delimitation still poses some problems, especially concerning *Osteospermum*. The predominant centre of speciation of the tribe is in South Africa (especially in the fynbos and karoo vegetation) although the Mediterranean is the centre of diversity for *Calendula*. Many taxa are widely cultivated throughout the temperate and subtropical zones, especially from *Calendula*, *Dimorphotheca* and *Osteospermum*. There are a few cultivated species in the flora area, all of which now belong in *Dimorphotheca*; no species of *Calendula* are apparently cultivated.

1. Achenes drupaceous, flesh distinctly black-blue or orange-red . . . . **114. Chrysanthemoides**
 – Achenes dry 2
2. Ray limbs white, pink or purple; disc achenes (when present) with thickened margins . . . . **115. Dimorphotheca**
 – Ray limbs yellow-orange; disc achenes absent. 3
3. Ray achenes 3-winged with apical cavity and 3-fenestrate . . . . **112. Tripteris**
 – Ray achenes winged or wingless, without apical cavity . . . . **113. Osteospermum**

112. **TRIPTERIS** Less.

Tripteris Less. in Linnaea **6**(1): 95 (1831). —Nordenstam in Bremer, Asterac. Cladist. Classif.: 376 (1994); in Compositae Newslett. **25**: 46–49 (1994); in Compositae Newslett. **44**: 38–49 (2006); Nordenstam in Kubitzki, Fam. Gen. Vasc. Pl. **8**: 244 [2006](2007); in Funk *et al.*, Syst. Evol. Biogeogr. Compositae: 527–529 (2009).
Tripterachaenium Kuntze, Revis. Gen. Pl. **3**(3): 182 (1893), as nom. nov. for Linnaeus' pre-1753 *Tripteris*.
Osteospermum L. subgen. *Tripteris* (Less.) Norl. sect. *Trifenestrata* Norl., Stud. Calenduleae **1**: 263–335 (1943).

Annual or perennial herbs, subshrubs or shrubs. Stems unarmed, rarely spiny. Leaves alternate or opposite, petiolate or sessile, entire or variously lobed, coriaceous or membranaceous, sometimes almost fleshy, glabrous or pubescent, oblong, elliptic oblanceolate or rhomboid. Inflorescence corymbose or of solitary capitula. Capitula heterogamous, radiate; involucre campanulate or turbinate; phyllaries imbricate or almost biseriate, rarely subequal and nearly uniseriate. Ray florets usually conspicuous, female, fertile, corollas yellow to orange, often tinged violet beneath; achenes usually homomorphic and 3-winged with an apical fenestrate air chamber, sometimes dimorphic with winged achenes and wingless rostrate achenes; carpopodium, if obvious, a narrow annulus concolorous with base of achene; pappus absent. Disc florets hermaphrodite, functionally male; corollas yellow; styles bifid with an apical ring of hairs; achenes absent.

A genus of c. 20 spp. from southern and tropical Africa N to Egypt, 'Arabia' and Jordan; four are currently recognized from the Flora area.

Often included as a subgenus of *Osteospermum* (e.g. Norlindh 1943, Beentje 2002), the main distinguishing character is the presence of an apical cavity in the achene which is absent in *Osteospermum*, and the achenes are usually markedly winged, whereas they may or may not be winged in *Osteospermum*; Nordenstam (2006) considered the achenes homomorphic. The genus is considered distinct in this Flora treatment. Nordenstam (2006) transferred several species into a number of new genera and has since indicated that further transfers out of *Tripteris* may well be necessary (Nordenstam 2009).

1. Basal, lower and intermediate cauline leaves opposite; achenes large (13–20 mm long), wings membranaceous . . . . **1.** *monocephala*
 – All leaves alternate, or only lowermost opposite and intermediate and upper alternate; achenes 5–12 mm long, wings semi-transparent 2
2. Annual to short-lived perennial herb . . . . **2.** *vaillantii*
 – Perennial herb or suffrutex 3
3. Achene body 9–12 mm long, wings c. 4.5 mm wide; perennial herb to subshrub to 40 cm tall . . . . **3.** *aghillana*
 – Achene body 5–6 mm long, wings 1–1.5 mm wide; suffruticose perennial with stems to c. 2 m . . . . **4.** *nyikensis*

1. **Tripteris monocephala** Oliv. & Hiern in F.T.A. **3**: 424 (1877). —Hoffmann in Pflanzenw. Ost-Afrikas C: 419 (1895). Types: [Malawi:] 'Mozamb. Distr. Manganja Hills, *Meller*! Zambesia [sic!], [1863,] *Stewart*!' *Meller* s.n. (K000307188 lectotype) lectotypified by Norlindh (1943: 288); *Stewart* s.n. (BM000924707 syntype).[2] FIGURE 6.5.**1**.

    Tripteris goetzei O.Hoffm. in Bot. Jahrb. Syst. **30**: 439, t. 22A (1902). Types: [Tanzania:] 'Kingagebirge: Dinda-Berg [, Njombe Distr, Ukinga (Kinga), Diuda (Dinda) Mt], auf rasigen Abhängen um 2400 m ([*Goetze*] n. 1288. – Blühend am 12. Sept. 1899). Usafua: Ngosi- oder Poroto-Berg [(Poroto Mts, Ngosi)], auf flachen, rasigen Abhängen um 2000 m ([*Goetze*] n. 1287. – Blühend am 25. Sept. 1899)' *Goetze* 1228 (B† syntype); *Goetze* 1287 (B†, BR887671, Z syntypes).

    Tripteris rhodesica R.E.Fr., Wiss. Ergebn. Schwed. Rhodesia-Kongo-Exped. **1**: 346, t. 22/1 (1916). Types: [Zambia:] 'Nordwest-Rhodesia: Chirukutu bei Broken Hill [(Kabwe)], in Trochenwald [in beginnender Blüte, aber einzelne Exemplare schon fruchttragend 8. Aug. [1911,] [*R. E. Fries*] –n. 282]; Bwana Mkubwa [?Kitzwe], auf abgebranntem Grasfeld [blühend 28. Aug. [1911] [*R. E. Fries*] – n. 464]' *R. E. Fries* 282 (B†, UPS syntypes); *R. E. Fries* 464 (UPS syntype).

    Tripteris gweloensis Mattf. in Notizbl. Bot. Gart. Berlin-Dahlem **8**: 180 (1922). Type: [Zimbabwe:] 'Südafrika: Matabeleland, Gwelo [(Gweru)], feuchte Stellen (*Klingberg* [s.n.] – VI. 1895)' (B† holotype, S 08-10946).

    Tripteris gossweileri Mattf. in Bot. Jahrb. Syst. **59** (Beibl. 133): 44 (1924). Type: 'Angola: Cuanza Norte: Malange, bei Gola Luije nahe dem Fluß Cole (Lucala), 1000 m ü. M. (*Gossweiler* n. 8871, blühend und fruchtend 5. September 1922)' (B† holotype, BM000924708, K000307187).

    Osteospermum monocephalum (Oliv. & Hiern) Norl., Stud. Calenduleae **1**: 288, fig. 28, map on 417 (1943). —Norlindh, Stud. Calenduleae **2**: fig. 2 (1946). —Norlindh in Kirkia **11**(1): 151, fig. 4 (1978). —Cribb & Leedal, Mountain Fl. S. Tanzania: 163, t. 44d (1982). —Lisowski, (Asterac. Fl. Afr. Cent. 2) Fragm. Flor. Geobot. **36** Suppl. 1: 497, fig. 101 (1991).

Perennial herb 5–70 cm high. Rootstock a woody xylopodium, 1–5(8) cm diam. Stems 1 to several, erect, rigid, slightly woody at base, few-branched, reddish, terete, striate, viscid-pilose; hairs short, glandular, rarely almost glabrous. Leaves distant, sessile or with a short petiole to 6 mm long, near base of plant opposite and obovate or narrowly obovate, more distally alternate, larger and elliptic to narrowly oblanceolate, 2–7 × 0.2–3 cm, short glandular-hairy and slightly scabrid, rarely subglabrous, base attenuate to cuneate; margins entire to sinuate-toothed, apex rounded to acute. Inflorescence of solitary capitula; pedicel 3–34 cm long, sweet-smelling at anthesis, nodding in fruit, slightly glandular-hairy, usually ebracteate. Capitula 5–8 mm long (to 22 mm with rays or fruit); phyllaries 12–18, ± uniseriate to biseriate, subequal, narrow-elliptic, 4–7 × 1–2 mm, sparsely hairy towards base, mid-portion green, margins white-scarious, apex acute or acuminate, often reddish or whole mid-portion becoming reddish to purplish in fruit. Ray florets same number as phyllaries, ray limbs 12–24 × 3–5 mm, yellow to orange, occasionally with brown or purple markings beneath. Disc florets numerous, corollas to 10 mm long, yellow; anthers yellow. Achenes 13–20 mm long, green, 3-angular, 3-winged, wings 3–6 mm wide, pale yellow turning crimson or purple, stiff and membranous, often undulate, distal cavity with 3 windows 1–2 mm long.

Zambia. N: Abercorn, 10.vii.1964, *Mutimushi* 865 (K). W: Solwezi, 16.viii.1953, *Fanshawe* F250 (K, USC, SRGH). C: Serenje, 2.xi.1972, *Fanshawe* 11662 (K, NDO). E: Great North Road, 138 km beyond Chipata, 20.ix.1935, *Galpin* 15118 (BOL, K, LD, PRE). S: Mumbwa, *Macaulay* 853 (K). **Zimbabwe**. N: Mazoe Dist., Birkdale Nine road, Umvukwes, at foot of 'Chrome' hills, c. 1600 m, 15.ix.1963, *Leach &Muller* 11721 (K, M, S, SRGH). W: Matobo Distr., Farm Quaringa, 1494 m, xii.1957, *Miller* 4809 (COI, K, LISC, SRGH). C: Makoni Dist., c. 8 km W of Rusape in woodland, 1500 m, 30.xi.1930, *Fries, Norlindh & Weimarck* 3300a (B, COI, LD, SRGH, WAG). E: Inyanga Distr., in open grassland above the Pungwe Falls, close by the river, c. 1700 m, 16.xii.1930, *Fries, Norlindh & Weimarck* 3737 (EA, LD, US). S: Victoria, 1909, *Monro* 727 (BM, SRGH). **Malawi**. N: Nyika Plateau, by main road, 3.4 km S from Rest House, 2150 m, 22.x.1958, *Robson & Angus* 238 (BM, CAH, K, LISC). C: summit

---

[2] Norlindh (1943: 289) cited the Meller collection as from Portuguese East Africa, not Malawi. Oliver & Hiern cited Stewart's 'Zambesi' on the sheet as 'Zambesia'; Norlindh in turn misinterpreted the small label on the sheet (as 'Zambesi Rev. ...') which actually reads 'Zambesi Revd Dr. Stewart 1863', from 'Livingston's Zambesi Expedition' – after the expedition had finished!

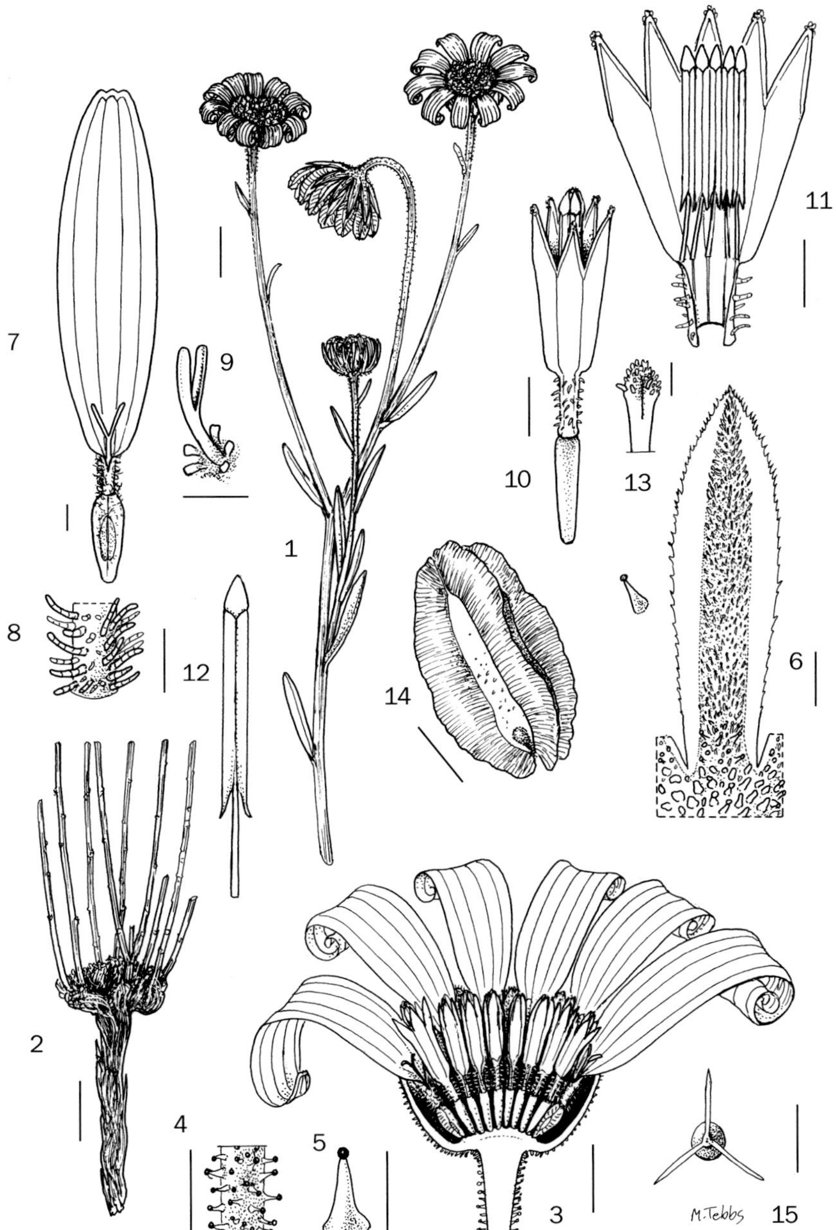

Fig. 6.5.**1**. TRIPTERIS MONOCEPHALA. 1, flowering stem; 2, rootstock, showing woody taproot, swollen apex and bases of several flowering stems; 3, l.s. capitulum; 4, detail of pedicel; 5. detail of glandular trichome on pedicel; 6, phyllary with detail of trichome; 7, ray floret; 8, detail of corolla tube of ray floret; 9, detail of style and surrounding staminodes; 10, disc floret; 11, disc floret corolla opened out showing attachment of filaments; 12, stamen; 13, apex of disc floret style; 14, achene; t.s. of achene showing wings. 1, 2 from *Jackson* 158, 3–7, 10–13 from *Eyles* 7196, 8–9, 14–15 from *Leach & Muller* 11721. Scale bars: 5 = 0.2 mm; 13 = 0.4 mm; 4, 6–12 = 1 mm; 3, 14–15 = 5 mm; 1 = 10 mm; 2 = 5 cm. Drawn by Margaret Tebbs.

of Mt Dedza, 2080 m, 15.x.1937, *Longfield* 42 (BM, SRGH). S: Foothpath Zomba to Zomba Plateau, 1070–1520 m, 27.xii.1936, *Lawrence* 233 (K). **Mozambique**. N: Niassa Distr., Vila Cabral, 12.x.1934, Torre 425 (COI, LISC). Z: Gurue, at the top of Gurué Mt, 25.ix.1944, *Mendonça* 2291 (LISC). T: Macanga, Furancungo, at 20 km on road to Angónia, 29.ix.1942, *Mendonça* 535 (LISC). MS: Manica, Rotanda, between R. Mussapa and the border, Tandara, 3 km from Tsetserra, 1750 m, 18.xi.1965, *Torre & Correia* 13114 (LISC).

Also in Angola, D.R. Congo and Tanzania; apparently the largest distribution in the genus after *T. vaillantii*. In open and frequently burnt grassland, and in moister grasslands of river valleys, also in miombo woodland and on sandy or on red soils; 1000–2700 (2950) m.

Norlindh (1978) proposed that the variation in vegetative parts was "predominantly due to different environmental conditions. The rainy and dry season and veld fires in particular greatly influence the appearance of the plant."

Conservation Status: Widespread species; Least Concern.

2. **Tripteris vaillantii** Decne. in Ann. Sci. Nat., Bot. sér. 2, **2**: 260 (1834). Type: [?Egypt:] 'Hab: le mont Sanaï [Sinai?]' *?Bové* (P00670372, holotype).[3]

    Tripteris cheiranthifolia Sch.Bip. in Flora **24** (Intelligentz. 1): 26 (1841), nom. nud. based on *Schimper* 345, a widely distributed collection.

    Tripteris cuneifolia Schweinf. & Asch. in Schweinfurth, Beitr. Fl. Aethiop.: 287 (1867), nom. nud.

    Tripteris rueppellii Schweinf. & Asch. in Schweinfurth, Beitr. Fl. Aethiop.: 287 (1867), nom. nud.

    Tripteris lordii Oliv. & Hiern in F.T.A. **3**: 424 (1877). Type: [?Sudan:], 'Nile Land, Upper Nubia, Hor Tamanib, [Red Sea, 17.ii.1869,] *J. K. Lord !* (with label stating that the flower is blue).' (K000307193 holotype).

    Tripteris lordii var. *racemosa* Balf.f. in Proc. Roy. Soc. Edinburgh **12**: 406 (1883)[?7.i.1884]. Types: [Yemen:] 'Socotra, prope Galonsir atque Tamarida abundans. [ii–iii.1880], B.C. S. No. 74. Schweinf. No. 443' *Balfour, Cockburn & Scott* 74 (B†, K000884966 syntypes); *Schweinfurth* 443 (B†, K000884966, P00128329, P00128330 syntypes).[4]

    Tripteris racemosa Wagner sensu Vierh. in Denkschr. Kaiserl. Akad. Wiss. Wien, Math.-Naturwiss. Kl. **71**: 483 (1907).

    Tripteris angustissima S.Moore in J. Bot. **38**(456): 460 (1900). Type: [Somalia:] 'Hab. Gan Liban, Somaliland, [5800', 25.ii.]1899 ; *Dr. Donaldson Smith*' (BM000924700 holotype).

    Osteospermum vaillantii (Decne.) Norl., Stud. Calenduleae **1**: 305 (1943).

    Osteospermum afromontanum Norl. in Svensk. Bot. Tidskr. **48**(1): 149 (1954). Type: [Tanzania:] 'Tanganyika Terr.: Mt Hanang, N. E. Slope, S. Mbulu Distr., c. 3300 m.s.m., *Greenway*, 1946, flor. et fruct. 8 Febr., n. 7665' (EA000003043 holotype, K000307189, PRE).

    Tripteris afromontana (Norl.) B.Nord. in Compositae Newslett. **25**: 47 (1994).

Annual to short-lived perennial herbs. Stems erect or ascending, (10)30–60(150) cm tall, often woody at base, to 15 mm diam., terete, striate or slightly sulcate, glandular-pubescent, branched, viscid and aromatic; internodes 5–30 mm. Leaves alternate or upper opposite, 15–115 × 0.4–3.5 mm, base attenuate, often semi-amplexical, surfaces glandular-pubescent; margins sinuate-dentate, 2–10 pairs of mucronate teeth, upper leaves sometimes entire; apex acute to obtuse and mucronate. Inflorescence of lax few-headed corymbs or racemes, sometimes a solitary capitula; capitula pedicellate, pedicels 1–14 cm long, densely glandular pubescent, bracteate, bracts scale-like, to 5 mm long. Capitula heterogamous and radiate, erect at anthesis, nodding in fruit; phyllaries (8)10–13, ± uniseriate, subequal, linear-lanceolate to lanceolate, (2)4–5 × 1–1.5 mm, midrib glandular-puberulous, green, margins broadly white-scarious, apex acute to acuminate. Ray florets female, as many as phyllaries, limb 5–12 ×1.5–3 mm, golden to lemon yellow, corolla tube pubescent. Disc florets 10–40, corollas 3–3.5 mm long, yellow, pubescent. Achenes obovoid, 6–11 mm long, green, sparsely hairy, 3-winged, wings 1–3 mm wide, tinged red or purple at maturity, apical window broad-elliptic, c. 1 mm long.

---

[3] Decaisne's discussion (Decaisne 1834: 261) indicated that he considered that this is a Bové collection, although the sheet is unmarked.

[4] The Balfour *et al.* syntype at K was determined as *T. vaillantii* var. *racemosa* by Norlindh (1938) but I have not been able to find this combination published anywhere. He failed to mention it in his tribal revision.

Zambia. N: Mafinga, 24.v.1973, *Fanshawe* 11974 (K, NDO). **Malawi**. N: Mafinga hilltop, Chitipa, 9°56'33.73"S 33°21'37.59"E, 2181 m, 5.viii.2007, *Chapama et al.* 708 (FRIM, K).

Also in Kenya, Tanzania and Uganda, N to Egypt. Largely in montane areas, amongst rocks or in grassland, sometimes occurring as a weed on cultivated ground; 900–3500 m.

Conservation Status: Although few collections in the Flora area it has a wide distribution; Least Concern.

I have followed Beentje (2002: 533–535) and accepted *Osteospermum afromontanum* as a synonym of *T. vaillantii*. In Tanzania alone there are a number of very woody-based shrub-like specimens that bridge the gap between Norlindh's concept of *O. afromontanum* and his concept of *T. vaillantii* (as *Osteospermum*). The position of *Angus* 843 (determined by Norlindh as 'aff. *afromontanum*') is commented on under *T. nyikensis* below. Clearly in the right environment this taxon can perennate and become quite woody at the base, not unlike material of *T. lordii* var. *racemosa* placed as a synonym of *O. vaillantii* by Norlindh himself.

3. **Tripteris aghillana** DC., Prodr. **6**: 457 (1838). —Harvey in Harvey & Sonder, Fl. Cap. **3**: 429 (1865). Type: [South Africa: Cape Province.] 'in Africa australi ad Cape Aghillas legit cl. *Drege* [3718]! … (v. s. comm. et seqq. à cl. inv.)' (G-DC-G00472382 holotype, P00132208, P00132209 – s.n., P00132210 – s.n.).

   Tripteris setifera DC., Prodr. **6**: 457 (1838). Type: [South Africa:] 'in Africa Capensi ad Zwarteberge [(Swartberg)] alt. 20–300 ped. legit. cl. *Drege* [6043]! … (v. s.)' (G-DC-G00472371 holotype, P00132217, P00132245 – s.n., P00132246 – s.n. number added later).[5]

   Osteospermum scariosum DC., Prodr. **6**: 464 (1838). —Norlindh, Stud. Calenduleae **1**: 311, 313, fig. 31 a-m (1943). —Acocks in Mem. Bot. Surv. South Africa **28**: 93, 131, 134 (1953). —Merxmüller, Prodr. Fl. Südwestafr. **139**: 128 (1967). Types: [South Africa: Cape Province.] 'in Africa Capensi ad Zwart Ruggens alt. 2–3000 ped. [1835] legit cl. *Drege*! et ad Graaf-Reynet cl. *Ecklon* [& *Zeyher* 641] ! … (v. s.)' *Ecklon & Zeyher* [88] 641 (G-DC-G00472555 lectotype, S-G-9767), effectively lectotypified by Norlindh (1943: 312); *Drège* 6114 (G-DC-G00472519 syntype); *Drège* 6381 (G-DC-G00472548 syntype).

   Osteospermum picridioides DC., Prodr. **6**: 465 (1838). —Harvey in Harvey & Sonder, Fl. Cap. **3**: 440 (1865). Type: [South Africa: Cape Province.] 'in Africa Capensis distr. Carro [(Karoo)] legit cl. *Drege* [6045] ! … (v. s.)' (G-DC-G00472510 holotype, P00132214, P00132215 – s.n., P00132216 – s.n.).

   Tripteris humilis Turcz. in Bull. Soc. Imp. Naturalistes Moscou **24**(1): 211 (1851). —Harvey in Harvey & Sonder, Fl. Cap. **3**: 430 (1865). Type: [South Africa: Orange Free State.] '*Ecklon* coll. Capensis n. 52. 114. 10.' (KW00100092, P00132231, P00132232, P00132234 syntypes).

   Tripteris glandulosa Turcz. in Bull. Soc. Imp. Naturalistes Moscou **24**(3): 93 (1851). —Harvey in Harvey & Sonder, Fl. Cap. **3**: 430 (1865). Type: [South Africa: Cape Province.] 'C. b. spei. *Zeyher* n. 3067' (KW001000923, K000307337, P00132230 syntypes).

   Tripteris aghillana var. *integrifolia* Harv. in Harvey & Sonder, Fl. Cap. **3**: 429 (1865). Type: cited merely as 'Hb. Sd.' without collector or locality.[6]

   Tripteris flexuosa Harv. in Harvey & Sonder, Fl. Cap. **3**: 429 (1865). —Compton in Trans. Roy. Soc. South Africa **19**(3): 324 (1931), nom. illeg. et superfl. pro *Osteospermum scariosum* DC., cited in synonymy.

   Tripteris natalensis Harv. in Harvey & Sonder, Fl. Cap. **3**: 430 (1865). Type: [South Africa:] 'Hab. Natal Country, *Miss Owen*! (Herb. D.)' (?TCD holotype).

   Tripteris glandulosa var. *dentata* Harv. in Harvey & Sonder, Fl. Cap. **3**: 430 (1865). Types: [South Africa:] 'Langehoogte, Caledon, *Ecklon*! Babylon'stoornberg, *Zey*[*her*]! 3067 [p.p.]' *Zeyher* 3067 p.p. (S holotype).

---

[5] Norlindh (1943: 322) cited the type as *Drège* 6043, but oddly from 600–900 m, considerably higher than that stated by Candolle, suggesting that the elevation should have been "2–3000 ped."

[6] Harvey cited 'Hab. Cape, L'Agulhas, *Drege*! Cape, E. & Z.! (Herb. Sond.)' beneath the diagnosis for the variety, without indicating which applied to the species and which to the variety. Nordenstam (1996) annotated S-G-9776 as having been Norlindh's 'lectotype' of the variety; this specimen is 'ex Herb. Sond.', but there is no indication of the collector. Norlindh's account (1978: 154) is typically brief adding no further information.

Tripterachaenium scariosum (DC.) Kuntze, Revis. Gen. Pl. **3**(3): 182 (1893).

Tripterachaenium humile (Turcz.) Kuntze, Revis. Gen. Pl. **3**(3): 182 (1893).

Osteospermum scariosum subsp. *setiferum* (DC.) Norl. in Kirkia **11**(1): 155 (1978), comb. inval. (post-1973, and not citing basionym reference).

Osteospermum scariosum subsp. *integrifolium* (Harv.) Norl. in Kirkia **11**(1): 155 (1978), comb. inval. (post-1973, and not citing basionym reference).

Perennial herb, sometimes becoming ± subshrubby, viscid and strongly aromatic, up to 40 cm tall. Stems arising from a woody rootstock to 2 cm diam., erect or ascending, branching from base and often flexuous, stems and branches terete, striate, usually woody in lower parts; ± densely glandular-puberulous (rough when dry) and besides ± hispid with whitish, multicellular swollen hairs, or older parts glabrous. Leaves alternate, somewhat fleshy, 2–7(12) × 0.5–2 cm including petiole, pale green or often glaucous, pubescence similar to that of stem and branches, rarely glabrate; basal leaves crowded, often ± rosulate, oblanceolate to oblong or a few obovate, tapering towards base into a petiole; apex obtuse to subacute, margins remotely and minutely toothed to coarsely sinuate-toothed, teeth with callose tips, rarely pinnatifid or entire, stem and branch leaves ± scattered, lower stem similar to basal, but usually narrower, upper smaller towards capitula, linear-lanceolate to oblong, sessile and often ± auriculate, apex obtuse to acute-acuminate, margins remotely toothed or sometimes entire, midrib whitish, prominent beneath, lateral veins usually inconspicuous. Inflorescence of solitary capitula at stem or branch apices, or rarely a lax corymb, capitula nodding in fruit; pedicels glandular-pubescent, 10–20 cm long, bracts few, scattered, acuminate, uppermost narrowly lanceolate or linear-subulate, margins entire. Phyllaries 12–15, ± uniseriate, ± equal, narrowly oblong or elliptic-oblong or lanceolate, 4–6 × 1–2.5 mm, apex obtuse to acute-acuminate, abaxially glandular-puberulent along midrib, margins ± broadly whitish-scarious, slightly lacerate-ciliate. Ray florets equal in number to phyllaries, limb yellow, rarely with a dark violet centre (eye) above, but often ± violet beneath, about twice as long as involucre. Disc florets yellow; apical anther appendages dark violet. Achenes 9–12 mm long (excluding wings), slightly incurved, 3-angular and broadly 3-winged, glabrous, sides often glandular, provided with a longitudinal, inconspicuous groove, transversely ribbed, scarcely spiny, pitted or rarely almost smooth, wings membranaceous, semi-transparent, whitish or stramineous, often with violet flecks, up to 4.5 mm wide, closed apical cavity 3-fenestrate with small (c. 1–2 × 0.5–1 mm) ovate-oblong windows.

Botswana. SE: Mochudi, 1914, *Harbor* in Herb. Rogers 6227 (Z).

Also widespread in South Africa. It has the widest range in southern Africa of all species belonging to the sub-genus *Tripteris* but only just gets into the Flora area in SE Botswana. In open often stony grassland, *Acacia* savanna, rocky hills with typical karroo shrublets, sandy loam, black clay soil, calcareous soil, grey sandy soil, by moist river-banks and in thorny bushes along watercourses or on vacant plots of ground; 300–1500 m.

Conservation Status: Although poorly collected in the Flora area it is very widely distributed in South Africa; Least Concern.

4. **Tripteris nyikensis** (Norl.) B.Nord. in Compositae Newslett. **25**: 48 (1994). Type: 'Malawi, Kawozya Hill, *Brummitt & Synge* WC211 211' (K000307185 holotype, MAL, S-G-9765, SRGH0106746-0).

Osteospermum nyikense Norl. in Kew Bull. **31**(1): 171 (1976).

Suffruticose perennial. Stems several, ascending at c. 45°, up to c. 2 m long, terete, glandular-pubescent, branched, lower part strongly lignified and up to 1 cm diam., bark brownish, upper part striate, green or dull-green, ending in inflorescences, lateral branches of stem very rarely flowering, internodes in middle part of stems 1–3 cm long or even longer in upper flower-bearing branches. Leaves (basal unknown) alternate, glandular-pubescent or nearly glabrous, elliptic or narrowly elliptic, gradually tapering into a petiole or sessile, variable in size, to c. 6 cm long including petiole and to 2 cm wide, acute or often obtuse, mucronate, margins distantly toothed to coarsely sinuate-toothed, teeth mucronate, up to 5 mm long, leaves on flowering branches narrowly elliptic or linear, midrib thick, prominent beneath, lateral veins 1–3 on each side, usually anastomosing, conspicuous or obscure in upper leaves. Inflorescences corymbose, with 3–6 capitula on each branch, fruiting capitula nodding; pedicels 1–6 cm long, densely glandular-pubescent (rough when dry), bracts few, linear-subulate. Phyllaries uniseriate, c. 13, ± equal, linear-oblong, acuminate, 3–4.5 × c. 1 mm wide, slightly glandular-hairy abaxially, margins whitish-scarious. Ray florets equal in number to phyllaries, ray limb yellow, 4–5 times as long as phyllaries. Disc floret corollas yellow. Achenes hard, sparsely glandular-papillose or almost smooth on tangential sides, triangular-obovoid, 5–6 mm long excluding wings, sides 2–2.5 mm wide, 3-winged, wings half-transparent, 1–1.5 mm wide, closed apical cavity with 3 small fenestrae, hardly more than 0.5 mm wide.

Zambia. N: Isoka Dist., Among rocks on summit of Mafingi Range above Chisenga, c. 2439 m, 22.xi.1952, *Angus* 843 (FHO, K). **Malawi**. N: Nyika Plateau, 18302–2135 m, vi.1896, *Whyte* 169 (K).

Only found in Zambia and Malawi. Amongst rocks on mountain summits; 1800–2440 m.

Conservation Status: With only *Angus* 834 giving any indication of population size, it is clear that its EOO and AOO would make it Critically Endangered/Endangered; for the present best recorded as Data Deficient. Clearly more detailed fieldwork needs to be done.

Norlindh's determination of *Angus* 834 (FHO, K) suggested '*Osteospermum* aff. *afromontanum*'; I feel that this is a third collection of *T. nykensis*, from Zambia, other than the type and paratype cited by Norlindh from Malawi.

113. **OSTEOSPERMUM** L.

Osteospermum L., Sp. Pl. **2**: 923 (1753); Gen. Pl., ed. 5: 395 (1754). —Bentham & Hooker f., Gen. Pl. **2**(1): 455 (1873). —Norlindh, Stud. Calenduleae **1**: 98–357 (1943); in Bot. Not. **113**: 385–399 (1960). —Dyer, Gen. S. Afr. Fl. Pl. **1**: 718 (1975). —Nordenstam in Bremer, Asterac. Cladist. Classif.: 375 (1994); in Kubitzki, Fam. Gen. Vasc. Pl. **8**: 244 [2006a](2007); in Compositae Newslett. **44**: 38–49 (2006b).

Annual or perennial herbs or shrubs. Leaves alternate or less often opposite, sometimes rosettiform, petiolate or sessile, entire or variously divided up to bipinnatipartite. Inflorescence of solitary terminal capitula or of terminal corymbose cymes or rarely racemes. Capitula heterogamous, radiate; involucre campanulate or hemispherical to turbinate; phyllaries free, 1–3(4)-seriate; receptacle flat or slightly convex, epaleate. Ray florets female, fertile, with short tube, ray limbs narrow oblong, minutely 3-dentate, yellow or orange; staminodes sometimes present; style linear, glabrous, style arms long, linear, obtuse with stigmatose swellings along margins. Disc florets hermaphrodite and functionally male, corolla with a gradually widening tube, 5-lobed; anther bases sagittate and ± tailed, apical appendage ovate; style shallowly cleft, style arms connate, with a collar of sweeping hairs at or slightly below bifurcation, papillae absent. Achenes hard, straight or slightly incurved, triangular or terete, sometimes 3–9-ridged, smooth, rugose or tuberculate, with or without wings (but never with an apical cavity), without or rarely with a rostrum; pappus absent.

A genus of probably less than 40 spp. from southern and tropical Africa, Somalia and SW Arabia; only two species are recognized from the Flora area now that many have been removed to *Tripteris*, and several cultivated species are now recognized as dimorphothecas.

Nordenstam (1994, 2006a) has suggested a split of the genus into *Osteospermum* sensu stricto and *Tripteris* based on the characters equality/inequality of phyllaries and presence/absence of sclerenchymatic strands in the corolla. In a recent account (Nordenstam 2006b) he included within the synonymy of *Dimorphotheca* the genus *Xenismia*, and *O.* sect. *Xenismia*, although later removing two of the taxa in this section to *Oligocarpus* Less. (Nordenstam 2006a). I have left the only representative of *Osteospermum* sect. *Xenismia* sensu Norlindh in *Osteospermum*, its sectional placement still needs to be resolved.

Leaves entire; achenes ± cylindrical and ± conspicuously ridged. . . . .
. . . .**1.** *imbricatum* subsp. *nervatum*
Leaves serrate-dentate or variously divided, even pinnatipartite, sometimes entire; achenes obovoid to obpyramidal, smooth or tuberculate . . . .**2.** *muricatum*

1. **Osteospermum imbricatum** L., Mant. Pl. Altera: 290 (1771). —Norlindh, Stud. Calenduleae **1**: 151, 145, fig. 10d, 153, fig. 12a-e (1943). Type: [South Africa: Cape of Good Hope.] 'Habitat ad Cap. b. spei.' *Tulbagh* 144 (LINN1037.16 lectotype), lectotypified by Norlindh (1943: 152).[7] FIGURE 6.5.**2**.

 Osteospermum glandulosum A.Spreng., Tent. Suppl.: 26 (1828), non sensu DC.

 Osteospermum corymbosum sensu DC., Prodr. **6**: 461 (1838), non L.

 Osteospermum corymbosum sensu DC. var. β *rotundifolium* DC., Prodr. **6**: 461 (1838), p.p. Types: [South Africa:] '… ad Vanstaadesrivier (*Drege*), in Stellenbosch et Caledon (*Eckl.* [& Zeyher]!). … (v. s.)' *Drège* s.n. (G-DC, K, L, P, Winsert: syntypes), following Norlindh (1943: 155); Stellenbosch, *Ecklon & Zeyher* [998] (G-DC-G00472496 syntype); Caledon, *Ecklon & Zeyher* [1452] (G-DC-G00472507 syntype), annotated as 'typus' by Norlindh in 1955.

 Osteospermum corymbosum sensu DC. var. γ *lasiocaulon* DC., Prodr. **6**: 462 (1838). Types: [South Africa:] '… ad Cap. Bonae-Spei (*Zeyh.*[*er*]! n. 78 in h. Moric. Eckl!), ad Zneenrobergen alt. 2–3000 ped. (*Drege*!). … (v. s.)' Zuurebergen, R IV, *Drège* 2125 (G-DC-G00472551); Zuurebergen, nördlichster Bergabhang, *Drège* 6119 (G-DC-G00472527, possibly HAL0113047 – s.n., possibly HBG504976 – s.n., P00127244 syntypes); Albany, [Zwartehoogde, Suurberg Range near Grahamstown, 2000′], *Ecklon* 374 (G-DC-G00472462); Uitenhage, [between Krakakamma and Vanstaadesberg], *Ecklon* [& *Zeyher*] 531 (G-DC-G00472476, possibly HBG504977 – s.n. syntypes); ibid., *Ecklon* [& *Zeyher*] 1243 (G-DC-G00472455 syntype); Uitenhage, [between Zwartkopsrivier and Zondagsrivier, under 1000′], *Ecklon* [& *Zeyher*] 375 (G-DC-G00472465 syntype); Uitenhage, [Koegarivier, near Winterhoeksberg], *Ecklon* [& *Zeyher*] 1101 (G-DC-G00472512 syntype); Uitenhage, [around Uitenhage], *Ecklon* [& *Zeyher*] 621 (G-DC-G00472546 syntype).

 Osteospermum dichotomum E.Mey ex DC., Prodr. **6**: 462 (1838). Types: [South Africa:] 'in Africa Capensi in Carro legit cl. *Drege* et ni fallor in Langekloof. … (v. s.)' Carro, *Drège* 510 (G-DC-G00472468, P00127238 – although this is localized to Lagenkloof, syntypes); Langekloof, *Drège* 6120 (G-DC-G00472498, P syntypes); s.loc., *Drège* 6122 (G-DC-G00472467 syntype).

 Osteospermum retirugum DC., Prodr. **6**: 462 (1838). Type: [South Africa:] '… ad Cap. Bonae-Spei legit cl. *Drege*. [6123] … (v. s.)' (G-DC-G00472487 holotype, G, L, P00127246, P00127247).

 Osteospermum amplexicaule Steud., Nomencl. Bot., ed. 2, **2**: 236 (1841), nom. nud.

 Osteospermum lanatum Spreng. ex Sch.Bip. in Flora **27**: 768 (1844), nom. nud.

 Osteospermum corymbosum var. γ *parvifolium* Harv. in Harvey & Sonder, Fl. Cap. **3**: 445 (1865). Type: [South Africa: Uitenhage: Grassrug, kalksteinhaltige, grassreiche Fläche und Hügel zwischen Koega und Zondagrivier, 150–300 m.] '*Zey*[*her*]! 3059.' (C, GRA, K, S, W, Z syntypes).[8]

Subsp. **nervatum** (DC.) Norl., Stud. Calenduleae **1**: 160, 161, fig. 13 (1943). —Acocks in Mem. Bot. Surv. South Africa **28**: 34 (1953).

 Osteospermum nervatum DC., Prodr. **6**: 462 (1838). —Harvey in Harvey & Sonder, Fl. Cap. **3**: 445 (1865). Types: [South Africa:] 'in Africâ Capensi ad Albany, Omtata et Omsamwubo legit cl. *Drege*! … (v. s.)' *Drège* [5106] (G-DC-G00472516 – s.loc., G-DC-G00472480 – 'Albany', G-DC-G00472464 – 'Omtata und Omsamwubo' syntypes).[9]

 Osteospermum helichrysoides DC., Prodr. **6**: 466 (1838). Type: [South Africa:] 'in Africa australi ad Omsanculo et Port-Natal legit cl. *Drege*! … (v. s.)' *Drège* [5104] (G-DC-G00473734 holotype, B†).

---

[7]   Norlindh merely specified the location, not the specific sheet in LINN.

[8]   Location of holotype unclear from protologue; Norlindh made no comment in his treatment.

[9]   There are several specimens in G-DC bearing the name *Osteospemum nervatum*. Three are numbered '5106', one attributed to 'Albany. R. IV.' (G-DC-G00472516), one to 'Omtata und Amsamwubo. R III.' (G-DC-G00472480) and the other unlocalized (G-DC-G00472464). There are also two other collections, one from/to 'M. Edm. Boissier 1833' (G-DC-G00472529) the other 'N. 1639. 85. Cap. M. Ecklon. 1805' (G-DC-G472515); the latter two are not considered type material. There are a number of potential isosyntypes although none have these localities marked on their labels or bear different sheet annotations (and none are numbered), although all are marked as this taxon in Drège's [?] hand. Norlindh (1943: 160) merely stated that the 'typus subspeciei' was *Drège* 5106 in G-DC, without specifying which of the specimens he was referring to, although by implication he lectotypified the name based on the second *Drège* 5106 mentioned above as this was illustrated in fig. 13 on p.161.

Osteospermum glaberrimum O.Hoffm. in Kuntze, Revis. Gen. Pl. **3**(2): 165 (1893). Type: [South Africa:] 'Natal: Krantzkloof [, 550 m, 12.iii.1894,]' *Kuntze* s.n. (NY00230832 holotype).[10]

Osteospermum imbricatum var. *helichrysoides* (DC.) Norl., Stud. Calenduleae **1**: 163 (1943).

Subshrub or shrub, up to c. 2 m tall. Rootstock a thick, woody xylopodium, crown woolly. Stem virgate, erect or rarely ascending, usually simple below and branched within inflorescence, but old (especially damaged) specimens developing lateral, semi-prostrate or ascending branches from rootstock; stem and branches densely leafy and strongly woody, glabrous or ± densely whitish-lanate when young, later glabrescent. Leaves alternate and imbricate, sessile and semi-amplexicaule, slightly fleshy and ± brittle or leathery when dry, glabrous, glaucous, margins entire and slightly cartilaginous, basal and lower stem leaves linear to linear-oblong, to 15(18) × 0.5–1.5(2) cm, obtuse to subacute, minutely mucronate, midrib whitish, prominent below and often with 1 or 2 inconspicuous longitudinal veins on each side of midrib, stem leaves smaller towards inflorescence, median and upper stem and branch leaves linear, linear-oblong or linear-oblanceolate, 0.6–3 × 0.2–0.5 cm, suberect, but hardly adpressed or imbricate, usually only midrib conspicuous. Inflorescence of terminal corymbose cymes or broad few-headed panicles, or capitula sometimes solitary at ends of lowermost lateral branches; pedicels ± rigid, ± glandular-muriculate, 1–5(10) cm long, bracts few, 3–8 mm long, lanceolate-linear to linear-subulate. Capitula 2–2.5 cm diam. across rays; phyllaries 10–14, ± uniseriate, elliptic-oblong or narrowly obovate-oblong, 5–6(7) × 1.5–3 mm, abaxially sparsely glandular-muriculate, acuminate or obtuse and mucronate, margins ± broadly whitish-scariose and slightly ciliate-lacerate. Ray florets equal in number to phyllaries, ray limbs yellow but often with violet flecks on beneath, exceeding involucre more than twice. Disc florets yellow, equalling involucre; apical anther appendages very dark violet. Achenes glabrous, almost cylindrical, 6–7 mm long and c. 2 mm diam., ridged, primary ridges 3, acute or often slightly winged, secondary ones 6, ± conspicuous, obtuse, in pairs between primary ridges, grooves between ridges foveolate.

Mozambique. M: Maputo (Lourenço Marques), Namaacha, near Cascata, 22.xii.1944, *Torre* 6943 (LISC).

Also in coastal and montane South Africa N to Natal. On grassy slopes and sand dunes near the coast, but also occurring a little inland in the mountains and in river-beds; up to 1500 m. Flowering almost throughout the year.

Conservation Status: The species and this subspecies are widespread in South Africa, especially up into Natal; Least Concern.

I am reliant upon Norlindh's record from Mozambique having seen no material from the Flora area.

Norlindh's var. *helichrysoides* is easily distinguished by its silky woolly leaves, but is known only from Natal.

2. **Osteospermum muricatum** E.Mey. ex DC., Prodr. **6**: 464 (1836). —Hilliard, Compositae Natal: 523 (1977). —Norlindh in Kirkia **11**(1): 150 (1978). Type: [South Africa:] 'in Africâ australi legit cl. Drege! … (v. s.)', *Drège* 869 (G-DC-G00472491 holotype, E00239856 – s.n., HBG504971 – s.n., K000307296 – s.n., MO391644 – s.n., P00132098, P00132099, P00132100, PRE0213043-0 – s.n., duplicate material from K presented to PRE, SAM0017114-0, TUB005831 – s.n.). FIGURE 6.5.**2**.

Osteospermum muricatum var. *asperum* Harv. in Harvey & Sonder, Fl. Cap. **3**: 441 (1865). Type: [South Africa:] '*Zey*[*her*]! 1005' (K000978061 – ex TCD post-1951, P00132109 syntypes).[11]

Osteospermum muricatum var. *glabratum* Harv. in Harvey & Sonder, Fl. Cap. **3**: 441 (1865). Type: [South Africa:] '*Zey*[*her*]! 1004', (K000307297, P00132108 syntypes).[12]

---

[10] See Wetter & Zanoni in Brittonia **37**: 334 (1985).

[11] Norlindh (1943: 187) noted "Orange Free State: Rhinosterkop near Vaalrivier" (a little different from that cited for *Zeyher* 1004) with duplicates in BM, K, SAM and W, but did not mention where the holotype might be.

[12] Norlindh (1943: 186) also noted "Beaufort West: Rhinosterkop" and isotypes in BM and W; Harvey did not annotate either syntype.

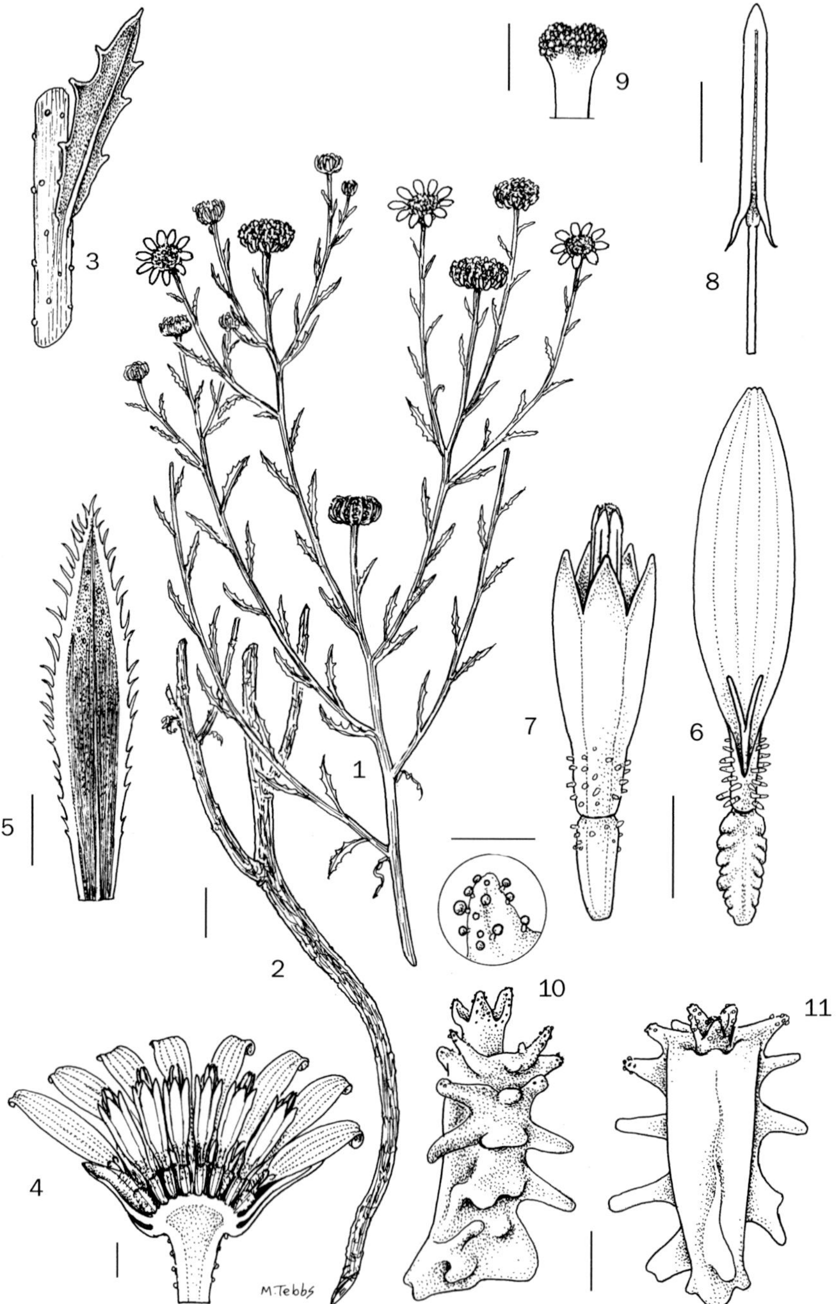

Fig. 6.5.2. **OSTEOSPERMUM MURICATUM**. 1, flowering branch; 2, rootstock and base of annual stems; 3, portion of stem and leaf; 4, l.s. capitulum; 5, phyllary; 6, ray floret; 7, disc floret; 8, stamen; 9, apical portion of disc floret style; 10, achene (abaxial view) with detail of protruberance; 11, achene, adaxial view; 12, detail of achene protruberance. 1, 3–9 from *Alemarumo et al.* MSB 518; 2, 10–11 from *Skarpe* S-396. Scale bars: 9 = 0.2 mm; 8, 12 = 0.5 mm; 4 – 7, 10, 11 = 1 mm; 1, 2 = 10 mm. Drawn by Margaret Tebbs.

Osteospermum hamiltonii S.Moore in J. Bot. **41**: 136 (1903). Types: [South Africa:] 'Hab. Griqualand West, between Fourteen Streams Station and Schmidt's Drift [*Capt. Barrett-Hamilton* s.n.] (Also Griqualand West; *Mrs. Barber* in herb. Kew [Comm. … 12/7/1]' *Barrett-Hamilton* s.n. (BM000924704 syntype); *Barber* 11 (K000978060 syntype).

Osteospermum muricatum subsp. *longiradiatum* Norl. in Bot. Not. **113**: 392 (1960). Type: [Namibia:] 'S. W. Africa: Reheboth: Hänge und Plateau des grossen Gamsberges (1900–2332 m), *Merxmüller*, 1958, flor. et fruct. 28/2, n. 962' (M0105084 holotype, LD, PRE0213075-0).

Perennial herb or small (sometimes 'gnarled') subshrub to 0.6 m tall, often forming plants c. 1 m diam.; whole plant sweetly resinous-aromatic and sticky (often leaving a black glue on fingers). Stems solitary or well branched from top of rootstock, usually woody in lower part, terete, striate, ± densely covered in stipitate-glandular hairs, later glabrescent. Leaves sessile, alternate, 5–50 × 4–20 mm, linear, linear-lanceolate or oblong, pale green or glaucous, glabrous or minutely stipitate-glandular pubescent, often scabrid when dry, margins entire, dentate or pinnatifid, longest lobes to 10 mm, apices acute. Inflorescences of solitary terminal capitula on ebracteate peduncles. Capitula heterogamous and radiate, c. 10 mm diam.; phyllaries 7–10, ± uniseriate or ± biseriate, subequal, lanceolate to narrow-ovate, 3–6(8) × 1–2.5 mm, abaxially glandular-puberulous, margins white-scarious, ciliate, apices acute or acuminate. Ray florets short, scarcely exceeding phyllaries by up to 2 mm (subsp. *muricatum*) or c. 4 mm (subsp. *longiradiatum*), ray limb yellow above and variously coloured beneath (greenish, violet or lilac). Disc florets shorter than phyllaries by up to 2 mm, corollas yellow. Achenes 3–5 × 1.5–2.5 mm, obovoid or pyriform, body hard when ripe, 3-angled, 2 outer angles forming blunt processes, outer face transversely furrowed, radial surfaces smooth or with a blunt longitudinal rib, apex glabrous or minutely glandular.

Botswana. SW: Ghanzi Dist., on old runway, ii.1952, *de Beer* D47 (K, SRGH). SE: Takatokwane Pan, 17.ii.1960, *Wild* 4991 (K, SRGH). **Zambia**. W: Kitwe, 5.viii.1971, *Fanshawe* 11238 (K, NDO). C: Lusaka, Mt Makulu Research Station, 2.vi.1956, *Angus* 1319 (K, SRGH). S: Monze, 7.vi.1962, *Fanshawe* 6863 (K, NDO). **Zimbabwe**. N: Mazoe, N of Jumbo, c. 1220 m, 13.i.1958, *Phipps* 842 (K, SRGH). W: Insiza, Fort Rixon, 8.ii.1974, *Mavi* 1515 (K, SRGH). C: Harare (Salisbury), Mt Pleasant, 1494 m, 14.x.1968, *Biegel* 2637 (K, SRGH). E: Nyanga Dist., Nyanga lake, c. 1700 m, 1930, *Norlindh & Weimarck* 3201 (LD). S: Masvingo Dist., Masvingo (Fort Victoria), road to Kyle Nat. Park, 9.ii.1972, *Gibbs Russell* 1455 (K, SRGH).

Also known from Angola, widespread in southern Africa, and apparently disjunct to the N in Yemen and Somalia. Grazed sandveld, pan margins, roadsides, fallow pastures, on granitic or red clayey soil, white sand or dolerite, or as a garden weed; 130–1900 m, but up to c. 2400 m in the Yemen.

Conservation Status: Widespread, sometimes weedy species throughout much of the Flora area, although apparently absent from Malawi and Mozambique; Least Concern.

If one cares to follow Norlindh's recognition of two subspecies – subsp. *longiradiatum* is endemic to Gamsberg in Namibia and is recognized by longer ray limbs.

114. **CHRYSANTHEMOIDES** Fabr.

Chrysanthemoides Fabr., Enum.: 79 (1759). —Norlindh, Stud. Calenduleae **1**: 367–403 (1943). —Dyer, Gen. S. Afr. Fl. Pl. **1**: 719 (1975). —Hilliard, Compositae Natal: 527 (1977). —Norlindh in Kirkia **11**(1): 157–161 (1978). —Nordenstam in Bremer, Asterac. Cladist. Classif.: 373 (1994). —Beentje in F.T.E.A., Compositae **2**: 535–537 (2002). —Nordenstam in Kubitzki, Fam. Gen. Vasc. Pl. **8**: 244 [2006](2007).

Unarmed or spinescent shrubs or small trees. Leaves alternate, pseudopetiolate, lamina coriaceous to subfleshy, glabrous or glabrescent or persistently arachnoid-tomentose, margins entire or denticulate, apices acute or obtuse. Inflorescence of solitary terminal capitula or of few-headed corymbs and terminal, subterminal or axillary, capitula pedicillate, pedicels arachnoid pubescent or glabrescent, bracteate, bracts scale-like. Capitula heterogamous and radiate; involucre broadly campanulate; phyllaries 2–3-seriate, margins scarious; receptacle glabrous or lanate, epaleaceous. Ray florets female, fertile, uniseriate, ray limb yellow, 4-veined, lobes 3, minute, tube short; staminodes 4; style filiform, style arm apices obtuse. Disc florets hermaphrodite and functionally male; tube glandular-hairy, lobes 5, ovate-triangular; anther bases sagittate, apical anther appendages ovate; style scarcely bifid, glandular, style arms pilose. Fruit a globose drupe, usually shiny blue or black, endocarp hard and globose or ellipsoid, surface smooth or with raised veins; pappus absent.

A genus of two, perhaps up to six, spp. in E and southern Africa and St Helena, introduced into Australia (for sand dune stabilization), New Zealand and California. Only one species is present in the Flora area, represented by two subspp. scarcely separable on the basis of measurements of the phyllaries. They are highly variable in terms of their leaf shape and size even on the same plant and certainly dependent upon the collecting season, but clearly separable in terms of their habitats. Norlindh's key is a polytomy with one couplet with five leads!

Chrysanthemoides monilifera (L.) Norl., Stud. Calenduleae **1**: 374, 399 (1943).— Hilliard, Compos. Natal: 527 (1977). —Coates Palgrave, Trees S. Afr.: 913, t. 314 (1977). —Norlindh in Kirkia **11**(1): 158 (1978). —Agnew & Agnew, Upland Kenya Wild Fl., ed. 2: 226 (1994). —Beentje in F.T.E.A., Compositae **2**: 535–537 (2002). Type: 'Habitat in Æthiopia.' Dillenius, Hort. Eltham.: t. 68, fig. 79 (1732), iconotype, lectotypified by Norlindh (1943: 375). FIGURE 6.5.**3**.

> *Osteospermum moniliferum* L., Sp. Pl. **2**: 923 (1753). —De Candolle, Prodr. **6**: 460 (1838).
>
> *Lepisiphon dentatus* Turcz. in Bull. Soc. Imp. Naturalistes Moscou **24**(1): 180 (1851). Type: [St Helena:] 'In insula Sanctae Helenae. *Cumming* coll. n. 2450.' (BM, G, ?K[13], KW001000913, P00132069, UPS syntypes).

Shrub 1–3 m high, densely branched; stems terete, branches whitish- greyish- or fawnish-arachnoid-pubescent when young but usually soon glabrescent. Leaves alternate, cuneate base/petiole 2–30 mm long, lamina slightly coriaceous, obovate, oblanceolate or elliptic, 1–7.5 × 0.6–6.5 cm, base cuneate to attenuate, midrib prominant and pale beneath, usually less prominent above, secondary venation inconspicuous, margins serrate-dentate to denticulate, apex acute to rounded and mucronate, mucro recurved, arachnoid-pubescent when young, glabrescent, hairs eglandular, leaf surface with slightly raised lighter-coloured areas of hair bases following hair loss. Capitula 5–7 mm tall (to 12 mm with erect rays), solitary or few; pedicels 1–6 cm long, arachnoid-pubescent, usually rapidly glabrescent, with a few linear bracts 2–6 mm long; involucre broadly campanulate, phyllaries ± 20, 2–3-seriate, spreading in fruit, reflexed after, outer linear-lanceolate, margins ± scarious, inner ovate or broadly lanceolate, 3–9 × 1.5–4.5 mm, acute to acuminate, margins broad, scarious; receptacle convex, glabrous. Ray florets 7–10, ray limbs 9–13 × 2–4 mm, yellow. Disc florets ± 20, corollas 5–6 mm long, lobes ± 1 mm, yellow, hairy; anthers brown; styles yellow. Drupes reddish purple to black; stone (sub-)globose, about as long as wide.

The two subspecies are scarcely separable, the most distinctive characters are provided in the following key (modified from Norlindh 1943) if the two are to be separated.

Inner phyllaries 4–5 mm long, inner lanceolate to narrow-ovate; leaves usually broad-ovate to rotund; plants of coastal areas at elevations below 1250 m . . . . b) subsp. *rotundata*
Inner phyllaries 6–8 mm long, inner ovate to lanceolate; leaves obovate to broad-elliptic; plants of montane areas at elevations above 1200 m . . . . a) subsp. *septentrionalis*

a) Subsp. **septentrionalis** Norl., Stud. Calenduleae **1**: 396, figs. 42m, 44e, 47 (1943). — in Kirkia **11**(1): 159, fig. 9 (1978). —Beentje in F.T.E.A., Compositae **2**: 535 (2002). Type: 'FRIES, NORLINDH et WEIMARCK n. 4166 in Herb. Lund.' [Zimbabwe, 'In campo fruticoso ad villam Maistone prope pagum Rusapi, c. 1450 m.s.m, NORLINDH et WEIMARK, 1931, n. 4166.'] (LD holotype, S04-1817, SRGH, US01848340).

> *Osteospermum moliniferum* subsp. *septentrionalis* (Norl.) J.C.Manning & Goldblatt in Strelitzia **29**: 801 (2012).

---

[13] Although a duplicate was cited by Norlindh as being in 'K', none has been found.

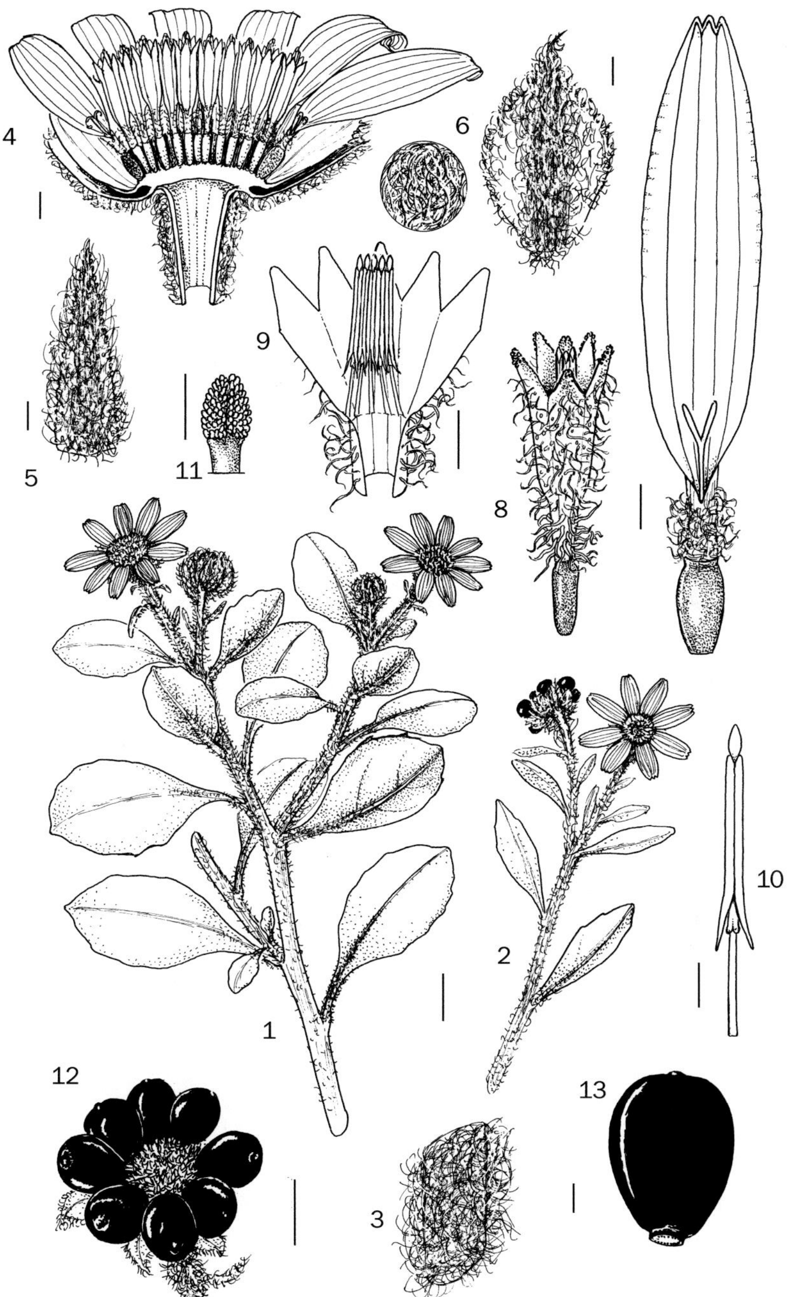

Fig. 6.5.**3**. CHRYSANTHEMOIDES MONILIFERA. 1, flowering shoot; 2, flowering and fruiting shoot; 3, detail of stem; 4, l.s. capitulum; 5, inner phyllary; 6, outer phyllary with detail of pubescence; 7, ray floret; 8, disc floret; 9, disc floret corolla opened out showing attachment of filaments; 10, stamen; 11, apex of disc floret style; 12, fruiting capitulum; 13, drupe. 1, 3–11 from *Wild* 1494; 2, 12–13 from *Hutchinson & Gillett* 3375. Scale bars: 11 = 0.4 mm; 10 = 0.5 mm; 3–9, 13 = 1 mm; 12 = 5 mm; 1–2 = 10 mm. Drawn by Margaret Tebbs.

Osteospermum moniliferum sensu Engler, Hochgebirgsfl. Afrika: 447 (1892). —Hoffmann in Pflanzenw. Ost-Afrikas C: 419 (1895). —Brenan in T.T.C.L.: 156 (1949). —Eyles in Trans. Roy. Soc. S. Africa **5**: 520 (1916), non L. s.str.

Shrub 0.5–2 m, sometimes described as scrambling, arachnoid pubescence on branches and pedicel persistent. Leaves obovate to elliptic, sometimes broadly so, 20–65 × 10–40 mm (excluding pseudopetiolate base to 15 mm), usually denticulate, teeth to 1 mm long to subentire. Phyllaries 3–9 × 2–4.5 mm wide, margins fimbriate, arachnoid-pubescent. Fruit ellipsoid-obovoid, 5–7 × 2.5–4 mm; stone slightly 3-angled with raised venation.

Zimbabwe. C: Harare Dist., Epworth, 1.iv.1956, *Wild* 4798 (K, PRE, SRGH). E: Nyanga Dist., Mt Nyangani (Nyanga Mt), Nyanga Nat. Park, 31.vii.1988, *Carter & Coates-Palgrave* 2054 (K). S: Ndanga Dist., near Bikita village, 20.x.1930, *Fries, Norlindh & Weimarck* 2123 (B, EA, K, LD, SRGH, UPS). **Malawi**. N: North Road between Mzimba and Kasungu, 23.viii.1946, *Brass* 17386 (K, NY, SRGH). C: between Bisiwini and Dedza, 4.1.1967, *Hilliard & Burtt* 4163 (E, K). S: Mt Mulanje, 2160 m, 6.vi.1970, *Brummitt* 11323 (K). **Mozambique**. T: Tete, between Vila Mouzinho and Zóbuè, 19.vii.1949, *Barbosa & Carvalho* 3696 (K, SRGH). MS: Serra Zuira, Tsetserra, c. 2100 m, 2.iv.1966, *Torre & Correia* 15577 (LISC).

Also present in Tanzania, Kenya and South Africa (Transvaal). Hillside grassland, bushed grassland, wooded grassland, in heath scrub, *Protea* scrub or montane forest margins; 1500–2350 m; flowering throughout the year.

Conservation Status: Widespread; not threatened; Least Concern.

The other subspecies of this species occur in South Africa and Mozambique. There seems to be quite some variation as well as intergradation, and it is quite possible they are all one and the same highly variable species. Norlindh (1943: 381) also pointed to the similarity with *C. monilifera* subsp. *canescens* (DC.) Norl. from the Drakensberg, which he called the Drakensberg race.

b) Subsp. **rotundata** (DC.) Norl., Stud. Calenduleae **1**: 391 (1943). —Hilliard, Compositae Natal: 527 (1977). Type: [South Africa:] 'in Africâ austr. propè Port Natal legit cl. *Drege*! [5051] … (v. s.)' (G-DC holotype).

> *Osteospermum rotundatum* DC., Prodr. **6**: 461 (1838).
>
> *Osteospermum macrocarpum* DC., Prodr. **6**: 461 (1838). Type: [South Africa:] 'in Africâ Capensis dist. Uitenhagen legit cl. *Ecklon*! … (v. s.)' *Ecklon* [& *Zeyher*] 9 (G-DC-G00472472 syntype), *Ecklon* [& *Zeyher*] 571 (G-DC-G00472563 syntype).[14]
>
> *Osteospermum moniliferum* var. *rotundatum* (DC.) Harv. in Harvey & Sonder, Fl. Cap. **3**: 436 (1865).

A well-branched bushy shrub to c. 3 m. Young stems and leaves arachnoid pubescent, later glabrescent. Leaves c. 80 × 15–35(45) mm, broadly obovate to rotund, base contracted into pseudopetiole or winged petiole c. 20 mm long, lamina coriaceous to subfleshy, shiny green, margins subentire, sometimes with few remote teeth, and ± revolute, apices obtuse and apiculate. Inflorescence of few-headed terminal corymbs, capitula pedicellate, pedicels 10–30 mm long, glabrous or arachnoid pubescent. Capitula c. 30 mm diam. across rays; phyllaries biseriate, 6–8 mm long, narrowly ovate or lanceolate, apices acute to acuminate. Ray florets yellow. Drupes subglobose, shiny, purplish-black when ripe, achene ellipsoid or obovoid, irregularly longitudinally ridged.

Mozambique. GI: Zavala (Quissico), viii.1936, *Gomes e Sousa* 1803 (COI, K, LISC). M: Maputo Dist., Maputo, Ponta Vermelho, 5.ix.1940, *Hornby* 2080 (K).

Widespread in Southern Africa from the coasts of Cape Province and Natal N to Mozambique. Now also naturalized, and an invasive weed, in Australia (New South Wales, Queensland) and Lord Howe Island; on sand dunes and in areas behind the main dune system, on cliffs, edges of *Typha* swamps, river margins, and forest margins at higher altitudes (according to Hilliard, 1977); 0–1250 m; flowering throughout the year.

---

[14] Norlindh (1943) listed two 'Ecklon & Zeyher s.n.' collections from Uitenhage District without further comment, although only one was cited for G-DC.

Conservation Status: Widespread; a somewhat weedy plant; Least Concern.

Beentje (2002: 537) suggested it was subsp. *septentrionalis* that was the weedy entity in Australia; Norlindh (1943) did not specify, citing Australian material under the species itself, although later (Norlindh, 1978: 159) confirming its Australian distribution.

Norlindh (1978) indicated *Gomes e Sousa* 1803 was from MS, which is contradicted by the label information; doubtless the taxon is widespread along the coast in suitable habitats, just that representative collections have not been made and/or seen – it is certainly common on Inhaca Isl.

115. **DIMORPHOTHECA** Moench

Dimorphotheca Moench, Methodus: 585 (1794). —Norlindh, Stud. Calenduleae **1**: 38 (1943). —Nordenstam in Kubitzki, Fam. Gen. Vasc. Pl. **8**: 245 [2006a](2007).
Gattenhoffia Neck., Elem. Bot. **1**: 39 (1790), nom. inval., published in opera utiq. oppr.
Meteorina Cass. in Bull. Sci. Soc. Philom. Paris **1818**: 167 (1818).
Blaxium Cass. in Dict. Sci. Nat., ed. 2, **30**: 328 (1824).
Arnoldia Cass. in Dict. Sci. Nat., ed. 2, **30**: 330 (1824).
Castalis Cass. in Dict. Sci. Nat., ed. 2, **30**: 332 (1824).
Dimorphotheca sect. *Meteorina* (Cass.) DC., Prodr. **6**: 70 (1838).
Dimorphotheca sect. *Blaxium* (Cass.) DC., Prodr. **6**: 71 (1838).
Dimorphotheca sect. *Castalis* (Cass.) DC., Prodr. **6**: 72 (1838).
Dimorphotheca sect. *Arnoldia* (Cass.) DC., Prodr. **6**: 73 (1838).
Acanthotheca DC., Prodr. **6**: 73 (1838).
Osteospermum sect. *Acanthotheca* (DC.) Norl., Stud. Calenduleae **1**: 191 (1943).
Osteospermum sect. *Blaxium* (Cass.) Norl., Stud. Calenduleae **1**: 237 (1943).
Dimorphotheca sect. *Acanthotheca* (DC.) J.C.Manning & Goldblatt in Strelitzia **29**: 794 (2012).

Annual or perennial herbs, subshrubs or shrubs. Leaves alternate, entire or variously divided. Inflorescences of solitary capitula or corymbose. Capitula pedicellate, radiate, heterogamous; involucre broad-campanulate; phyllaries uniseriate; receptacle epaleaceous. Ray florets female and fertile or female and sterile or neuter, corollas yellow or orange, purple or white, sometimes darker beneath; style fertile or sterile, sometimes absent; achenes (when present) triquetrous or subterete, sometimes winged or with numerous wing-like protuberances, smooth or tuberculate, straight or slightly curved; pappus absent. Disc florets hermaphrodite or functionally male; style short-bilobed with an annular collar of hairs; achenes (when present) flattened with thickened margins, circular or elliptic to obcordate or obovate; pappus absent.

About 20 spp., predominantly in Southern Africa, and N to Angola and Zimbabwe. Only two spp. recorded for the Flora area.

Dimorphotheca barberae Harv. (*Osteospermum barberae* (Harv.) Norl.), *Dimorphotheca ecklonis* DC. (*O. ecklonis* (DC.) Norl.) and *Dimorphotheca fruticosa* (L.) Less. (*O. fruticosum* (L.) Norl.) are cultivated and do not appear to have naturalized in the Flora area.

Shrub or subshrub to 1 m; leaves pinnatifid-dentate to denticulate, 1–2(3) cm long; disc floret achenes c. 10 mm long; phyllaries 6–9 mm long. . . . . **1.** *cuneata*
Perennial herbs 5–25 cm tall; leaves entire, dentate, pinnatifid (to pinnatipartite), 5–8 cm long; disc florets achenes c. 5 mm long; phyllaries 5–8 mm long . . . . **2.** *zeyheri*

1. **Dimorphotheca cuneata** (Thunb.) DC., Prodr. **6**: 72 (1838). Type: not cited. [South Africa. Sutherland: Roggeveld, *Thunberg*] (UPS-THUNB holotype).
Calendula cuneata Thunb., Prodr. Pl. Cap. **2**: 164 (1800).
Calendula viscosa Andrews, Bot. Repos. **6**: t. 412 (1804). Type: 'The figure annexed [Plate CCCCXII] represents a new and very ornamental species of the genus *Calendula*, which we lately

discovered flowering beautifully, amongst many other rare plants in the charming collection of exotics at George Hibbert's, esq. Clapham; where it was lately raised from seeds, sent to him, we believe, from the Cape of Good Hope, of which country we understand it is a native'. Herbarium material unknown.

Arctotis glutinosa Sims in Bot. Mag. **33**: t. 1343 (1811). Type: 'Communicated by Messrs. LEE and KENNEDY, who raised it from Cape seeds about five years ago.' Herbarium material unknown.

Erect shrub to c. 1 m tall. Stems well-branched, young stems glabrous, glandular-viscid, densely leafy, internodes 1–5 mm long, older stems becoming woody. Leaves alternate, somewhat fleshy or rigid membranaceous or subcoriaceous, ± glandular and viscid, obovate to oblanceolate, rarely ± linear, 1–2(3) × 0.3–1 cm, base narrowly cuneate, margins pinnatifid-dentate to gross-dentate or denticulate, teeth 1–3, midrib conspicuous above, prominent beneath. Capitula solitary, terminal, pedicellate, pedicels 1–4(7) cm long, short stipitate-glandular; capitula nodding in fruit; phyllaries c. 13, uniseriate, subequal, lanceolate or linear-lanceolate, 6–9 × 1–2 mm, stipitate glandular outside, margins scarious, white, ciliate, apices acuminate. Ray florets similar in number to phyllaries, ray limb white above blue to violet (or reddish) and often copper-infused beneath (sometimes with white), 2–3 times longer than phyllaries. Disc florets numerous, yellow, equalling involucre. Ray achenes 5 6 × c. 2 mm, 3-angled to obpyramidate, incurved, angles slightly or markedly crenate, ± stipitate-glandular; disc achenes 10–12 × 8–12 mm, flattened obcordate to suborbicular, ± stipitate-glandular, margins inflated.

Botswana. SE: Southern Dist., Morapedi Ranch, 15.ix.1978, *Hansen* 3453 (BM, C, GAB, K, PRE, SRCH, UPS, WAG).

Widespread in South Africa; common in highveld and karroid plains; 30–1800 m.

Conservation Status: Although clearly undercollected in the Flora area, Norlindh (1943) indicated that it has the widest distribution in the genus; Least Concern.

2. **Dimorphotheca zeyheri** Sond. in Harvey & Sonder, Fl. Cap. **3**: 421 (1865). Types: [South Africa:] 'HAB. Zuurepoost, Stormberg, Sep. *Zey*[*her*]! Near Albert, *T. Cooper*! 771 (Herb. Sond., D.)' *Zeyher* Comp. 16 (S lectotype[15], B, P, W), lectotypified by Norlindh (1943: 69); *Cooper* 771 (Herb. Sond., D. syntype). FIGURE 6.5.**4**.

Perennial herb 5–25 cm tall; rhizome woody 2–6 mm diam. Stems solitary or few, simple or few-branched, terete, striate, stipitate-glandular. Basal leaves conspicuously rosulate, stem leaves moderately dense, upper leaves remote, alternate, linear-lanceolate or linear, 50(80) × 5(8) mm, rigidly membranaceous, subglabrous, margins callose, stipitate glandular, entire to remotely dentate or pinnatifid or rarely pinnatipartite, midrib prominent beneath, primary veins conspicuous. Capitula solitary, terminal; phyllaries c. 15, uniseriate, subequal, 5–8 × 1–1.5 mm, lanceolate or linear-lanceolate, stipitate-glandular, apices acuminate, ciliate. Ray florets equal in number to phyllaries, ray limb white above, purple to bluish or pale pink beneath, often infused brownish, 2–3 times length of involucre. Disc floret corollas yellow. Ray achenes 4–5 mm long, triangular-obpyramidal, angles irregularly crenate or subentire, glabrous, body ± smooth or rugose, apex truncate; disc achenes 5–6 × 4–5 mm, obcordate to suborbicular, laterally compressed, glabrous, margins inflated, apices emarginate.

Botswana. SW: Kgalagadi Dist., S Kgalagadi, 962 m, 18.xi.2009, *Gwafila et al.* MSB 632 (BNHBG, BNPGRC, K). SE: Kanye Dist., 17.xi.1948, *Hillary & Robertson* 592 (K, PRE). **Zimbabwe**. W: Bulawayo, 1220 m, ?1912, *Rogers* 13750 (K).

Also widespread in South Africa. Common in grassy veld; 600–1600 m.

Conservation Status: Relatively widespread, although poorly collected in the flora area; Least Concern.

---

[15] Norlindh (1943: 69) cited the 'Typus speciei: in Herb. Sonder, Stockholm', but only cited one relevant *Zeyher* collection from S, 'Wodehouse: Cis-Garipina, vom nördlichen Fuss der Sormbergen bis Buffelvalei am Garip, 1200–1500 m, [119.0] ZEYHER, Comp. 16'.

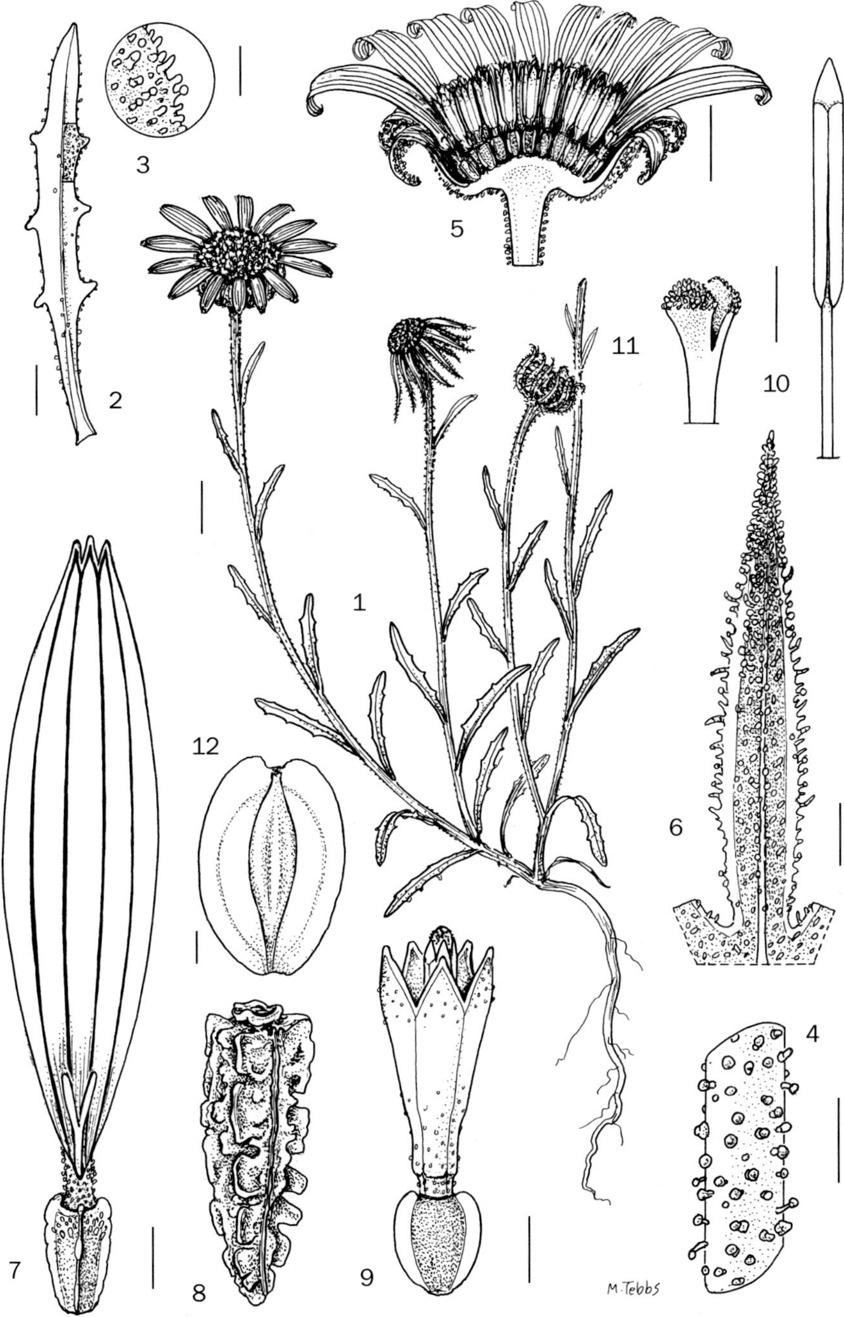

Fig. 6.5.4. DIMORPHOTHECA ZEYHERI. 1, branch of flowering plant showing rootstock; 2, leaf (showing detail of part of surface); 3, detail of leaf margin; 4, detail of stem; 5, l.s. capitulum; 6, phyllary; 7, ray floret; 8, ray floret achene; 9, disc floret; 10, stamen; 11, style apex of disc floret; 12, disc floret achene. 1, 8, 12 from *Gwafila et al.* MSB 632; 2–3 from *Rogers* 3750; 4–7, 9–11 from *Hillary & Robertson* 592. Scale bars: 3–4, 6–12 = 1 mm; 2, 5 = 5 mm; 1 = 10 mm. Drawn by Margaret Tebbs.

Tribe 13. **HELIANTHEAE** Cass.[16]

Heliantheae Cass. in J. Phys. Chim. Hist. Nat. Arts **88**: 189 (1819). —Wild in Kirkia **6**(1): 1–62 (1967). —Pope in Kirkia **10**(1): 101–121 (1975). —Robinson, Smithsonian Contr. Bot. **51**: 1–102 (1981). —Lisowski, (Asterac. Fl. Afr. Cent. 1) Fragm. Flor. Geobot. **36** Suppl. 1: 108–236 (1991). —Hind in Fl. Masc., Composées **109**: 168–226 (1993). —Karis & Ryding in Bremer, Asterac. Cladist. Classif.: 521–624 (1994) [Helenieae & Heliantheae]. —Beentje & Hind in F.T.E.A., Compositae **3**: 702–818 (2005). —Panero in Kubitzki, Fam. Gen. Vasc. Pl. **8**: 391–510 [2006](2007). —Baldwin in Funk *et al.*, Syst. Evol. Biogeogr. Compositae: 689–711 (2009) [Heliantheae alliance]. —Crawford *et al.* in Funk *et al.*, Syst. Evol. Biogeogr. Compositae: 713–730 (2009) [Coreopsideae].

Annual or perennial herbs, subshrubs, shrubs, rarely trees or vines. Rootstock fibrous, tuberous, or a woody or fleshy xylopodium. Stems prostrate, procumbent or erect, terete or distinctly angled, sometimes winged from decurrent leaf bases, usually unarmed, very rarely armed with nodal spines. Leaves opposite or sometimes alternate, or opposite below and alternate above in flowering parts, sessile, pseudopetiolate or petiolate, petioles winged or unwinged, lamina linear to ovate, deltoid or suborbicular, entire, lobed, variously dissected, or pinnatifid, single-veined to pinnately or palmately veined, usually pubescent with coarse, often scabrid, indumentum, sometimes conspicuously gland-dotted and often aromatic, usually unarmed, very rarely spiny. Inflorescences of solitary capitula, rarely scapiform, or panicles of few to many capitula, variously arranged into cymes, racemes or corymbs, sometimes of secondary aggregations into glomerules or synflorescences, surrounded by a secondary involucre. Capitula either heterogamous radiate, heterochromous or homochromous, rarely disciform, or homogamous, discoid, homochromous, florets usually hermaphrodite, rarely unisexual, rarely radiate and disciform capitula existing in one species; involucre usually herbaceous, rarely membranous or scarious; phyllaries in 1 to many series, when uniseriate often gland-dotted, subequal to unequal gradate and imbricate, occasionally distinctly biseriate with a distinct outer series; receptacle flat or convex, sometimes distinctly conical or even columnar, paleaceous, paleae carinate or flat, sometimes conspicuously conduplicate about achene and becoming ornamented with age, sometimes developing into a spiny bur (perygynium) enclosing achene, sometimes epaleaceous. Florets 1–∞. Ray florets, when present, female, sterile or neuter, rarely all florets female and fertile in dioecious capitula, uniseriate in radiate capitula, limb entire or 1–3-lobed, deciduous or rarely persistent, commonly yellow, but ranging from greenish, white, orange, and red to purple or atropurpureous. Disc florets usually numerous, hermaphrodite or functionally male, rarely all florets male in dioecious capitula, corollas greenish through white, yellow, orange, red to purple or atropurpureous, corolla lobes usually short, usually 5, but sometimes reduced to (2–3)4, glabrous, papillose or rarely pubescent; anther thecae often black, sometimes yellow; basal anther appendages often short-hastate, never tailed; apical anther appendages ovate, often with a constricted base; filaments usually glabrous, very rarely papillose or short-pubescent; style base sometimes enlarged, usually glabrous, style-arms truncate or appendiculate; pollen mostly echinate, but smooth in some wind-pollinated genera. Achenes few- (2–5) ribbed, terete, prismatic, compressed or obcompressed, often variously winged, sometimes apically awned, body usually black or dark brown with phytomelanin in the walls, glabrous or variously setuliferous, very rarely existing in a calvous form with both setuliferous and pappose and glabrous and epappose achenes in one capitulum, rarely achenes conspicuously bimorphic or very rarely heteromorphic in one capitulum, rarely achenes enclosed by accompanying phyllaries with surface variously ornamented or armed with spines and forming a distinctive bur (perygynium), sometimes achenes with a conspicuous neck, sometimes distinctly beaked; pappus usually of awns or scales, sometimes coroniform, absent, rarely of hairs, hairs barbellate or very rarely plumose, white or off-white, or variously coloured.

A tribe of over c. 280–300 genera, with c. 3400 spp., predominantly from the New World but with many genera in the Eastern hemisphere; 31 genera are recorded for the Flora area. The tribe includes several plants of horticultural and agricultural use, but also several pantropical weeds and a few species that are the main cause of hay fever in N America. Recent phylogenetic analyses are at odds with previous treatments that include *Welwitschiella*

---

[16] By D.J.N. Hind (except *Bidens* by Mesfin Tadesse, edited by D.J.N. Hind).

within the Heliantheae (incl. the Helenieae) (q.v. Hoffmann in Nat. Pflanzenfam. **4**, 5(54): 210–283, 1894; Pope 1975; Robinson 1981; Karis & Ryding 1994); Panero (2006) placed it in the Athroismeae; Brouillet *et al.* (in Funk *et al.* Syst. Evol. Biogeogr. Compositae: 589–630, 2009a; in Kew Bull. **64**(4): 589–630, 2009b) placed the genus in the Astereae subtribe Grangeinae – it is treated within the tribe Astereae, adjacent to *Grangea*.

Robinson's early work on the tribe (Robinson 1981) recognized 35 subtribes, and placed the Helenieae within a broadened tribal concept; this broad concept is accepted here. It should be noted, however, that Panero (2006) recognized some 12 tribes (in place of the broad Heliantheae) within the 'Heliantheae alliance', together with the associated 53 subtribes (still with several unassigned genera within some tribes). He also positioned the Eupatorieae within this complex. Jeffrey (in Bot. Zhurn. **89**(12): 1817–1822, 2004) provided an extreme view in recognizing the tribes as 'supersubtribes', still recognizing the Eupatorieae as part of this alliance. This was matched by Robinson (in Phytologia **86**(3): 116–120, 2004) in using the rank of 'supertribe' to include the Heliantheae. This has not helped the taxonomic hierarchy in this part of the family, neither view serving any useful purpose, especially considering the wholly impracticable key to the family that resulted (Jeffrey in Kubitzki, Fam. Gen. Vasc. Pl. **8**: 61–86, [2006] 2007), where several of the tribes in question had multiple entries.

Species of several genera are widely cultivated in the Flora Zambesiaca area including: **Coreopsis lanceolata* L., *Dahlia imperialis* Roezl ex Ortgies and **Dahlia* sp. (cultivar), *Dyssodia tenuiloba* (DC.) B.L.Rob., *Gazania rigens* var. *leucolaena* (DC.) Roessler, **Helianthus angustifolius* L., *Rudbeckia hirta* L., *R. laciniata* L., **Sphagneticola trilobata* (L.) Pruski, *Tagetes erecta* L. and *Zinnia* 'elegans'. Those with '*' are expected to escape and/or naturalize, or have already naturalized. When relevant they are keyed out under native or naturalized taxa in appropriate genera. Only one specimen of an apparently naturalized *Dahlia* has been seen; this is closest to a cultivar of *Dahlia coccinea* Cav. and would key out alongside *Coreopsis* in the generic key below but has considerably larger capitula and leaves, the plant perennating from tubers.

1.   Capitula unisexual, with separate male and female capitula on same plant 2
–   Capitula heterogamous, each capitulum with male and female florets, or florets hermaphrodite 3
2.   Fruit a bur covered in short, straight spines or tubercles; female capitula single-flowered . . . . **134. Ambrosia**
–   Fruit a bur (achenes enclosed by indurate and hardened phyllaries) covered in hooked spines; female capitula 2-flowered . . . .**135. Xanthium**
3.   Pappus of retrorsely barbed awns 4
–   Pappus lacking retrorsely barbed awns, pappus setae or elements capillary or of scales, antrorsely barbellate or plumose, or pappus absent 5
4.   Anther filaments pubescent; outer phyllaries usually erect or ascending; achene apices tapering or rostrate, achene body evenly furrowed . . . .**118. Cosmos**
–   Anther filaments glabrous; outer phyllaries usually recurved; achenes usually erostrate, achene body irregularly furrowed or terete . . . .**119. Bidens**
5.   Disc florets functionally male and achenes undeveloped 6
–   Disc florets hermaphrodite producing fertile achenes 7
6.   Ray floret achenes enclosed by indurate and spiny phyllaries forming a bur covered in hooked spines . . . . **143. Acanthospermum**
–   Ray floret achenes enclosed by accompanying unarmed palea that also encloses 2 adjacent disc florets . . . . **133. Parthenium**
7.   Phyllaries connate; oil glands present on leaves and phyllaries . . . . **122. Tagetes**
–   Phyllaries free to base; oil glands absent although plants may be variously glandular. 8
8.   Leaves alternate 9
–   Leaves opposite (although upper and inflorescence leaves or leaf-like bracts may be alternate) 13

9.   Leaves entire, simple or lobed. 10
–   Leaves pinnatisect, pinnate or bipinnate 12
10. Phyllaries uniseriate; paleae indurate about accompanying achene and falling with achene as a dispersal unit; pappus absent; ray limb small . . . . **138. Sclerocarpus**
–   Phyllaries few-seriate; paleae only conduplicate about accompanying achene and not falling with achene; pappus of persistent or caducous scales or awns; ray limb conspicuous, often showy 11
11. Leaves usually 3–5-lobed; pappus of persistent scales; pedicels inflated below involucre . . . . **139. Tithonia**
–   Leaves entire; pappus of 2 caducous/deciduous setae or awns; pedicels of uniform diameter beneath involucre . . . . **140. Helianthus**
12. Pappus absent; receptacle paleaceous, paleae flat; achenes compressed and conspicuously winged; leaves simple to bipinnatisect, lobes narrow-lanceolate to linear . . . . **116. Chrysanthellum**
–   Pappus of few scarious squamellae; receptacle epaleaceous; achenes usually quadrangular; leaves pinnate to bipinnate, lobes linear to filiform . . . **124. Schkuhria**
13. Receptacle conical to almost columnar . . . . **136. Acmella**
–   Receptacle flat to convex 14
14. Ray florets absent (in spp. in the Flora area); at least some pappus setae uncinate (sometimes straight setae alternating with uncinate setae); corollas orange to red. . . . . **123. Hypericophyllum**
–   Ray florets present (or sometimes sporadically and irregularly radiate and capitula in congested scorpioide cymes or glomerules in *Flaveria*); none of pappus setae uncinate; corollas white, yellow, pink or red. 15
15. Pappus absent. 16
–   Pappus of setae, scales, awns or coroniform (and with 1 or 2 fine setae or awns) . . . 23
16. Phyllaries conspicuously stipitate-glandular . . . . **146. Sigesbeckia**
–   Phyllaries lacking stipitate glands but sometimes variously pubescent or glandular-punctate 17
17. Ray florets white. 18
–   Ray florets (when present) yellow 19
18. Annual herbs; ray limbs 1–2 mm long; phyllaries erect in fruit and spreading after achenes have fallen; paleae filiform and weak, 1–2 mm long. . . . . **128. Eclipta**
–   Shrubs or small trees; ray limbs 3–35 mm long; phyllaries reflexed in fruit; paleae deltoid to triangular and conspicuous in fruit, to 22 mm long. . . . . . . **125. Montanoa**
19. Capitula in dense scorpioides cymes or glomerules, irregularly radiate in outer capitula. . . . . **121. Flaveria**
–   Capitula solitary or in lax corymbs or cymes, usually regularly radiate, sometime inconspicuously so 20
20. Phyllaries 4 in opposite pairs of 2; capitula axillary and sessile or subsessile; plants of wet habitats, perennial . . . . **120. Enydra**
–   Phyllaries > 4, biseriate and not in decussate pairs; inflorescences terminal and corymbose or cymose, rarely of solitary capitula. 21
21. Disc floret corolla tube glabrous. . . . . **130. Wollastonia**
–   Disc floret corolla tube pubescent. 22
22. Ray limb conspicuous, > 8mm long; ray floret achenes 3-angled, disc floret achenes 4-angled; leaves sessile and often connate at base . . . . **145. Guizotia**
–   Ray limb inconspicuous, c. 1 mm long; achenes monomorphic and 4-angled; leaves petiolate or pseudopetiolate and gradually narrowing to connate base **144. Micractis**
23. Ray floret corollas persistent on achenes; pappus of 1 or 2 awns . . . . **137. Zinnia**
–   Ray floret corollas deciduous; pappus of setae, scales, awns (and then ray limbs small) or coroniform (with or without awns). 24

24. Achenes dimorphic, ray floret achenes winged or obcompressed or obconical, disc floret achenes wingless and angled, or ray floret achenes trigonous or unevenly winged and disc floret achenes conspicuously winged; pappus of awns when present 25
–   Achenes monomorphic, obcompressed or winged; pappus of setae, scales, awns or coroniform with awns or fine setae (sometimes ray floret achenes epappose) 27
25. Ray floret achenes winged (wings notched or lobed) and with a 2-awned pappus, disc floret achenes wingless but with pappus of 2 conspicuous awns; inflorescences of several sessile axillary capitula . . . . **126. Synedrella**
–   Ray floret achenes obcompressed or obconical, or 3-angled; inflorescences corymbose and terminal (other spp. outside of the Flora area), scapiform on long pedicels or cymose (or paniculate) and sometimes with solitary axillary short-pedicellate capitula 26
26. Disc floret achenes winged, wings often corky and adnate to base of awns; ray limbs conspicuous . . . . **132. Verbesina**
–   Disc floret achenes usually 3-angled, wingless, 1–3 weak, barbellate awns arising from a narrow cupule; ray limbs inconspicuous . . . . **127. Blainvillea**
27. Pappus of numerous plumose setae . . . . **141. Tridax**
–   Pappus of scale, awns or coroniform with awns 28
28. Pappus of scales, scales with or without terminal awns . . . . **142. Galinsoga**
–   Pappus of awns and/or a laciniate crown 29
29. Pappus a laciniate crown with 1 or 2 setae or awns arising from crown . . **131. Aspilia**
–   Pappus of awns or teeth 30
30. Achenes distinctly flattened with spreading lateral wings, not corky (some spp. with wingless achenes outside Flora area); phyllaries in 2 distinct series, outer appressed or spreading, herbaceous, inner appressed, scarious; pappus of persistent fimbriate teeth (or absent outside of Flora area); paleae flat or slightly concave . . . . . **117. Coreopsis**
–   Achenes slightly compressed, few-angled or cylindrical, wingless; phyllaries 2–3-seriate, gradate, imbricate; pappus of few to several caducous antrorsely barbed setae/awns; paleae conduplicate about achenes . . . . **129. Lipotriche**

116. **CHRYSANTHELLUM** Rich.

Chrysanthellum Rich. in Pers., Syn. Pl. **2**: 471 (1807). —Wild in Kirkia **6**(1): 34–35 (1967). —Turner in Phytologia **51**(4): 291–293 (1982). —Ariza Espinar & Cerana in Bol. Soc. Argent. Bot. **22**(1–4): 267–273 (1983). —Turner in Phytologia **64**(6): 410–444 (1988). —Lisowski, (Asterac. Fl. Afr. Cent. 1) Fragm. Flor. Geobot. **36** Suppl. 1: 173 (1991). — Mesfin Tadesse, Fl. Ethiopia & Eritrea **4**(2): 338 (2004). —Beentje & Hind in F.T.E.A., Compositae **3**: 810–811 (2005).

Sebastiania Bertol., Lucubr. Re Herb.: 37 (1822), nom. illeg., non Spreng. (1820: 118) [EUPHORBIACEAE].

Collaea Spreng., Syst. Veg., ed 16 **3**: 622 (1826), nom. illeg., non DC. (1825: 96) [LEGUMINOSAE].

Chrysanthellina Cass. in Dict. Sci. Nat., ed. 2, **25**: 391 (1822), nom. illeg. superfl.

Adenospermum Hook. & Arn. in J. Bot. (Hooker) **3**: 318 (1841).

Hinterhubera Sch.Bip. in Flora **24**(1), Intell. 3: 42 (1841), nom. nud.

Annual, or rarely perennial, prostrate to erect, mostly glabrous herbs. Leaves predominantly alternate, rarely superficially opposite, simple to bipinnatisect. Capitula solitary, terminal or axillary, radiate, heterogamous, pedicellate; involucre biseriate, series subequal; receptacle flat to slightly convex, paleaceous; paleae membranous, persistent, linear. Ray florets female, fertile, ray limb yellow, 2- or 4-dentate; style arms filiform, apices acute. Disc florets hermaphrodite, fertile or functionally male throughout; corollas 4–5 lobed, tube and limb well-marked, funnel-shaped, outer floret corollas usually yellow about ½ size of inner, inner corollas usually brown or reddish brown; anther thecae small, apical anther appendages well developed, apices acute; style arms well-developed, linear, apices penicillate,

2–3 × length of stigmatic lines in fertile florets, or fused and bifid in functionally male florets. Achenes dimorphic, ray achenes usually clavate, wingless, straight or variously recurved or coiled, epappose; disc achenes absent or flattened, prominently winged, epappose or very rarely with 2 deciduous awns.

A genus of 11 spp., mostly from Mexico, but with one widespread pantropic weed, *Chrysanthellum indicum* DC. subsp. *afroamericanum* B.L.Turner, also present in the Flora area.

Chrysanthellum indicum DC., Prodr. **5**: 631 (1836). Types: 'in Indiâ orient. ad Gojpur et Sukanaghur. Bidens? Bardana Wall.! cat. et herb. n. 3231. comp. n. 401. Achillea? Bardana Hamilt. herb. ex Wall. … (v.s. comm. ab hon. Cur. Ind. or. et à cl. Royle.)' *Hamilton* 401 (G-DC-G00456653, K-W syntypes); *Royle* 174 (G-DC-G00456566 syntype).

Subsp. **afroamericanum** B.L.Turner, Phytologia **51**(4): 291 (1982). —Lisowski, (Asterac. Fl. Afr. Cent. 1) Fragm. Flor. Geobot. **36** Suppl. 1: 173–175 (1991). —Mesfin Tadesse, Fl. Ethiopia & Eritrea **4**(2): 338 (2004). —Beentje & Hind in F.T.E.A., Compositae **3**: 810–811 (2005). Type: 'ARGENTINA. Prov. Cordoba, Dcpt. Colon; Rio Ceballos, 15 Mar 1944, *C. A. O'Donell & J. M. Rodriguez* V. 501' (A holotype, F, LIL001588, UC). FIGURE 6.5.5.

    Adenospermum tuberculatum Hook. & Arn., J. Bot. (Hooker) **3**: 318 (1841). Types: 'Adenocarpum tuberculatum, *Don, mst.*–Province of Cordova [Saladillo & Los Caldenes]; *Dr Gillies* [73]. Cordova; *Tweedie*, (n. 1109)', 'ARGENTINA, Prov. Cordoba, Cordoba. "On hillsides and hard dry soils", w/o date, *L. Tweedie* 1107 (not 1109 as cited).' (K000497550 lectotype[17]), lectotypified by Turner (1988: 432); *Gillies* 73 (K000497551 syntype).

    Hinterhubera kotschyi Sch.Bip. in Flora **24**(1), Intell. 3: 42 (1841), nom. nud..[18]

    Plagiocheilus erectus Rusby, Mem. Torrey Bot. Club **4**: 212 (1895). Type: [Bolivia:] '[Cochabamba, 1891][*Bang*] 965.' (NY00232688 holotype[19], E00413898, GH00011308, GH00011311, K000221591, M0030215, MICH1107626, MIN1002054, US49209, US00145520, Z000003799).

    Chrysanthellum boliviense Sch.Bip., Bull. Soc. Bot. France **12**: 82 (1865). —Linnaea **34**(5): 536 (Feb. 1866), nom. nud., based on *Mandon* 295.

    Chrysanthellum weberbaueri Chung, Phytologia **14**(6): 321 (1967). Type: 'Peru, *Weberbauer* 6465.' (F629085 holotype, GH00004801, S, US1876711).[20]

    Chrysanthellum tuberculatum (Hook. & Arn.) Cabrera, Bol. Soc. Argent. Bot. **15**(1): 117 (1973).

    Chrysanthellum indicum var. *afroamericanum* B.L.Turner, Phytologia **51**(4): 291 (1982). Type: see subspecies.

    Chrysanthellum argentinum Ariza & Cerana, Bol. Soc. Argent. Bot. **22**(1–4): 267 (1983). Type: 'ARGENTINA. Prov. La Rioja: Dpto. San Martín: Estancia El Tala, Bajo Hondo, potrero 21 *Anderson* 2629, 27-XI-1972.' (CORD holotype).

Glabrous branching annual herb 3–45 cm tall. Stems prostrate, ascending or erect, striate. Leaves alternate (radical leaves usually absent in older specimens), petiolate, petiole 10–30 mm long, ± widened and clasping stem at base, lamina slightly fleshy, 10–30(50) × 10–40 mm, ovate in outline, 2–3-pinnatifid, segments often lobed linear or very narrowly obovate, apices apiculate. Inflorescences of solitary capitula or of few-headed cymes, capitula pedicellate, pedicels to 6 cm long, sulcate. Capitula radiate, heterogamous, to c. 1 cm diam.; involucres short-cylindrical, 3–4 mm tall; phyllaries 3–4 mm long, oblong, obtuse, striate,

---

[17] Although determined on the sheet as 'Holotype'.

[18] Turner (1988: 433) cited a type (*Kotschy* 175) for this name giving a page reference in Flora 24(2) which is a prose text '*Zusätze und Verbesserungen zur Synopsis Florae Germanicae et Helveticae*'. Where the name does appear, q.v. there is no associated description, nor specimen citation. The binomial, followed by the generic name, also appeared in Flora 25(2): 419 (1842), although neither appeared with a validating description or diagnosis in Latin or German.

[19] Turner (1988: 433) oddly cited the holotype as in US; the material in NY is undoubtedly the holotype material.

[20] In Brako & Zarruchi, Monogr. Syst. Bot. Missouri Bot. Garden **45** (1993), this was accepted as a distinct species in opposition to Turner's revision.

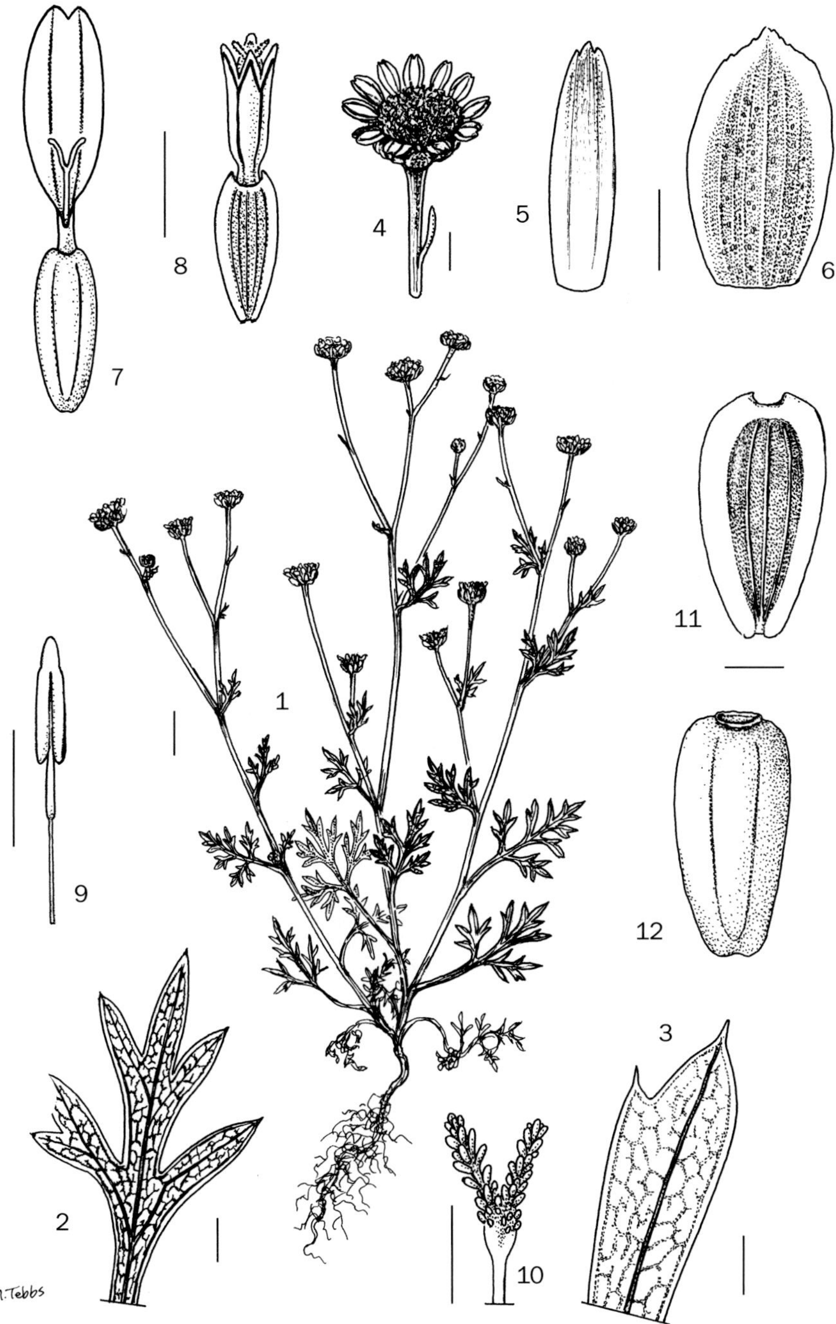

Fig. 6.5.**5**. CHRYSANTHELLUM INDICUM ssp. AFROAMERICANUM. 1, flowering shoot; 2, adaxial view of leaf apex; 3, detail of abaxial view of leaf lobe; 4, capitulum; 5, palea; 6, phyllary; 7, ray floret; 8, disc floret; 9, stamen; 10, style of disc floret; 11, disc floret achene; 12, ray floret achene. 1, 6, 11 from *West* 2686; 2–5, 7–10 from *Bingham* 8857; 12 from *Harder et al.* 2514. Scale bars: 10 = 0.25 mm; 9 = 0.5 mm; 2–8 = 1 mm; 1 = 10 mm. Drawn by Margaret Tebbs.

margins scarious; receptacle shallow concave; paleae persistent, 3–3.5 mm long, linear, obtuse at apex, 3-veined. Ray florets yellow or orange-yellow, ray limb 1–2.5 mm long, oblong, 2-lobed, lobes sometimes emarginate, tube 0.5–0.7 mm long; style arms very short. Disc florets (8)13–34, corollas orange, 0.8–1.8 mm long, corolla tube c. 0.6 mm long; style arms c. 0.7 mm long. Achenes 2–6 mm long, ray achenes oblong-cylindrical ± 3-angled, disc achenes compressed, apex emarginate, setuliferous; epappose.

Zambia. N: North Luanga National Park, 700 m, 25.i.1994, *Smith* 157 (K). C: Iolanda, near Kafue Town, 14.iii.1965, *Robinson* 6425 (K, SRGH). S: Livingstone, 3.ii.1961, *Fanshawe* 6174 (K, SRGH). **Zimbabwe**. W: Wankie Game Reserve, 15.ii.1956, *Wild* 4728 (K, SRGH). C: Salisbury, 29.ii.1960, *Phipps* 2530 (BM, COI, EA, K, LD, SRGH). S: Livingstone Dist., ?1911, *Gairdner* 563 (K). **Malawi**. N: Karonga Dist., Sangilo Point, 705 m, 24.ii.1978, *Pawek* 13881 (K, MA, MO). C: Kota Kota, 1.iii.1944, *Benson* 77 (PRE). S: Zomba, 23.ii.1956, *Banda* 198 (BM, SRGH). **Mozambique**. N: Maniamba, 27.v.1948, *Pedro & Pedrógão* 3885 (EA, LMJ). ?N or ?T: Ngame Valley, 3.ii.1942, *Hornby* 3562 (PRE). Z: Mocuba, Namagoa, iii. 1944, *Faulkner* 49 (EA, K, PRE).

A widespread tropical weed throughout tropical Africa from the Sudan and W Africa southwards to Zimbabwe, but apparently absent from Botswana; also widely distributed in S America. Turner (1988: 433) suggested that it was possibly introduced into Africa in 'relatively recent times'. Often common in damp situations, a weed of gardens, roadsides, heavily grazed grassland, or in dry soil. Common in the Flora area; 60–1900 m.

Conservation Status: Least Concern.

This taxon has been widely cited under the name *Chrysanthellum americanum* (L.) Vatke in Abh. Naturwiss. Vereins Bremen **9**(2): 122 (1885) and its synonyms. —Moore in Fawc. & Rendle, Fl. Jamaica **7**: 256, fig. 87 (1936). —Andrews, Fl. Pl. Sudan **3**: 17, fig. 2 (1956). —Adams in F.W.T.A., ed. 2, **2**: 234 (1963). The name *Chrysanthellum procumbens* Pers., Syn. Pl. **2**: 471 (1807), nom. illeg., has also been cited in synonymy, but this is a synonym of *Ch. americanum* only. —Oliver & Hiern in F.T.A. **3**: 395 (1877). —Eyles in Trans. Roy. Soc. S. Afr. **5**: 515 (1916). —Mendonça, Contrib. Conhec. Fl. Angola **1** (Compositae): 106 (1943). — Lisowski, (Asterac. Fl. Afr. Cent. 1) Fragm. Flor. Geobot. **36** Suppl. 1: 173 (1991).

117. **COREOPSIS** L.

Coreopsis L., Sp. Pl. **2**: 907 (1753). —Sherff, Publ. Field Mus. Nat. Hist., Bot. Ser. **11**(6): 279–475 (1936). —Robinson in Fl. Ecuador. **77**(1), 190(6): 113–116 (2006). —Mesfin Tadesse & Crawford, Syst. Bot. Monogr. **114**: 1–283 (2023).

Acispermum Neck., Elem. Bot. **1**: 34 (1790), nom. inval., published in opera utiq. oppr.

Coreopsoides Moench, Methodus [Moench]: 394 (1794).

Anacis Schrank in Denkschr. Königl. Akad. Wiss. München **5**: 6 (1817).

Leachia Cass. in Dict. Sci. Nat., ed. 2, **25**: 388 (1822), nom. superfl., incl. type of *Coreopsis* L.

Calliopsis Rchb., Icon. Descr. Pl. Cult. (Mag. Aesth. Bot.) **1**: t. 70 (1823).

Diplosastera Tausch, Hortus Canal. **1** (Decas Prima): [unpaginated] t. 4 (1823).

Chrysomelea Tausch, Hortus Canal. **1** (Decas Prima): [unpaginated] sub t. 4 (1823), nom. illeg., based on *Coreopsoides* Moench.

Chrysostemma Less., Syn. Gen. Comp.: 227 (1832).

Conopsis Nutt. ex Less., Syn. Gen. Comp.: 228 (1832), orth. err. pro *Coreopsis* L.

Leptosyne DC., Prodr. **5**: 531 (1836).

Electra DC., Prodr. **5**: 630 (1836), nom. illeg., non Panz. (1813: 299) [= *Schismus* P.Beauv., POACEAE].

Epilepis Benth., Pl. Hartw.: 17 (1839).

Coreopsis L. sect. *Gyrophyllum* Nutt. in Trans. Amer. Philos. Soc., ser. 2, **7**: 358 (1841).

Coreopsis L. sect. *Eublepharis* Nutt. in Trans. Amer. Philos. Soc., ser. 2, **7**: 359 (1841).

Coreopsis L. sect. *Calliopsis* (Rchb.) Nutt. in Trans. Amer. Philos. Soc., ser. 2, **7**: 360 (1841).

Tuckermannia Nutt. in Trans. Amer. Philos. Soc., ser. 2, **7**: 363 (1841).

Pugiopappus A.Gray in Pacif. Rail. Rep. **4**: 104 (1857).

Coreopsis L. sect. *Leptosyne* (DC.) O.Hoffm. in Nat. Pflanzenfam. **4**, 5(54): 243 (1894).
Coreopsis L. sect. *Tuckermannia* (Nutt.) S.F.Blake in Proc. Amer. Acad. Arts **49**: 340 (1913).
Coreopsis L. sect. *Pugiopappus* (A. Gray) S.F. Blake in Proc. Amer. Acad. Arts **49**: 340 (1913).
Selleophytum Urb. in Repert. Spec. Nov. Regni Veg. **13**: 483 (1915).

Erect annual or perennial herbs, subshrubs or shrubs, poorly- to well-branched. Rootstock a taproot. Stems ± terete to strongly ribbed, glabrous to pilose. Leaves usually opposite and narrowly connate across node or sometimes alternate above especially within inflorescence, sessile to narrowly petiolate or pseudopetiolate, lamina simple and margins entire to dentate, to tripartite or ternately divided with pinnae dissected, in outline linear oblong, obovate or triangular, with lobes linear to ovate or obovate, apices obtuse to acute, surfaces without evident glandular dots. Inflorescences of solitary capitula or many-headed in simple to corymbose panicles, capitula pedicellate, pedicels glabrous or pilose to tomentose. Capitula heterogamous and radiate; involucres campanulate; phyllaries biseriate, rarely 3–4-seriate, outer herbaceous to submembranaceous, appressed to spreading, inner usually larger, scarious, membranous, brown to yellowish or reddish, margins yellowish; receptacle slightly to strongly convex, paleaceous, paleae flat or slightly concave, easily deciduous, lanceolate to linear, apices obtuse to acute or filiform, glabrous or sericeous along midrib. Ray florets 4–18(20), uniseriate, usually sterile, sometimes fertile, some sterile without style; corollas yellow, tube narrow, glabrous to pilosulous, limb oblong to oblanceolate, apex 3- or 4-lobed, papillose above. Disc florets 10–65, hermaphrodite; corollas yellow, glabrous to pilosulous, basal tube short, throat narrowly campanulate, lobes 4 or 5, triangular-oblong, mamillose inside; filaments glabrous and smooth; anther thecae and usually apical anther appendages blackish, anther bases scarcely to strongly hastate, without glandular dots; style base not inflated; style arms with paired stigmatic lines, sometimes with short caudate appendage and with longer sweeping hairs at base of appendage. Achenes obcompressed, biconvex or plano-convex, obovate to oblong, or elliptical, straight-sided or evenly narrowed above, erostrate, with or without lateral wings, glabrous to densely setuliferous on margins and midrib of adaxial surface with biseriate setulae, body black, usually with pale striae, base narrowed; carpopodium short, broad or sometimes indistinct; pappus of 2 awns, sometimes antrorsely barbed, or absent.

A genus of 18 spp. predominantly from N America, but with several cultivated species, two of which are widely cultivated and one naturalizing in the Flora area. *Coreopsis tinctoria* Nutt. (an annual with 1–3-pinnate leaves, and ray floret limbs often with a red basal blotch) is cultivated (as *C. atkinsoniana* Douglas ex Lindl.) but does not appear to have become naturalized. Similarly, a plant called '*Coreopsis grandiflora* Hogg ex Sw.' is cultivated and possibly escaped in Nkana township according to *Mutimushi* JMM 48 (K) – but this specimen is exceptionally long-petiolate for this species! *Coreopsis grandiflora* is a perennial with sessile or very short-petiolate 1(2)-irregularly pinnate or pedately-lobed leaves (rarely simple), and has much shorter disc floret corollas than *C. lanceolata* (3.3–4.8 vs. 6–7.5 mm).

Coreopsis lanceolata L., Sp. Pl.: 908 (1753). —Hind in Fl. Masc., Composées **109**: 196 (1993). —Beentje & Hind in F.T.E.A., Compositae **3**: 706 (2005). —Hind, Lambkin & Tebbs in Curtis's Bot. Mag. **40**(4): 463–478 (2023) [2024]. Type: 'Habitat in Carolina.' 'Bidens caroliniana, florum radiis latissimis, insigniter dentatis, semine alato per maturitatem convoluto' in Martyn, Hist. Pl. Rar., **26**, t. 26 (1728)[21], iconotype, lectotypified by Green, Prop. Brit. Bot.: 183 (1929). FIGURE 6.5.**6**.
   Coreopsis crassifolia Dryand. ex Aiton, Hort. Kew **3**: 253 (1789). Type: 'Nat. of Carolina. Introd. 1786, by Mr. Adam Robson. Fl. August–October.' Herbarium material unknown.
   Coreopsoides lanceolata (L.) Moench, Methodus [Moench]: 594 (1794).
   Coreopsis lanceolata var. α *glabella* Michx., Fl. Bor.-Amer. (Michaux) **2**: 137 (1803). Type: 'Hab. in Carolina' [but no distinction made between either variety.]
   Coreopsis lanceolata var. *villosa* Michx., Fl. Bor.-Amer (Michaux) **2**: 137 (1803). Type: 'Hab. in Carolina' [but no distinction made between either variety.]
   Leachia lanceolata (L.) Cass. in Dict. Sci. Nat., ed. 2, **25**: 389 (1822).

---

[21] Smith (Sida **6**: 141, 1976) also designated this plate as the neotype. A number of authors have also 'cited' the Herb. Clifford material (BM-HC 420 Coreopsis 1) as the lectotype.

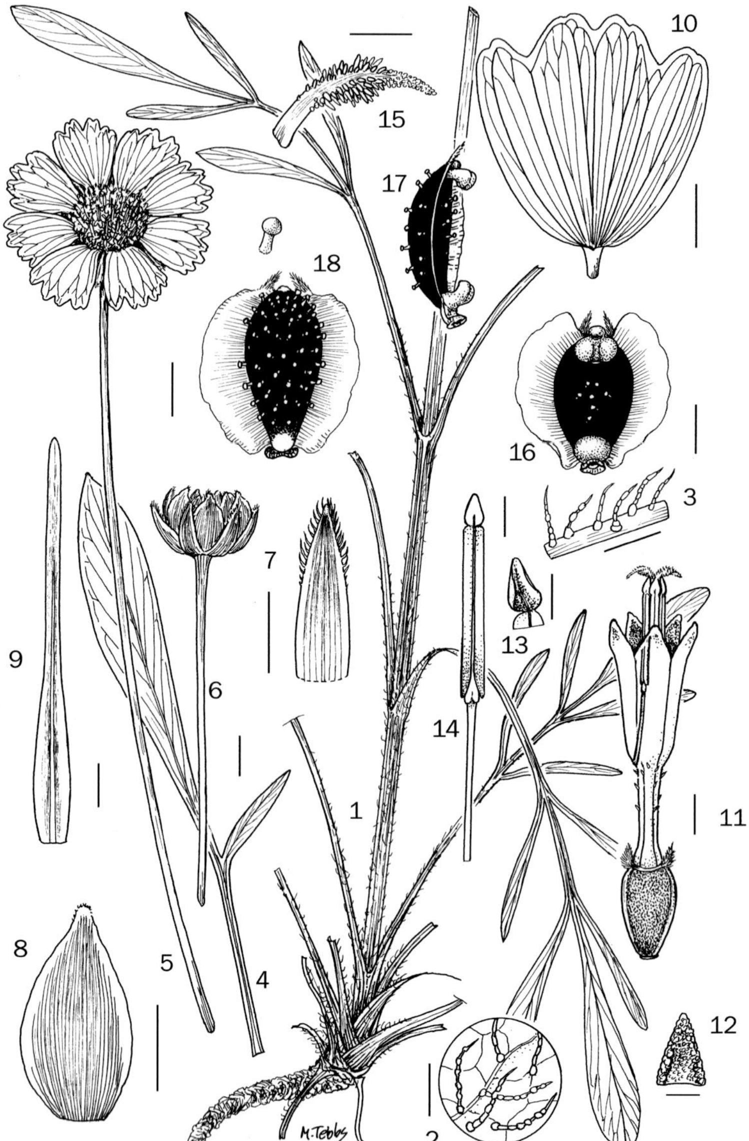

Fig. 6.5.**6**. COREOPSIS LANCEOLATA. 1, rootstock and pinnate basal leaves of flowering shoot; 2, detail of pubescence of abaxial leaf surface; 3, detail of leaf margin pubescence; 4, variant leaf form with a single-lobed lamina; 5, flowering capitulum and upper part of scape; 6, involucre post-flowering after loss of ray florets; 7, outer phyllary; 8, inner phyllary; 9, palea; 10, ray floret corolla; 11, disc floret, showing cut-away detail of attachment of filament within corolla; 12, detail of apex of corolla lobe of disc floret showing short-papillose margins; 13, detail of keeled apical anther appendage; 14, stamen; 15, detail of style arm of disc floret; 16, adaxial view of disc floret achene; 17, side view of disc floret achene, with closest wing removed, showing apical and basal 'bladders'; 18, abaxial view of disc floret achene with detail of knob-like projection. 1, 2 from *Bryant* 11; 3–5, from *Lye & Katende* 6506; 6–14 from *274-69-02341 – ex Cult. Kew*; 15–17 from *Sherff* 5014. Scale bars: 12 = 0.25 mm; 14, 15 = 0.5 mm; 13, = 0.8 mm; 2, 3, 9, 11, 16 – 18 = 1 mm; 7, 8, 10 = 5 mm; 1, 4 – 6 = 10 mm. Drawn by Margaret Tebbs. Repoduced from Curtis's Botanical Magazine (2023) [2024] with permission of the artist.

Coreopsis oblongifolia Nutt. in J. Acad. Nat. Sci. Philadelphia **7**: 76 (1834). Type: 'Hab. In North Carolina and Georgia.' (?GH00006181 holotype).[22]

Coreopsis lanceolata var. γ *crassifolia* (Dryand. ex Aiton) Heynh., Nomencl. Bot. Hort. **1**: 219 (1840).

Coreopsis lanceolata var. α *succisifolia* DC., Prodr. **5**: 570 (1836). Type: 'Haec varietas hortensis videtur typus speciei et generis. Dill. elth. 55. t. 48. f. 56. Linn. hort. cliff. 420. ... (v. s. c.)'.[23]

Chrysomelea lanceolata (L.) Tausch, Hort. Canal. **1** (Decas Prima): [unpaginated] sub t. 4 '*Diplosastera tinctoria*' (1823).

Coreopsis lanceolata var. β *angustifolia* Torr. & A. Gray, Fl. N. Amer. **2**: 344 (1842), nom. illeg.[24]

Coreopsis heterogyna Fernald in Rhodora **40**(480): 475 (1938). Type: 'VIRGINIA: rich alluvial woods and thickets back of sand-beach of James River, below Sunken Meadow Beach, Surry County, June 14, 1938, *Fernald & Long*, no. 8506' (GH00006175 holotype, PH00006711, US00128135).

Coreopsis lanceolata var. *pumila* Moldenke in Phytologia **26**(4): 224 (1973). Type: 'The type of this variety was collected by *Alma L. and Harold N. Moldenke* (no. 26471) on dry sandy road shoulders and in fields at Bostwick, Putnam County, Florida, on March 26, 1973, and is deposited in the Lundell Herbarium at the University of Texas.' (LL00373807 holotype).[25]

Perennial herbs, (10)20–60 cm tall. Stems erect or ascending, terete or scarcely angled, glabrous, branched, leafy towards base. Leaves opposite, lower leaves petiolate, petioles long and slender (1–8+ cm), upper sessile, lamina 5–15 × 0.8–1.8 cm, spathulate or linear to linear-lanceolate, usually simple, rarely 1–2 lobed, glabrous or ciliate at base. Pedicels (8)12–20(35) cm. Capitula 3–6 cm diam. at anthesis, 1–1.4 cm tall; outer phyllaries 8–10, 4–8 mm long, lanceolate or oblong-ovate, glabrous or apices pubescent, inner phyllaries 8–12 mm long, lanceolate-ovate or oblong-ovate; paleae 4–6 mm long, base oblong-linear, apex filiform. Ray florets c. 8, sterile, ray limbs 13–30 mm long, obovate or cuneate, yellow, apex 3-lobed. Disc floret corollas 6–7.5 mm long, yellow; style arm apical appendages caudate/cuspidate. Achenes 2.3–3(4) mm long, body black, obcompressed, orbiculate, winged, wings spreading, entire; pappus of 2 short fimbriate scales or awns (sometimes described as 'teeth').

Zimbabwe. C: there is no herbarium material of this species from the Flora area available for study[26].

A species native of central and southeastern USA, now frequently cultivated and with several well-known cultivars. In the FTEA region it was only recorded as a cultivated plant, not having apparently escaped; it was not recorded by Wild, or Pope, in earlier accounts of the Heliantheae for our Flora area although now apparently naturalized.

Conservation Status: Least Concern.

118. **COSMOS** Cav.

Cosmos Cav., Icon. **1**: 9, t. 14, 79 (1791). —Sherff in Publ. Field Columb. Mus., Bot. ser. **8**(6): 401–447 (1932). —McVaugh, Fl. Novo-Galiciana **12**: 262 (1984). —Hind in Fl. Masc., Composées **109**: 198 (1993). —Mesfin Tadesse, Fl. Ethiopia & Eritrea **4**(2): 335 (2004). —Beentje & Hind in F.T.E.A., Compositae **3**: 808 (2005).
Cosmea Willd., Sp. Pl., ed. 4, **3**: 2250 (1803), orth. var.
Cosmus Pers., Syn. Pl. **2**: 477 (1807), orth. var.

---

[22] This material is labelled 'N. Carol.'

[23] The Hort. Clifford material was also one of the elements cited by Linneus in Species Plantarum, as was the Dillenius plate.

[24] Citing *Coreopsis lanceolata* L. var. β [sic!] *glabella* Michx. (based on the Greek lettering in de Candolle's Prodromus) and *Coreopsis lanceolata* L. (based on Hooker's Fl. Boreali-Americana).

[25] Sherff (1936: 341) also listed '*Bidens caroliniana* Hemsl.' as apparently provided in the combined 'Index to the Botanical Magazine 1–130' (1906) on p. 27. The attribution of this name to Hemsley is in error, since Hemsley was merely taking the first two 'words' from the polynomials provided, in Curtis's Bot. Mag. **51**: Plate 2451 (1824), as synonyms of *Coreopsis lanceolata*; he was not providing a new name or combination.

[26] Originally a garden plant and now naturalized (possibly since the 1980s), and apparently locally common in the Harare and Vumba area on roadside verges.

Annual or perennial herbs, rarely subshrubs, erect, glabrous to hispidulous, sparsely branched. Rootstock a taproot and fibrous roots or of fasciculate fleshy tubers. Stems pale greenish to brown or reddish-tinged, terete to quadrate or hexagonal, often striate, pith solid. Leaves opposite, mostly ternately or quadriternately compound, dissected, petiolate, segments filiform to lanceolate, distal segments sometimes lobes or pinnatifid, sometimes with only central vein, lacking glandular dots or with minute dots. Inflorescence terminal and from axils of upper leaves, becoming long-pedicellate, sometimes with reduced bracteoles and alternate branching, capitula solitary or of few-headed lax corymbs. Capitula radiate, heterogamous; involucre broadly campanulate or cup-shaped; phyllaries biseriate, outer, usually 8, uniseriate, herbaceous, subulate to filiform or broadly elliptical, shorter than to longer than inner, with or without dark longitudinal lines, inner phyllaries c. 8, chartaceous, subequal, broadly oblong, dark or with dense dark longitudinal lines formed by resin ducts, margins narrow, scarious, apices short-acute, phyllaries glabrous to hispidulous; receptacle paleaceous, paleae similar to phyllaries, thinly chartaceous, glabrous, flat, oblong, margins scarious, apices short-acute. Ray florets 5 or 8, uniseriate (more in some cultivars), neuter; ray limb white, yellow to orange, pink or reddish-violet to atropurpureous, tube short, limb oblong, with or without numerous hairs along veins below, papillose above, apices blunt or sharply 3-lobed; style absent. Disc florets 13–100, hermaphrodite; corollas yellow to reddish-violet or atropurpureous, tube cylindrical to funnelform, upper tube and lower throat glabrous to short-pubescent, throat cylindrical, lobes 5, oblong-ovate to triangular, sometimes with hairs distally outside with fringe of dense hair-like papillae along inside margin, smooth to papillose on inner surface; anther filaments puberulous; anther thecae usually blackish; apical anther appendages ovate; style base not broadened, upper style shaft pilose; style arm apices flagelliform, densely short-pubescent with long papillae. Achenes ± quadrangular to slightly obcompressed, constricted above and becoming long-rostrate, lines along side scabrid, body blackish or dark brown; carpopodium broad, annular; pappus usually of 1–3 stout, sometimes reflexed, usually retrorsely barbed awns, sometimes absent.

A genus of c. 30 spp., predominantly Mexican, with three species frequently cultivated, these (*C. bipinnatus* Cav., *C. caudatus* Kunth, *C. sulphureus* Cav.) often escaping and naturalizing. Only two spp. are present in the Flora area, *Cosmos caudatus* has not yet been recorded but may well be found (keying out alongside *C. sulphureus* but with white, pink or purple rays).

Leaves pinnatisect, segments narrowly linear; paleae lanceolate, apices filiform or subulate; body and beak of achene not usually exserted from capitulum at maturity; ray limbs white, pink or lilac . . . . **1.** *bipinnatus*
Leaves pinnatisect, segments lanceolate to elliptic-ovate, rarely linear; paleae narrowly oblong, apices obtuse or attenuate, sometimes acuminate; body and beak of achene well exserted from capitulum at maturity; ray limbs yellow, sulphur yellow or orangeish . . . . **2.** *sulphureus*

1. **Cosmos bipinnatus** Cav., Icon. [Cavanilles] **1**: 10, pl. 14 (1791). —Martineau, Rhod. Wild Flowers: 90, t. 32 fig. 2 (1953). —Cronquist in Gleason, New Illustr. Fl. NE. U.S. & Adj. Canada **3**: 358 cum fig. (1955). —McVaugh, Fl. Novo-Galiciana **12**: 264–265 (1984). —Mesfin Tadesse, Fl. Ethiopia & Eritrea **4**(2): 336 (2004). —Hind in Fl. Masc., Composées **109**: 199 (1993). —Beentje & Hind in F.T.E.A., Compositae **3**: 808 (2005). Type: 'Habitat in Mexico. ⊙. Floruit in Regio horto Matritensi annis 1789 et 1790 mensibus Octob. Novemb. et Decemb.' (MA475586, MA475586 [sic!], MA475587, MA475588 syntypes).[27] FIGURE 6.5.**7A**.

---

[27] The original material in MA is represented by a few sheets. MA (475588 – Fiche 27/A4) is simply labelled *Cosmos bipinnatus* on a type written label; MA (475587 – Fiche 27/A5) bears three handwritten labels, the top 'Cosmos bipinnatus Cav./H. M. M. 1829', the middle 'Cosmos bipinnatus/H. B. M.', the bottom 'Cosmos bipinnatus Cav. ic. 7a. 14/Cosmea bipinnata Willd. Syng. pa 2250/ex Herb. Botani. Hispalens. anno 1797.', together with a typewritten label 'Cosmos bipinnatus Cav./M'ejic 1789-1790'; MA (475586 – Fiche 27/A6) bears just a typewritten label with the same text as MA 475587; MA (475586 [sic!] – Fiche 27/A7) bears a handwritten label 'Cosmos bipinnatus Cav. Icon. tab. 14/ Mexico 1789-1790' and a typewritten label with the name and 'Horto B. Hispalensis.' Sherff (1932) noted that the description was based on several specimens but did not select a lectotype.

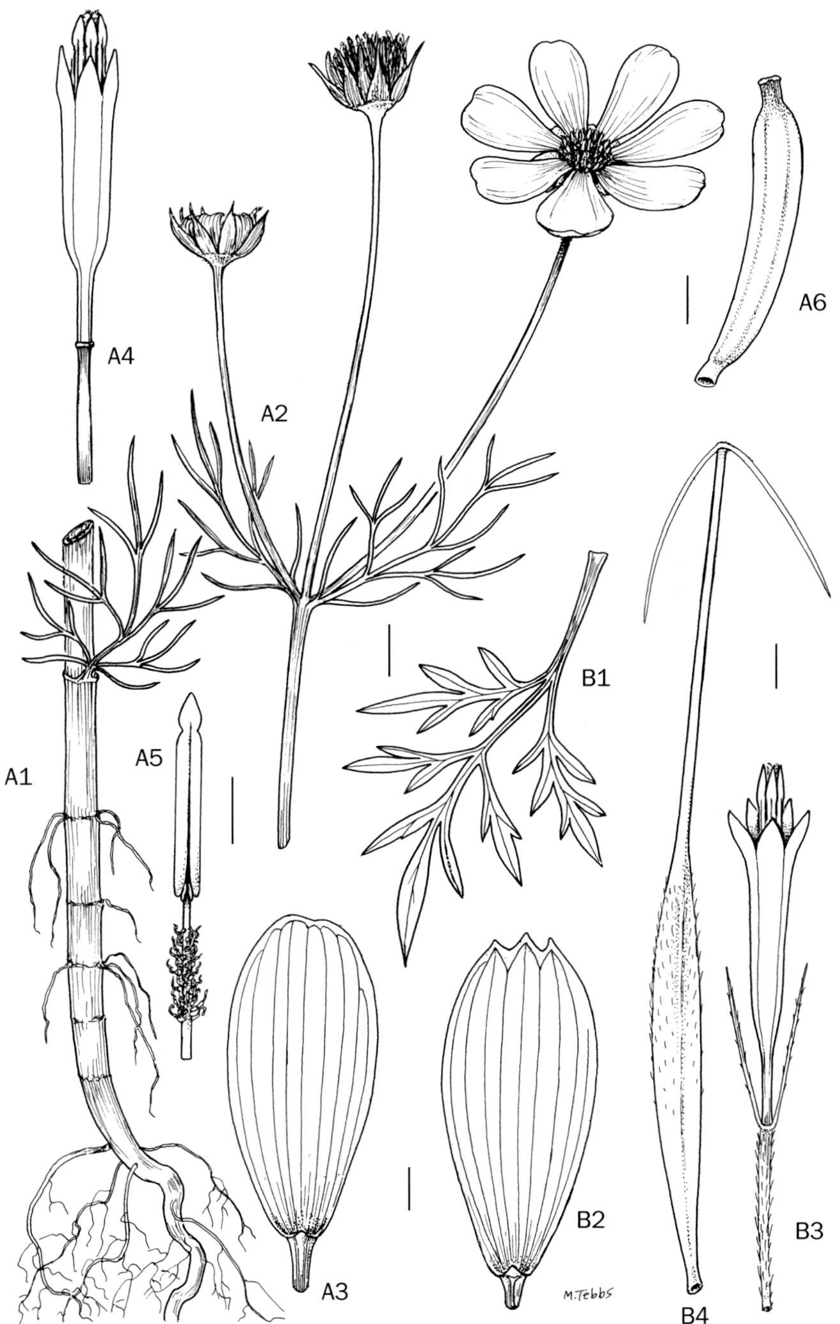

Fig. 6.5.7. A. —COSMOS BIPINNATUS. A1, base of plant showing rootstock; A2, flowering branch; A3, ray floret; A4, disc floret; A5, stamen showing pubescent filament; A6, epappose ray floret achene. B. —COSMOS SULPHUREUS. B1, leaf; B2, ray floret limb; B3, disc floret; B4, ray floret achene. A1–A5 from *Brummitt* et al. 14997A; A6 from *Brummitt* 9586; B1–B3 from *Drummond* 7409; B4 from *Chase* 7682. Scale bars: A3–A6, B2–B4 = 1 mm; A1–A2, B1 = 10 mm. Drawn by Margaret Tebbs.

Cosmos formosa Bonato, Pisaura automorpha e Coreopsis formosa: 22 (1793). Type: '..., è finalmente annua ed indigina del Perù, vive all' aria aperta, ...'. It is unclear if Bonato preserved material, nor where his herbarium might be located.

Cosmea bipinnata (Cav.) Willd., Sp. Pl., ed 4 **3**(3): 2250 (1803).

Georgia bipinnata (Cav.) Spreng., Syst. Veg., ed. 16, **3**: 611 (1826).

Cosmos bipinnatus Cav. var. β *exaristatus* DC., Prodr. **5**: 606 (1836). Types: '① in Mexico (*Alam.*!) ad Pezcuaro (*H. et B.*), Oaxaca (*Andr.*! pl. exs. n. 307). S. Angel. (*Berl.*! pl. exs. n. 938). Haec var. sola in hortis mihi notis nunc colitur et sola provenit in herb. Mexic. (v. s. sp. et v. c.)' *Alaman* s.n. (G-DC-G00456074 syntype); *Humboldt & Bonpland* 4321 (P0032436 syntype); *Andrieux* 307 (G-DC-G00456976, K000373019, K000497467, P02140492 syntypes); *Berlandier* 938 (G-DC-G00456975, P02140493, P02140494, P02140495 syntypes).

Cosmos tenuifolius Lindl., Edwards's Bot. Reg. **23**: pl. 2007 (1837). Type: 'A beautiful annual Mexican plant, ... The drawing was made in the garden of the Horticultural Society, where it has been raised from seeds presented by George Frederick Dickson, Esq. F. H. S. ...' (?CGE holotype).

Cosmea tenuifolia (Lindl.) J.W.Loudon, Ladies' Flower-Gard.: 185 (Feb. 1840); cf. Mabberley in Taxon **32**(1): 87 (1983).[28]

Cosmea tenuifolia (Lindl.) Heynh., Nomencl. Bot. Hort. **1**: 223 (Sept.-Oct. 1840), comb. superfl.[28]

Bidens formosa (Bonato) Sch.Bip in Seem., Bot. Voy. Herald: 307 (1856). —Kuntze, Revis. Gen. Pl. **3**(3): 137 (1898). —Lisowski, (Asterac. Fl. Afr. Cent. 2) Fragm. Flor. Geobot. **36** Suppl. 1: 172 (1991).

Bidens lindleii Sch.Bip. in Seem., Bot. Voy. Herald: 307 (1856), nom. nov. pro *Cosmos tenuifolius* Lindl.

Bidens bipinnata Baill., Hist. Pl. **8**: 50 (1882), nom. illegit. non L. (1753).

Cosmos hybridus Goldring, Gard. & Forest. **1**: 474 (1888). Type: not cited, although Goldring mentioned only that 'Though a Mexican plant ...'[29]

Cosmos ×*spectabilis* Carrière, Rev. Hort. **64**: 372 (1892).[30]

Cosmos ×*spectabilis* Carrière var. *alba* Carrière, Rev. Hort. **64**: 373 (1892).[30]

Cosmos ×*spectabilis* Carrière var. *rosea* Carrière, Rev. Hort. **64**: 374 (1892).[30]

Bidens lindleyii Sch.Bip. ex B.D.Jacks., Index Kew. **1**: 301 (1895), nom. nud. et superfl.

Annual herbs c. 1(2) m tall; taprooted. Stems ± 4-angled, sulcate, simple, usually well-branched in inflorescence, glabrous or minutely scaberulous-pubescent. Leaves sessile or with 2 ± widened and clasping petioles 4–7 mm long, lamina 6–11 × 3–5 cm, ovate in outline, 2- or 3- pinnately dissected, deeply so, ultimate segments 0.4–1.2 mm. wide, linear or linear-filiform, glabrous. Capitula to c. 2 cm diam. (excluding rays, to 8 cm including), solitary on terminal and side branches; pedicels 5–20 cm long or often much less, striate, scaberulous-pubescent or glabrescent. Involucre cup-shaped; outer phyllaries herbaceous, c. 1 cm long, ovate with a ± long-acuminate scaberulous apex, inner phyllaries membranous, ovate-oblong, obtuse and shorter than outer; paleae c. 7 mm long, oblong, with a linear appendage c. 2 mm long. Ray florets c. 8, limbs to 4 cm long, to 1.8 cm wide, oblong-obovate, white, pink or lilac, apex bluntly 3-dentate. Disc florets numerous (c. 60–100), c. 7 mm long, corollas cylindrical but widening slightly near apex, tube c. 2 mm long, yellow, corolla lobes c. 1.5 mm long; anther thecae dark-coloured; style arms with caudate appendages. Achenes blackish, 7–16 mm long, sulcate, beak 1–6 mm long, antrorsely scabrous; pappus of 2 or 3 short retrorsely-barbed awns or rarely absent (especially in cultivated material).

Zambia. S: Mumbwa, 20.iii.1963, *van Rensburg* 1730 (K, SRGH). **Zimbabwe**. W: Khami, iii.1944, *Martineau* 809 (SRGH). C: Gwelo, iv.1934, *Goldschmidt* (BM). E: Inyanga, 12.i.1931, *Norlindh & Weimarck* 4233 (LD). **Malawi**. N: Mzimba Dist., 2 mi. on Musese-Lwanjati road, near Champira, 1550 m, 24.iii.1978, *Pawek* 14114 (K, MA, MO). C: Dedza Dist., Massasa village, 1530 m, 27.iii.1977, *Brummitt et al.* 14997A-C (colour variants) (all K).

---

[28] Although Mabberley cited this as 1840, a date supported in TLII, the latter also notes a problem with dates, especially since the title page indicates 1842.

[29] Oddly, this name is attributed to 'Hort.', although it is quite clear that the article describing this plant was by William Goldring, as 'W.G.'. It is unclear if herbarium material was preserved.

[30] Although of hybrid origin the live material upon which Carrière based his descriptions was cited as 'Disons d'abord que les plantes que nous avons étudiées provenaient de graines que avaient été semées en février dernier dans les cultures de M. Forgeot, à Vincennes, ...'

From Arizona in the U.S.A. to central Mexico. Cultivated widely but not nowadays in our area since it became a widespread road side weed, although not appearing to extend beyond disturbed ground. Also a weed in the Transvaal. Cultivated for seed for export to a limited extent in Zimbabwe. 0–2400 m.

Conservation Status: LC (Least Concern).

2. **Cosmos sulphureus** Cav., Icon. **1**: 56, pl. 79 (1791). Type: 'Habitat in Mexico. 0. Floruit in Regio horto Matritensi mense Novembri.' (MA holotype).[31] —Andrews, Fl. Pl. Sudan **3**: 1 (1956). —Cronquist in Gleason, New Illustr. Fl. NE U.S. & Adj. Canada **3**: 358 (1955). —McVaugh, Fl. Novo-Galiciana **12**: 280–281 (1984). —Hind in Fl. Masc., Composées **109**: 199, t. 68 (1993). —Mesfin Tadesse, Fl. Ethiopia & Eritrea **4**(2): 336–337 (2004). —Beentje & Hind in F.T.E.A., Compositae **3**: 808–810 (2005). FIGURE 6.5.**7B**.

    Coreopsis artemisiifolia Jacq., Icon. Pl. Rar. [Jacquin] **3**: 16, pl. 595 (1793), as '*artemisiaefolia*'. Type: not cited.

    Cosmea sulphurea (Cav.) Willd., Sp. Pl., ed 4 **3**(3): 2250 (1803).

    Bidens sulphureus (Cav.) Sch.Bip. in Seem., Bot. Voy. Herald: 308 (1857). —Wild in Kirkia **6**(1): 32 (1967). —Lisowski, (Asterac. Fl. Afr. Cent. 2) Fragm. Flor. Geobot. **36** Suppl. 1: 171 (1991). —Maroyi in Kirkia **18**(2): 198 (2006).

    Cosmos aurantiacus Klatt in Leopoldina **25**: 105 (1889). Type: 'Crescit in campis Tacontenango pro Guatemala, flor. Dec., leg. *Gust. Bernoulli* 1865.' (B† holotype).[32]

    Bidens artemisiifolia (Jacq.) Kuntze, Rev in Gen. Pl. **1**: 321 (1891), comb. illeg. non Poepp. (1843) (= *Bidens triplinervia* Kunth).

    Cosmos sulphureus Cav. var. *exaristatus* Sherff in Publ. Field Columbian Mus., Bot. ser. **8**(6): 411 (1932). Type: 'Collected by *C. A. Purpus*, No. 6793, Sierra de Tonalá, State of Chiapas, Mexico, September, 1913 (Calif.).' (UC172476 holotype, F415644, GH53187, NY00076721, US567132).

    Cosmos artemisiifolius (Jacq.) M.R.Almeida, Fl. Maharashtra **3A**: 93 (2001).

Annual herb 0.6–2.1 m tall; taprooted. Stems ± 4-gonous, simple below, well-branched in inflorescence, sulcate pilose-pubescent or glabrescent. Leaves subsessile or petiolate, petiole 2–7 cm long (shorter on upper leaves), widening and clasping at base sparsely pilose or pubescent, lamina 7–15 × 5 cm, ovate, 2- to 3-pinnatisect, deeply so, ultimate segments 2–4(8) mm wide, narrowly oblong and apiculate, glabrous. Capitula 3–6.5 cm diam. (across rays), solitary on terminal and side-branches; pedicels 10–22 cm long but often much less, striate, glabrous. Involucre cup-shaped or campanulate; outer phyllaries herbaceous, c. 1.2 cm long, narrowly ovate, tapering to a scaberulous acuminate apex, inner phyllaries with membranous margins ± same length as outer narrowly oblong obtuse or subacute at apex; paleae c. 9 mm long, narrowly oblong, margins membranous, apices acute or subobtuse. Ray florets c. 10, limb rich orange or sulphur yellow, 1.8–3 × 1–1.7 cm, oblong-obovate, apex 3-dentate. Disc floret corollas orange-yellow, c. 7 mm long, narrowly funnel-shaped, corolla lobes c. 1.5 mm long; anther-thecae dark-coloured. Achenes blackish, 16–28 mm long, beak 5–12 mm long, antrorsely scabrid; pappus of 2 horizontally spreading retrorsely barbed awns, 4.5–7 mm long.

Zambia. B: Balovale, Makondu R., 25.iii.1961, *Drummond & Rutherford-Smith* 7321 (SRGH). W: 51 km. N. of Kabompo, 27.iii.1961, *Drummond & Rutherford-Smith* 7409 (SRGH). C: Lusaka, 3.iii.1957, *Noak* 138 (K, PRE, SRGH). **Zimbabwe**. N: Eldorado, 28.v.1941, *Hopkins* in GHS 8061 (K, SRGH); Mrewa Dist., Chivake River, 15.iii.1978, *Pope* 1664a (BR, K, LISC, and at least 5 other institutes!). C: Marandellas, iii.1949, *Dehn* in GHS 23086 (SRGH). E: Umtali, 1098 m, 5.iv.1951, *Chase* 3735 (BM, K, SRGH). **Malawi**. N: Mzimba Dist., Mzimba,

---

[31] There is one sheet in MA (475589 – Fiche 27/A8) which is taken as the holotype. There is one handwritten label on the sheet 'Cosmos sulphureus M./Icon. Tab. 79/Coreopsis artemisiefolia Jacq. Ic./tab. 525 Coll. 5. p. 155./Habitat in Mexico H. B. Mt.' and one typewritten label 'Cosmos Coreopsis sulpureus Cav./Méjico.'

[32] *Klatt expressly described this taxon in material cited as 'Compositae Guatemalenses et Costaricenses ex Herb. Mus. Berol. determinatae et novae descriptae'.*

c. 1370 m, 22.iii.1975, *Pawek* 9170 (K, MO, SRGH, UC). S: Chiradzulu Dist., 5 km from Chiradzulu on road to Limbe, c. 1015 m, 11.iv.1970, *Brummitt* 9791 (K).

Native in Mexico but occurs in the U.S.A. as a weed. Also recorded as a weed in Tanganyika. Stated to be cultivated in the Sudan (Andrews, loc. cit.) but is apparently cultivated no more in Zimbabwe where it has become a common roadside weed amongst long grass, especially in deciduous woodland. 0–3600 m.

Conservation Status: LC (Least Concern).

119. **BIDENS** L.[33]

Bidens L., Sp. Pl. **2**: 831 (1753) & Gen. Pl.: 362 (1754). —Sherff, Publ. Field Mus. Nat. Hist., Bot. Ser. **16**(1–2): 1–709 (1937). —Wild in Kirkia **6**(1): 1–62 (1967). —Mesfin Tadesse in Symb. Bot. Upsal. **24**(1): 1–138 (1984). —Rayner in Phytologia **73**(2): 77–97 (1992). —Mesfin Tadesse in Kew Bull. **48**(3): 437–516 (1993). —Mesfin Tadesse in F.T.E.A., Compositae **3**: 778–807 (2005). —Panero in Kubitzki, Fam. Gen. Vasc. Pl. **8**: 411 [2006](2007).

Annual or perennial herbs or shrubs. Rootstock fibrous or a taproot, sometimes stoloniferous. Stems terete or tetragonal, variously branched, glabrous or pubescent, hairs simple or rarely stellate. Leaves opposite, sessile or petiolate, lamina simple or variously compound, segments filiform to broadly ovate, margins entire, denticulate, serrate or variously incised. Inflorescences of solitary terminal capitula or cymose, capitula pedicellate. Capitula heterogamous and radiate or homogamous and discoid; involucre campanulate to hemispherical; outer phyllaries 1–2(3)-seriate, green, 1– to many-veined, linear to spathulate, herbaceous to coriaceous; inner phyllaries 1–2-seriate, oblong to ovate, membranous to subcoriaceous, grey- or orange-striate, margins scarious, connate below or free; receptacle flat to short-conical, paleaceous; paleae 2– to many-striate, membranous, white to yellowish-brown, oblong-linear, persistent. Ray florets when present uniseriate, neuter with aristate or exaristate ovary or pistillate and fertile with well-developed achenes and aristae, ray limbs yellow, golden-yellow, orange, rarely pink, purple, crimson, white, lemon-yellow or lilac, striate. Disc florets hermaphrodite, corollas yellow, 5-lobed; anthers brown to black or purplish, base caudate to sagittate, anther filaments glabrous; style bifurcate, style arm apices penicellate. Achenes obcompressed and flat or 4-angled, striate-sulcate, ribbed, unmargined or margins narrowly winged or callose-thickened, apex with or without aristae, aristae with or without antrorse and/or retrorse barbs.

A genus with 63 spp. in Africa and about 280 spp. worldwide, but predominantly from Mexico, Central and S America; 19 species are treated for the Flora area.

Wild (1967), and Mesfin Tadesse (1984), included all African species of *Coreopsis* within their concept of *Bidens*, a position accepted here, although Wild also included *Cosmos*, which is generically distinct and recognized separately in this Flora treatment.

1. Ray floret limbs pink, purple, crimson or lilac, rarely white but then leaf lobes linear to linear-oblong or linear-oblanceolate 2
  – Ray floret limbs yellow, golden-yellow, orange or if white, then leaflets broadly ovate or ovate-lanceolate, rarely rays absent 3
2. Outer phyllaries subulate, glabrous or glabrescent; achenes with pale or dark-brown tubercles on inner suface, outer surface glabrous, usually exaristate, occasionally with 2 entire, short (c. 0.3 mm long) aristae; inner phyllaries 8–16 × 4–8 mm in fruit . . . . **15.** *rubicundula*
  – Outer phyllaries linear-lanceolate, hispidulous; achenes setuliferous on both surfaces, biaristate, aristae 1–2 mm long, retrorsely barbed; inner phyllaries up to 10 × 3.5 mm in fruit . . . . **14.** *urceolata*

---

[33] By Mesfin Tadesse; edited by D.J.N. Hind.

3.   Annual herbs 4
–   Perennial herbs, subshrubs or shrubs 13
4.   Pappus (aristae) antrorsely barbed, rarely barbs absent or few 5
–   Pappus (aristae) retrorsely barbed or, rarely exaristate 6
5.   Leaves glandular-punctate (use ×10 lens); achenes narrowly obovoid, 1–1.7 mm wide,
      apex truncate; capitula 5–6.5 cm diam. at anthesis . . . . **9.** *steppia*
–   Leaves without clearly visible glands; achenes orbicular or broadly ovoid, 3–4 mm wide,
      apex 'V'-shaped; capitula 7–8 cm diam. at anthesis. . . . . **10.** *oligoflora*
6.   Leaf segments linear to filiform, 0.5–3 mm wide 7
–   Leaf segments ovate to lanceolate, 5–30 mm wide or more 9
7.   Ray floret limbs 6–7 × 1–1.5 mm; achenes beaked, with beak up to 24 cm long . . . . .
      . . . . **12.** *acuticaulis*
–   Ray floret limbs 9–20 × 3–6 mm; achenes not beaked 8
8.   Achenes without aristae, 3–3.5 mm long; ray floret limbs 10–15 mm long . . . .
      . . . . **11.** *malawiensis*
–   Achenes with retrorse barbs, 6–14 mm long; ray floret limbs 15–20 mm long . . . .
      . . . . **13.** *diversa*
9.   Achenes without aristae, 3–3.5 mm long . . . . **11.** *malawiensis*
–   Achenes with retrorsely or antrorsely-barbed aristae, 4–16 mm long 10
10.   Leaves glandular-punctate (using ×10 lens); capitula 2.5–4 cm diam.; achenes oblong-
      elliptic, flat or compressed, bifacial; ray limbs yellow, 6–20 × 5–7 mm . . . .
      . . . . **16.** *schimperi*
–   Leaves without visible glands; capitula, if radiate, 0.5–2 cm diam.; achenes linear-
      oblong, tetragonal; ray limbs yellow and 3–6 × 1–2.5 mm, or white and 7–15 ×
      3–4.5 mm, or absent 11
11.   Ray floret limbs yellow or yellowish-white; outer phyllaries linear, lanceolate or
      oblanceolate, usually much longer than inner phyllaries, marginal hairs up to 1 mm
      long; inner achenes much longer than others, exceeding phyllaries and paleae by half
      when fully mature 12
–   Ray floret limbs, when present, white or creamy-white; outer phyllaries spathulate, not
      exceeding inner phyllaries, marginal hairs up to 0.3 mm long; achenes more or less
      uniform. . . . . **19.** *pilosa*
12.   Leaves pinnately (3)5–9-lobed with ovate-lanceolate segments; outer phyllaries with
      long (0.5–1 mm) white hairs along margins; achenes densely setuliferous . **17.** *biternata*
–   Leaves 2- or rarely 3- pinnatisect with ovate to deltoid segments; outer phyllaries with
      minute (c. 1 mm) white hairs near apex and rarely also along margins; achenes hairy
      or glabrous . . . . **18.** *bipinnata*
13.   Leaves simple, rarely lobed or bilobed[34] 14
–   Leaves variously compound 17
14.   Capitula 6–8.5 × 1–1.8 cm; ray floret limbs 3–4.5 cm long; outer phyllaries (5)10–30 mm
      long . . . . **2.** *kilimandscharica*
–   Capitula 3.5–6(7) × 0.5–1 cm; ray floret limbs 1.5–3(4.5) cm long; outer phyllaries
      3–10(15) mm long 15
15.   Pappus retrorsely barbed; decumbent herb or shrub, 30–60 cm tall with 1–2.5 mm
      wide stem(s) from tuberous rootstock; leaves oblanceolate or elliptic, margins entire or
      regularly coarsely dentate from middle to apex . . . . **5.** *moorei*
–   Pappus antrorsely barbed or barbs absent; erect herb or shrub, up to 2 m high, with
      3–6 mm wide stem(s) from tuberous rootstock; leaves ovate; margins irregularly
      lobulate-serrate 16

---

[34] If both simple and compound leaves are on the same specimen or plant, try both leads of the couplet.

16. Achenes 9–11 × 3–4.5 mm, brown; outer phyllaries 7–11(15) mm long; ray floret limbs 8–13, 15–45 mm long . . . . **3.** *oblonga*
 —   Achenes 6–11 × 1–1.7 mm, black; outer phyllaries 6–7 mm long; ray floret limbs 6–8, 15–18 mm long . . . . **4.** *baumii*
17. Leaf segments linear to linear-elliptic, 1.5–8 mm wide 18
 —   Leaf segments ovate to ovate lanceolate, 10–40 mm wide 21
18. Capitula 1–3 cm diam. at anthesis; achenes 9–18 mm long . . . . **6.** *crocea*
 —   Capitula (3)3.5–7.5 cm diam. at anthesis; achenes 4–10 mm long 19
19. Ray floret limbs orange or golden yellow; achenes 2.5–3.5 mm wide . . . . . . **7.** *ochracea*
 —   Ray floret limbs yellow; achenes 1.0–1.7 mm wide 20
20. Plant with tuberous rootstock; leaves membranous, minutely glandular-punctate (using ×10 lens), hispid, rarely glabrous; capitula 3.5 –7 cm diam. at anthesis . . . . . **4.** *baumii*
 —   Plant rhizomatous; leaves thick, coriaceous; capitula 3–4.5 cm diam. at anthesis . . . . **8.** *kirkii*
21. Achenes 9–11 × 3–4.5 mm; pappus of outer achenes 2, naked except at base . . . . . . **3.** *oblonga*
 —   Achenes 4.5 × 10 × 1–2.5 mm; pappus of outer achenes absent, rarely 2 naked 22
22. Ray limbs 23–30 × 5–7 mm; leaves subtending inflorescence branches overtopping capitula . . . . **1.** *pinnatipartita*
 —   Ray limbs 20–45 × 10–15 mm; capitula above foliage . . . . **2.** *kilimandscharica*

1. **Bidens pinnatipartita** (O.Hoffm.) Wild in Kirkia **6**(1): 18 (1967). —Mesfin Tadesse in Kew Bull. **48**(3): 449 (1993). Types: [Tanzania:] 'Usafua: Poroto-Berg, Abhänge um 2300 m ([*Goetze*] n. 1041. –Blühend am 17. Juni 1899). ... Ein von *Buchanan* im Jahre 1891 im Nyassaland gesammeltes Exemplar (n. 380) gehört offenbar zu derselben Art.' *Goetze* 1041 (P00091460 lectotype, B†, BM000924446, BR8677761, K000410579, L0001853, P00091460, Z), lectotypified by Rayner (1992: 93); *Buchanan* 380 (K syntype).[35]

   Guizotia bidentoides Oliv. & Hiern in F.T.A. **3**: 386 (1877). Type: [Malawi:] 'Mozamb. Distr. Manganja Hills, 500–3,000 ft. alt., *Kirk*!'[36] (K00109233[37] lectotype – upper collection, the apical portion of branch (possibly with an associated capsule – although this may be for the whole sheet!, K001092339) lectotypified by Rayner (1992: 94).

   Coreopsis pinnatipartita O.Hoffm. in Bot. Jahrb. Syst. **30**(3–4): 432 (1901).

   Coreopsis lupulina O.Hoffm. in Bot. Jahrb. Syst. **30**(3–4): 432 (1901). Type: [Tanzania:] 'Usafua: Abhänge des Beya-Berges, um 2400 m ([*Goetze*] n. 1069. –Blühend und fruchtend am 27. Juni 1899).' (B† holotype, P00091461 lectotype, BM000924447, BR8677778, E00239250, K000410578, L0012056), lectotypified by Rayner (1992: 92).

   Coreopsis whytei S.Moore in J. Linn. Soc., Bot. **35**: 348 (1902). Type: [Malawi:] 'Mount Milanji, Nyassaland; *A. Whyte*, [Oct.] 1891, no. 35.' (BM000924448 holotype).

---

[35] Wild (1967: 18) selected *Buchanan* 380 in K as the neotype.

[36] The sheet in K is an interesting sheet with 4 shoots mounted on it from at least 2, possibly 3, collections that probably correspond to type material. There are 2 collections with 'Livingston's [sic!] Zambesi Expedition' printed labels; these have been interpreted as *Kirk's* collections (by Oliver & Hiern 1877, and Mesfin Tadesse 1993), even though they have 'Coll. C. J. Meller' printed on the bottom right hand corner. The supplementary handwritten labels give the altitudes – 'from a hill 500 ft. – 3 miles from river Shire.' (which is an excellent match for Meller's handwriting!) and 'Karizakwwo. Entr. to Banque Pass. 3000 ft.' (which is an excellent match for Kirk's handwriting!) It is the latter to which 'Maganja Hills' is printed on the associated 'Meller' label found on the bottom right hand corner of the sheet. If these are relevant to Oliver & Hiern's name there are at least 2 syntypes – this would correspond with the two blue stamps ('Herbarium Hookerianum 1867') which appear on the sheet.

[37] It is this specimen that has Kirk's handwriting on the strip of paper above the printed Meller label. The remaining pencil notes, and sketches, on the sheet are those of Oliver. The sheet has previously been determined by Brenan as '*Coreopsis pinnatipartita* O.Hoffm. (*Guizotia bidentoides* Oliv. & Hiern)', and by Wild as *B. pinnatipartita*. Rayner (1992: 94–95) misinterpreted the labelling and their authors.

Perennial herb or soft-wooded shrub 0.5–3 m tall (once described as a tree to 4.5 m); stems usually several from a thick rootstock, often tinged reddish, tomentose to glabrous. Leaves petiolate, petiole to 50 mm long, lamina pinnatisect to pinnatipartite, ovate in outline, 6–19 × 6–19 cm, leaf segments narrowly ovate or ovate, margins serrate to deeply lobed, pubescent above, pubescent to tomentose beneath. Inflorescence of solitary capitula solitary or capitula up to 5 together; pedicels to 3 cm long. Capitula radiate; involucre cylindrical, 7–15 mm tall; outer phyllaries leafy, reflexed at anthesis, 7–15 mm long, pubescent to tomentose, inner phyllaries yellow, 7–16 mm long, spreading, pubescent; paleae 6–15 mm long, usually with 3–6 purple lines. Ray florets neuter, 6–13, ray limbs yellow, 23–30 × 5–7 mm, tube 3–3.3 mm long, pubescent. Disc floret corollas yellow or orange-yellow, 5.5–8 mm long, sparsely puberulous in lower parts. Achenes shiny dark brown or black, obovoid, 4.5–7 × 1–2 mm, glabrous; pappus usually absent, rarely with two entire or distally barbed yellow aristae to 2.5 mm long.

Zambia. N: Nyika Plateau, side of road near Zambia Rest House, 2 miles from house, 2250 m, 14.xi.1967, *Richards* 22528 (K). E: Lundazi Dist., upper slopes of Kangampande Mt., Nyika Plateau, 2134 m, *White* 2747 (K). **Malawi**. N: Vipya, 10.vii.1952, *Jackson* 961 (K). C: Dedza Mt., 5.i.1967, *Hilliard & Burtt* 4177 (K). S: Mlanje Mt., west slope, 1850 m, 18.vii.1946, *Brass* 16864 (K, NY). **Mozambique**. N: Niassa, Litunde, a 2 kms a oeste da aldeia na base do monte, 900 m, 30.ix.1983, *Groenendijk & Dungo* 660 (K, LMU). Z: Zambezia, Serra do Guruè, 18.x.1949, *Barbosa & Carvalho* 4518 (K, LMJ).

Also in Tanzania and Angola. Forest margins, bushland, bushy grassland, secondary bushland; 900–2440 m.

Conservation Status: Relatively widespread, with many collections from the Flora area, although mostly from the same districts and division, and predominantly from Malawi; LC (Least Concern).

2. **Bidens kilimandscharica** (O.Hoffm.) Sherff in Bot. Gaz. **59**(4): 309 (1915). —Sherff, Publ. Field Mus. Nat. Hist., Bot. Ser. **16**(2): 606, fig. 174 (1937). —Wild in Kirkia **6**(1): 31 (1967). —Agnew, Upland Kenya Wild Fl.: 467 (1974). —Blundell, Wild Fl. E. Afr.: pl. 354 (1987). —Mesfin Tadesse in Kew Bull. **48**(3): 451 (1993). —Agnew & Agnew, Upland Kenya Wild Fl., ed. 2: 218 (1994). Types: 'Kilimandscharo (*Abbot* a. 1890); Landschaft Uschiri, häufig auf einem grasigen Plateau oberhalb des Lumiflusses (*Volkens* n. 398 – 14. Juni 1893); Landschaft Marangu, im Gebüsch am Wege zur Steppe (*Volkens* n. 537 – 6. Juli 1893).' *Volkens* 398 (BM lectotype, B, G), lectotypified by Rayner (1992: 90–91); *Abbot* s.n. (B† syntype); *Volkens* 537 (B†, HBG504239 syntypes).[38]

   Coreopsis kilimandscharica O.Hoffm. in Bot. Jahrb. Syst. **20**(1 & 2): 234 (1894).

   Bidens volkensii O.Hoffm. in Pflanzenw. Ost-Afrikas C: 415 (1895). —Sherff, Publ. Field Mus. Nat. Hist., Bot. Ser. **16**: 610, fig. 175 (1937). Type: [Tanzania:] '15.(Kl., am Quarefluss unterhalb Madwame – *Volk*[*ens*] n. 1694).' (B† holotype).

   Coreopsis crataegifolia O.Hoffm. in Bot. Jahrb. Syst. **30**(3–4): 431 (1901). Type: [Tanzania:] 'Livingstone-Gebirge: Yawulanda-Berg, auf unbewaldeten Abhängen, um 1800 m ([*Goetze*] n. 851. – Blühend am 18. April 1899).' (B† holotype, BM lectotype, BR8872142), originally selected by Sherff (1937), reselected by Rayner (1992: 89).

   Bidens robustior S.Moore in J. Linn. Soc., Bot. **35**(245): 349 (1902). —Sherff, Publ. Field Mus. Nat. Hist., Bot. Ser. **16**(2): 581, Pl. 158 (1937). —Wild in Kirkia **6**(1): 31 (1967). Type: [Kenya:] 'Hab. Masailand, Elmentaita at 6000 feet; *G. F. Scott Elliot*, no. 6846.' (BM000924458 holotype, K000410577).

   Bidens ukambensis S.Moore in J. Linn. Soc., Bot. **35**: 350 (1902). —Sherff, Publ. Field Mus. Nat. Hist., Bot. Ser. **16**: 609, Pl. 170 a–i (1937). Type: [Kenya:] 'Hab. Ukamba at 5–6000 feet; *G. F. Scott Elliot*, no. 6462.' (BM000924457 holotype, K000410576).

   Bidens crataegifolia (O.Hoffm.) Sherff in Bot. Gaz. **76**(2): 158 (1923). —Sherff, Publ. Field Mus. Nat. Hist., Bot. Ser. **16**(2): 605, Pl. 173 (1937).

---

[38] Sherff (1937) originally cited the *Abbott* collection as 'type', but specified that this 'was the first one cited by Hoffmann', something suggested by Rayner as 'artificial requiring re-lectotypification'.

Coreopsis leptoglossa Sherff in Bot. Gaz. **76**(2): 88 (1923). —Sherff, Publ. Field Mus. Nat. Hist., Bot. Ser. **11**(6): 382 (1936). Type: [D.R. Congo:] '*T. Kassner* 2871, on mountain slope, Lofuku River, Congo Free State, May 25, 1908' (B† holotype, BM000924451, K000410542, P00086757, P00086818, Z).[39]

Bidens kilimandscharica var. *retrorsa* Sherff in Bot. Gaz. **92**(2): 202 (1931). —Sherff, Publ. Field Mus. Nat. Hist., Bot. Ser. **16**(2): 607 (1937). Type: [Tanzania:] '*A. E. Haarer* 1472, at altitude of about 4000 feet, Doloti, Moshi District, German East Africa, August, 1928' (K000410574 holotype).

Bidens insignis Sherff in Publ. Field Mus. Nat. Hist., Bot. Ser. **17**(6): 591 (1939). Type: [Kenya:] '*Mrs. Brodhurst-Hill* 612, a very handsome plant, growing in large clumps, 5 feet tall, alt. about 7,000 feet, Menengai, Kenya Colony' (K(000311746 holotype, F0075217 – several fragments from the holotype).

Bidens kilimandscharica var. *oxymera* Sherff in Amer. J. Bot. **34**(3): 155 (1947). Type: [Kenya:] '*Miss E. R. Napier* 6,085, alt. 6,000 feet, Donyo, Sabuk, Kenya Colony, January 6, 1933 (type, Amani, ex Kew; ...)' (EA holotype, K000311745).

Bidens meruensis Sherff in Bot. Leafl. **5**: 18 (1951). Type: [Tanzania:] '*C. G. Van Someren*, alt. 4,000 ft., Meru [just west-southwest of Mt. Kilimanjaro, extreme northeastern Tanganyika], 1932' (EA holotype).

Bidens cuspidata Sherff in Bot. Leafl. **9**: 12 (1954). Type: [Mozambique:] '*I. B. Pole Evans & J. Erens* 483, herb under trees, on road to Vila Gauveia, Portuguese East Africa, May 31, 1938' (SRGH holotype, K000410672, P00086765, PRE0194317).

Bidens crataegifolia var. *burttii* Sherff in Amer. J. Bot. **42**(6): 561 (1955). Type: [Tanzania:] '*Bidens D. Burtt* 4,117, up to 6 feet tall, common locally, edge of herbaceous thicket and grass plateau, associated with vernonias, *Hagenia* zone, alt. 7,500 feet, west side of Mt. Meru, Arusha District, Tanganyika Territory, Sept. 28, 1932' (K000410575 holotype, EA).

Bidens leptoglossa (Sherff) Lisowski, (Asterac. Fl. Afr. Cent. 1) Fragm. Flor. Geobot. **36** Suppl. 1: 146 (1991).

Perennial herb 0.6–3 m high. Stems slightly woody at base, little branched, pilose to thinly pubescent. Leaves petiolate, petiole 20–40 mm long, lamina variable, pinnatipartite to bipinnatisect or uppermost and lowermost ones sometimes simple, 5–15 cm long, up to 10 cm wide, when lobed with ovate segments, when simple with serrate to lobed margins, teeth mucronate to apiculate, apex acute to acuminate, tomentose or hispid-pubescent on both surfaces. Inflorescences of terminal, solitary capitula or of few-headed leafy cymes, erect; peduncle 1–5 cm long, often much shorter, densely hairy near receptacle. Capitula radiate; involucre 6–14 × 8–12 mm; outer phyllaries green, 9–13, linear oblong or narrowly ovate to ovate, 5–14 × 1–4.5 mm at anthesis, pubescent on outside and on upper half on inside; inner phyllaries with scarious margins, pilose to pubescent, narrowly oblong to ovate-oblong, about same length as outer; paleae 5.5–14 mm long, linear or linear-oblong, obtuse. Ray florets neuter, 8–13, ray limbs 20–40 × 10–15 mm, yellow or orange, tube 3.5 mm long, pubescent. Disc floret corollas yellow, 5–6 mm long. Achenes black or dark brown, 5.5–10 × 1.5–2.5 mm, flat, setuliferous; aristae 2, divergent, to 2.5 mm long, naked or with 1 or 2 antrorsely- or retrorsely-set barbs at apex or at base.

Zimbabwe. E: Inyanga Dist., Stapleford, Hope Patrol, 1830 m, 5.iv.1962, *Wild* 5716 (K, MO, SRGH). **Malawi**. N: Karonga to Kondowe, 610–1830 m, vii.1896, *Whyte* s.n. (K). **Mozambique**. MS: Bracken/grassland W face Gorongosa Mt., 1220–1525 m, 10.vii.1969, *Leach & Cannell* 14297 (K, SRGH).

Also in Uganda, Kenya, Tanzania, D.R Congo, Burundi, and Angola. Grassland in rocky sites, scattered tree grassland, bushed or wooded grassland, bushland, forest margins; 900–3050 m.

Conservation Status: A widespread species, although still relatively poorly collected and apparently with a limited distribution in the Flora area; LC (Least Concern).

Leaf variation is considerable as to dissection and indumentum: 3-lobed leaves are very common, with the terminal segments the largest, but up to 11 lobes per leaf do occur.

---

[39] Rayner (1992: 91) considered only the material in Z fits the description, the right hand specimen being selected as the lectotype; Mesfin Tadesse (1993: 452) selected the K isotype as the lectotype.

3. **Bidens oblonga** (Sherff) Wild in Kirkia **6**(1): 30 (1967). —Mesfin Tadesse in Kew Bull. **48**(3): 455 (1993). Type: '*Alex. Carson* 106, Lake Tanganyika, East Equatorial Africa, September 1893' (K000410667 holotype).

 Coreopsis oblonga Sherff in Bot. Gaz. **76**(2): 80, fig. 7 (1923). —Sherff, Publ. Field Mus. Nat. Hist., Bot. Ser. **11**(6): 374, fig. 2 (1936).

 Bidens nyikensis Sherff in Bot. Gaz. **81**(1): 50 (1926). —Sherff, Publ. Field Mus. Nat. Hist., Bot. Ser. **16**(2): 613, Pl. 178 (1937). Type: [Malawi:] '*A. Whyte* 191, at altitude of 6000–7000 feet, Nyika plateau, British Central African Protectorate, June, 1896' (K000410671 holotype).

 Bidens rhodesiana Sherff in Bot. Gaz. **91**: 309 (1931). —Sherff, Publ. Field Mus. Nat. Hist., Bot. Ser. **16**: 608, Pl. 170 figs. j–p (1937). Type: [Zimbabwe:] '*A. J. Teague* 226, in the Odzani River Valley, District of Manica, District of Umtali, Southern Rhodesia, 1914' (K000410666 holotype, BOL139030).

 Coreopsis exilis Sherff in Bull. Jard. Bot. État Bruxelles **13**(4): 288 (1935). —Sherff, Publ. Field Mus. Nat. Hist., Bot. Ser. **11**(6): 385 (1936). Type: [D.R. Congo:] 'District of Upper Katanga: Katuba-Elisabethville, on farm Droogmans, May 1927, *Quarré* 419bis' (BR8677792 holotype).

 Coreopsis goffardi Sherff in Bull. Jard. Bot. État Bruxelles **13**(4): 290 (1935). —Sherff, Publ. Field Mus. Nat. Hist., Bot. Ser. **11**(6): 390 (1936). Type: [D.R. Congo:] 'District of Upper Katanga: Kipila-Elisabethville, on Goffard's farm, May 1929, *Quarré*, 1616' (BR8873071 holotype).

 Bidens dielsii Sherff var. *intermedia* Sherff in Amer. J. Bot. **42**(6): 562 (1955). Type: [Zambia:] '*A. A. Bullock* 2,740, Tanganyika Territory, 1949-1951' (K000410670 holotype, K)[40].

 Bidens richardsiae Sherff in Amer. J. Bot. **42**(6): 562 (1955). Type: [Zambia:] '*Mrs. Richards* 1,546, growing everywhere in the bush, a tall, handsome plant 3–4 ft., Chilongowelo, Abercorn District, Northern Rhodesia, Apr. 24, 1952' (K000410668 holotype).

 Bidens exilis (Sherff) Lisowski, (Asterac. Fl. Afr. Cent. 1) Fragm. Flor. Geobot. **36** Suppl. 1: 140 (1991). Type as for *Coreopsis exilis*.

Perennial herb, 1–2 m high, with tuberous rootstock; stem 4-angled, pilose to glabrous. Leaves petiolate, petiole to 50 mm long, lamina ovate, leathery, simple to pinnately 3–5-lobed, 4–22 × 3–15 cm, margins coarsely crenate to irregularly lobed, apex acute, hispid to glabrous. Inflorescence of solitary terminal capitula or capitula in few-headed cymes; peduncles 0.5–12(20) cm long, densely pilose to glabrous. Capitula radiate; involucre 6–10 × 8–10 mm; outer phyllaries 6–12, green, linear-spathulate, 7–11(15) × 1–2(3) mm at anthesis, up to 17 × 4 mm in fruit, pubescent at base and at margins in lower half to densely hispid-pilose throughout; inner phyllaries 8–10, oblong to oblong-obovate, 8–10 × 3–5 mm at anthesis, up to 15 × 5 mm in fruit, pilose, appearing much longer than outer ones in fruit; paleae 8–10 mm long. Ray florets neuter, 8–13, ray limbs yellow or orange, 15–45 × 5–10 mm wide, tube puberulous. Disc floret corollas yellow. Achenes brown, oblong, 9–11 × 3.5–4.5 mm, flat, setulose; aristae 2, 1–3 mm long, usually with a few antrorse barbs near base, otherwise nude.

Zambia. N: Abercorn Dist., Lake Chila, 1650 m, 22.iv.1962, *Richards* 16380 (K). **Zimbabwe**. E: Umtali Dist., Odzani Heights, 1525 m, 24.iii.1957, *Chase* 6369 (K, SRGH). **Malawi**. *Whyte* 191 – type of *B. nyikensis*, q.v.

Also in Tanzania, D.R. Congo and Angola. Rough grassland; 1000–1650 m.

Conservation Status: A widespread species, although relatively poorly collected in the Flora area with several of the collections the types of many of the synonyms; LC (Least Concern).

4. **Bidens baumii** (O.Hoffm.) Sherff in Bot Gaz. **59**(4): 309 (1915). —Sherff, Publ. Field Mus. Nat. Hist., Bot. Ser. **16**(2): 589, Pl. 163 (1937). —Lisowski, (Asterac. Fl. Afr. Cent. 1) Fragm. Flor. Geobot. **36** Suppl. 1: 163, fig. 37 (1991). —Mesfin Tadesse in Kew Bull. **48**(3): 460 (1993). Type: [Angola:] 'Zwischen Kuma und Kutsi auf sandigem Lehmboden in lichtem Wald sporadisch, 1300 m ü. M. ([*Baum*] Nr. 883, blühend und fruchtend am 2, Dezember 1900.)' (B† holotype, W19010009347

---

[40] The main collection, not the duplicate, with minimal and incorrect information on its label, was sent on loan to Sherff! The main label reads: '[Zambia:] *A. A. Bullock* 2740. Abercorn. 27.3.1950. alt. 5,500 ft. Occasional in *Brachystegia* woodland. An escape? Erect to 3 ft. Flowers to 3 ins. diam., sunflower yellow.'

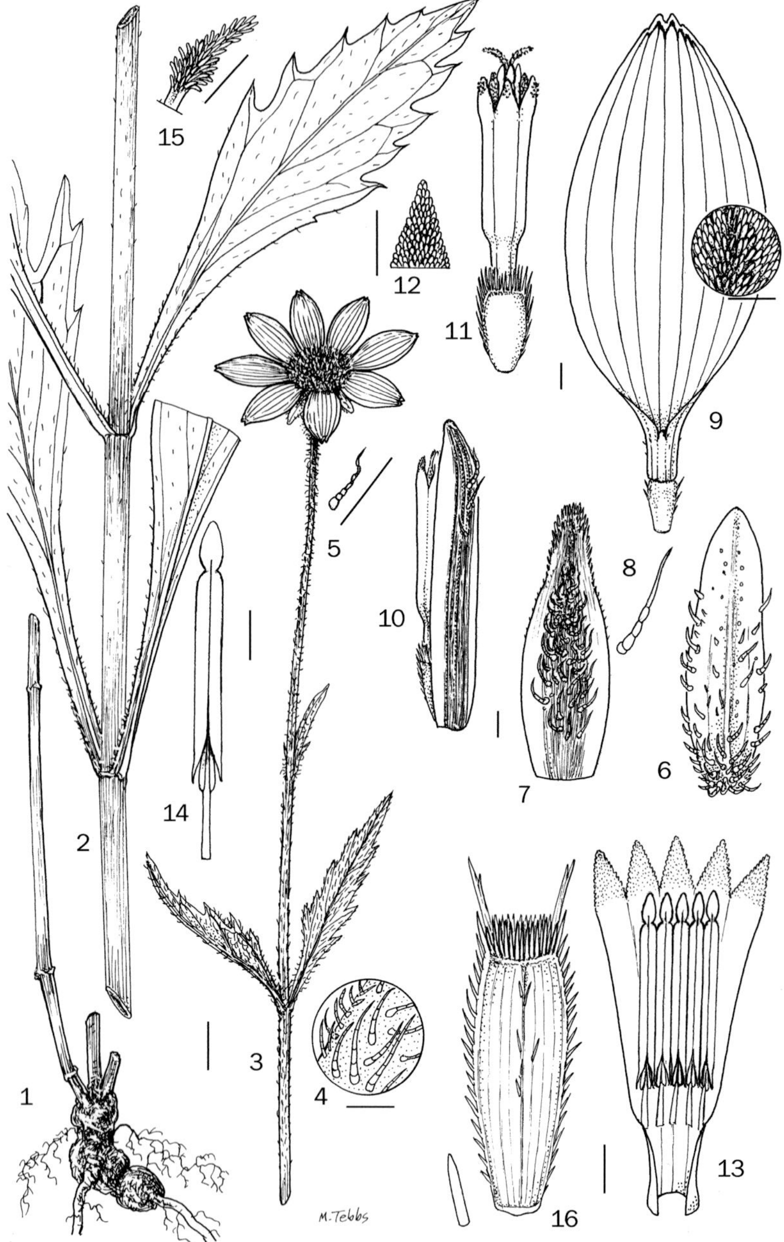

Fig. 6.5.**8**. BIDENS BAUMII. 1, Xylopodiaceous rootstock and current year's stem and remains of past seasons, stems; 2, portion of leafy mid-stem; 3, flowering stem; 4, detail of leaf surface; 5, detail of leaf hair; 6, outer phyllary; 7, inner phyllary; 8, detail of inner phyllary hair from midrib; 9, ray floret with inset detail of ray limb surface; 10, disc floret with accompanying palea; 11, disc floret; 12, detail of corolla lobe; 13, disc floret corolla opened out showing attachment of filaments; 14, stamen; 15, detail of apex of style arm; 16, disc floret achene with accompanying detail of setula from achene surface. 1, 3–4, 16 from *Brummitt* 10923; 2, 5–15 from *Phillips* 1740. Scale bars: detail on 9, 15 = 0.5 mm; 4–9, 11–16 = 1 mm; 1–3 = 10 mm. Drawn by Margaret Tebbs.

lectotype, BM000924475, COI, E00239240, G00018023, HBG504243, M0105197), lectotypified by Rayner (1992: 89). FIGURE 6.5.**8**.

    Coreopsis baumii O.Hoffm. in Warburg, Kunene-Sambesi-Exped.: 419 (1903).

    Coreopsis scabrifolia Sherff in Bot. Gaz. **76**(2): 86 (1923). —Sherff, Publ. Field Mus. Nat. Hist., Bot. Ser. **11**(6): 397 (1936). Type: [D.R. Congo:] '*T. Kassner* 2776, under trees, Kundelungu, Congo Free State, May 15, 1908' (B† holotype, BM000924477 lectotype, BM000924476, HBG504242, K000410540, K000410541, P00073759, P00073761, P00418384), lectotypified by Rayner (1992: 93).

    Bidens ruandensis Sherff in Bull. Jard. Bot. État Bruxelles **13**(4): 285 (1935). —Sherff, Publ. Field Mus. Nat. Hist., Bot. Ser. **16**(2): 590 (1937). Type: [Rwanda:] 'District of Ruanda-Urundi: Ruanda, Ruhengeri, *H. Scaetta* 427' [Route Kigali-Rubonito-Kilali-Ruhengeri (Ruanda), 1928, *Scaetta* 427] (BR8872470 holotype).

    Bidens somaliensis Sherff var. *bukobensis* Sherff in Publ. Field Mus. Nat. Hist., Bot. Ser. **17**(6): 584 (1939). Type: [Tanzania:] 'in a rather remote Bukoba Province of northwestern Tanganyika Territory: [Largely growing in country which reminded me of alpine meadows, Bugufi, 6000 ft. *Miss G. A. Chambers*] K21, [January 1936]' (K000410563 holotype).

Perennial herb, 0.6–1.3 m high, erect to decumbent. Stem single from a woody tuberous rootstock or several from root apex, few-branched, glabrous. Leaves pseudopetiolate, pseudopetiole to 35 mm long, lamina lanceolate to ovate, simple to pinnately 3–7-lobed, 5–15 × 1–3 cm, base long attenuate but wider near stem, margins coarsely dentate to shallowly pinnately lobed, apex acute to attenuate, glandular-punctate, scabrid on both surfaces. Inflorescences of solitary terminal capitula or capitula few in lax cymes, 3.5–7 cm wide at anthesis; peduncle up to 16 cm long. Capitula radiate; involucre 8–11 mm long; outer phyllaries 8, green, 6–7 × 1–2 mm, pilose on outside and also on upper half on inner surface; inner phyllaries 8, linear-lanceolate, 8–11 × 2.5–4 mm, sparsely pilose on outer surface; paleae 7–8.5 mm long. Ray florets neuter, 6–8, ray limbs ovate to elliptic, 15–18 × 7–8 mm, yellow, tube to 3 mm long. Disc floret corollas yellow, 5.5–6 mm long, pilose. Achenes black, linear, 6–11 × 1–1.7 mm, setuliferous, c. 16-striate-sulcate; aristae 2, without barbs or with barbs connate to arista, 1–2.5 mm long, rarely absent.

Zambia. W: Mufulira, 5.iii.1957, *Fanshawe* 3021 (K, NDO). **Malawi**. N: Rumphi Dist., Ascent to Nyika Plateau from Katumbi, 4 km outside National Park gate, 1860 m, 19.v.1970, *Brummitt* 10923 (K).

Also in Tanzania, D.R. Congo, Rwanda, Burundi and Angola. Grassland, bushed grassland, Brachystegia-Uapaca woodland; 1000–2200 m.

Conservation Status: LC (Least Concern).

5. **Bidens moorei** Sherff in Bot. Gaz. **81**(1): 25 fig. 1/a–g (1926). —Sherff, Publ. Field Mus. Nat. Hist., Bot. Ser. **16**(2): 591, Pl. 164 figs. a–g (1937). —Liswoski, (Asterac. Fl. Afr. Centr. 1) Fragm. Flor. Geobot. **36** Suppl. 1: 166 (1991). Type: [Angola:] '*John Gossweiler* 3339, in woods, Angola, April, 1906' (BM000924453 holotype, K000410665, LISC014897).

    Bidens modesta Sherff in Publ. Field Mus. Nat. Hist., Bot. Ser. **17**(6): 602 (1939). Type: [Angola:] '*John Gossweiler* 11,899, a perennial much dispersed in hiemifruticeta, alt. 1,700 meters, Huambo, New Lison, Angola, [1700 metros] 1937' (K000410664 holotype, F0075221, US00406457).

    Bidens moorei [var.] *verrucosa* Sherff in Bot. Gaz. **81**(1): 26 (1926). Type: [Angola:] '*John Gossweiler* 3021, Angola, April, 1906' (BM holotype, F0075222).

    Bidens bampsiana Lisowski in Bull. Jard. Bot. Natl. Belg. **57**(3–4): 464 (1987). Type: [D.R. Congo:] 'Zaïre: Shaba, plaine à l'embouchure de la Kapanda, déc. 1912, *Homblé* 1011'. [Region du Lualaba (Kat) (Vallée de la Kapanda) plaine à l'emb. de la Kapanda. Fleur jaune. Déc. 1912. *Homblé* 1011.] (BR8872852 holotype, BR8872128).

Decumbent perennial herb with woody rootstock and tuberous roots, 30–60 cm high. Stems several from rootstock, subtetragonal, glabrous. Leaves petiolate, petiole up to 3 cm long simple, lamina oblanceolate or elliptic, leathery, glabrous, 6–14 × (0.2)0.5–3 cm, base cuneate, margins recurved, glabrous, entire below middle, entire or minutely or coarsely dentate above with 2–4 teeth on each side. Inflorescences of solitary terminal capitula, or 4–5 per stem; peduncle glabrous or apex ciliate, 6.5–27 cm long. Capitula radiate; involucre c. 4 × 1 cm at anthesis, up to 13 mm tall in fruit; outer phyllaries green, glabrous or sparsely ciliate at base, 8, oblong, 4–7.5 × 1–3 mm at anthesis, not altered in fruit and much shorter than inner, leathery, apex obtuse or rounded, callose-thickened, inner yellowish-green (brown or dark brown after drying), c. 8, ovate-oblong, leathery, glabrous except at base and apex where sparsely ciliate, 7–8 × 2–2.5 mm at anthesis, up to 11 mm tall in fruit. Ray florets 8, neuter, ray limbs lemon-yellow, 20–26 × 6–8 mm, linear-oblong, thickish, 8–9 × 2–2.5 mm, glabrous, apex acute. Disc floret corollas lemon-yellow, 5-lobed; anthers brown; styles yellow, bifurcate. Achenes dark brown (immature), linear-oblong, 2–4-striate on each surface, flat, glabrous except at margins where setuliferous, 7–12 × 1–1.5 mm; apex with a crown of bristles and 2 (rarely 3–5) aristate; aristae retrorsely barbed, 2–3 mm long.

Zambia. W: Mwinilunga Dist., Near source of R. Matonchi, 16.ii.1939, *Milne-Redhead* 4604 (K).

Also in Angola, D.R. Congo. *Brachystegia* woodland and grassland; 1200–2400 m.

Conservation Status: Poorly collected in the Flora area, but apparently widespread outside; LC (Least Concern).

6. **Bidens crocea** Welw. ex O.Hoffm. in Bol. Soc. Brot. **10**: 177 (1892). —Sherff, Publ. Field Mus. Nat. Hist., Bot. Ser. **16**(2): 585, Pl. 161 figs. a–g (1937). —Lisowski, (Asterac. Fl. Afr. Cent. 1) Fragm. Flor. Geobot. **36** Suppl. 1: 160 (1991). —Mesfin Tadesse in Kew Bull. **48**(3): 463 (1993). Types: 'Angola ([ad 4500 ped., Febr. 1860] *Welwitsch*, n.º 3964). Huilla (*Antunes*).' *Welwitsch* 3964 (B† – originally lectotypified by Sherff (1937: 586), BM000924470 lectotype, BR8873187, C10000322, COI, G, K000410662, LISU218460, LISU218461, M, P00086709), re-lectotypified by Rayner (1992: 82).

    Bidens crocea var. *verrucifera* S.Moore in J. Linn. Soc., Bot. **37**(260): 322 (1906). Type: 'Angola, Catombe near Mlange; *Gossweiler*, 1210.' (BM holotype, F0075207, P00086707).

    Bidens ambigua S.Moore in J. Linn. Soc. Bot. **37**(260): 322 (1906). —Sherff, Publ. Field Mus. Nat. Hist., Bot. Ser. **16**: 582, Pl. 148 figs. h–m (1937). Type: [Angola:] 'Hab. Angola, in open forests on the left bank of the river Quanze et Kiambella; *Gossweiler*, 1189.' (BM000924469 holotype, K000410660, P00086708).

    Bidens bequaertii De Wild. in Repert. Spec. Nov. Regni Veg. **13**(359–362): 204 (1914). —Sherff, Publ. Field Mus. Nat. Hist., Bot. Ser. **16**(2): 573, Pl. 153 (1937). —Lisowski, (Asterac. Fl. Afr. Cent. 1) Fragm. Flor. Geobot. **36** Suppl. 1: 153 (1991). Type: [D.R. Congo:] 'Katanga: Élisabethville, mars 1912 (*J. Bequaert*, no. 270); Welgelegen, 1912 (Corbisier, coll. *Homblé*, no. 605).' *Bequaert* 270 (BR8873194 lectotype[41], BR8873163, F0075200 – fragments from BR8873194), first-step lectotypification by Sherff (1937: 574), second-step lectotypification by Rayner (1992: 81); *Homblé* 605 (? syntype).

    Bidens palustris Sherff in Bot. Gaz. **76**(2): 148 (1923). —Sherff, Publ. Field Mus. Nat. Hist., Bot. Ser. **16**(2): 568, Pl. 150 figs. a–g (1937). —Lisowski, (Asterac. Fl. Afr. Cent. 1) Fragm. Flor. Geobot. **36** Suppl. 1: 152 (1991). Type: [D.R. Congo:] '*T. Kassner* 2599, in swamps. Kundelungu, Congo Free State, March 13, 1908'. (B† holotype, K000410539 lectotype, BM000924468, BR887869, HBG504236, P00086867, Z), lectotypified by Rayner (1992: 84).

    Bidens phalangiphylla Sherff in Bot. Gaz. **76**(2): 152 (1923). —Sherff, Publ. Field Mus. Nat. Hist., Bot. Ser. **16**(2): 579, Pl. 156 figs. a–h (1937). Type: [Tanzania:] '*C. Holst* 2967, at altitude of 30 m., Doda, Usambara, German East Efrica, June 28, 893' (B† holotype).

    Bidens bequaertii var. *amplior* Sherff in Publ. Field Mus. Nat. Hist., Bot. Ser. **17**(6): 595 (1939). Type: [Tanzania:] '*Read Admiral H. Lynes* no. I. h. 263a, alt. 6,200–6,400 feet, Mt. Mferu, Iringa, central Tanganyika Territory, March 22, 1932.' (K000974634 holotype).

    Bidens gossweileri Sherff in Publ. Field Mus. Nat. Hist., Bot. Ser. **17**(6): 604 (1939). Type: [Angola:] '*John Gossweiler* 11,845, dispersed in *hiemifruticeta*, alt. 1,000 meters, Xassengue-Caiango, near Cuango River, Lunda Region, Angola, April, 1937' (K000410661 holotype, F0075211 – fragments from holotype).

---

[41] First-step lectotypification appears to have been undertaken by Sherff (1937: 574) in mentioning as 'type' the 2 sheets of *Bequaert* 270 in BR. Rayner (1992: 81), at length, provided second-step lectotypification, selecting BR8873194, albeit omitting to point out that there are two specimens mounted on the sheet, one with the rootstock, the other without; this sheet now has a full capitulum dissection mounted on it beneath the two specimens.

Bidens crocea var. *ornata* Sherff in Ann. Mag. Nat. Hist., ser. 12, **10**(2): 42 (1957). Type: [Angola:] '*Gossweiler* 4,304, ... stems procumbent-ascending, florets beautifully golden-yellow, in open mumua woods between Kaconda and Bissapa, Angola, Mar. 5, 1907' (BM000924471 holotype).

Bidens palustris var. *cubangona* Sherff in Ann. Mag. Nat. Hist., ser. 12, **10**(2): 43 (1957). Type: [Angola: Benguela – Huambo, Capango – Posto do Sambo, proximum flumen Cubango, alt. 1.700 m. app., 2/III/1950, *Brito Teixeira* 291. Vivaz, até 1 m. de altura. Flores amarelas. Frequente, na anhara da margen direita do Rio Cubango.] (BM000924471 holotype).

Bidens kasaiensis Lisowski in Bull. Jard. Bot. Natl. Belg. **58**(1–2): 259 (1988); (Asterac. Fl. Afr. Cent. 1) Fragm. Flor. Geobot. **36** Suppl. 1: 159 (1991). Type: [D.R. Congo:] 'Zaïre: Kwango, rivière Tshilualua, forêt claire, mars 1956, *Devred* 3523' (BR8872135 holotype).

Perennial herb, 0.5–1.2 m high, rarely annual. Stem single or several, glabrous, from thickened woody rootstock. Leaves petiolate, petiole 0.5–7 cm long, lamina pinnatifid with 3–5 segments or bipinnatifid with few side lobes, 5–15 cm long, segments linear, 1–6 cm × 0.5–2 mm, glabrous. Inflorescences of solitary terminal capitula; peduncle 12 cm long. Capitula radiate; involucre 4–8 × 8–10 mm; outer phyllaries 6–8, green, linear, 3–8 × 0.2–0.5 mm at anthesis, glabrous or sparsely pubescent; inner phyllaries 8, oblong, glabrous except for apex 8–10 × 1.5–2.5 mm at anthesis; paleae linear. Ray florets neuter, 5–8, ray limbs 10–25 × 3–5 mm, golden yellow, tube glabrous. Disc floret corollas yellow, 3.8–5 mm long, glabrous or sparsely puberulous. Achenes dark brown or black, flat, linear, 9–18 × 1.2–1.5 mm, glabrous or setuliferous, curved near apex when fully mature; aristae 2(–3), retrorsely barbed, to 4.3 mm long.

Zambia. N: Mpika, 2.ii.1955, *Fanshawe* F.1943 (K, NDO). C: Serenje Dist., 24 km NE of Mkushi River, Uapaca woodland, 16.ii.1970, *Drummond & Williamson* 9613 (K, SRGH).

Also in Tanzania, D.R. Congo and Angola. Open woodland; 500–1900 m.

Conservation Status: LC (Least Concern).

7. **Bidens ochracea** (O.Hoffm.) Sherff in Bot. Gaz. **76**(2): 158 (1923). —Mesfin Tadesse in Kew Bull. **48**(3): 465 (1993). Type: [Tanzania:] 'Uhehe: Bweni, auf rotem Laterit eines hügeligen Plateaus im lichten Busch, um 1700 m ([*Goetze*] n. 731. – Blühend am 11. März 1899).' (B† holotype, BM000924450 lectotype), lectotypified by Rayner (1992: 92).

Coreopsis ochracea O.Hoffm. in Bot. Jahrb. Syst. **30**(3–4): 431 (1901). —Sherff in Bot. Gaz. **80**(4): 375, fig. 19 (1925). —Sherff, Publ. Field Mus. Nat. Hist., Bot. Ser. **11**(6): 380, fig. 3 (1936).

Coreopsis cosmophylla Sherff in Bot. Gaz. **76**(2): 90, fig. 9/h–n (1923). Type: [Tanzania:] '*Muenzer* 159, Msamvia, Lake Tanganyika District, German East Africa, February 24, 1909' (B† holotype).

Coreopsis ochraceoides Sherff in Amer. J. Bot. **42**(6): 564 (1955). Type: [Tanzania:] '*E. T. Ward, Esq.* U31, Uhinga and Ubena areas, Iringa Province, Tanganyika Territory, March, 1937.' (K000410562 holotype).

Perennial herb, up to 1.5 m high; stems several from thickened or tuberous rootstock, hardly branched, glabrous. Leaves petiolate, petiole to 5 cm long, lamina ovate in outline, deeply pinnatipartite with 3–7 segments or bipannatipartite, simple in upper leaves, 4–10(18) × 2–7 cm, leaf segments linear, thick, 1.5–7 mm wide, margins scabrid or pubescent and slightly revolute, apex acute or mucronate. Inflorescences of few (1–3) capitula at apex of branches; peduncle up to 15 cm long. Capitula radiate, up to 7.5 cm diam. at anthesis; involucre 8–13 mm long; outer phyllaries 8–10, green, 8–13(19) × 1–2.5(4) mm at anthesis, pilose at base; inner phyllaries 8–9, lanceolate, ovate-lanceolate, or narrowly oblong, 8–13 × 2–4 mm, yellow with a red stripe, glabrous except for apex; paleae to 10 mm long. Ray florets neuter, 8–9, ray limbs 20–35 × 10–14 mm, orange or golden yellow, tube ± 3 mm long, pubescent. Disc floret corollas pale cream yellow, sparsely pubescent near base. Achenes black, narrowly obovoid, 6–7 × 2.5–3.5 mm, narrowly callose-thickened, glabrous or sparsely bristled on margins; aristae 2, naked or sparsely antrorsely barbed, to 2 mm long.

Zambia. N: Abercorn Dist., Lake Chila, Abercorn, 1650 m, 6.iii.1959, *Richards* 11107 (K). **Malawi**. N: Karonge Dist., nr. Kayelekera, 960 m, 8.vi.1989, *Brummitt* 18440 (K, MO).

Also in Tanzania. Miombo woodland, *Brachystegia* woodland, grassland, on a variety of soils; 960–2550 m.

Conservation Status: There appear to be a number of collections from each of the northern divisions of Zambia and Malawi, with the species found through into neighbouring Tanzania; LC (Least Concern).

8. **Bidens kirkii** (Oliv. & Hiern) Sherff in Bot. Gaz. **59**(4): 309 (1915). —Sherff, Publ. Field Mus. Nat. Hist., Bot. Ser. **16**(2): 561, Pl. 145 (1937). —Wild in Kirkia **6**(1): 29 (1967). Type: [Mozambique:] 'Mozamb. Distr. Moramballa, alt., 3,000 ft., [30 Dec. 1858] *Kirk* [s.n.]!' (K000410659 holotype).

    Coreopsis kirkii Oliv. & Hiern in F.T.A. **3**: 390 (1977).

    Coreopsis insecta S.Moore in J. Bot. **46**(542): 42 (1908). Type: [Zimbabwe:] 'Hab. Mazoe, Bernheim Hill, *F. Eyles* 266.' (BM000924455 holotype, BOL139029, SRGH).

    Bidens insecta (S. Moore) Sherff in Bot. Gaz. **59**(4): 309 (1915).

    Bidens rufovenosa Sherff in Bot. Gaz. **59**(4): 301 (1915). —Sherff, Publ. Field Mus. Nat. Hist., Bot. Ser. **16**(2): 546, Pl. 138 figs. a–i (1937). —Mendonça, Contrib. Fl. Angola **1** (Compositae): 101 (1943). Type: [Angola:] '*John Gossweiler* 4176, among the ferruginous rocks near the fort at Kabango, Princip. Amelia, Africa, Dec. 1907' (BM000924456 holotype).

    Bidens schlechteri Sherff in Bot. Gaz. **76**(2): 146 (1923). —Sherff, Publ. Field Mus. Nat. Hist., Bot. Ser. **16**(2): 570, Pl. 151 figs. a–i (1937). Type: [South Africa:] '*R. Schlechter* 4745, at altitude of 5000 feet, on slopes of Hortbosch Mountain, Transvaal, March 30, 1894 (type in Herb. Berl., two sheets).' (B† holotype, G00022138 lectotype, C10000323, K000410644, P00086766), lectotypified by Rayner (1992: 86).

    Coreopsis curtisiae Sherff in Bot. Gaz. **96**(1): 146 (1934), as '*curtisii*'. —Sherff, Publ. Field Mus. Nat. Hist., Bot. Ser. **11**(6): 384 (1936). —Mesfin Tadesse in Compositae Newslett. **33**: 29 (1999). Type: '*Mrs. Richard C. Curtis*, cultivated in 1930, from seeds obtained in Angloa' (F! holotype, photos: F719934, F765708).

    Coreopsis schlechteri (Sherff) Burtt Davy in Bull. Misc. Inform. Kew **1935**: 568 (1935).

    Bidens schlechteri var. *wildii* Sherff in Amer. J. Bot. **41**(9): 762 (1954). Type: [Zimbabwe:] '*Hiram Wild* 4,320, bushy, straggling perennial, among rocks, alt. 5,000 feet, Mt. Buhwe, District of Belingwe, Southern Rhodesia, Dec. 10, 1953' (F holotype, K000410644, PRE0194316-0, SRGH).

Perennial herb, up to 1 m tall, glabrous. Stem irregularly branched, terete or ribbed-tetragonal, several from a thick rhizome. Leaves petiolate, petiole up to 3.5 cm long, lamina ovate in outline, deeply pinnately or sub-bipinnately 3–7-lobed, up to 10 cm long, segments glabrous, linear or narrowly elliptic, rather thick and coriaceous, apex acute to acuminate, indurate, margins entire, subrevolute. Inflorescences of few 1–2(3) capitula, terminal on peduncles; peduncle up to 16 cm long, glabrous. Capitula radiate, 3–4.5 cm diam. at anthesis; involucre 3–4.5 × 7–9 mm at anthesis; outer phyllaries green, 4–8, linear or narrowly elliptic, 4–14 mm long, sparsely pubescent at base; inner phyllaries c. 8, oblong, glabrous except for pubescent apex, 6–10 mm long, greyish-orange, margins scarious; paleae thin, 7–10 mm long, linear. Ray florets neuter, 5–8, ray limbs yellow, 1.5–3 cm long narrowly elliptic-obovate with 13–20 dark brown striae. Disc floret corollas yellow. Achenes black or dark brown, linear-oblong, flat, hispidulous or glabrescent with a few short setulae near apex and margins, 6–7 × 1.2–1.6 mm; aristae 2, glabrous except for few bristles at base and a single antrorse or retrorse barb at apex, c. 2 mm long.

Zimbabwe. N: Mazoe, Bernheim Hill, *F. Eyles* 266 (BM, BOL, PRE). C: About 10 miles E of Marandellas, 1677 m, 29.iv.1969, *Plowes* 3201 (K, SRGH). S: Belingwe Dist., Mount Buhara, 1500 m, 3.v.1973, *Biegel* et al. 4273 (K, SRGH). **Mozambique**. MS: see type of *Coreopsis kirkii*.

Also in Angola and S Africa (Transvaal). Grassland on mountain slopes and in *Brachystegia* woodland; 600–1800 m.

Conservation Status: Apparently poorly collected in the southern and eastern areas of the Flora, but also present outside; LC (Least Concern).

9. **Bidens steppia** (Steetz) Sherff in Bot. Gaz. **76**(2): 82 (1923). —Sherff, Publ. Field Mus. Nat. Hist., Bot. Ser. **16**(2): 542, Pl. 137 figs. j–r (1937). —Wild in Kirkia **6**(1): 19 excl. syn. p.p. (1967). —Mesfin Tadesse in Kew Bull. **48**(3): 487 (1993). Type: [Mozambique:] 'Standort: Mossambique, an verschiedenen Orten, neben feuchten Stellen, Rios de Sena. [*Peters* 58]' (B† holotype); 'Mozambique. Mocuba, Lugela District, Namagra, 1946-7, *H. G. Faulkner* Kew 10' (K neotype, BM, BR, COI, EA, FT, G0007741, G00015715, K(× 2), P, PRE, S-G-1592, SRGH), neotypified by Rayner (1992: 94). FIGURE 6.5.**9**.

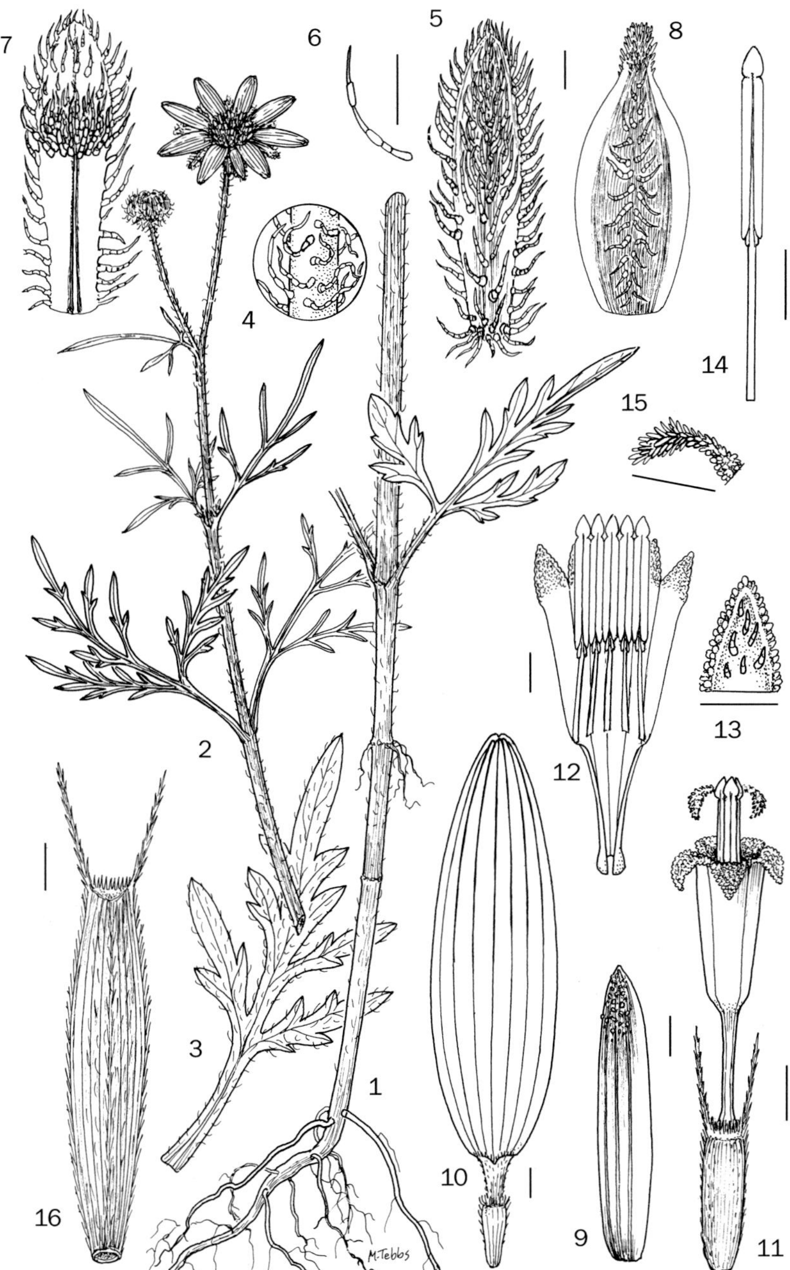

Fig. 6.5.**9**. BIDENS STEPPIA. 1, stem base; 2, apex of flowering shoot; 3, leaf; 4, detail of stem pubescence; 5, abaxial surface of inner phyllary; 6, hair from abaxial surface of outer phyllary; 7, adaxial surface of outer phyllary; 8, inner phyllary; 9, palea; 10, ray floret; 11, disc floret; 12, disc floret corolla opened out showing attachment of filaments; 13, detail of abaxial surface of apex of corolla lobe of disc floret; 14, stamen; 15, detail of style arm apex; 16, mature achene. 1 from *Pawek* 4763; 2, 8–15 from *Pawek* 3432; 3–7, 16 from *Kwatha et al.* 180. Scale bars: 13, 15 = 0.5 mm; 4–12, 14 = 1 mm; 1–3 = 10 mm. Drawn by Margaret Tebbs.

Coreopsis steppia Steetz in Peters, Naturw. Reise Mossambique **6** (Bot. 2): 496 (1864).

Coreopsis ambacensis Hiern, Cat. Afr. Pl. **1**(3): 586 (1898). Type: [Angola:] 'Ambaca. –At the left bank of the river Caringa; fr. June 1855. [*Welwitsch*] No. 3272, partly.' (BM000924449 holotype, F0075249, LISU218452, P00091575).

Bidens ambacensis (Hiern) Sherff in Bot. Gaz. **59**(4): 309 (1915).

Bidens steppia var. *ambacensis* (Hiern) Sherff in Bot. Gaz. **90**(4): 392 (1930). —Sherff, Publ. Field Mus. Nat. Hist., Bot. Ser. **16**(2): 546 (1937).

Bidens steppia var. *leptocarpa* Sherff in Bot. Gaz. **90**(4): 392 (1930). Type: [Tanzania:] '*Ad. Stolz* 729, growing 1.5 m. high at alt. 1350 m., Langenburg, German East Africa, May 26, 1911' (B† holotype, G – the sheet bearing the barcode label and Stolz's 'Flora Africae orientali' label, C10000326, G00007740 × 2, K000410607, L0001857, LE, LU, M0105181, S-G-962, STU, W, WAG0000524, Z), lectotypified by Rayner (1992: 87).

Bidens uhligii Sherff in Bull. Jard. Bot. État Bruxelles **13**(4): 286 (1935). —Sherff, Publ. Field Mus. Nat. Hist., Bot. Ser. **16**(2): 542 (1937). Type: [Tanzania:] 'Tanganyika Territory: Near Lake Victoria Nyanza, Ukerewe near edge of forest, April 1904, *Uhlig* V. 46' (B† holotype).

Bidens steppia var. *elskensii* Sherff in Bull. Jard. Bot. État Bruxelles **13**(4): 286 (1935). —Sherff, Publ. Field Mus. Nat. Hist., Bot. Ser. **16**(2): 545 (1937). Type: [Rwanda:] 'District of Ruanda-Urundi: Urundi, hills at Kitega, in places long cultivated, Dec. 1922, *O. Elskens* 257' (BR8872395 holotype).

Coreopsis vulgaris Sherff in Bull. Jard. Bot. État Bruxelles **13**(4): 291 (1935). —Sherff, Publ. Field Mus. Nat. Hist., Bot. Ser. **11**(6): 391 (1936). Type: [D.R. Congo:] 'District of Middle Katanga: Mukishi-Lomami, alt. 1,000 m., common on savanna, flowering entire year except dry season, Sept. 1927, *A. Becquet* 62' (BR8677808 holotype).

Coreopsis multiflora Sherff in Bull. Jard. Bot. État Bruxelles **13**(4): 292 (1935). —Sherff, Publ. Field Mus. Nat. Hist., Bot. Ser. **11**(6): 383 (1936). Type: [D.R. Congo] 'District of Upper Katanga: Plateau of Bianos, Kansenia, Febr. 1911, *Dom Benoît Thoreau* [11]' (BR8677815 holotype).

Coreopsis injucunda Sherff in Bull. Jard. Bot. État Bruxelles **13**(4): 293 (1935). —Sherff, Publ. Field Mus. Nat. Hist., Bot. Ser. **11**(6): 393 (1936). Type: [D.R. Congo] 'District of Upper Katanga: Kipila-Elisabethville, on Goffard's farm, May 1929, *P. Quarré* 1662' (BR8872845 holotype, BR8872401).

Bidens steppia var. *inarmata* Sherff in Amer. J. Bot. **34**(3): 152 (1947). Type: [Tanzania:] '*Amani Inst. no.* 7,951, Neulangenburg, southwesternmost Tanganyika Terr., June' (EA holotype).

Bidens steppia var. *kalamboensis* Sherff in Amer. J. Bot. **42**(6): 562 (1955). Type: [Zambia:] '*Mrs. Richards* 1,342, in bush close to river, Kalambo Falls, North Rhodesia, March 30, 1952' (K000410658 holotype, BR9862029, F).

Bidens steppia var. *garusonis* Sherff in Ann. Mag. Nat. Hist., Ser. 12, **10**(2): 42 (1957). Type: [Mozambique:] '*H. B. Gilliland* 1,847, on the ridge ascending Garuso, savannah, Portuguese East Africa, April, 1935' (BM000924441 holotype[42], K000410657).

Annual herb 7–200 cm high, whole plant with long, silvery or shiny hairs; stem often reddish, terete or tetragonal, branching towards apex, often reddish, angular, pubescent. Leaves sessile or petiolate, petiole to 9 cm long, lamina deltoid in outline, incised to bi- or tri-pinnatisect or pinnatifid, 2.5–35 × 2–26 cm, segments linear to narrowly ovate, variously lobed or cleft, glandular-punctate and pilose to glabrous. Inflorescence of solitary terminal capitula or in few-headed lax cymes; peduncle to 12 cm long. Capitula radiate; involucre 5–6.5 × 7–10 mm; outer phyllaries 8–16, linear or oblanceolate, 5–15 × 1–3 mm at anthesis, up to 30 × 5 mm in fruit, pilose on upper half, inner phyllaries ± 8, oblong or narrowly ovate, 5–10 × 2–3.5 mm at anthesis, yellow with red striations, pilose; paleae thin, membranous, oblong-linear, 5–10 mm long. Ray florets neuter or pistillate, 8–13, ray limbs yellow or orange, often two-toned, 25–35 × 5–12 mm, apex minutely 3-fid or entire. Disc floret corollas yellow or orange-yellow, 2.8–4 mm long. Achenes black, narrowly obovoid, 8-striate-sulcate, 6–10 × 1–1.7 mm, flat, setuliferous; aristae 2, antrorsely barbed, 1–2.5 mm long, rarely without aristae or aristae nude.

Zambia. N: By Furrow South Farm to Kasulo, 1525 m, 2.iv.1952, *Richards* 1592 (K). W: Ndola Dist., 26.v.1953, *Fanshawe* F 33 (K, NDO). C: Mpika Dist., 4.6 mi. E of Mfuwe, 7.v.1965, *Mitchell* 2864 (MO). E: Lundazi, 1100 m, 1.vi.1954, *Robinson* 802 (K). S: Mazabuka Dist., Mvuma hill, 22 km S of Kafue town on road to Mazabuka, 1160–1200 m, 25.iii.1972, *Kornas* 1450 (MO). **Zimbabwe**. N: vicinity of Umvukwe mts., 5 miles N of Banket,

---

[42] Rayner (1992) determined the holotype as *B. schimperi* Sch.Bip.

23.iv.1948, *Rodin* 4395 (K, MO). C: 31 km SE of Lusaka, Nachitete River, Fulawula village, 15°36'S 28°31'E, 1220 m, 10.iv.1993, *Bingham* 9062 (K). **Malawi**. N: Chitipa Dist., 7 mi. N of cross roads to Misuku Hills, 3500 ft., 25.iv.1977, *Pawek* 12663 (MO). C: Ntchisi Dist., Ntchisi Forestry Reserve, 11.v.1984, *Banda & Kaunda* 2180 (MO). S: Zomba Dist., along Mulunguzi R., 11.viii.1978, *Msiska* 92 (MO). **Mozambique**. N: near Vila Cabral, iv.1934, *Gomes Sousa* 4 (MO). T: Zobue District, 914 m, 17.vi.1947, *Hornby* 2759 (K). C: Harare, Msasa woodland, 4500 ft., 5.iv. 1984, *Bayliss* 10085 (MO).

Also in Uganda, Tanzania, Cameroon, Central African Republic, D.R. Congo, Angola and Burundi. Grassland, floodplain vegetation, bushed grassland, woodland, weed of cultivation; 280–2100(2740) m.

Conservation Status: LC (Least Concern).

Leaves are very variable in dissection of the lamina, varying from deeply incised to 2- or 3- pinnatisect or pinnatifid, with linear, narrowly oblong-linear or narrowly ovate segments, segments variously lobed or cleft, glandular-punctate, terminal segment usually long with entire margins. The ray florets are often described by collectors as 'two-toned' or 'darker yellow in the lower parts and lighter in the upper parts'.

Vernacular name: "Kisosogwe" (Chinyaanja – Malawi).

10. **Bidens oligoflora** (Klatt) Wild in Kirkia **6**(1): 20 (1967). —Lisowski, (Asterac. Fl. Afr. Cent. 1) Fragm. Flor. Geobot. **36** Suppl. 1: 143, t. 33 (1991). —Mesfin Tadesse in Kew Bull. **48**(3): 487 (1993). Type: [Angola:] 'Hab. Malange (Angola) in virgultis, leg. *Dr. Buchner* 1879, No. 32.' (B† holotype, GH6212 lectotype – a fragment of the holotype and a drawing of the holotype), lectotypifieded by Wild (1967: 20).

    Coreopsis oligoflora Klatt in Leopoldina **25**: 107 (1889), p. 4 of a preprint in K. —Sherff, Publ. Field Mus. Nat. Hist., Bot. Ser. **11**(6): 386 (1936).

    Coreopsis oligantha Klatt in Ann. Nat. Hofmus. Wien. **7**: 103 (1892). Type: [Angola:] 'Hab.: Angola, Malange, leg. Mart. 1880, [*Mechow*] No. 459.' (B† holotype, M lectotype, JE, W), lectotypified by Mesfin Tadesse (1993: 489).[43]

    Coreopsis mattfeldii Sherff in Bot. Gaz. **76**(2): 83 (1923). —Sherff, Publ. Field Mus. Nat. Hist., Bot. Ser. **11**(6): 392 (1936). Type: [Zambia:] '[N.] *Carson* 75, Fwamba, German East Africa, in 1894' (B† holotype, K000410656 lectotype), lectotypified by Mesfin Tadesse (1993: 489).

    Coreopsis oligoflora var. *robusta* Sherff in Bot. Gaz. **90**(4): 386 (1930). —Sherff, Publ. Field Mus. Nat. Hist., Bot. Ser. **11**(6): 386, 388 (1936). Type: [D.R. Congo:] '*Büttner* 408' (B† holotype).

    Bidens onisciformis Sherff in Bot. Gaz. **96**(1): 144 (1934). —Sherff, Publ. Field Mus. Nat. Hist., Bot. Ser. **16**(2): 559 (1937). Type: [D.R. Congo:] '*Ven den Houdt* 211, alt. about 2200 m., northeastern Belgian Congo, 1932' (BR8873095 holotype).

    Coreopsis isokoensis Sherff in Amer. J. Bot. **42**(6): 565 (1955). Type: [Zambia:] '[N.] *Mrs. Richards* 1,108, in open bush, alt. 4,000 feet, annual? Isoko Valley, Northern Rhodesia, Mar. 19, 1952' (K000410654 holotype, F, SRGH0106737).

    Bidens drummondii Wild in Kirkia **6**(1): 17 (1967). Type: [Zambia:] '[W.] Zambia, 11 km. W. of Chizera, Brachystegia woodland, 23/3/1961, *Drummond & Rutherford-Smith* 7204' (SRGH0106864 holotype, K000410655, LISC002776, M0105187).

Annual herb to 1 m tall, whole plant with long, shiny or silky hairs. Stem erect, pilose to glabrous, tetragonal. Leaves petiolate, petiole to 70 mm long, lamina pinnatipartite or bi-pinnatipartite, 3–20 × 2–6 cm, segments ovate to linear (narrower in upper leaves), sparsely pilose on both surfaces. Inflorescences terminal, of few-headed, lax corymbose cymes, 7–8 cm wide at anthesis; peduncle sparsely pilose, up to 13 cm long. Capitula radiate; involucre 7–8 × 8–10 mm; outer phyllaries 8–10, pilose, 8–13 × 1–1.5 mm at anthesis, linear or linear-lanceolate, outer surface sparsely pilose, inner surface pilose on upper half; inner phyllaries 8, brown with yellow margins, 8–10 × 2–4 mm, oblong, pilose on outer surface; paleae membranous, linear-oblong, 8–10 mm long, glabrous. Ray florets

---

[43] However, annotations on the material in W-Rchb. (1889-0054613) suggest that this sheet is the holotype of Klatt's name.

neuter, 8, ray limbs golden yellow to orange, 10–30 × 2–12 mm. Disc floret corollas yellow, pubescent in middle part; anthers brown. Achenes black or dark brown, broadly obovoid or orbicular, 6–9(12) × 3–4 mm, scabrid, margins thickened, light brown, apex 'V'-shaped; aristae 2, antrorsely barbed, to 2 mm long, rarely absent.

Zambia. N: Luangwa Valley, Lupande Munkanya, 19.iv.1968, *Phiri* 179 (K). W: Mufulira, 1220 m, 2.iv.1948, *Cruse* 309 (K). C: 28 m SW of Broken Hill, 1159 m, 13.vii.1930, *Hutchinson & Gillett* 3626 (K). **Zimbabwe**. E: Odzi riverside, 18.iv.1948, *Chase* 667 (K, SRGH).

Also in Tanzania, Cameroon, D.R. Congo and Angola. Bushy grassland; 600–1828 m.

Conservation Status: LC (Least Concern).

Similar to *Bidens steppia* and may be easily distinguished by its achenes which are broadly obovate or orbicular with notched or broadly 'V'-shaped apices and usually also cartilaginous-thickened margins.

11. **Bidens malawiensis** Mesfin in SINET: Ethiop. J. Sci. **12**: 125 (1989). —Mesfin Tadesse in Kew Bull. **48**(3): 492 (1993). Type: [Malawi:] '*E.A. Banda* 494. Malawi, Dedza Mountain forest 30.IV.1963' (K000410653 holotype, M).

   Bidens setigera sensu auct. non (Sch.Bip.) Sherff. —Wild in Kirkia **6**(1): 16 (1967).

Annual herb, 20–75 cm high. Stem subtetragonal, 1.5–4 mm wide at base. Leaves 1- to 2-pinnatisect, up to 10 cm long, segments linear to lanceolate, 1–3(8) mm wide, densely glandular punctate and pilose on both surfaces, rarely subglabrous, petiole up to 25 mm long. Inflorescences of solitary terminal capitula or in axillary or terminal open corymbiform cymes; peduncle up to 8 cm long, often less than 1 mm wide, hairy. Capitula radiate, 2–3 cm diam. at anthesis, involucre 4–6 mm tall; outer phyllaries (5)8, oblanceolate, 4–6 × 0.5–1.2 mm, glabrous; inner phyllaries 8, oblong-ovate, 4–6 × 1.5–2 mm, densely pilose on outer surface; paleae 4.5–5.5 × 0.5–1 mm, membranous, glabrous. Ray florets 8, neuter with exaristate ovary, ray limbs yellow, with c. 5 striae, 10–15 × 4–6 mm, apex minutely 2–3-fid. Disc floret corollas yellow, 2–2.5 mm long, 5-lobed; anthers black, 2.5–4 mm long. Achenes black, oblanceolate, glabrous, shiny, 3–3.5 × 0.8–1.1 mm, outer surface convex, c. 8-striate-sulcate, tuberculate at apex, inner surface concave and medially ridged, often tuberculate all over, margins not winged, apex exaristate, truncate.

Malawi. N: Dedza Dist., Dedza Mountain Forest, 10.iv.1980, *Banda* 1645 (K, SRGH). C: Ncheu Dist., 13 mi N of Ncheu junction, 1430–2000 m, 4.iv.1978, *J. Pawek* 14200 (K, MO).

Montane grassland, margins of forests, *Brachystegia* woodland; 550–2000 m.

Conservation Status: Apparently common in grassland and colouring whole hillsides; LC (Least Concern).

Bidens malawiensis resembles *Bidens occidentalis* (Hutch. & Dalziel) Mesfin of W Africa, *Bidens negriana* (Sherff) Cufod. and *Bidens setigera* (Sch. Bip.) Sherff subsp. *setigera* of E and NE Africa in leaf morphology. From particularly *Bidens setigera*, with which it was compared by Wild (1967), it differs in having campanulate corolla, non-setiferous foliar teeth, and neuter ray florets.

12. **Bidens acuticaulis** Sherff in Bot. Gaz. **59**(4): 301 (1915). —Sherff, Publ. Field Mus. Nat. Hist., Bot. Ser. **16**(2): 347, Pl. 104 figs. j–r (1937). —Mesfin Tadesse in Kew Bull. **48**(3): 492 (1993). Type: [Angola:] '*John Gossweiler* 4052, in herb-grown woods, Angola, April 4, 1906 (type in Herb. Brit. Mus.)' (BM000924478 holotype, F0075198, LISC014895).

   Bidens paupercula Sherff in Bot. Gaz. **76**(2): 158, fig. 12/a–g (1923). —Sherff, Publ. Field Mus. Nat. Hist., Bot. Ser. **16**(2): 380. Pl. 94 (1937). —Lisowski, (Asterac. Fl. Afr. Cent. 1) Fragm. Flor. Geobot. **36** Suppl. 1: 133, fig. 31 (1991). Type: [Tanzania:] '*Ad. Stolz* 1442, in forest at altitude of 900 m., Kyimbila, Nyassaland, July 22, 1912 (two type sheets in Herb. Berl.)' (B† holotype, M0105198 lectotype, B100093284, C10000321, G00018024, K000410606, L0001842, LD1095139, S-G-957, W, WAG0000516, WAG0000517), lectotypified by Rayner (1992: 85).

   Bidens ciliata De Wild. in Repert. Spec. Nov. Regni Veg. **13**(359–362): 203 (1914), non Hoffmanns. ex Fisch. & C.A.Mey. (1839), nom. illeg. Type: [D.R. Congo:] 'Katanga: Elisabethville, avril 1912 (*J. Bequaert*, no. 302)' (BR8872456 holotype).

Annual herb, 0.14–1 m tall. Stem erect, tetragonal, simple or branched above, reddish or speckled with reddish, glabrous or pilose in upper parts. Leaves petiolate/pseudopetiolate, (pseudo)petiole to 50 mm long, lamina narrowly winged pinnate to bipinnatifid, 2–13 × 6 cm, segments 3–5, narrowly linear to filiform, to 50 by 2 mm, minutely ciliate and sometimes sparsely hispidulous. Inflorescences of solitary terminal capitula; peduncle to 15 cm long, glabrous to sparsely pilose. Capitula radiate; involucre 4–8 mm long lengthening to 13 mm in fruit; outer phyllaries 4–9, green to reddish, linear, 4–8 × 0.3–0.5 mm, often reflexing, glabrous or ciliate near apex; inner phyllaries ± 8, pale brown with dark red or purple margins or red all over, 4–4.5 × c. 1 mm, glabrous or pubescent. Ray florets neuter, 5–8, ray limbs 6–7 × 1–1.5 mm, striate, apex acute to minutely 2-fid, yellow, upper part paler. Disc florets corollas yellow, campanulate, 5-lobed, lobes glabrous or puberulous; style bifid, c. 6 mm long, penicillate at apex; anthers brown, c. 2 mm long. Achenes black or dark brown, narrowly ellipsoid, appressed setuliferous, 8-striate, not winged, beaked, beak up to 2.5 (var. *acuticaulis*) or up to 24 cm long (helically twisted when immature, straight and widely spreading when ripe: var. *filirostris*), with 2–3 aristae, 1–2.5 mm long, retrorsely barbed.

Achene beak to 2.5 cm long, golden when ripe; aristae retrorsely barbed . . . . a) *acuticaulis*
Achene beak to 24 cm long, pendulous, golden brown, twisted when immature, straight in
    mature fruit; aristae only retrorsely barbed at apex . . . . b) *filirostris*

a) Var. **acuticaulis**. FIGURE 6.5.**10A**.

Zambia. N: Samfya Dist., Samfya Forestry Office, saw mill yard, 29°33'E, 11°21'S, 21.iv.1989, *Pope et al.* 2192 (BR, K, LISC, NDO, SRGH). W: Chingola Dist., 18.iv.1954, *Fanshawe* F1111 (K, NDO). C. Mumbwa, 21.iii.1963, *van Rensberg* 1763 (K). S: Choma Dist., 15.iv.1963, *van Rensburg* 1938 (K). **Malawi**. N: Rumphi Dist., Livingstonia Escarpment above bend 1, 1159 m, *Pawek* 2332 (K). C: Dedza Dist., Research Station, Chongoni, 18.iv.1961, *Banda* 418 (K, PRE, SRGH). **Mozambique**. *Pedro & Pedrogão* 3506 – collection details unavailable.

Also in Tanzania, D.R. Congo and Angola. *Brachystegia* woodland; 900–2050 m – for species.

Conservation Status: Although apparently scattered in parts of the Flora area it is also known from a wide area outside; LC (Least Concern).

Philcox et al. 9825, from Ndola Dist., would to appear to be a rayless form of this taxon.

b) Var. **filirostris** (P.Taylor) T.G.J.Rayner in Phytologia **72**(2): 101 (1992). Type: [Tanzania:]
    'TANGANYIKA. Southern Province: Songea District, about 140 km. east of Songea in Brachystegia-Uapaca woodland on sand, 900 m., alt. ... 4 June 1956, *Milne-Redhead & Taylor* 10547' (K000410603 – 'Sheet 1', K000410604 – 'Sheet 2', K000410605 – 'Sheet 3' holotype,[44] B100093283, BR8677785, EA, LISC002775, P00086869, S-G-958, SRGH). FIGURE 6.5.**10B**.
    Bidens paupercula Sherff var. *filirostris* P.Taylor in Hooker's Icon. Pl. **36**: tab. 3580 (1962).

Zambia. N: Shiwa Ngandu, 1555 m, 4.vi.1956, *Robinson* 1580 (K).
Also in Tanzania. *Brachystegia* woodland; 850–1555 m.

Conservation Status: Very poorly collected in the Flora area (two collections from the N Province), and little collected outside apparently (three collections from Tanzania); clearly more fieldwork is needed: DD (Data Deficient).

---

[44] It is quite clear that the holotype consists of 3 sheets, the first of which ('Sheet 1') has the original full label, which is not present on the other two sheets, as well as a duplicated continuation label, which is the one marked 'Sheet 1', 'Sheet 2' and 'Sheet 3' on the respective sheets. It is clear that this is one gathering and as such, since the protologue is quite specific, that all 3 sheets represent the holotype. Further discussion of lectotypification is unnecessary.

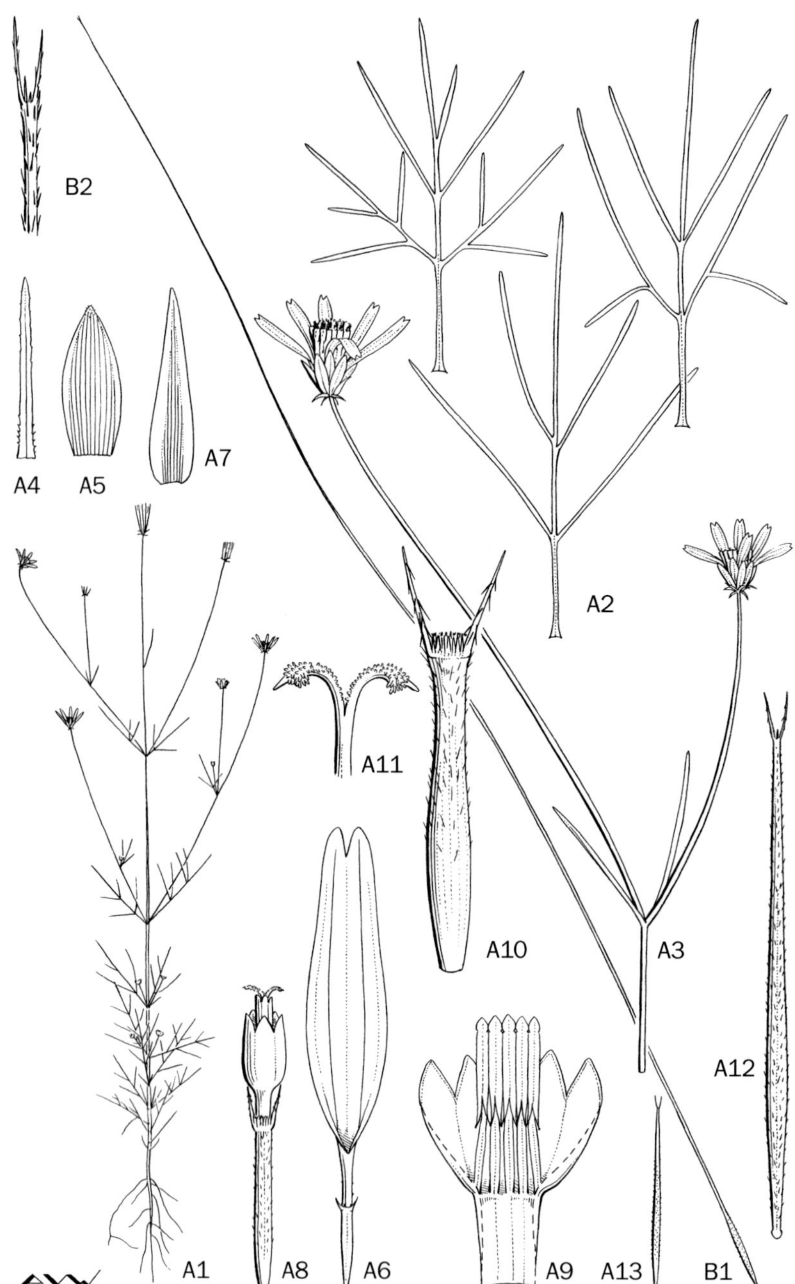

Fig. 6.5.**10**. A. —BIDENS ACUTICAULIS var. ACUTICAULIS. A1, habit (× ¹/₆); A2, leaves (× ²/₃); A3, flowering branch (natural size); A4, outer phyllary (× 4); A5, inner phyllary (× 4); A6, ray floret (× 4); A7, palea (× 4); A8, disc floret (× 4); A9, disc floret corolla opened out showing attachment of filaments (× 8); A10, immature achene (× 8); A11, style arms (× 12); A12, achene, enlarged view (× 3); A13, achene (natural size). B. — BIDENS ACUTICAULIS var. FILIROSTRIS. B1, achene (natural size); B2, achene apex (× 8). A1–A11 from *Milne-Redhead & Taylor* 9887; A12–13 from *Milne-Redhead & Taylor* 10582; B1–2 from *Milne-Redhead & Taylor* 10547. Drawn by 'A.V.W.' Reproduced from Hooker's Icones Plantarum (1962).

13. **Bidens diversa** Sherff in Bot. Gaz. **76**(2): 159 (1923). —Sherff, Publ. Field Mus. Nat. Hist., Bot. Ser. **16**(2): 329, Pl. 75 figs. a,b,d–i (1937). —Mesfin Tadesse in Kew Bull. **48**(3): 493 (1993). Type: [Angola:] '*Antunes* 315, in forest at altitude of 1760 m., Mounyino, Portuguese West Africa, March 1901 (two type sheets in Herb. Berl.)' (B† holotype); 'Angola. Huilla, Lubango, Tundavala, at 12 km, source of the Inhames, 30 Apr. 1971, *A. Borges* 167.' (LISC002693 neotype, LISC0014896, M0105192, P, PRE0798429, SRGH), neotypified by Rayner (1992: 82).

 Bidens filiformis Sherff in Publ. Field Mus. Nat. Hist., Bot. Ser. **17**(6): 600 (1939). —Wild in Kirkia **6**(1): 21 (1967). Type: [Zambia:] '*B. D. Burtt* 6,269, alt. 4,500–5,000 feet, Lake Chila, Northern Rhodesia, April, 1936' (F0075210 holotype, BM0924466, BR8873156, K00410651, K00410652).

 Bidens cochlearis Merxm. in Mitt. Bot. Staatssamml. München **2**(heft 11): 33 (1954). Type: 'Angola, Provinz Huila: Berg Eyvila, 40 km südlich Quilengues; Grannit, ca. 1150 m. Blüten gelb. 4.5.1952 leg. *H. Hess* nr. 52/1531.' (Herb. Hess, now in ZT7979 holotype, M0105193, ZT7980, ZT7981).

 Bidens diversa subsp. *filiformis* (Sherff) T.G.J. Rayner in Phytologia **75**(2): 156 (1993).

Annual herb, 13–50(100) cm tall. Stem tetragonal, slender, only 2–4 mm wide at base, glabrous, usually dichotomously branched above. Leaves petiolate, petiole to 3 cm long, lamina pinnate, to 10 cm long, with a filiform rachis and 3–5 filiform segments to 5 cm long, glabrous or sparsely setulose. Inflorescences of solitary terminal capitula; peduncle to 12 cm long, glabrous. Capitula radiate, 4–4.5 cm diam. at anthesis; involucre 4–5 mm long; outer phyllaries 8, narrowly linear-oblong, 2–5 × 0.3–0.5 mm at anthesis, glabrous or sparsely ciliate at margins toward apex, apex acute-apiculate, base slightly dilated; inner phyllaries reddish brown (after drying) with scarious margins, oblong or ovate-oblong, 3–5 × c. 1 mm at anthesis, up to 9 mm long in fruiting heads, outer surface glabrous or sparsely ciliate, apex obtuse, usually purple-tipped. Ray florets 8, neuter with exaristate ovary or rarely minutely aristate ovary, ray limbs yellow with orange base, oblong to narrowly obovate, with 6–8 purplish striae, 15–20 × 3–4 mm, apex minutely 2-fid or entire, tube c. 1.8 mm long, glabrous. Disc florets corollas yellow or orange, glabrous to sparsely ciliate above, lobes papillate; anthers brown, 1–3 mm long; style 2.3–6 mm long. Achenes black, narrowly ellipsoid, c. 8-striate-sulcate and setuliferous, margins with appressed tuberculate bristles, graded monomorphic, 6–14 × 0.7–1 mm, apex yellowish, biaristate and with dense erect setulae; aristae 2, yellowish, retrorsely barbed at apex, ± 1 mm long.

Zambia. N: Abercorn Dist., Chiyanga stream near Kalambo Road, 1500 m, 9.v.1966, *Richards* 21471 (K, MO). **Malawi**. N: Mzimba Dist., Champira Forest, 1464 m, 20.iv.1974, *Pawek* 8412 (CA, K). **Mozambique**. Namplua Prov., Nampula, 15.iv.1937, *Torre* 1375 (COI, LISC).

Also in Tanzania and Angola. Tall grassland, woodland on rocky outcrops; gregarious elsewhere; 825–2000 m.

Conservation Status: Although patchy in its distribution in the Flora area it is a relatively widespread species, and usually recorded as common on labels'; LC (Least Concern).

Bidens diversa varies in the manner of leaf dissection, size of phyllaries and in the texture of the achenes. It has affinities with *Bidens acuticaulis* from which it can be distinguished on the basis of the size and shape of the achene, size of rays and diameter of the capitulum.

14. **Bidens urceolata** De Wild. in Ann. Mus. Congo Belge sér. 4: 167 (1903). —Sherff, Publ. Field Mus. Nat. Hist., Bot. Ser. **16**(2): 550, Pl. 140 figs. a–g (1937). —Wild in Kirkia **6**(1): 16 (1967). —Lisowski, (Asterac. Fl. Afr. Cent. 1) Fragm. Flor. Geobot. **36** Suppl. 1: 148 (1991). —Mesfin Tadesse in Kew Bull. **48**(3): 495 (1993). Type: [D.R. Congo:] 'Katanga, Lukofu, April 1900, *Verdick* 464.' (BR8677693 holotype). FIGURE 6.5.**11**.

 Bidens rubra De Wild. in Repert. Spec. Nov. Regni Veg. **13**(359–362): 203 (1914). —Sherff, Publ. Field Mus. Nat. Hist., Bot. Ser. **16**(2): 548, Pl. 140 figs. h–o (1937), non Sch.Bip., nom. nud., in sched. (cf. Sherff 1937: 481) [= *Bidens gardneri* Baker (1884)]. Types: [D.R. Congo:] 'Katanga: Welgelegen, mai 1912 (*J. Bequaert*, no. 389 et coll. *Homblé*, no. 563).' *Bequaert* 389 (BR8677686[45]

---

[45] Marked as holotype, although clearly a syntype!

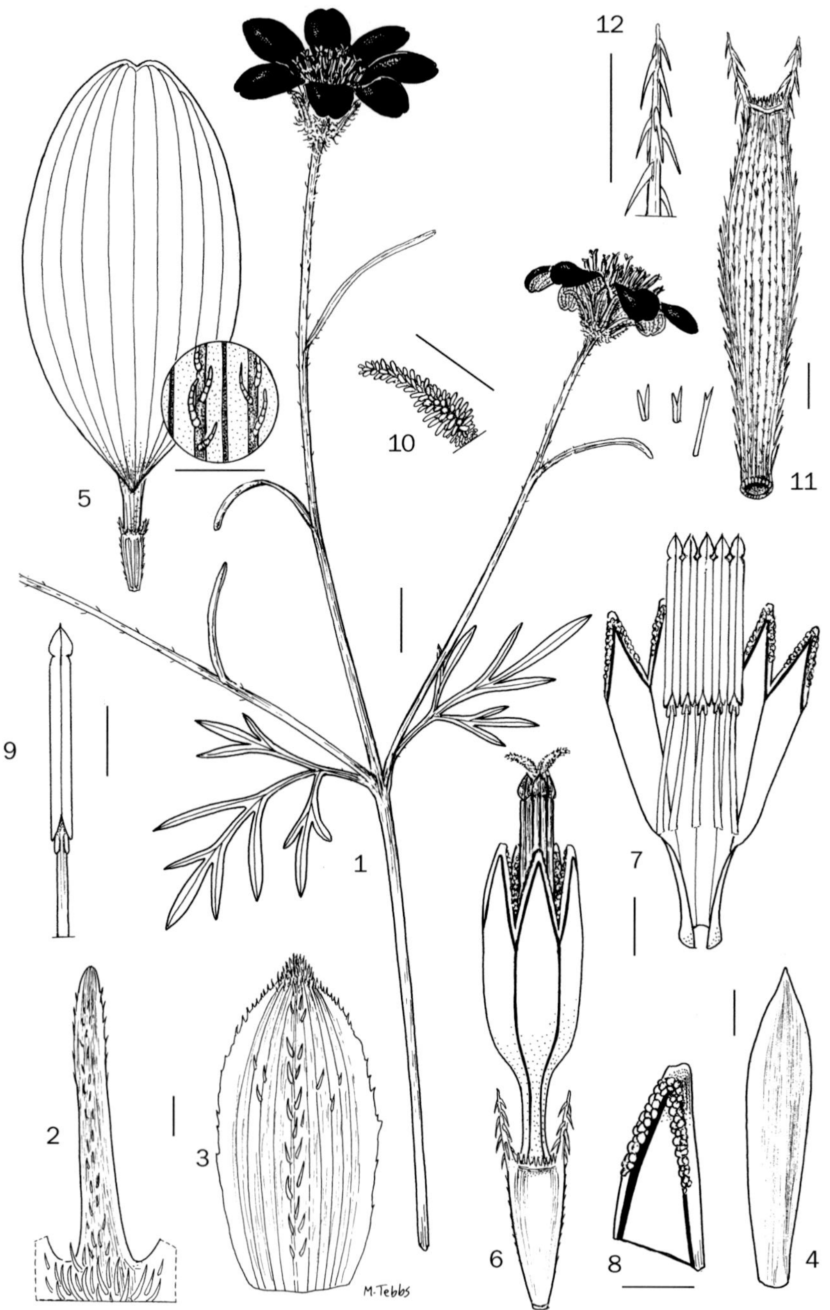

Fig. 6.5.**11**. BIDENS URCEOLATA. 1, flowering branch; 2, outer phyllary; 3, inner phyllary; 4, palea; 5, ray floret with inset detail of abaxial surface showing pubescence on venation; 6, disc floret; 7, disc floret corolla opened out showing attachment of filaments; 8, detail of apex of corolla lobe; 9, stamen; 10, detail of style arm apex; 11, mature achene with detail of setulae; 12, detail of awn from disc floret achene. 1–10 from *Robinson* 6651; 11–12 from *Mutimushi* 3240. Scale bars: 8, 10 = 0.5 mm; 2–7, 9 = 1 mm; 1 = 10 mm. Drawn by Margaret Tebbs.

lectotype), lectotypified by Sherff (1937: 549); *Homblé* 563 (?BR syntype).[46]

    Bidens leptolepis Sherff in Bot. Gaz. **76**(1): 85, fig. 9/a–g, f (1923). —Sherff in Bot. Gaz. **85**(1): 12 (1928). —Sherff, Publ. Field Mus. Nat. Hist., Bot. Ser. **16**(2): 55, Pl. 138 figs. f, j–p, (1937). —Lisowski, (Asterac. Fl. Afr. Cent. 1) Fragm. Flor. Geobot. **36** Suppl. 1: 150–152 (1991). Type: [D.R. Congo:] '*T. Kassner* 2725, under trees, Mt. Kundelungu, Congo Free State, May 10, 1908' (B† holotype; Z lectotype, BM00924437, E00239246, HBG504238, K00410534, P0086769, P0091616), lectotypified by Rayner (1992: 84).

    Bidens leptolepis f. *pallida* Sherff in Amer. J. Bot. **42**(6): 563 (1955). Type: [Zambia:] '*E. G. Walter* 21, common in plains and shady places, alt. 3,000–4,000 feet, Kawambwa and Chienje districts (Lake Mweru), Northern Rhodesia, May, 1931 (type, K, my photograph n. 4,411)' (K holotype).

    Bidens urceolata var. *leptolepis* (Sherff) T.G.J. Rayner, Phytologia **77**(6): 444 (1995).

Erect annual herb, up to 70 cm tall. Stem simple or branched towards apex, tetragonal, glabrous to sparsely pilose. Leaves petiolate, petiole up to 3 cm long, glabrous to hairy, lamina pinnatipartite to bi-pinnatipartite with 3–7 slender, up to 5 cm long and 1–4 mm wide segments, segments linear-oblong to sub-spathulate, glabrous, margins sparsely hispid to glabrous. Inflorescences of solitary terminal capitula or of open corymbose cymes; peduncles up to 15 cm long, glabrous or pilose. Capitula erect, 3.5–7.5 cm diam. at anthesis; involucre 6–10 mm long lengthening to 14 mm in fruit, c. 5 mm diam.; outer phyllaries green, 7–18, linear-lanceolate, slightly dilated at base, 6–10 × 0.5–1 mm, outer surface and margins hispidulous, apex acute; inner phyllaries dark-purple with light purplish or scarious margins, c. 8, oblong to narrowly ovate-oblong, 6–10 × 1.5–3.5 mm, outer surface pilose mainly medially or at base, rarely glabrous. Ray florets 5–8, neuter with exaristate ovary, limbs crimson (deep purple after drying) or pink (pale purple after drying), narrowly obovate with 6–8 purple striae, 25–38 × 8–10 mm, apex entire or minutely 3-fid, tube and basal part puberulent. Disc floret corollas red or pink, 5-lobed, glabrous; style 6.8–10.3 mm long; anthers red or pink. Achenes black, linear-oblong or elliptic, 8–15 × c. 1 mm, 8-striate-sulcate, setuliferous on both surfaces; outer surface flat, inner surface more or less concave, apex setose and biaristate; aristae 1–2 mm long, yellowish, retrorsely barbed, base hyaline.

Zambia. N: Kawambwa, 25.viii.1957, *Fanshawe* F 3614 (K, NDO). W: 80 km W of Solwezi, 1350 m, 15.iv.1960, *Robinson* 3541 (K).

Also in D.R. Congo. Dry *Brachystegia* woodland on sandy soil; 900–1350 m.

Conservation Status: Relatively well-collected from the Flora area, albeit from the same two divisions, where it is often locally common or frequent, forming extensive, quite spectacular, populations; LC (Least Concern).

15. **Bidens rubicundula** Sherff in Amer. J. Bot. **41**(9): 762 (1954). —Wild in Kirkia **6**(1): 15 (1967). —Mesfin Tadesse in Kew Bull. **48**(3): 495 (1993). Type: [Zambia:] '*D. Bidens Fanshawe* F64, "flowers pale mauve pink," Chizera, Northern Rhodesia, June 11, 1953 (two type sheets, Kew, my photographghs no. 4,390 and no. 4391...)' (K00410649 –'Sheet 1', K00410650 –'Sheet 2' holotype,[47] BR8677754, F, SRGH).

    Bidens rubicundula f. *alba* T.G.J.Rayner in Phytologia **77**(6): 437 (1995). Type: 'Zambia. North-Western Province – Kalenda Plain, Matonchi [11°39'S, 24°06'E], alt. 1400 m, 16 Apr. 1960, *E. A. Robinson* 3634' (K holotype, BR, EA, M0105184, SRGH).

---

[46] Sherff (1937: 549) also indicated that there was a *Bequaert* 563 (BR15413390) and a *Homblé* 563 collection in BR, which is probably a mistake.

[47] The K material consists of two sheets of material, marked clearly as 'Sheet 1' and 'Sheet 2', with 'Sheet 1' possessing the full label information, and it is clear that Sherff saw both (as both were sent on loan, and determined by Sherff), and was well aware that the preparation of the type specimen consisted of the two; there is a third sheet which is simply of detailed line drawings by Peter Taylor of floral parts. Rayner (1992: 85) was of the opinion that Sherff's 'designation is contrary to the definition of a holotype and must be considered ineffective.'; this is incorrect as the ICBN (2006) Art. 8.2 and 8.3 allow for this situation. Unnecessarily, 'Sheet 2' was selected by Rayner as the lectotype – which merely clarifies which material represents the name, albeit unnecessarily.

Annual herb, up to 60 cm high. Stem erect, simple or branched from base, subtetragonal, glabrous, woody at base. Leaves petiolate, petiole up to 2 cm long, lamina deeply 3-lobed or irregularly bipinnatifid, narrowly winged upper leaves often simple, sometimes 2-lobed and often lobes secondarily lobed, segments up to 9 cm long, linear to narrowly elliptic-linear, apex acuminate to subacute, glabrous, margins sparsely ciliate or glabrous. Inflorescences of solitary, terminal capitula or of open corymbose cymes; peduncles up to 9 cm long, glabrous. Capitula erect, 4.5–5 cm diam. at anthesis; involucre 4–9 mm tall, lengthening to 20 mm in fruit; outer phyllaries green, (2)4–9 × 1–2 mm, up to 20 × 3 mm in fruit, usually much narrowed for remaining 2/3 of length, subulate, reflexed in fruit, 3-veined, veins purplish, glabrous to glabrescent, margins ciliate; inner phyllaries reddish-purple (dark-purple after drying), 8–10(12) × 3–4.5(6) mm at anthesis, up to 17 × 9 mm in fruiting heads, broadly oblong, glabrous, apex acute, puberulous; paleae thin, glabrous, narrowly oblong, 8–10 × 1–2 mm at anthesis. Ray florets 5–6, neuter with exaristate ovary, ray limbs pink (pinkish purple after drying) or rarely white, elliptic to narrowly obovate-elliptic with 6–8 purple striae, 30–35 × 6–9 mm, apex entire or 3-fid, tube sparsely hairy. Disc floret corollas purplish or yellowish, 5-lobed, glabrous or sparsely hispid; anthers purple. Achenes black, narrowly oblong-elliptic, 10–14 × 0.8–1 mm, c. 8-striate-sulcate, outer surface convex, glabrous, inner surface flat with muricately arranged pale or dark brown tubercles and setulae towards apex, rarely tubercles absent, apex setose, usually exaristate, occasionally with 2 entire, 0.3 mm long aristae.

Zambia. W: Kabompo Gorge, Mwinilunga Dist., 18.iv.1965, *Robinson* 6628 (EA, K, M). Also in southern D.R. Congo. *Brachystegia* woodland; 650–1400 m.

Conservation Status: Apparently restricted to the Western Province of Zambia within the Flora area, with very few collections, and only one from D.R. Congo. Only one collection from Zambia (the holotype) gives any indication of abundance where it is noted as occasional – clearly more fieldwork is necessary; DD (Data Deficient).

16. **Bidens schimperi** Sch.Bip. ex Walp. in Repert. Bot. Syst. **6**: 168 (1846). —Oliver & Hiern in F.T.A. **3**: 393 (1877). —Sherff in Bot. Gaz. **81**(1): 53 (in note under *Bidens schimperi* var. *leptocera*), fig. 3 (1926). —Sherff in Bot. Gaz. **85**(1): 16 (1928). —Sherff, Publ. Field Mus. Nat. Hist., Bot. Ser. **16**(2): 554, Pl. 142 figs. i–q, 143, 144 (1937). —Andrews, Fl. Pl. Sudan **3**: 11 (1956). —Wild in Kirkia **6**(1): 23 (1967). —Agnew, Upland Kenya Wild Fl.: 466 (1974). —Cribb & Leedal, Mountain Fl. S. Tanzania: 151, fig. 40b (1982). —Mesfin Tadesse in Symb. Bot. Upsal. **24**(1): 110, fig. 56 (1984). —Mesfin Tadesse in Kew Bull. **48**(3): 497 (1993). —Agnew & Agnew, Upland Kenya Wild Fl., ed. 2: 217, t. 89 (1994). Type: [Ethiopia:] 'Schimp. it. Abyss. sect. III. no. 1429. ... Crescit in Abyssiniae vallibus prope Djeladjeranne.' (P00091511 holotype, BR8360670, FI000794, G00015760, G00015761, G00015762, GH00004221, GOET001096, K000410586, K000410587, K000410588, M0105183, MO391525, P00091512, S-G-960, TUB005544, TUB005545). FIGURE 6.5.**12**.

Coreopsis exaristata O.Hoffm. in Pflanzenw. Ost-Afrikas C: 414 (1895), non DC. Types: [Tanzania:] '(Usb., auf Wiesen der Gebirge und der gerodeten Urwälder, am Rande von Planzungen – *Holst* n. 102, 207, 5002). Gebirgstropenwald, um 1100 m.' *Holst* 102 (B lectotype), lectotypified by Sherff (1937: 567); *Holst* 207 (BM syntype); *Holst* 5002 (B syntype).[48]

Bidens prolixa S.Moore in J. Linn. Soc., Bot. **40**(275): 116 (1911), as '*prolixus*'. Type: [Zimbabwe:] 'Melsetter 200–600 ft.; in fl. & fr. April; [*Swynnerton*] n. 1884.' (BM000924442 holotype, K000410647).

Bidens punctata Sherff in Bot. Gaz. **59**(4): 302 (1915). Type: [Malawi:] '*W. P. Johnson* 343, Tumbi (Makapula), April 27 (1901?)' type in Herb. Kew' (K000410647 holotype).[49]

---

[48] Sherff (1937: 567) effectively lectotypified the name based on *Holst* 102 in B, noting a duplicate of *Holst* 207 was in BM, and a duplicate of *Holst* 5002 was in B. Rayner (1992: 90) suggested that this was a mechanical selection to be rejected, and there is no indication that this is what Sherff did, although the plate was drawn from *Holst* 207. Rayner proposed the duplicate of *Holst* 5002 in WU as the lectotype.

[49] Sherff (1915) specified *Johnson* 343 as the type, but the description was also clearly based on *Johnson* 341 as declared by Sherff.

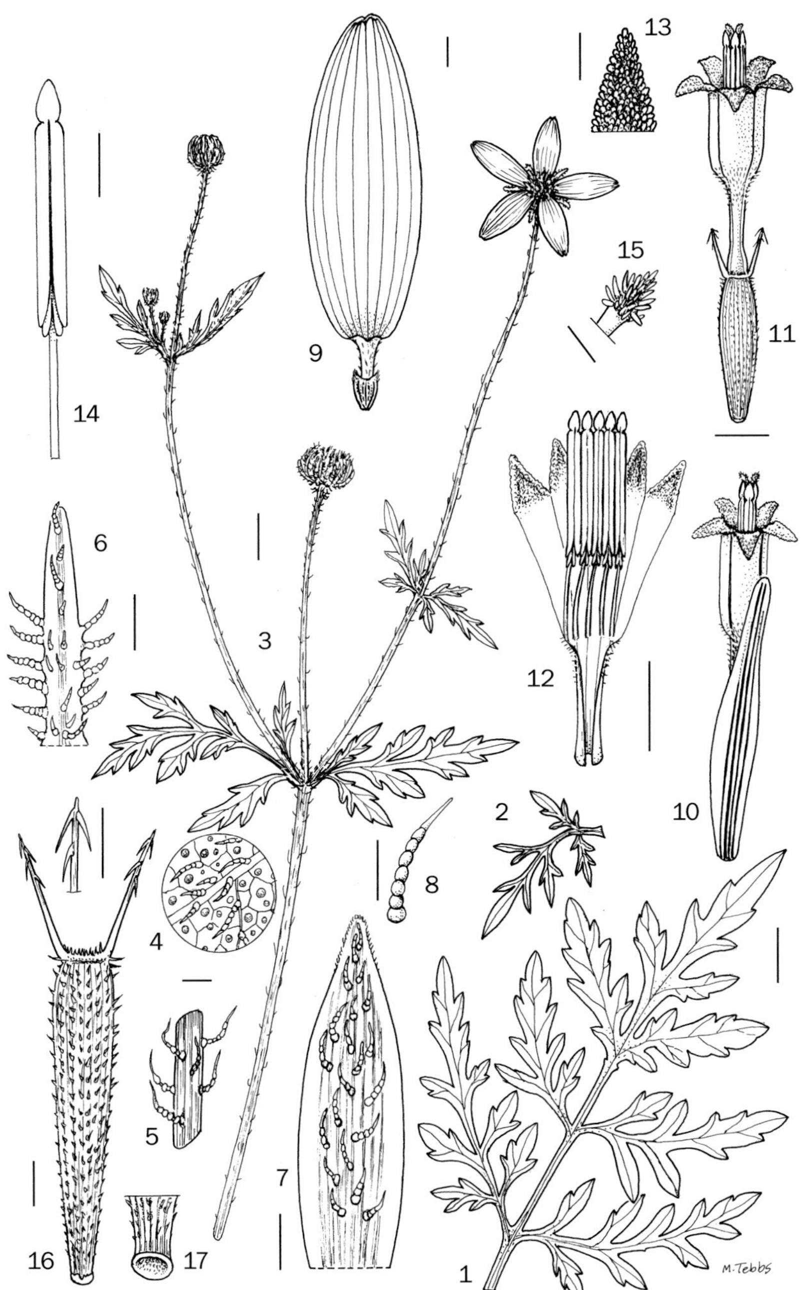

Fig. 6.5.**12**. BIDENS SCHIMPERI. 1–2, leaf variation; 3, flowering branch; 4, detail of lower leaf surface; 5, detail of stem pubescence; 6, outer phyllary; 7, inner phyllary; 8, detail of inner phyllary hair; 9, ray floret; 10, disc floret with accompanying palea; 11, disc floret; 12, disc floret corolla opened out showing attachment of filaments; 13, detail of apex of corolla lobe; 14, stamen; 15, detail of apex of style arm of disc floret; 16, disc floret achene; 17, detail of base of disc floret achene. 1 from *Chase* 8542; 2 from *Pawek* 9492; 3–14 from *Zimba* 882; 15–16 from *Smith* 0547. Scale bars: 13 = 0.25 mm; 8, 14–15 = 0.5 mm; 4–7, 9–12, 16 = 1 mm; 1–3 = 10 mm. Drawn by Margaret Tebbs.

Bidens microcarpa Sherff in Bot. Gaz. **76**(2): 84 (1923), as nom. nov. pro *Coreopsis exaristata* O.Hoffm. —Sherff, Publ. Field Mus. Nat. Hist., Bot. Ser. **16**(2): 567, Pl. 149 figs. a–h (1937). Type as for *Coreopsis exaristata* O.Hoffm.

Bidens acutiloba Sherff in Bot. Gaz. **76**(2): 147 (1923). Type: [Tanzania:] '*G. Volkens* 384, at altitude of 1500 m., Kwa Kinabo, Mt. Kilimanjaro, German East Africa, June 1893' (B† holotype, BM lectotype, G00015759, K000410600), lectotypified by Mesfin (1993: 497).

Bidens schimperi var. *leptocera* Sherff in Bot. Gaz. **81**(1): 52 (1926). —Sherff, Publ. Field Mus. Nat. Hist., Bot. Ser. **16**(2): 554 (1937). Type: [Tanzania:] '*C. F. M. Swynnerton* 845 A. Kilossa, German East Africa, May 8, 1921' (BM holotype).

Bidens schimperi var. *punctata* (Sherff) Sherff in Bot. Gaz. **85**(1): 17 (1928).[50]

Bidens schimperi var. *brachycera* Sherff in Publ. Field Mus. Nat. Hist., Bot. Ser. **17**(6): 585 (1939). Type: [Tanzania:] '[*Romala D.*] *Bax* 206, growing 1.5 feet tall, in grass, Shinyanga, [northern Tanganyika Territory], May, 1933' (K000410602 holotype, F0075235 – fragments and photograph of holotype, US00406073).

Bidens schimperi var. *leiocera* Sherff in Publ. Field Mus. Nat. Hist., Bot. Ser. **17**(6): 586 (1939). Type: [Zambia:] '*J. D. Martin* [s.n.], Mazabuka, northern Rhodesia' (K000410645 holotype, F0075236 – fragments and photograph of holotype).

Bidens schimperi var. *greenwayi* Sherff in Amer. J. Bot. **42**(6): 563 (1955). Type: [Tanzania:] '*Greenway* 6,876, a common yellow-flowered much-branched annual herb up to 2 feet tall, growing in *Senecio goetzei*, *Erlangea calycina*, *Aneilema vigna* spp. ass. with *Acalypha indica*, *Berkheyopsis diffusa*, *Crotalaria*, and *Polygala fischeri* on a black, cracking clay of volcanic derivation, alt. 5,000 feet, Engari (Engare) Nairobi, west slope of Kilimanjaro, Tanganyika Terr., June 19, 1944' (K000410601 holotype).

Bidens grantii (Oliv.) Sherff. var. *stapfioides* Sherff in Kew Bull. **11**(3): 445 (1957). Type: [Tanzania:] 'Tanganyika. Tabora, 1200 m., 12 April 1954, *F. G. Smith* 1134. [tall erect herb up to 1.5 m., on fold of rich ground near hilltop.]' (EA holotype).

Annual herb, 7–150 cm tall. Stem erect, 4-angled, simple to much branched, often with red or purple markings, sparsely to densely pubescent. Leaves petiolate, petiole 10–55 mm long broadly ovate in outline, tripartite with lobed parts, rarely simple or 2- to 3-pinnatisect, 2.5–20(30) cm long, 1.5–10(15) cm wide, leaf segments ovate to narrowly ovate, margins lobed, entire, incised-dentate to crenate-serrate, pubescent, especially above, rarely almost glabrous, densely glandular-punctate. Inflorescences of lax terminal corymbose cymes; peduncle 2–18 cm long, capitula erect or nodding. Capitula radiate, 2.5–4 cm diam.; involucre 4–8 mm tall; outer phyllaries 8, linear, oblanceolate to sub-spathulate, 2.5–10 × 0.5–1.2 mm, lengthening to 16 mm in fruit, sparsely pubescent; inner phyllaries 8, oblong-ovate, 3.5–11 × 1.2–3.5 mm, pubescent or glabrous, light brown to red with wide yellow margins. Ray florets neuter with aristate ovary, 6–8, ray limbs 6–20 mm long, yellow or occasionally orange, tube ±1.5 mm long. Disc floret corollas dull yellow to orange, ± 4 mm long; anthers brown, c. 1 m long; style c. 3.5–4.5 mm long. Achenes dark brown to black, 3–10 mm long (outer) to 5–16 mm long (inner), striate, setuliferous, oblong-elliptic, compressed or flat, sometimes margins callose thickened; aristae 2, retrorsely barbed, 0.2–4 mm long.

Botswana. N: Matlapeneng, Thamalakane River bridge, 8 km NE of Maun, 19.iii.1965, *Wild & Drummond* 7172 (K, SRGH). **Zambia**. B: Barotseland Dist., 45 miles W of Nangweshi, 1037 m, 6.viii.1952, *Codd* 7421 (K, PRE). E: Chizombo, Chipata Dist., Luangwa Valley, 698 m, 20.ii.1969, *Astle* 5501 (K). S: Kafue Mission, 35 miles south of Lusaka, 976 m, 30.iv.1955, *Best* 85 (K). **Zimbabwe**. N: Mutoko Dist., Nyamahere Hill, 17°23'S, 32°13'E, 1340 m, 18.ii.1978, *Pope* (K). C: Gwelo head waters, 1433 m, iv.1918, *Eyles* 1302 (K). E: Inyanga Dist., slopes of Nyahokwe Mountain, Ziwa Farm (path to ancient village), 1372 m, 27.ii.1966, *Chase* 8389 (K, SRGH). S: Buhera Dist., near Mudanda Rest Camp, 800–1000 m, *Plowes* 3497 (K, SRGH). **Malawi**. N: Karonga Dist., 22 mi. west of Karonga, Stevenson Road, 820 m, 22.iv.1975, *Pawek* 9492 (K, MAL, MO). C: Dedza Dist., 1.iv.1961, *Chapman* 1227 (K, SRGH).

---

[50] Mesfin Tadesse (1993: 497) was in error citing 'Malawi, Thumbi Hill, Kabata, *Johnston* [sic!] 341 (K!, holotype)' where he assumed *Johnson* 341 was the holotype. It is quite clear that Sherff made a combination in 1928; *Johnson* 341 is a paratype of *Bidens punctata* Sherff.

Mozambique. T: Tete Dist., entre o Inhacapirire e a Chicoa, a 10.8 km de Inhacapirire, 19.ii.1972, *Macedo* 4862 (K, LMA).

Also in southern Egypt, Sudan, Ethiopia, Djibouti, Somalia, Kenya, Tanzania, Angola, Namibia, S Africa. Abundant in disturbed, overgrazed or burnt soil, and may form stands in wheat fields; also in grassland, bushy or wooded grassland (especially on black cotton soils), woodland, evergreen bushland and secondary thicket; 150–2600 m.

Conservation Status: A widespread species; LC (Least Concern).

Leaf dissection is very variable, as is the length of the ray florets limbs and the inner achenes.

17. **Bidens biternata** (Lour.) Merr. & Sherff in Bot. Gaz. **88**(3): 293 (1929). —Sherff, Publ. Field Mus. Nat. Hist., Bot. Ser. **16**(2): 388, Pl. 99 figs. a, c–m (1937). —Adams in F.W.T.A., ed. 2, **2**: 234 (1963). —Wild in Kirkia **6**(1): 26 (1967). —Agnew, Upland Kenya Wild Fl.: 466 (1974). —Mesfin Tadesse in Symb. Bot. Upsal. **24**(1): 115, fig. 58 (1984). —Lisowski, (Asterac. Fl. Afr. Cent. 1) Fragm. Flor. Geobot. **36** Suppl. 1: 135 (1991). —Hind in Fl. Masc., Composées **109**: 195, t. 66 (1993). —Mesfin Tadesse in Kew Bull. **48**(3): 498 (1993). —Agnew & Agnew, Upland Kenya Wild Fl., ed. 2: 217, t. 89 (1994). Type: [China:] 'Habitat agrestis prope Cantonem Sinarum' *Loureiro* s.n. (?)[51]; 'China, Kwangtung Prov., Canton (Honam Island), 13.X–9.XI.1916, *Merrill* 10122' (UC neotype), neotypified by Mesfin Tadesse (1984: 115).

Coreopsis biternata Lour., Fl. Cochinch., ed. 1, **2**: 508 (1790) & ed. 2, **2**: 622 (1793).

Bidens abyssinica Sch.Bip. ex Walp. in Repert. Bot. Syst. **6**: 167 (1846). Type: [Ethiopia:] 'in Schimp. Hb. it. Abyss. sect. I. no. 337. ... Crescit in Abyssinia in saxosis montium et vallium prope Adoam et ubique pro Djeladjeranne.' (P00073876, P00073877, P00073878 lectotypes, BM, BR8360335, BR8872494, FT, G00022139, 00022140, HAL0111895, HAL0111896, HBG504241, HOH009462, K000410582, K000410583, LG90025051, M0105196, MO357031, NY00162602, S-G-950, TUB005541, TUB005542, UPS), first-step lectotypification by Sherff (1937: 402).[52]

Bidens kotschyi Sch.Bip. ex Walp. in Repert. Bot. Syst. **6**: 168 (1846).[53] Type: [Sudan:] 'Crescit in Nubia ad stagna pluvialia in radice orientali monti Arasch-Cool et in paludosis Cordofanis.' *Kotschy* 79 (P lectotype, BR8872500, G, K, L, M0105412, M0105413, MO256891, S, STU, UOS, W, WAG, WU), first-step lectotypifications by Sherff (1937: 369) and Mesfin Tadesse (1993: 498);[54] *Kotschy* 91 (K 2×, W syntypes).

Bidens abyssinica var. *glabrata* Vatke in Linnaea **39**(6): 500 (1875), as '*glabratus*'. Types: [Ethiopia:] '([*Schimper*] n. 285.) Adest in coll. a. 1854 n. 105. e Gaha Meda prope Dschadscha, a Schweinfurthio

---

[51] Mesfin Tadesse (1984: 115) noted that Loureiro cited no material, and none could be found.

[52] Mesfin Tadesse (1993: 498) claimed he had selected the lectotype in 1984, yet in that work he referred back to Sherff (1937); it is quite clear that Sherff lectotypified the name.

[53] Sherff (1937: 366) placed *B. kotschyi* into the synonymy of *B. bipinnata* but Mesfin Tadesse (1993: 498; 2005: 806) placed it into the synonymy of *B. biternata*.

[54] Two collections, *Kotschy* 79 (ad stagna pluvialia in radice orientali montis Arusch-Cool, 30. Sept. 1839), and *Kotschy* 91 (in paludosis Cordofanis, Aug. 1837) represent syntypes. Sherff (1937) cited *Kotschy* 79 as the type collection mentioning many duplicates; this is accepted as lectotypification, by specifying one of the syntypes. Rayner (1992: 83) considered perfectly valid Sherff's lectotypification as 'ineffective', and pointed out that Mesfin's citation of 'holotype' in P also as 'ineffective', together with an explanation suggesting a misunderstanding of the Code. Mesfin's assigning of material in P, is effectively first step lectotypification (the term holotype being correctable to lectotype). Second step lectotypification is therefore required. Rayner (1992: 83) selected the lectotype as from P, but noting only two sheets. However, there are 3 sheets in P representing 'Kotschyi iter Nubicum 79'; P00073858 is labelled as the lectotype – an odd choice, especially since it was not in Schultz Bipontinus possession and not annotated by him – and P00073859 and P00073860 are labelled as isolectotypes. It is worth noting that P00073860 is from Herb. Schultz Bipontinus, was one of the two isolectotypes determined by Mesfin Tadesse in 1988, and is clearly annotated by Schultz Bipontinus as *Bidens kotschyi* – it would have been a better choice as lectotype.

Beitr. 142 cum. var. altera (quadriaristata Hochst. fide ejusdem) n. 305 e Gageros confusa, a qua promo intuitu diversissima; nostra transitum praebere videtur ad B. bipinnatum L. a Kotschyo in Nubia repertum, cui forte stirps abyssinica reducenda.' *Schimper* 285 (Z lectotype, BM, PRE), lectotypified by Rayner (1992: 80–81); *Schimper* 105 (G, Z syntypes).[55]

Bidens quadriseta Hochst. ex Oliv. & Hiern in F.T.A. **3**: 393 (1877). Type: [Ethiopia:] 'Nile Land. Abyssinia, *Schimper*! alt. 1–7000 ft.' *Schimper* 2181[56] (B† holotype, G00018021 lectotype, BR8872487, G00018022, GOET001090, K000410581, MO357631, P00073894, S-G-959, Z), lectotypified by Mesfin Tadesse (1984: 115).

Bidens pilosa L. var. *quadriseta* (Hochst. ex Oliv. & Hiern) Engl. in Abh. Königl. Akad. Wiss. Berlin **1891**: 437 (1892).

Bidens pilosa L. var. *glabrata* (Vatke) Engl. in Abh. Königl. Akad. Wiss. Berlin **1891**: 437 (1892).

Bidens abyssinica var. *incisifolia* Hochst. ex Chiov. in Annuario Reale Ist. Bot. Roma **8**: 186 (1904). Types: 'Schimper fl. Abyss. ed. Hohn. (1854) n. 2324; Oliver a Hiern l.c. Assaorta: Cualo-Enrot, 500 m. circa, 18.III. 1893 (P. n. 305). Oculè Cusai: Valle Damos, 600 m. circa, 14.IV.92 (P. n. 4128).' *Schimper* 2324 (K lectotype, BM8360656, G, P, S), lectotypified by Mesfin Tadesse (1984: 115).[57]

Bidens pilosa L. var. *abyssinica* (Sch.Bip. ex Walp.) Fiori in Nuovo Giorn. Bot. Ital. n.s., **20**: 390 (1913).[58]

Bidens chinensis (L.) Willd. var. *abyssinicus* (Sch.Bip. ex Walp.) O.E.Schulz in Bot. Jahrb. Syst. Suppl. **50**: 180 (1914).

Bidens lasiocarpa O.E.Schulz in Bot. Jahrb. Syst. Suppl. **50**: 185 (1914). Type: 'Hab. in India orientale, distr. Scinde: *Stocks* n. 608' (B† holotype).

Bidens cylindrica Sherff in Bot. Gaz. **81**(1): 28, fig. 91/g–l (1926). Type: [Zambia:] '*Menyhart* 1110, not abundant, in shady places at St. Joseph, Boruma ("Boroma"), Northern Zambesia, Rhodesia, April, 1892' (WU0038579 holotype, WU0038581).[59]

Bidens biternata var. *glabrata* f. *abyssinica* (Sch.Bip.) Sherff in Bot. Gaz. **90**(4): 389 (1930).

Bidens biternata var. *glabrata* f. *lasiocarpa* (O.E.Schulz) Sherff in Bot. Leafl. **9**: 10 (1954).

Bidens bipinnata sensu auct. mult. non L.; e.g., —Oliver & Hiern in F.T.A. **3**: 393 (1877), p.p. —Andrews, Fl. Pl. Sudan **3**: 11 (1956). —Cufodontis, (Enum. Pl. Aethiop.) Bull. Jard. Bot. Natl. Belg. **37** (3, suppl.): 1135 (1967). —Lisowski, (Asterac. Fl. Afr. Cent. 1) Fragm. Flor. Geobot. **36** Suppl. 1: 133 (1991).

Annual herb, erect, 0.05–2 m high; stem reddish tinged, 4-angled, simple or branched, pubescent or glabrous. Leaves petiolate, petiole to 60 mm long, opposite or rarely alternate near apex, lamina broadly ovate in outline, pinnately (3)5–9-lobed or almost pinnatisect, 3–20 × 2.5–12 cm, leaf segments lobed or bilobed at base, ovate to lanceolate, margins crenate-serrate to almost lobulate, apices acute. Inflorescences of terminal lax paniculate cymes. Capitula radiate; involucre cup-shaped; outer phyllaries 4–8(15), linear to oblanceolate, 4–6(12) mm long, to 20 mm long in fruit, pubescent; inner phyllaries 8–11, yellowish green, margins paler, 3.5–7 mm long, pubescent; paleae 4.5–6 mm long. Ray florets neuter or with staminodes, (1)2–5, ray limbs 3–6 × 1.3–2.5 mm, striate, yellow, rarely orange, tube 0.8–1.3 mm long, pubescent. Disc floret corollas yellow to orange-yellow, 3–3.7 mm long, glabrous. Achenes black, 4–8-ribbed, linear-tetragonal, 6–16(20) mm long, setuliferous in upper half; aristae (2)3–4(5), yellow-brown, 2–4 mm long, retrorsely barbed.

---

[55] Mesfin Tadesse (1984: 115) apparently lectotypified this name based on material of *Schimper* 105 in Z, with a duplicate in G, having noted that the 'holotype' in B had been destroyed. Rayner (1992: 80–81), at length, quite clearly indicated that Vatke had been precise in associating the name with *Schimper* 285, and had merely discussed the other *Schimper* collections. Rayner re-lectotypified the name based on *Schimper* 285 in Z; isolectotypes: BM, PRE.

[56] Mesfin Tadesse (1993: 498) gave the type collection as *Schimper* 2181.

[57] Mesfin Tadesse (1984: 115) suggested there was a holotype, citing *Schimper* 2324, and selecting a lectotype in K; this ignored the fact that clearly Chiovenda had cited several other collections.

[58] Mesfin Tadesse (1993: 498) was of the opinion there was/were types of this name, selecting a lectotype in FT. It is quite clear from the source references it was a combination.

[59] There is an unnumbered, undated *Menyhart* collection also in WU (WU0038580) which was considered type material by Sherff who determined it as '2nd type sheet' – it is from the type locality.

Botswana. N: Northern Dist., Sigera Pan, 30 mls. W of mouth of Nata River, 868 m, 26.iv.1957, *Drummond & Seagrief* 5255 (K, SRGH). **Zambia**. W: Kitwe Dist. 22.iv.1963, *Fanshawe* F7782 (K, NDO). S: Kaloma Dist., Machipapa, near Choma, 1220 m, 10.iii.1962, *Astle* 1488 (K). **Zimbabwe**. N: Darwin Dist., near Mutepatepa, on granite kopje, 14.iii.1963, *Wild* 6078 (K, SRGH). C: Makoni Dist., Chiduku, 1372 m, iv.1955, *Davies* 1172 (K, SRGH). E: Umtali Dist., 1098 m, 6.iii.1956, *Chase* 6002 (K, SRGH). **Malawi**. N: Mzimba Dist., Mzambazi Mission, 4 mi. N of Euthini, c. 1220 m, 18.iv.1974, *Pawek* 8362 (K, MAL, MO, SRGH, UC). C: Dedza Dist., Dedza Mountain Forest, 9.iv.1968, *Jeke* 173 (K, SRGH). S: Mt. Mulanje, Likabula Valley, Chapaluka Path, 1300 m, 17.iv.1988, *Chapman & Chapman* 9053 (K, MO). **Mozambique**. M: Vila Luísa, andado c. 15 km. para Manhiça próximo da Reserva do Bobole, 16.xii.1974, *Marques* 2514 (K, LMU).

Widespread in tropical and subtropical Africa, Asia and Australia. Forest margins, weed of cultivation (especially maize), ruderal sites, woodland, evergreen bushland, river-banks, grassland; 500–2500 m.

Conservation Status: A very widespread and somewhat weedy species; LC (Least Concern).

18. **Bidens bipinnata** L., Sp. Pl. **2**: 832 (1753). —Oliver & Hiern in F.T.A. **3**: 393 (1877). —Sherff, Publ. Field Mus. Nat. Hist., Bot. Ser. **16**(2): 366, Pl. 89 figs. l–s (1937). —Adams in F.W.T.A., ed. 2, **2**: 234 (1963). —Lisowski, (Asterac. Fl. Afr. Cent. 1) Fragm. Flor. Geobot. **36** Suppl. 1: 133 (1991). Type: [USA:] 'Habitat in Virginia.') (LINN – Herb. Linn. No. 975.12. lectotype), lectotypified by Mesfin Tadesse (1993: 499).[60]

Bidens fervida hort. ex Colla, Herb. Pedem. **3**: 306 (1834), nom. illeg. non Lam. (1785). Type: '(Patria?). Habeo hanc plantam ex hb: BIROLI; …'. Herbarium material unknown.

Bidens myrrhidifolia Tausch in Flora **19**(2): 394 (1836). Type: not cited, but based on cultivated material – probably in LZ.

Bidens cicutifolia Tausch in Flora **19**(2): 395 (1836). Type: not cited, but based on cultivated material – probably in LZ.

Bidens elongata Tausch in Flora **19**(2): 395 (1836). Type: not cited, but based on cultivated material – probably in LZ.

Bidens decomposita Wall. ex DC., Prodr. **5**: 602 (1836). Types: '(Wall.! cat. et herb. n. 298) … in India orient. ad Monghir prov. Bahar et in Nepalia. *Coreopsis corymbifolia* Hamilt.! ex Wall. (v. s. comm. ab hon. aula merc. angl. Ind. or.)' (G-DC 2×, K-W, K).[61]

Coreopsis corymbifolia Buch.-Ham. in Wall. ex DC., Prodr. **5**: 602 (1836), nom. nud. pro syn.

Kerneria bipinnata (L.) Gren. & Godr., Fl. Fr. **2**: 169 (1850).

Bidens pilosa var. *bipinnata* (L.) Hook.f., Fl. Brit. India **3**: 309 (1881).

Bidens pilosa var. *decomposita* (Wall. ex DC.) Hook.f., Fl. Brit. India **3**: 309 (1881).

Bidens bipinnata var. *minor* Memm. in J. Elisha Mitchell Sci. Soc. **30**(3): 148 (1915). Type: not cited.

Erect annual herb, 20–150(250) cm high. Stem tetragonal, glabrous or minutely setose-hispid in young plants, sometimes densely branched above. Leaves petiolate, petiole ebracteate, 1–6 cm long, lamina bipinnatisect, usually pubescent in lower parts, ovate, up to 20 cm long, segments membranous, ovate to deltoid or terminal segment lanceolate, ciliate, margins crenate-serrate, ciliate. Inflorescences of open paniculate cymes; peduncle glabrous, up to 10 cm long. Capitula radiate; involucre 4–7 × 4–6 mm; outer phyllaries green, linear or lanceolate, 7–10, glabrous or ciliate at base, 3-striate, 3–5 × 0.5 mm at anthesis, slightly dilated towards apex, shorter than inner phyllaries, apex acute; inner phyllaries dark brown (after drying) with yellowish-white scarious margins, ciliate at base, 7–8, linear, 3–6 × 1 mm at anthesis, margins ciliate near apex; receptacle sparsely pubescent, hairs white; paleae membranous, pale

---

[60] Sherff also noted the existence of this specimen but he did not lectotypify the name.

[61] There are two sheets in G-DC representing this material. The number '298' is the number applied to this collection by de Candolle when he 'numbered' the Compositae; the Catalogue number in the Herbarium of the East India Company is 3188, and there are clearly four collections, A, B, C, D, mentioned under Wallich's herbarium name, the material of which is represented in K-W, together with a duplicate in K (apparently of B – ex HBC).

brown, linear-lanceolate, c. 3–5 × 1 mm, up to 10 mm long in fruit, glabrous. Ray florets neuter, 3–5, ray limbs yellowish-white, oblong, c. 5 × 1.5 mm, apex entire or 3-fid, with 5 orange striae, tube c. 1 mm long. Disc floret corolla yellow, c. 4 mm long, 5-lobed; anthers brown, c. 1 mm long; style c. 3.5 mm long. Achenes black, linear-fusiform, tetragonal, glabrous or sparsely tuberculate-hispid towards apex, 7–18 × 1 mm, apex truncate; aristae 4 (rarely 2 or 3) yellow, slightly divergent, retrorsely barbed, 2–4 mm long.

Zimbabwe. W: Bulawayo, Waterford at the end of Dold Avenue past Waterford School, 28.i.1974, *Norrgrann* 461 (K, SRGH). C: Gwelo Dist., Mlozu School Farm, 18 m. SSE of Que Que, 1281 m, 26.ii.1966, *Biegel* 951 (K, SRGH). **Malawi**. *Brown* 8272 (collection details unavailable).

A widespread weedy species particularly in N and S America. Less widespread in Europe, Asia and Africa; 450–1400 m.

Conservation Status: Surprisingly, very infrequently collected in the Flora area; LC (Least Concern).

19. **Bidens pilosa** L., Sp. Pl. **2**: 832 (1753). —Oliver & Hiern in F.T.A. **3**: 392 (1877). — Sherff, Publ. Field Mus. Nat. Hist., Bot. Ser. **16**(2): 412, Pl. 99 fig. b, Pl. 102 figs. 1–b, e–j (1937). —Wild in Kirkia **6**(1): 24, excl. syn. (1967). —Agnew, Upland Kenya Wild Fl.: 466 (1974). —Mesfin Tadesse in Symb. Bot. Upsal. **24**(1): 122, fig. 61, fig. 58 (1984). —Ballard in Amer. J. Bot. **73**(10): 1452–1465 (1986). —Agnew & Agnew, Upland Kenya Wild Fl., ed. 2: 217, t. 89 (1994). —Mesfin Tadesse in Kew Bull. **48**(3): 500 (1993). —Blundell, Wild Fl. E. Afr.: pl. 90 (1987). —Lisowski, (Asterac. Fl. Afr. Cent. 1) Fragm. Flor. Geobot. **36** Suppl. 1: 136, fig. 32 (1991). —Hind in Fl. Masc., Composées **109**: 195 (1993). Type: 'Habitat in America.' (LINN – Herb. Linn. No. 975.8 lectotype),[62] lectotypified by D'Arcy in Woodson & Schery, Ann. Missouri Bot. Gard. **62**(4): 1178 (1975).

    Coreopsis leucanthema L., Cent. I Pl.: 29 (1755). Type: 'Habitat in America. Miller.' (LINN – Herb. Linn. No. 1026.5 lectotype, BM), lectotypified by Ballard (1986: 1464).

    Coreopsis leucantha L., Sp. Pl., ed. 2, **2**: 1282 (1763), orth var., type as above.

    Bidens hispida Kunth in Humb., Bonpl. & Kunth, Nov. Gen. Sp. Pl. **4** (ed. folio): 186 (1818). Type: 'Crescit locis siccis, prope La Venta de Sanchorquiz, alt. 760 hex. (Prov. Venezuelæ.). Floret Januario.' [Humboldt & Bonpland 'mss. n. 698. Caracas'; B-W: 'Caracas. Jan. 1800 __?__ 698. in Siccis La Venta'] (P-Bonpl. holotype, B-W).

    Bidens reflexa Link, Enum. Pl. Hort. Berol. Alt. **2**: 306 (1822). Type: 'Hab. in Mexico K.' (B† holotype).

    Bidens adhaerescens Vell., Fl. Flumin.: 348 (1825) [7 Sept.–28 Nov. 1829]; Fl. Flumin. Icon. **8**: tab. 88 (1827)[29 Oct. 1831]. Type: 'Habitat et frequenter, et ubique.' Herbarium material unknown.

    Bidens sundaica Blume, Bijdr. Fl. Ned. Ind. **13**: 913 (1826). Type: 'Crescit: in graminosis humidis prope Buitenzorg. Floret: omni tempore. Nomen: Harruga.' (L0001851 – 'Herb. Lugd. Bat. No. 900, 146 ... 70' holotype).[63]

    Bidens sundaica var. *minor* Blume, Bijdr. Fl. Ned. Ind. **13**: 914 (1826). Type: 'Crescit: cum praecedente.' [see *Bidens sundaica* Blume] (L0484314 – 'Herb. Lugd. Bat. No. 900, 146 ... 72' ?holotype).[64]

    Bidens decussata Pav. ex DC., Prodr. **5**: 599 (1836), nom. nud. pro syn. based on *B. hispida*.

    Bidens californica DC., Prodr. **5**: 599 (1836). Type: '① in Californià legit cl. *Douglas* [56]. ... (v.s. comm. ab hon. soc. hortic. Londin.)' (G-DC-G00454684 holotype, K001065520, K001065519 – mounted with what is probably a *Coulter* 295 collection, although the barcode is closer to the Coulter collection).

---

[62] Although Sherff (1937) mentioned the material he did not select a lectotype for the name.

[63] Mesfin Tadesse (1984, 1993) has repeatedly cited 'Bijd. Nat. Wetens. **1**: 913 (1826)' for the location of the protologue. The 'Bijdragen tot de Natuurkundige Wetenschappen. Amsterdam', to which this refers, is a multi-subject journal. However, I can find nowhere, in both physical and electronic copies, where the Flora is published. It is possible that the Flora was published as a separate (pre- or re-print) but no other reference work has any indication of this.

[64] L0484315 – 'Herb. Lugd. Bat. 900, 146 ... 77' is also marked as an isotype.

Bidens hirsuta Nutt., Trans. Amer. Phil. Soc. ser. 2, **7**: 369 (1841). Type: [Hawaii:] 'Hab. In Atooi.' (BM000810489 holotype).

Bidens pilosa f. *discoidea* Sch.Bip. in Webb & Berthelot, Hist. Nat. Iles Canaries **3**(2, sect. 2): 242 (1844), nom. illeg. based on *Bidens pilosa* L.

Bidens pilosa f. *radiata* Sch.Bip. in Webb & Berthelot, Hist. Nat. Iles Canaries **3**(2, sect. 2): 242 (1844), nom. illeg. based on *Coreopsis leucantha* L./*Bidens leucantha* (L.) Willd.

Bidens leucantha (L.) Willd. f. *discoidea* (Sch.Bip.) Krauss, Beitr. Fl. Cap. Natal.: 77 (1846), nom. inval.

Bidens pilosa var. *radiata* (Sch.Bip.) J.A. Schmidt, Beitr. Fl. Cap. Verd. Ins.: 197 (1852), comb. illeg.

Bidens leucantha var. *pilosa* (L.) Griseb., Cat. Pl. Cub.: 155 (1866).

Kerneria pilosa (L.) Lowe, Man. Fl. Madeira **1**: 474 (1868).

Kerneria pilosa var. α *radiata* (Sch.Bip.) Lowe, Man. Fl. Madeira **1**: 474 (1868), comb. illeg., citing *Coreopsis leucantha* L. in synonymy.

Kerneria pilosa var. β *discoidea* (Sch.Bip.) Lowe, Man. Fl. Madeira **1**: 474 (1868), comb. illeg., based on *Bidens pilosa* L.

Bidens pilosa var. *leucantha* (L.) Kuntze [forma] 1. *subsimplicifolia* Kuntze, Revis. Gen. Pl. **1**: 322 (1891). Type: 'Portorico. Hongkong.' ['PUERTO RICO. Caguay, 8 Mar 1874, *Kuntze* s.n.' – according to Wetter & Zanoni in Brittonia **37**: 328, 1985] *Kuntze* s.n. ex Puerto Rico (NY00115509 holotype).

Bidens pilosa var. *leucantha* Sch.Bip. [forma] 2. *ternata* Kuntze, Revis. Gen. Pl. **1**: 322 (1891). Types: 'Porto Rico. Hongkong. Sikkim. Gran Canaria.' Hong Kong (NY00162606, NY00162607 syntypes); Sikkim (NY00162610 syntype).[65]

Bidens pilosa var. *leucantha* Sch.Bip. [forma] 4. *subbiternata* Kuntze, Revis. Gen. Pl. **1**: 322 (1891). Type: 'Caracas. [*Kuntze* 1462b]' (NY00162594 holotype).

Bidens pilosa var. *discoidea* Sch.Bip. [forma] 3. *pinnata* Kuntze, Revis. Gen. Pl. **1**: 322 (1891). Type: 'Colon. Macao. Turong.' ['CHINA. Macao, 7 Feb 1875, *Kuntze* 3541. VIETNAM. Turong [Da Nang], Feb 1875, *Kuntze* 3791.' – according to Wetter & Zanoni, 1985: 328] *Kuntze* 3541 (NY 00162608 syntype); *Kuntze* 3791 (NY00162605 syntype).

Bidens pilosa var. *discoidea* Sch.Bip. [forma] 4. *subbiternata* Kuntze, Revis. Gen. Pl. **1**: 322 (1891). Type: 'St. Thomas. Birma.' ['BURMA. Maulmein, 18 Oct 1875, *Kuntze* 6248. VIRGIN ISLANDS. St. Thomas, 26 Feb 1874, *Kuntze* 213.' – according to Wetter & Zanoni, 1985: 328] *Kuntze* 6248 (NY 00162609 syntype); *Kuntze* 213 (NY00115508 syntype).

Bidens pilosa subvar. *radiata* (Sch.Bip.) Pit. in Pit. & Proust, Iles Canaries Fl.: 226 (1908).

Bidens pilosa subvar. *discoidea* (Sch.Bip.) Pit. in Pit. & Proust, Iles Canaries Fl.: 226 (1908).

Bidens pilosa var. *minor* (Blume) Sherff in Bot. Gaz. **80**(4): 387 (1925). —Sherff, Publ. Field Mus. Nat. Hist., Bot. Ser. **16**(2): 421, Pl. 102 figs. c–d, k–r (1937).

Annual herb, erect, 0.1–1.5 m tall; stem much branched or sometimes hardly branched, pale green or sometimes reddish, 4-angled, sparsely pubescent to glabrous. Leaves petiolate, petiole to 70 mm long, lamina pinnately 3–5-lobed or occasionally simple in upper or lowermost part of stem, up to 15(20) cm long, segments ovate to ovate-lanceolate, margins serrate or crenate-serrate, sparsely pubescent to glabrous, terminal segment 5–10 × 2.5–5 cm, lateral segments 2–5 × 1–2 cm and asymmetric, often lobed or bilobed. Inflorescences of solitary, terminal capitula or of few capitula; peduncles to 10 cm long. Capitula usually radiate; outer phyllaries 7–10, spathulate, 3–4 mm long, reflexed at anthesis; inner phyllaries 5–8, yellow-green to pale brown, 3–4.5 mm long with yellow scarious margins, glabrous except for apex; receptacle club-shaped in fruit; paleae light brown, striate, 3–5 mm long. Ray florets white or cream, sometimes absent, neuter or with pistillodes or staminodes, 4–8, rays (when fully developed) 7–15 × 3–4.5 mm, tube 0.5–0.8 mm long. Disc floret corollas yellow or yellow-orange, ± 4 mm long, pubescent at base. Achenes black, 4–6-ribbed, linear-tetragonal, 4–12 mm long, setuliferous or verrucose; aristae 2–3(4), 2–4 mm long, retrorsely barbed.

Botswana. N: Maun, 19°29'S, 23°26'E, 21.iii.1975, *Smith* (K, SRGH). **Zambia**. B: Kalabo, near resthouse Kalabo, 13.xi.1959, *Drummond & Cookson* 6415 (K, SRGH). N: Chilongowelo Bush and Garden, 1464 m, 16.ii.1952, *Richards* 971 (K). W: Mwinilunga Dist.,

---

[65] Wetter & Zanoni (1985) noted that no material had been located in NY. There are 2 sheets in NY (NY00162606, NY00162607) that correspond to two collections from Hong Kong. There is one sheet in NY (NY00162610) corresponding to a collection from Sikkim.

by R. Matonchi, 11.xi.1937, *Milne-Readhead* 3193 (K). C: Mumbwa, on the banks of the Lutale River between Mumbwa and Kasempa, 4.i.1963, *van Rensburg* 1157 (K). **Zimbabwe**. C: Salisbury, Agricultural Experimental Station, 1494 m, 15.viii.1957, *Drummond* 5820 (K, SRGH). **Malawi**. N: Nyika Plateau, juniper forest edge, 1799 m, 28.xii.1975, *Phillips* 713 (K, MO). S: Bvumbwe, 8.ii.1985, *Croix* 2634 (OS). **Mozambique**. M: Inhaca Island, 23 mls E of Lourenco Marques, 0–200 m, 27.ix.1958, *Mogg* 28373 (K).

Pantropical and pan-subtropical, also extending into some temperate areas. A weed of cultivation and ruderal sites, often on alluvial sites, also in forest margins and grassland, in swamps and along streams; may be locally common; (0)60–2500 m.

Conservation Status: A pantropical and pansubtopical weed, albeit surprisingly poorly collected from many areas – perhaps because it is such a common weed!; LC (Least Concern).

120. **ENYDRA** Lour.

Enydra Lour., Fl. Cochinch. **2**: 510 (1790). —Wild in Kirkia **6**(1): 35–36 (1967). —Dyer, Gen. S. Afr. Fl. Pl. **1**: 696 (1975). —Gibbs Russell in Kirkia **10**(2): 501 (1977). —Lack in Willdenowia **10**(1): 3–12 (1990). —Lisowski, (Asterac. Fl. Afr. Cent. 1) Fragm. Flor. Geobot. **36** Suppl. 1: 176–177 (1991). —Mesfin Tadesse, Fl. Ethiopia & Eritrea **4**(2): 314 (2004). —Beentje & Hind in F.T.E.A., Compositae **3**: 768–770 (2005). —Robinson in Harling & Andersson, Fl. Ecuador, **77**(1), 190(6): 151–154 (2006).
Enhydra DC., Prodr. **5**: 636 (1836), orth. var.
Phyllimena Blume ex DC., Prodr. **5**: 636 (1836), nom. nud. pro syn.
Meyera Schreb., Gen. Pl., ed. 8[a] **2**: 570 (1791), non Adans. (1763).
Sobreyra Ruiz & Pav., Fl. Peruv. Chil. Prodr.: 109, t. 23 (1794).
Sobreya Pers., Syn. Pl. **2**: 473 (1807), orth. var.
Cryphiospermum P.Beauv., Fl. Oware **2**: 24, t. 74 (1810).
Tetraotis Reinw., Syll. Pl. Nov. **2**: 8 (1825)[1826].
Wahlenbergia Schumach., Beskr. Guin. Pl.: 387 (1827), non Schrad. ex Roth. (1821), nom. cons. [CAMPANULACEAE].
Hingtsha Roxb., Fl. Ind., ed. 2, **3**: 448 (1832).

Erect or prostrate perennial herbs of wet places. Leaves opposite, sessile or subsessile. Capitula axillary, and overtopped by lateral innovations, sometimes appearing terminal, solitary, sessile or subsessile, hemispherical, heterogamous, radiate; phyllaries 4, in opposite pairs, foliaceous, broad, biseriate; receptacle convex, paleaceous, paleae (and inner phyllaries) sheathing florets. Ray florets numerous (c. 20–30), female, 1–3-seriate, fertile, limb very short, broad, 3–4-dentate. Disc florets several to numerous (10–30), hermaphrodite, fertile, or innermost sterile, corollas tubular but widening above, 5-lobed; anther thecae ± black, apical anther appendages ovate, glandular-punctate outside, base obtuse; style base thickened, style arms linear, puberulent with separate stigmatic lines. Achenes surrounded by persistent paleae, oblong or fusiform, terete or usually ± compressed, base short- to long-stipitate; pappus absent.

Although most floras have recorded 10 spp. I think this number is too high and that there are really only a handful of species found between both hemispheres. Only one species is found in the Flora area, although most probably native of the New World but now widespread in the tropics.

Enydra fluctuans Lour., Fl. Cochinch. **2**: 511 (1790). —Oliver & Hiern in F.T.A. **3**: 372 (1877) (as '*Enhydra*'). —Adams in F.W.T.A., ed. 2, **2**: 242 (1963). —Wild in Kirkia **6**(1): 35–36 (1967). —Lisowski, (Asterac. Fl. Afr. Cent. 1) Fragm. Flor. Geobot. **36** Suppl. 1: 176–177 (1991). —Mesfin Tadesse, Fl. Ethiopia & Eritrea **4**: 314 (2004). —Beentje & Hind in F.T.E.A., Compositae **3**: 768–770 (2005). —Chen Yousheng & Hind in Fl. China **20-21**: 861 (2011). Type: 'Habitat spontanea in paludibus Cochinchinae.' (BM holotype). FIGURE 6.5.**13**.

Fig. 6.5.**13**. ENYDRA FLUCTUANS. 1, habit (× ²/₃); 2, capitulum (× 4); 3, ray floret (× 10); 4, disc floret and accompanying palea (× 4); 5, disc floret (× 10); 6, disc floret with corolla opened out showing attachment point of filaments (× 8). 1–2 from *Richards* 21619; 3–6 from *Purseglove* 3350. Drawn by Juliet Williamson. From Flora of Tropical East Africa.

Cryphiospermum repens P.Beauv., Fl. Oware **2**: 24 (1810). Type: 'Cette plante croît sur les bords du fleuve Formose: c'est une de celles dont les naturels du pays font usage pour la guérison des plaies. …' (?G holotype).[66]

Meyera fluctuans (Lour.) Spreng., Syst. Veg., ed. 16, **3**: 602 (1826).

Meyera guineensis Spreng., Syst. Veg., ed. 16, **3**: 602 (1826), nom. illeg., citing both *Caesulia radicans* Willd. and *Cryphiospermum repens* P.Beauv.

Tatraotis paludosa Reinw. in Blume, Bijdr. Fl. Ned. Ind. **15**: 892 (1826). Type: 'Crescit: in paludosis prope Bataviam.' Type material may well be in L.

Tetraotis longifolia Reinw. in Blume, Bijdr. Fl. Ned. Ind. **15**: 892 (1826). Type: 'Crescit: cum priori.', q.v. *T. paludosa.* Type material may well be in L.

Hingtsha repens Roxb., Fl. Indica, ed. 2, **3**: 448 (1832). Type: 'Beng.[al] Hingtsha./A native of Bengal, delighting in a moist soil, and often extending itself considerably over the surface of adjoining pools of water.' Location of type material unknown.[67]

Enydra [sub *Enhydra*] *longifolia* (Reinw.) DC., Prodr. **5**: 637 (1836).

Enydra [sub *Enhydra*] *paludosa* (Reinw.) DC., Prodr. **5**: 637 (1836).

Enydra [sub *Enhydra*] *heloncha* DC., Prodr. **5**: 637 (1836). Types: 'in aquosis Indiae or. in prov. Silhet (*Wall.*[*ich*]), ad Dumdumma et Golpara (*Ham.*). Meyera Heloncha Wall.! cat. n. 3195. comp. 305. Hingcha repens Roxb. cat. calc. 62. … (v.s.)' (E, G-DC, K, K-W syntypes).[68]

Enydra anagallis Gardner in London J. Bot. **7**: 409 (1848). Type: [Brazil:] '[*Gardner*] 5522. … Hab. In ditches at the Laranjeiras, near Rio de Janeiro. Jan. 1841.' (BM, E00433364, GH00006580, K000895507, K000895508 – mounted with *Traill* s.n., NY00168363, P02535047 syntypes).

Enydra woollsii F.Muell., Fragm. **3**: 139 (1863). Type: [Australia:] 'In paludibus prope Manly Beach portus Jacksonii. *W. Wools.*' (MEL2159823 holotype, K001065502).

Perennial herb; stems fleshy, c. 5 mm diam., hollow, often rooting at nodes, elongate, prostrate, ascending or erect, pilose-pubescent or glabrescent, 20–30(100) cm long, internodes 5–10 cm long. Leaves sessile or subsessile, petiole (when present) to 2 mm long and ± dilated, lamina 5–7(12) × 0.5–1 cm, narrow-oblong, base subhastate, subcordate or broadly cuneate, flat, margins entire or serrate, sparsely pubescent or glabrescent on both surfaces, minutely glandular-punctate, apex acute or subacute. Capitula to c. 2 cm diam.; phyllaries c. 1 cm long, broadly ovate, outer pair completely obscuring capitulum at anthesis, many-veined, very sparsely pubescent or glabrous, apex obtuse; paleae 5–5.5 mm long, completely enfolding disc florets, with a keel and 2 strong lateral veins, ciliate towards apex, longitudinally veined. Ray floret corollas greenish-yellow, c. 2.2 mm long, limb 0.7–1 × 0.3 mm wide, broadly oblong, apex 3-lobed, tube 1.5 mm long; style exserted for c. 1.5 mm. Disc florets greenish-yellow, c. 3 mm long, tubular but widening above, corolla lobes c. 1 mm long. Achenes brown, c. 2-3.7 mm long, narrowly oblong, very short-stipitate, body striate, covered with very minute black or golden dots.

Zambia. B: Mongu, 12.xi.1959, *Drummond & Cookson* 6364 (K, LD, PRE, SRGH). **Zimbabwe.** N: Mana Pools, iii.1971, *Guy* 1626 (K, SRGH). **Mozambique.** GI: Chibuto, Maniquenique, 10.x.1957, *Granvaux Barbosa & de Lemos* 7981 (COI, K).

Widespread in tropical Africa to the N of our area; also in SE Asia, the E Indies and Australia (New South Wales), and is widespread in S America from Argentina N to Colombia; introduced into Mexico. A marsh plant of flood plains, at the edge of lagoons and in soak zones.

Vernacular name: 'Tchau' (Mozambique).

Conservation Status: Most probably under-collected in all suitable marshy places; probably not threatened, but best recorded as DD (Data Deficient).

---

[66] Robinson (2006: 152) indicated that the holotype was in BM where there may well be a duplicate.

[67] Robinson (2006: 152) queried whether there may well be type material in K. I have been unable to locate any amongst the appropriate regional material, although clearly that name appeared in Wallich's 'Numerical List' based on Roxburgh's mss name.

[68] There are three specimens numbered 305 in G-DC (G00457153, G00457202, G00457209) and one other numbered 3195 (G00457202 – second sheet).

121. **FLAVERIA** Juss.

Flaveria Juss., Gen. Pl.: 186 (1789). —Pope in Kirkia **10**(1): 111–114 (1975). —Powell in Ann. Missouri Bot. Gard. **65**: 590–636 (1978). —Lisowski, (Asterac. Fl. Afr. Cent. 2) Fragm. Flor. Geobot. **36** Suppl. 1: 490–491 (1991). —Hind in Fl. Masc., Composées **109**: 220–222 (1993). —Beentje & Hind in F.T.E.A., Compositae **3**: 724–725 (2005).
Vermifuga Ruiz & Pav., Fl. Peruv. Prodr.: 114, t. 24 (1794).
Brotera Spreng. in J. Bot. (Schrader) **4**[1800]: 184, t. 5 (1801), non *Brotera* Cav. (1799).
Nauenburgia Willd., Sp. Pl., ed 4 **3**(3): 1489, 2393 (1803/4), based on *Brotera* Spreng.
Dilepis Suess. & Merxm. in Mitt. Bot. Staatssamml. München **1**: 14 (1950).

Annual or perennial, delicate to rather robust herbs, very rarely small trees. Stems glaucous, glabrous to densely short-pubescent, branches opposite and decussate, stems often clothed in sheathing bases of dead leaves. Leaves opposite and decussate, simple, petiolate or sessile, leaf bases connate, sometimes weakly so, very rarely perfoliate, lamina linear, lanceolate, ovate, elliptic or oblanceolate, glabrous or puberulous, rarely villous or sparsely pilose, margins entire or serrate, apices acute. Inflorescences of laxly or densely aggregated capitula in flat-topped corymbose panicles or axillary glomerules. Capitula radiate and heterogamous (radiate capitula usually peripheral in inflorescences) or discoid and homogamous, both types usually present within each cluster/glomerule of capitula, sessile or short-pedicellate; involucre ± cylindrical or urceolate; phyllaries 1- or 2-seriate, few (2–4), navicular, herbaceous or hyaline, very rarely corky, apices acute; receptacle minute, glabrous or rarely setose (sometimes regarded as modified paleae), epaleaceous. Ray florets usually 1, rarely 2, if present, female, fertile; ray limb yellow or cream-yellow, ovate or elliptic, usually small and inconspicuous, corolla tube glabrous or pubescent. Disc florets 1–15, hermaphrodite, fertile; corollas yellow, tubes glabrous or puberulous towards base, throats gradually or abruptly expanded, narrow funnelform to campanulate, lobes short, acute; apical anther appendages ovate-lanceolate; basal anther appendages obtuse; style short, glabrous, style arms flattened, truncate, apices short-pubescent. Achenes 10-ribbed, body black, glabrous; carpopodium a narrow, inconspicuous annulus; pappus absent or rarely of 2–4 hyaline squamellae or a short crown of united scales.

A genus of 22 spp.; a New World genus with 1 sp. apparently endemic to Australia and 2 widespread pantropic weeds.

Capitula scorpioid but contracted (secund on branches of pedunculate clusters); ray limbs narrow, erect; receptacle glabrous; phyllaries 3 (or 4); florets (2)3–8 per capitulum . . . . **1.** *bidentis*
Capitula in dense sessile axillary and terminal glomerules; ray limbs suborbicular recurved; receptacle setose; phyllaries usually 2; floret usually 1, rarely 2 per capitulum . . . . **2.** *trinervia*

1. **Flaveria bidentis** (L.) Kuntze, Revis. Gen. Pl. **3**: 148 (1898). —Rydberg in N. Amer. Fl. **34**: 143 (1915). —Cabrera, Fl. Prov. Buenos Aires **6** Compuestas: 237, t. 70 (1963). —Henderson & Anderson, Common Weeds S. Afr.: 370, fig. 184 (1966). –Merxmüller, Prodr. Fl. Südwestafr. **139**: 71 (1967). —Hilliard in Ross, Fl. Natal: 367 (1972). Type: 'Habitat in India?' (LINN – Herb. Linn. No. 977.4 lectotype), lectotypified by Howard, Fl. Lesser Antilles **6**: 566 (1989).
Ethulia bidentis L., Mant. Pl.: 110 (1767).
Eupatorium chilense Molina, Sagg. Stor. Nat. Chile: 142, 354 (1782/9). Type: '… *Feuil.*' Feuillée s.n. Herbarium material unknown.
Flaveria chiloensis Juss., Gen. Pl.: 187 (1789). Type: '... (Milleria Chiloensis H. R. P.) capitalis, in Peruvianâ à *Dombeyo* datâ spicatis.' (?P-JU holotype).
Milleria chiloensis Juss., Gen. Pl.: 187 (1789), nom. nud. pro syn.
Flaveria chilensis J.F.Gmel., Syst. Nat., ed. 13, **2**: 1269 (1791)[1792], nom. illeg. based on *Flaveria chiloensis* Juss.
Milleria contrayerba Cav., Icon. **1**: 2, pl. 4 (1791). Type: 'Habitat in Huanuco imperii Peruani. 0. Vidi floridam in Regio hortu Matritensi a mense Septembri usque ad Decembrem.' (MA holotype).
Vermifuga corymbosa Ruiz & Pav., Syst. Veg. Fl. Peruv. Chil. **1**: 216 (1798), based on *Milleria contrayerba* Cav.

Flaveria contrayerba (Cav.) Pers., Syn. Pl. **2**: 489 (1807).

Flaveria capitata Juss. ex Sm. in Rees, Cycl. **14**: *Flaveria* no. 1. (1810), nom. illeg. (incl. *Ethulia bidentis* L. in synonymy).

Flaveria bonariensis DC., Prodr. **5**: 635 (1836). Type: '① circa Bonarium legit cl. *Bacle* [147]. Flores ign. (v.s.)' (G-DC holotype). G-DC (G00457188)

Flaveria contrayerba var. *latifolia* Phil. in Anales Univ. Chile **36**: 185 (1871).[69] Type: 'Mendoza.' Herbarium material unknown.[70]

Flaveria bidentis [forma][71] β *angustifolia* Kuntze, Revis. Gen. Pl. **3**(3): 148 (1898). Type: 'Argentina: Cordoba ([15 Mar 1878] *Galander* [s.n.]), westliche Pampas.' (NY00169353 holotype).

Erect semi-herbaceous divaricately-branched annual up to c. 1 m tall; branches opposite, decussate; stem and branches ± ribbed, sparsely puberulent to glabrescent, green or sometimes purple-tinged, drying yellowish. Leaves simple, sessile, 1.5–9.5 × 0.3–2.4 cm, very narrowly elliptic or narrowly oblanceolate to narrowly lanceolate, apex acute; base narrowed and pseudo-petiolate stem clasping in lower leaves, ± narrowed and amplexicaul in upper leaves ± villous where leaf or petiole clasps stem, margins serrulate to serrate, 3-veined from base, outer pair of veins running ± parallel to the leaf-margins nearly to the apex, midrib and 2 lateral veins prominent below, puberulent to glabrescent. Capitula secund, closely spaced in many pedunculate congested glomerules, c. 1–2.5 cm wide; capitula 3-5 mm long, narrowly obovoid c. 6–14 together, apical 2–4 usually containing female floret only, basal 4–9 containing 1 female plus 1–5 hermaphrodite florets. Phyllaries 4–5 mm long, glabrous, the outer narrowly cymbiform and enclosing narrowly oblanceolate phyllaries, with c. 2 narrowly lanceolate subsidiary bracts up to 3 mm long inserted just below phyllaries. Marginal florets female, rayed, ray limb yellow, c. 0.5–1 mm long, narrow, upright, apex emarginate, glabrous, exceeding phyllaries; corolla tube c. 1 mm long, base sparsely pubescent. Hermaphrodite florets 2.5–3 mm long, slightly exceeding phyllaries, lower part c. 1.5 mm long, tubular, ± sparsely pubescent, upper part yellow, 0.75–1 mm long, campanulate, lobes 5, spreading. Achenes black, 2–2.5 mm long, narrowly obovoid, slightly compressed, ribbed, glabrous; pappus absent.

Botswana. N: Maun, near the river, ii.1967, *Lambrecht* 67 (COI, K, PRE, SRGH). SE: 5 miles S of Palapye road at Serowe turn-off, 20.i.1960, *Leach &Noel* 268 (K, LISC, PRE, SRGH). **Zimbabwe**. S: Chiredzi, Mtilikwe River, 427 m, 26.i.1949, *Wild* 2784 (K, LISC, SRGH). **Mozambique**. M: Maputo, Matutunuíne, Catuane, Empresa Agro-Pecuária de Catuane, 29.ix.1983, *Daniel* et al. 565 (K, LMU).

A native of tropical America where it is particularly widespread; also recorded from Natal, the Orange Free State, Transvaal, Northern Cape, eSwatini and SW Africa. Apparently, an uncommon weed in the Flora area, although almost certainly under-collected and thus under-recorded, usually at roadsides, and preferring moist situations, but also in dry cultivated areas, roadsides, scrub, and dry woodland; 50–2800 m; probably flowering throughout the year.

Conservation Status: A widespread weed although not well-recorded in the Flora area; LC (Least Concern).

2. **Flaveria trinervia** (Spreng.) C.Mohr in Contr. U.S. Natl. Herb. **6**: 810 (1901). — Johnston in Proc. Amer. Acad. **39**: 284 (1903). —Rydberg in N. Amer. Fl. **34**: 142 (1915). —Andrews, Fl. Pl. Sudan **3**: 29 (1956). —Lisowski, (Asterac. Fl. Afr. Cent. 2) Fragm. Flor. Geobot. **36** Suppl. 1: 491 (1991). Type: A cultivated specimen from Halle Botanic Garden – 'Ipse *Oederam* prolificam a *Thunbergio* ad Caput bonae spei lectam sedulo examinavi, eiusdemque generis cum nostra planta esse, facile cognovi. Planta ipsa in area ad Octobrim usque bene viget, set floret tantummodo in hypocausto, cuius temepries 66 Gr. *Fahr.* est.' FIGURE 6.5.**14**.

---

[69] In a separately paginated reprint/preprint in K this appeared on p. 27.

[70] This was not listed by Muñoz Pizarro in his list of Philippi collections: Las especies de plantas descritas por R. A. Philippi, durante el siglo XIX - Santiago de Chile (1960).

[71] Kuntze actually used the word 'Form' not 'Varietäten'.

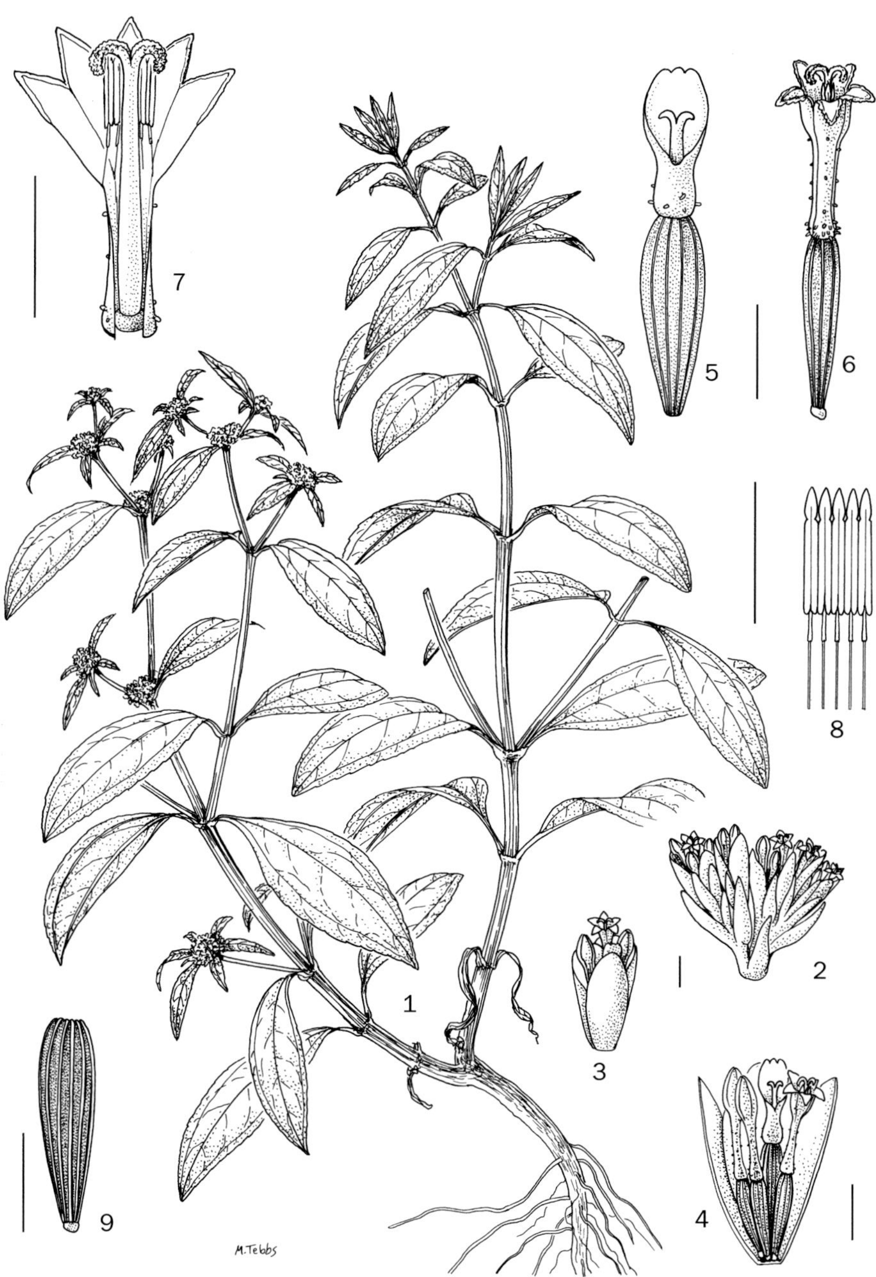

Fig. 6.5.**14**. FLAVERIA TRINERVIA. 1, flowering plant; 2, terminal glomerule; 3, capitulum; 4, l.s. capitulum; 5, ray floret; 6, disc floret; 7, corolla of disc floret opened out to show style and partial view of anther cylinder and filament attachments; 8, anther cylinder opened out; 9, achene. Scale bars: 2–9 = 1 mm; 1 = 10 mm. All from *Hind et al.* in PCD 3373. Drawn by Margaret Tebbs.

Oedera trinervia Spreng. in Bot. Gard. Halle: 63 (1800).

Brotera contrayerva Spreng. in J. Bot. (Schrader) **4**[1800]: tab. 5 (1801), non *Flaveria contrayerba* Pers. (1807)[= *Flaveria bidentis*]. Type: 'Diese merkwürdige Pflanze wächst um Huanuco in Peru wild.'[72]

Nauenburgia trinervata Willd., Sp. Pl., ed. 4, **3**: 2393 (1803), nom. illeg., based on *Brotera contrayerva* Spreng.

Brotera trinervata (Willd.) Pers., Syn. Pl. **2**: 498 (1807), comb. illeg.

Flaveria repanda Lag., Gen. Sp. Pl.: 33 (1816). Type: 'Azuminates vulgo in Nova Hispania, ubi sponte nascitur. Colitur in R. M. H. uni Autumno floret.' (?MA holotype).

Brotera sprengelii Cass. in Dict. Sci. Nat., ed. 2, **34**: 306 (1825), nom. illeg., based on *Brotera trinervata* Pers., *Nauenburgia trinervata* Willd., and *Brotera contrayerva* Spreng.

Flaveria trinervata (Willd.) Baillon, Hist. Pl. **8**: 55, Fig. 94, 95 (1882).

Dilepis dichotoma Suess. & Merxm. in Mitt. Bot. Staatssamml. München **1**: 14 (1950). Type: 'Venezuela, Caracas, San José de Avila, cr. 1000 m. – Leg. *M. Zehentner* nr. 91 (14.IX.1925). – Typus -3 Bogen- im Herb. München.' (M0030099, M0030100, M0030101 syntypes).

Flaveria australasica sensu Wild, Common Rhod. Weeds, fig. 69 (1955).

Erect semi-herbaceous divaricately branched annual 10–75(200) cm tall; branches opposite and decussate; internodes usually much longer than leaves, slightly compressed; stems and branches ± ribbed, sparsely pubescent to glabrescent, green or purple-tinged. Leaves simple, sessile, 1.5–7 × 0.3–2 cm, narrowly elliptic to oblanceolate, upper leaves smaller and very narrowly elliptic or lorate to linear, apex acute to obtuse, base narrowed and pseudopetiolate widening to become stem clasping in lower leaves ± narrowed and amplexicaul in upper leaves, margin serrulate to serrate, teeth ± remote, 3-veined from base, outer pair of veins ± parallel to leaf-margins nearly to apex, midrib and lateral veins prominent below, puberulent to glabrescent. Inflorescence glomerules congested, sessile, 1–2.5 cm wide, axillary and terminal; bracts of glomerules leaf-like, to c. 6 cm long. Capitula numerous, 4–5 mm long, narrowly oblong-cylindrical to ellipsoid, capitula with 1 female floret only, or less often 1 female plus 1 hermaphrodite floret, or of 2–4 hermaphrodite florets only. Phyllaries green, sometimes yellow towards apex, 4–5 mm long, glabrous; 2–3 outer phyllaries narrowly cymbiform enclosing 0–2 narrowly oblanceolate to lanceolate inner phyllaries, with 0–3 subsidiary phyllaries to 3 mm long, linear or widened near apex, inserted just below main phyllaries. Marginal florets, female, rayed, ray limb yellow, 0.5–1 mm long, obovate to suborbicular, emarginate at apex, glabrous, exceeding phyllaries, usually recurved; corolla tube green, c. 1 mm long, pubescent in lower part. Hermaphrodite florets 2–4; corolla 2–2.5 mm long, slightly exceeding phyllaries, tube base green, c. 1.5 mm long, tubular, pubescent particularly toward base, hairs eglandular, upper part yellow, shortly campanulate, with 5 acute spreading lobes, glabrous. Achenes black, 2–2.5 mm long, those of female florets usually longer than those of hermaphrodite florets, narrowly obovoid, slightly compressed, ribbed, glabrous; pappus absent.

Botswana. SE: Orapa Game Park, 21°20.041'S, 25°19.821'E, 982 m, *Smith et al.* (K). **Zambia.** C: Lusaka, 22.vi.1970, *Anton-Smith* in GHS 211774 (SRGH). S: Monze, 7.vi.1962, *Fanshawe* 6873 (K, SRGH). **Zimbabwe.** N: Shamva, 17.ii.1961, *Rutherford-Smith* 550 (K, M, SRGH). W: Wankie, Kennedy Siding, 12.xii.1968, *Rushworth* 1362 (COI, K, M, PRE, SRGH). C: Gatooma, 1.vii.1932, *Eyles* 7013 (K, SRGH). E: Umtali, Aerodrome Road, 13.v.1963, *Chase* 8015 (K, M, PRE, S, SRGH). S: Gwanda, Umzingwani River, Doddieburn Ranch, 9.v.1972, *Pope* 689 (K, LISC, PRE, SRGH). **Mozambique.** LM: Lourenco Marques, Costa do Sol, 28.ix.1959, *Mogg* 29750 (SRGH).

Tropical American in origin, now a widespread as a weed in Africa, and is recorded from the D.R. Congo, Kenya, Tanzania, Ethiopia and the Sudan. Ruderal, locally abundant on industrial sites, mine dumps, railway yards, cattle kraals and roadsides; 0–1150 m.

Vernacular name: 'Gaika weed' (Zimbabwe).

Conservation Status: A widespread weed; LC (Least Concern).

---

[72] Only the plate was titled *Brotera contrayerva*, the text, containing a full Latin description (referring to the plate), was titled '*Millera contrayerva* Cavan. neu untersucht und bestimmt.' [= *Millera contrayerva* Cav. re-examined and determined].

122. **TAGETES** L.[73]

Tagetes L., Sp. Pl. **2**: 887 (1753). —Pope in Kirkia **10**(1): 115–119 (1975). —Lisowski, (Asterac. Fl. Afr. Cent. 2) Fragm. Flor. Geobot. **36** suppl. 1: 491–496 (1991). ——Hind in Fl. Masc., Composées **109**: 222–224 (1993). —Soule in Hind & Beentje, Proc. Intl. Compositae Conf., Kew, Compositae Syst. **1**: 435–443 (1996). —Beentje & Hind in F.T.E.A., Compositae **3**: 725–728 (2005). —Panero in Kubitzki, Fam. Gen. Vasc. Pl. **8**: 429 [2006](2007).

Diglossus Cass. in Dict. Sci. Nat., ed 2 **13**: 241 (1819).

Enalcida Cass. in Bull. Sci. Soc. Philom, Paris **1819**: 31 (1819).

Solenotheca Nutt. in Trans. Amer. Phil. Soc., ser. 2, **7**: 371 (1841).

Annual or perennial herbs, sometimes shrubby; plants strongly aromatic especially when fresh, either pungent or of anise/aniseed. Rootstock a taproot, rhizomatous or woody. Stems terete, usually glabrous, smooth or striate. Leaves opposite or alternate, petiolate, often connate across node, connate margins with glandular foliaceous appendages, lamina usually pinnate, sometimes simple, gland-dotted, margins variously serrate. Inflorescences of solitary terminal or axillary capitula or many-headed dense terminal corymbs, or of dense clusters, pedicels short or long, slender, sometimes enlarge and fistulose beneath involucre. Capitula radiate and heterogamous or discoid and homogamous, small, medium or relatively large; involucres narrowly cylindrical to broadly campanulate, often inflated and fistulose beneath involucre; phyllaries uniseriate, margins connate, sometimes splitting at maturity, variously conspicuously glandular; receptacle convex or flat, epaleaceous, glabrous or ciliate. Ray florets few (5 or 8), rarely absent, uniseriate, female; ray limb 2- or 3-lobed, papillose above, white, cream, yellow, orange, maroon or red to brownish, sometimes bicoloured, sometimes variously marked with variously shaped coloured blotches on limb base, often lost on drying; corolla tube glabrous or short-pubescent; style base with conspicuous node, style arms filiform to attenuate, papillae long, stigmatic lines separate. Disc florets few to many, rarely lacking, hermaphrodite, corollas tubular, corolla lobes usually 5, rarely 4+1, inner surface long-papillate; apical anther appendages triangular, acute or rarely acuminate, anther bases rounded; style branches subacute and pilose. Achenes narrowly cylindrical to ± prismatic, body black, glabrous or densely setuliferous, setulae of twin-hairs; carpopodium interrupted; pappus of few (1–3) setae and more numerous (7–9) much shorter scales with laciniate margins, sometimes variable even within one capitulum, pappus rarely absent.

A genus of at least 50 spp. from tropical and subtropical America, especially in Central America, with several spp. as yet undescribed. One species, *Tagetes minuta*, is found in temperate and tropical regions of the Old World; *Tagetes erecta* is widely cultivated but seldom escapes and naturalizes from cultivation unlike *T. patula* which is a more common escape.

Capitula narrowly cylindrical, 1–3 mm diam.; inflorescences of dense corymbs; ray limbs <3 mm long . . . . **1.** *minuta*

Capitula ± campanulate, 15–40 mm diam.; inflorescences of lax corymbs; ray limbs >5 mm long . . . . **2.** *patula*

1. **Tagetes minuta** L., Sp. Pl. **2**: 887 (1753). —Rydberg in N. Amer. Fl. **34**(2): 156 (1915). —Eyles in Trans. Roy. Soc. S. Afr. **5**: 516 (1916). —Mendonça, Contrib. Conhec. Fl. Angola **1** (Compositae): 108 (1943). —Merxmüller in Proc. & Trans. Rhod. Sci. **43**: 145 (1951). —Martineau, Rhod. Wild Flowers: 90, t. 31 (1953). —Wild, Common Rhod. Weeds: fig. 77 (1955). —Cabrera, Fl. Prov. Buenos Aires **6** Compuestas: 250, t. 75 (1963). —Henderson & Anderson, Common Weeds in S. Afr.: 400, fig. 199 (1966). —Merxmüller, Prodr. Fl. Südwestafr. **139**: 176 (1967). —Binns, Check List Herb. Fl. Malawi: 37 (1968). —Guillarmod, Fl. Lesotho: 285 (1971). —Biegel & Mavi in Wild,

---

[73] As yet there is no complete revision of the genus although there are two unpublished PhD theses, one by Neher (1966 – Indiana Univ.) and the other by Soule (1993 – Univ. of Texas); Neher's is by far the most widely cited and used. Soule (1996) has since formalized her concepts on the infrageneric classification.

Rhod. Bot. Dict. African & English Plant Names: 254 (1972). —Hilliard in Ross, Fl. Natal: 367 (1972). —Lisowski in (Asterac. Fl. Afr. Cent. 2) Fragm. Flor. Geobot. **36** Suppl. 1: 493–495 (1991). —Beentje & Hind in F.T.E.A., Compositae **3**: 726–727 (2005). Type: 'Habitat in Chili.' 'Tagetes multiflora, minuto flore albicante' in Dillenius, Hort. Eltham. **2**: 374, t. 280, f. 362 (1732), iconotype, lectotypified by Delgado-Montaño in Jarvis & Turland, Taxon **47**(2): 368 (1998). FIGURE 6.5.**15**.

    Tagetes bonariensis Pers., Syn. Pl. **2**: 459 (1807). Type: 'Hab. in Bonaria. *Commers.*' (?L holotype).[74]

    Tagetes glandulifera Schrank, Pl. Rar. Hort. Monac. **2**: pl. 54 (1819). Type: 'PATRIA Brasilia. Dr. Martius./Colitur in Caldario.' (M holotype).

    Tagetes glandulosa Schrank ex Link, Enum. Hort. Berol. Alt. **2**: 339 (1822), nom. illegit. in error pro *Tagetes glandulifera* Schrank

    Tagetes porophyllum Vell., Fl. Flum. Icones 8: t. 116 (1827) [29 Oct. 1831]. Type: not stated. ['Habitat nudequaque, hoc etst, mediterraneis aeque, ae maritimis, sed his frequentius. Floret Jan.' see Arch. Mus. Nac. Rio de Janeiro **5**: 336 (1881).]

    Tagetes montana hort. ex DC., Prodr. **5**: 644 (1836), nom. nud. pro syn.[75]

    Tagetes tinctoria Hornsch., Delect. Sem. Hort. Gryph., Ann. **1845** Coll.: 3 (1845).[76] Original work not seen.

    Tagetes arvensis Rojas Acosta, Cat. Hist. Nat. Corrientes: 68 (1897). Type: 'Chincilla. ... Corrientes i otras provincias. ... Es quzá la chilca de Córdoba, eto es, el tagetes glandulifera de Entre Rios i Tucumán.'[77]

    Tagetes riojana M.Ferraro in Bol. Soc. Argent. Bot. **6**(1): 34 (1955). Types: [Argentina:] 'La Rioja: Cuesta de Miranda, 1900 m. s. m., leg. *J. Frenguelli*, No 511, 15-III-1943 [(]Co-tipo: LP.); Cuesta de miranda, 1900 m. s. m., leg. *J. Frenguelli*, No 510, 15-III-1943 (Co-tipo: LP).' *Frenguelli* 511 (LP syntype); *Frenguelli* 510 (LP syntype).

Erect, strongly aromatic annual herb, (10)70(250) cm tall. Stems simple, or branched in large plants, terete, ribbed, glabrous, dotted with elongate glands. Leaves sessile, mostly opposite, often alternate in inflorescence branches and upper part of stem, 3–30 × 0.7–8 cm wide, elliptic, pinnatisect, rhachis channelled above and narrowly-winged, wings widest near apex, lobes 7–17, 0.7–11.2 × 0.12–1.4 cm, very narrowly elliptic to linear-oblong, apex acute mucronate, margins sharply serrate with teeth well spaced, main vein channeled above and prominent below, lamina conspicuously dotted with linear glands particularly near margins, glands orange, glabrous on both surfaces, dark green above, paler below, basal lobes filiform usually laciniate up to 2.3 cm long. Inflorescences branched, densely corymbose, capitula numerous, pedicellate, pedicels short, 1–5 mm. Capitula 8–12 × 1–3 mm; involucre narrowly cylindrical; phyllaries 3–4(5), connate, 8–12 × 1.5–3 mm, drying yellow-green, apices ± finely ciliate, glabrous, with 16–25 large linear oil glands, glands brown or orange. Ray florets 2–3, limbs suborbicular or oblate to very broadly obovate, c. 2 × 2.0–3.5 mm wide, apices 1–2-lobed, pale yellow or cream to lemon-yellow; tube 2.5–3.8 mm long, ± pubescent. Disc florets (3)4–7; corollas yellow or dark yellow, 4–5 mm long, slightly exceeding involucre, ± funnel-shaped, pubescent outside, 5-lobed, margins ± ciliate. Achenes black, (5)6–7 mm long, fusiform, moderately, and conspicuously, adpressed and antrorsely setuliferous, setulae of pale straw-coloured twin-hairs, apices apparently connate or finely bifid and long-acute; pappus of 1–2 subulate to narrowly lanceolate ciliate setae c. 3 mm long and 3–4 lanceolate scales less than 1 mm long, apices obtuse, ciliate.

---

[74] Robinson & Soule in Harling, Andersson, Fl. Ecuador, **77**(1), 190(6): 142 (2006) placed this synonym under *Tagetes patula*, without explanation. Persoon's description, albeit minimal, would suggest a plant typical of *T. minuta*.

[75] Robinson & Soule (2006) suggested that this name has a type, but it is quite clear that this was an unpublished name in synonymy.

[76] Robinson & Soule (2006) have 'Delect. Sem. Hort. Greiswald, Coll. p. 3 (1845). – Type: 'Germany, Wolgast [harbor town 30 km ESE of Greifswald], Sep. 1845, HORNSCHUCH 28' (GFW holotype).

[77] Robinson & Soule (2006) suggested 'Holotype: AS' which is highly dubious since it was implicit that the species was widely distributed (and probably represented by several collections). I have my doubts as to whether this was a valid description as it was effectively only a comment on the size of the plant, which was not particularly diagnostic.

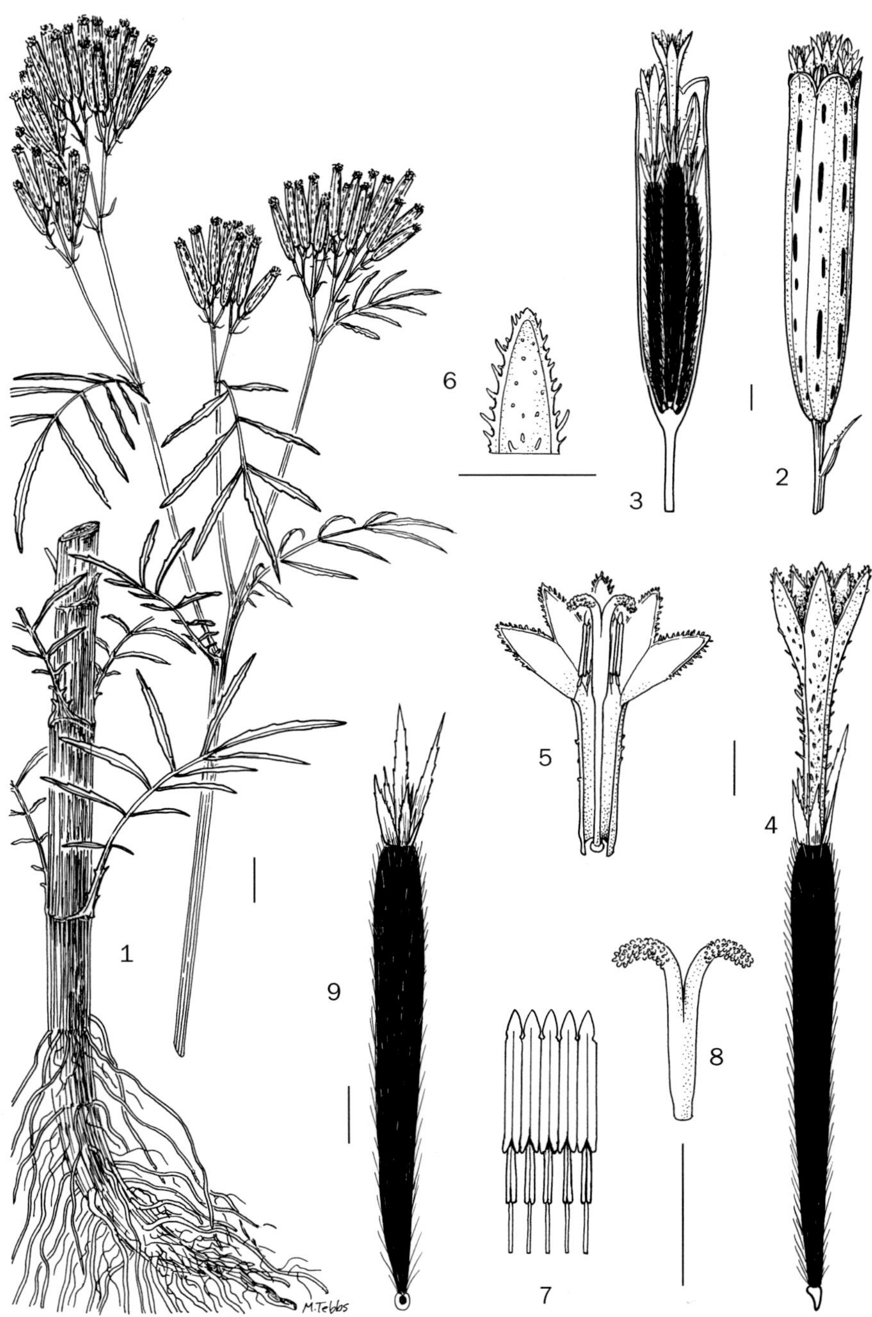

Fig. 6.5.**15**. TAGETES MINUTA. 1, apex of flowering stem with stem base and roots; 2, capitulum; 3, l.s. capitulum; 4, floret; 5, corolla opened out; 6, detail of corolla lobe; 7, anther cylinder opened out; 8, style arms; 9, achene. Scale bars: 2–9 = 1 mm; 1 = 10 mm. All from *Hind et al.* in PCD 3339. Drawn by Margaret Tebbs.

Botswana. N: Chobe Nat. Park, 23.iv.1970, *Mahundu* CNP 366 (SRGH). SE: Mahalapye, 976 m, vii. 1960, *Yalala* 110 (K, SRGH). **Zambia**. N: Mbala, Kawimbi Mission, 1525 m, 14.iii.1952, *Richards* 991 (K). W: Kitwe, 25.ii.1964, *Mutimushi* JMM 629 (K, SRGH). C: Chilanga, Mt. Makulu Res. Sta., 9.vi.1956, *Angus* 1329 (K, SRGH). S: Choma, Muckle Neuk, 1281 m, 11.x.1954, *Robinson* 909 (K, SRGH). **Zimbabwe**. N: Mazoe Citrus Estate, 10.xi.1970, *Searle* 33 (SRGH). W: Bubi, Inyati Mission, 29.iv.1947, *Keay* in FHI 21242 (K, SRGH). C: Salisbury, Nat. Bot. Gard. 21.ii.1957, *Phipps* 531 (LISC, SRGH). E: Umtali, 22.vi.1946, *Chase* 217 (SRGH). S: Lundi River, upstream of Hippo Pools, 2.v. 1962, *Drummond* 7849 (K, LISC, M, MO, PRE, SRGH). **Malawi**. N: Mzimba Dist., Mzuzu, Marymount, 1370 m, 13.vi.1975, *Pawek* 9736 (K, MA, MO, SRGH, UC). C: Nchisi, Chintembwe Mission, 418 m, 22.ii.1959, *Robson* 1716 (BM, K, LISC, PRE, SRGH). S: Blantyre, Ndirande Forest, 13.v.1968, *Banda* 1036 (SRGH). **Mozambique**. MS: Chimoio, 19.x.1944, *Mendonça* 2523 (LISC). GI: Vila de João Belo, Limpopo Riverbank, 8.x.1945, *Pedro* 245 (PRE). M: Namaacha, 30.vi.1961, *Balsinhas* 502 (BM, K, LISC, PRE, SRGH).

Native to S America but now introduced in southern Europe, Australia, New Guinea, Kenya, Angola and throughout southern Africa. A widespread weed of cultivated and disturbed ground, occasionally established in natural open woodland, roadsides; 350–2500 m.

Conservation Status: A widespread weed in the Flora area; LC (Least Concern).

2. **Tagetes patula** L., Sp. Pl. **2**: 887 (1753). —de Candolle, Prodr. **5**: 643 (1836). —Hiern, Cat. Afr. Pl. **1**(3): 589 (1898). —Hoffmann in Warburg, Kunene-Sambesi-Exped.: 420 (1903). —Rydberg in N. Amer. Fl. **34**(2): 154 (1915). —Mendonça, Contrib. Conhec. Fl. Angol. **1** (Compositae): 107 (1943). —Koster in Backer, F. Java **2**: 416 (1965). — Adams, Flower. Pl. Jamaica: 740 (1972). —Lisowski, (Asterac. Fl. Afr. Cent. 2) Fragm. Flor. Geobot. **36** Suppl. 1: 495–496 (1991). —Beentje & Hind in F.T.E.A., Compositae **3**: 728 (2005). Type: 'Habitat in Mexico.' (LINN – Herb. Linn. No. 1009.1, a cultivated specimen grown from seed from Central America), lectotypified by Hind in Jarvis *et al.*, Regnum Veg. **127**: 92 (1993).

 Tagetes corymbosa Sw., Brit. Fl. Gard., pl. 151 (1816). Type: 'Corymbus-flowered Tagetes. ... Our drawing of this new and beautiful species of Tagetes was taken from some fine plants at the Nursery of Mr. Colvill, in October last, where it had been raised from seed kindly presented to him by Mrs. Sutton the Lady of the Archbishop of Canterbury, who had received them from Mexico, ...'[78]

 Tagetes tenuifolia Kunth in Humb., Bonpl. & Kunth, Nov. Gen. Sp. Pl. **4** (ed. folio): 153 (1818). Type: [Mexico:] 'Crescit prope Santa Rosa et in amoena convalli Sancti Jacobi, inter Guanaxuato et Valladolid, alt. 900 – 1200 hex. (Regno Mexicano.). Floret Septembri.' (B-W holotype, P-Bonpl. – '4290. S. Rosa').

 Tagetes remotiflora Kunze in Linnaea **20**(1): 23 (1847). Type: 'Ex seminibus Mexicanis a. 1844 enatae plantae Octobri sub dio floruerunt, semina non maturarunt; indcirco in horto evanuit.' (?LZ holotype).

 Tagetes macroglossa Pol. in Linnaea **41**(7): 580 (1877). Type: [Costa Rica:] 'Ad vias propse San josé. Augusto. ([*Polakowsky*] No. 372 coll.)' (B†–F0BN0125488 holotype, BM000796234, GOET002092).[79]

---

[78] Two unranked taxa (= forma?) were listed at the beginning of the account, 'α. *purpurea*. Supra' (referring to the purplish colour on the upper side of the ray limb in the normal plants), and 'β *lutea*' where the ray was entirely yellow. It is unlikely that any herbarium material of this plant was preserved. Robinson & Soule (2006: 145) quaintly referred to the Mrs Sutton as 'Lady Archbishop of Canterbury' – something not yet attained in the early 21st century! They also declared that the holotypes of all three [!] entities were in CGE, clearly not having carefully read the protologue.

[79] Also referable to the synonymy are several names originally of horticultural origin ascribed by Robinson & Soule (2006: 145–146) as either considered as authored by Rümpler or Voss. Rümpler, Vilm. Ill. Blumengärtn. **2**: 990 (1879) provided several names as varieties of *Tagetes patula*, including: '*patula nana*', '*patula pumila*', '*pumila lutea*', and '*variegata*'. Voss, Vilm. Ill. Blumengärtn., ed. 3, **1**: 498 (1894) was quite clear that his names were 'Gruppen', as he certainly did not use the term variety – 'a) *simplex*' (which he equated to f. *variegatus* hort.), 'b) *tubulosus*', and 'c) *ligulosus* (which he equated to *nanus* hort. and *pumilus* hort.). All of these names were based on live material where it is unlikely that any herbarium material was preserved.

Erect much branched annual up to c. 0.6 (0.8) m tall, strongly aromatic when bruised. Stems and branches terete, ± strongly ribbed, glabrous. Leaves sessile, opposite, becoming alternate above, 3–19 cm long, elliptic, pinnatisect, basal lobes filiform, usually laciniate with teeth often ending in long setae, midrib narrowly winged with wings widest near apex, lobes 0.5–5 cm long, narrowly elliptic to lanceolate or sometimes linear, margins sharply serrate with scattered sessile orbicular pellucid glands on both surfaces, glabrous on both surfaces, apex acute or sometimes with an elongate seta up to 12 mm particularly in upper leaves. Inflorescence a lax corymb, 1.5–4 cm diam., capitula pedicellate, pedicels 2.5–15(20) cm long, inflated beneath involucre, bracteate near base, bracts up to c. 1.5 cm long, pinnatisect with filiform lobes. Capitula heterogamous, radiate; involucre campanulate or cylindrical, 1.5–2 × 0.5–1.5 cm; phyllaries 5–7, connate, apices dentate, 5–7-lobed, ± longitudinally folded, glandular, glands mostly linear or very narrowly elliptic becoming broadly elliptic near apex, glabrous except for lobes which are tomentose near apex and on margins. Ray florets female, limbs 5–15 mm long, broadly obovate to broadly oblong with a ± emarginate apex and ± erose margin, yellow, orange, yellow with an orange base, or with brownish-red markings, or nearly completely brownish-red, tube 5–8 mm long, glabrous. Disc florets numerous, hermaphrodite, corollas 10–16 mm long, narrowly funnel-shaped, ± 5–ribbed and glabrous outside, ± pilose in throat, apex deeply 5-lobed, lobe margins ± cucullate, ciliate and ± densely hispidulous at apex of inner surface. Achenes black, 7–10 mm long, ± spindle-shaped or tapering slightly to base, ± 5-ribbed, antrorsely adpressed setuliferous (predominantly on ribs), setulae of amber-coloured twin hairs, apices bifid, long-acute); carpopodium conspicuous; pappus of 5 or 6 ± united paleaceous setae, 1 or 2 setae 6–10 mm long with attenuate apices, remainder 3–5 mm long with obtuse apices, margins minutely ciliate.

Zambia. W: Ndola, 8.iv.1972, *Fanshawe* F11410 (K, NDO). C: Lusaka, open bush, 12.iv.1956, *Noak* 199 (K, SRGH). S: Choma, 21.ii.1963, *van Rensburg* 1382 (K, SRGH). **Zimbabwe**. N: Bindura, 16.vii.1969, *Mogg* 34309 (SRGH). **Malawi**. C: Dedza, Chongoni Forest, 24.iii.1968, *Salubeni* 1276 (SRGH). S: Near Chiradzulu, 21.x.1905, *Cameron* 183 (K).

A cultivated ornamental introduced from Central America, sometimes becoming established as a garden escape (usually as depauperate forms). *Tagetes erecta* L. is also grown as an ornamental in the Flora area (Zambia and Zimbabwe), but has so far not been found as an escape. *Tagetes patula* is distinguished from *T. erecta* in having a lower more branching habit, capitula usually in corymbs (solitary in *T. erecta*), and linear or narrowly elliptic glands on the phyllaries instead of the broadly elliptic or rotund glands of *T. erecta*. *T. patula* has been treated as a synonym of *T. erecta* in some recent floras (Chen Yousheng & Hind in Fl.

China **20-21**: 854, 2011; Ghazanfar, Edmondson & Hind in Fl. Irak **6**: 424, 2019) or as a distinct species (Koyama *et al.* in Fl. Thailand **13**(2): 414, 2016).

Conservation Status: As a cultivated plant, albeit escaping occasionally; it is best recorded as LC (Least Concern).

123. **HYPERICOPHYLLUM** Steetz

Hypericophyllum Steetz in Peters, Naturw. Reise Mossambique **6**, Bot. 2: 498 t. 50 (1864). —Brown in J. Linn. Soc., Bot. **35**(244): 120–123 & pl. 6 (1902). —Pope in Kirkia **10**(1): 104–111 (1975). —Lisowski, (Asterac. Fl. Afr. Cent. 2) Fragm. Flor. Geobot. **36** Suppl. 1: 482–490 (1991). —Beentje & Hind in F.T.E.A., Compositae **3**: 719–722 (2005). —Panero in Kubitzki, Fam. Gen. Vasc. Pl. **8**: 437 [2006](2007). —Hind in Kew Bull. **69**(2)-9500: 1–8 (2014).

Jaumea Pers., Syn. Pl. **2**: 397 (1807), sensu Oliver & Hiern in F.T.A. **3**: 395 (1877), p.p. —Mattfeld in Bot. Jahrb. Syst. **59**(4, Beibl. 133): 20–24 (1924).

Perennial herbs, glabrous, ± pilose or scabrid; stems terete, solitary or few from rootstock, simple or branched in upper part, branches opposite. Rootstock rhizomatous, roots usually fleshy. Leaves simple, opposite or subopposite, sessile or pseudopetiolate, glandular-punctate, margins entire or ± undulate, ± scabrid. Capitula large, solitary, terminal on stems and branches, usually homogamous and discoid or sometimes heterogamous and radiate; peduncles long, usually thickened towards apex; involucre broadly turbinate-campanulate; phyllaries 3–5-seriate, outer ± broadly ovate, ± striate, inner longer and narrower; receptacle ± convex, epaleaceous. Ray florets, when present, several, female, ray limb conspicuous, whitish or whitish-yellow; style long-exserted. Disc florets numerous, hermaphrodite; corollas cylindrical, orange or reddish-orange, sometimes white or whitish-yellow, widening gradually from base, apex deeply 5-lobed; anthers ± exserted, apical anther appendage narrowly ovate; anther bases obtuse; style glabrous; style arms papillate outside, obtuse or truncate. Achenes tapering to base, longitudinally ribbed, sometimes 4- or 5-angled, hirsute or strigose to glabrous, setulae of twin-hairs, apices unequal, usually scarcely separate; pappus uniseriate, setae numerous, ± paleaceous and laterally winged at base or throughout, or terete, apices attenuate and often uncinate.

A genus of about 10 spp. restricted to tropical Africa; four spp. are found in the Flora area; the higher than previously recorded number of spp. has resulted from a number of necessary combinations needed to recognize two names left in *Jaumea*, and other taxa that have been overlooked. Since Oliver & Hiern merely expanded the concept of the monospecific *Jaumea* (as they did not exclude the type), no attribution to Oliver & Hiern can be made, cf. Lisowski (1991), etc.

Robinson (Smithsonian Contrib. Bot. **51**, 1981) placed *Hypericophyllum* in the Heliantheae subtribe Chaenactidinae Rydb.; Baldwin (in Syst. Bot. **27**(1): 161–198, 2002) and Panero (2006) placed the genus in an essentially New World tribe, the Bahieae B.G.Baldwin (a refined group of genera from the subtribe Chaenactidinae sensu Robinson); *Apostates* Lander is endemic to Rapa Island and *Hypericophyllum* to tropical Africa. The reason given by Baldwin was of long-distance dispersal – presumably the uncinate pappus setae were caught in bird feathers!

1. Pappus setae 6–9 mm long, ± hispidulous, fimbriately winged for half or more of length, straight and uncinate setae within same pappus; leaves glabrous with scabrid margins, ± coriaceous, mostly wider at base than at middle . . . . **2.** *compositarum*
 – Pappus setae 2–7 mm long, glabrous, with or without fimbriate wings in lower third, all setae uncinate or short and straight setae ± alternating with longer uncinate setae; leaves scabrid, pubescent or glabrous, usually narrower at base than middle 2
2. Pappus of 5–8 terete uncinate setae 3.5–5 mm long ± alternating with c. 6 setae without uncinate apices, setae c. 1 mm long without fimbriate wings or if present wings poorly developed; leaves oblong-elliptic . . . . **4.** *tessmannii*
 – Pappus of 15–20, setae 2–7 mm long, all uncinate and ± fimbriately winged; leaves broadly elliptic to lanceolate or oblanceolate 3
3. Leaves glabrous or pubescent but not scabrous, usually elliptic; capitula usually 3–12 . . . . **3.** *elatum*
 – Leaves scabrid or hispidly pilose, mostly oblanceolate to spathulate; capitula usually 1–3 . . . . **1.** *angolense*

1. **Hypericophyllum angolense** (O.Hoffm.) N.E.Br. in J. Linn. Soc., Bot. **35**: 122 (1902). —Mendonça, Contrib. Conhec. Fl. Angola **1** (Compositae): 106 (1943). —Pope in Kirkia **10**(1): 106–107 (1975). —Lisowski in (Asterac. Fl. Afr. Cent. 2) Fragm. Flor. Geobot. **36** Suppl. 1: 484–486 (1991). —Beentje & Hind in F.T.E.A., Compositae **3**: 719–720 (2005). Type/s?: 'Angola, Humpata, Huilla, Chella (*Newton*). Huilla (*Antunes*). Malange (*Teuscz* [sic!] na expedição de v. Mechow, no. 470)' *Theush* [sic!] 470 (K000410637 syntype); *Welwitsch* 3965 (BR9864153, K000410636, P0092969 syntypes).[80,81] FIGURE 6.5.**16**.

---

[80] The *Newton* and *Antunes* syntypes appear to have been cited by Mendonça (1943: 107), but not as syntypes, and are in COI.

[81] The handwriting on the label is probably that of Vatke. Other material in K has the collector's name as Theusch, sometimes corrected to Teusz, although the collector was H. Teucsz, the official collector of the Mechow expedition to the Congo in 1884–6. At the point that Baker described *J. johnstoni* he compared his new species with '*J. Oliveri*, Vatke, et *J. Compositarum*, Benth. et Hook. fil., majis accedit.' This refers to the ined. name on the top of the label of the *Teucsz* 470 collection, a syntype of *Jaumea compositarum*. The writing on the K specimen, probably by N. E. Brown, suggests that in March 1884 he received the *Teucsz* 470 material from Vatke – K000410637.

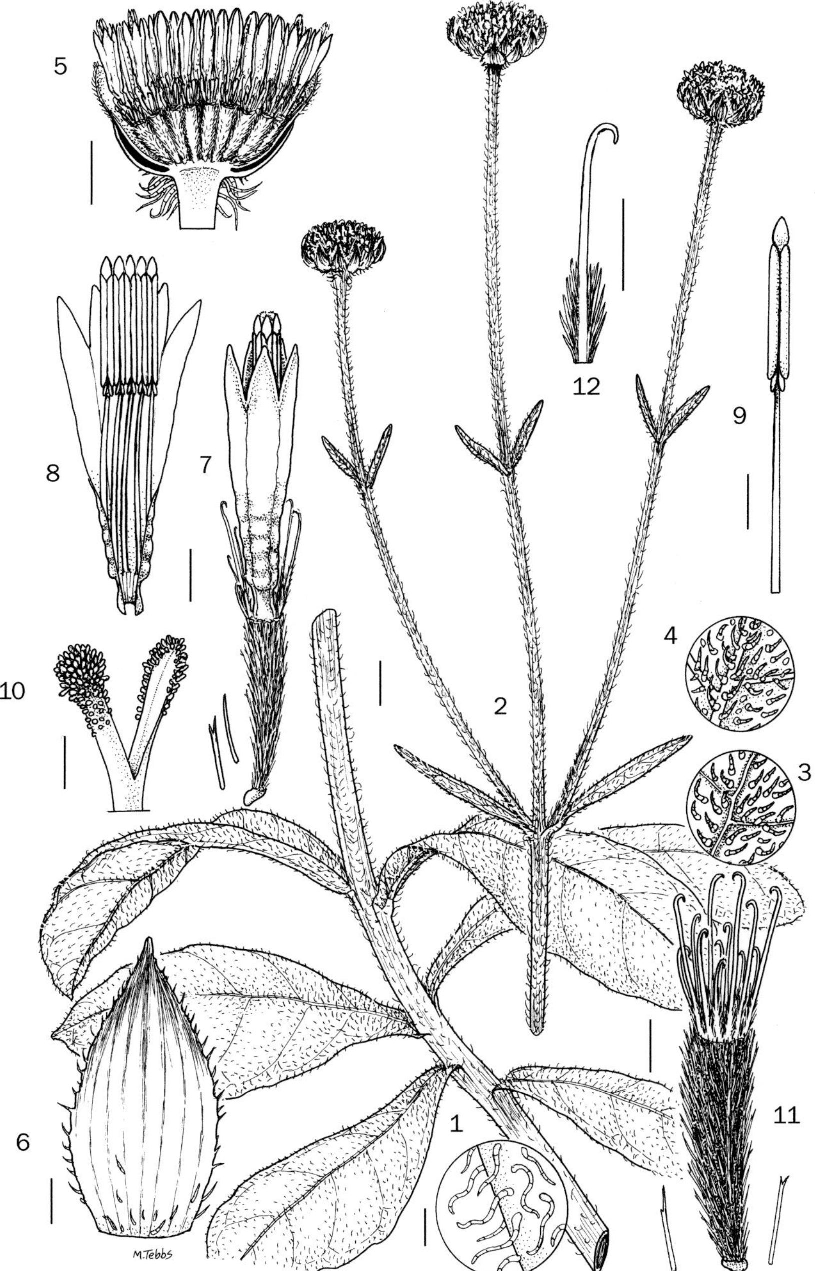

Fig. 6.5.**16**. HYPERICOPHYLLUM ANGOLENSE. 1, lower portion of stem showing mature leaves and inset detail of stem pubescence; 2, apex of flowering stem; 3, detail of upper leaf surface; 4, detail of lower leaf surface; 5, l.s. capitulum; 6, phyllary; 7, floret with detail of two setulae from immature achene; 8, corolla opened out showing attachment of filaments; 9, stamen; 10, style arms; 11, mature achene with details of two setulae; 12, detail of uncinate pappus setae with winged base. 1, 11–12 from *Smith* 0496; 2–4 from *Fanshawe* F18; 5–10 from *Stohr* 108. Scale bars: 10 = 0.5 mm; 1(inset), 3–4, 6–9, 11–12 = 1 mm; 5 = 5 mm; 1–2 = 10 mm. Drawn by Margaret Tebbs.

Jaumea angolensis O.Hoffm. in Bol. Soc. Brot. **10**: 178 (1893). —Hiern, Cat. Afr. Pl. **1**(3): 589 (1898). —Hoffmann in Warburg, Kunene-Sambesi-Exped.: 420 (1903).

Hypericophyllum scabridum N.E.Br. in J. Linn. Soc., Bot. **35**: 122 (1902). Types: 'British Central Africa. –Nyasaland: between Kondowe and Karonga, 2000–6000 ft., *Whyte*! Mangaja Hills, 1000 ft. alt., *Kirk*! Shire Highlands, near Blantyre, not at all plentiful, *Buchanan*, 73! 439!' '[Malawi:] between Kondowe and Karanga, *Whyte* s.n.' (K001097347 syntype); 'Manganja Hills, 1000 ft. alt., *Kirk* s.n.' (K000410739 syntype); 'Shire Highlands, *Buchanan* 73' (K000410737, K000410738 syntypes); '*Buchanan* 439' (K000410735, K000410736 syntypes).

Jaumea scabrida (N.E.Br.) Lawalrée in Bull. Jard. Bot. État Bruxelles **19**: 224 (1949).

Erect perennial herb, usually c. 80 cm tall with extreme forms from c. 20–100 cm tall. Roots fleshy. Stems terete, ribbed, ± densely hirsute becoming scabrid; simple or with a pair of opposite branches in upper part, rarely branching more than once; branches 10–30 cm long with opposite reduced narrowly lanceolate leaves 0.8–2 cm long. Leaves simple, opposite, sessile or pseudopetiolate, lower leaves 7.5–17 × 1–7 cm, narrowly oblanceolate or oblanceolate to spathulate or more rarely narrowly ovate to elliptic, upper leaves becoming smaller and very narrowly elliptic to lanceolate, scabrous or ± hispidly-pilose on both sides, sessile-glandular, apex obtuse rounded, margins entire sometimes undulate shortly scabrous, base cuneate to attenuate subauriculate in pseudopetiolate leaves semi-amplexicaul, bases connate, midrib and lateral veins ± prominent and hirsute or pilose below, reticulation ± conspicuous below. Inflorescence few-headed (1–3), cymose, frequently reduced to a solitary terminal capitulum; capitula, homogamous, c. 1.5(2) × 2.5–3 cm, long-peduncled (or pedicellate) with peduncle fistulose towards apex; involucre broadly turbinate; phyllaries 3–4-seriate, 4–15 mm long, gradate, narrowly ovate to narrowly lanceolate, glabrous or sometimes pubescent on exposed surfaces, apex obtuse in outer phyllaries becoming acute in inner, mucronate, margin ciliate. Florets numerous, exceeding involucre; corollas reddish-orange c. 10 mm long, with a constricted basal portion c. 2 mm long widening gradually to deeply 5-lobed apex, glabrous. Achenes 4–8(9) mm long, very narrowly turbinate, ± 4-angled, ribbed, ± densely hispid to strigose or rarely glabrescent, glandular-punctate; pappus of rigid bristles 3–6(7) mm long, ± paleaceous at base, mostly with uncinate apices, laterally membranously and ± fimbriately winged in basal $^{1}/_{3}$, glabrous or rarely with a few hairs near base.

Zambia. N: Luwingu, iv.1922, *Jeff* 9 (BM). W: 6 km west of Solwezi, 18.iii.1961, *Drummond & Rutherford-Smith* 6993 (K, LISC, M, PRE, SRGH). C: 56 km south of Ndola, 27.iii.1955, *Exell, Mendonça & Wild* 1225 (BM, LISC, SRGH). **Malawi**. N: Karonga, 60 km west of Karonga, 27.iv.1972, *Pawek* 5290 (SRGH). S: Shire Highlands, x.1881, *Buchanan* 439 (E). **Mozambique**. N: Niassa, 39 km west of Ribáuè, 16.v.1961, *Leach & Rutherford-Smith* 10891 (M, SRGH).

Also in Angola, D.R. Congo and Tanzania. An uncommon plant of miombo woodland and occasionally vlei grassland; 950–1950 m.

Conservation Status: Relatively widely distributed and considered LC (Least Concern).

2. **Hypericophyllum compositarum** Steetz in Peters, Naturw. Reise Mossambique **6**, Bot. 2: 499, t. 50 (1864). —Brown in J. Linn. Soc., Bot. **35**: 122 (1902). —Verdcourt in Kew Bull. **6**: 363 (1952). —Martineau, Rhod. Wild Flowers: 86, t. 31 (1954). —Andrews, Fl. Pl. Sudan **3**: 37 (1956). —Binns, Check List Herb. Fl. Malawi: 34 (1968). —Linley & Baker, Fls. Veld: 116, fig. 71 (1972). —Biegel & Mavi in Wild, Bot. Dict. African and English Plant Names (1972). —Pope in Kirkia **10**(1): 107–108 (1975). —Lisowski in (Asterac. Fl. Afr. Cent. 2) Fragm. Flor. Geobot. **36** Suppl. 1: 486–488 (1991). —Beentje & Hind in F.T.E.A., Compositae **3**: 720 (2005). Type: [Mozambique:] 'Standort: Boror. Rios de Sena. [*Peters*]' (B† holotype – no isotypes have been located); Taf. 50 in Peters, Naturw. Reise Mossambique **6**, Bot. 2 (1864), iconotype, lectotypified by Hind (2014: 3); [Mozambique:] 'Manica e Sofala: Tete, entre Casula e Furancungo, a 22.8 km de Casula, *L. A. Grandvaux Barbosa* et *M. Carvalho* no. 3514 in G. B. col.' (K001097345 epitype, LMJ, SRGH), epitypified by Hind (2014: 3).

Jaumea compositarum (Steetz) Oliv. & Hiern in F.T.A. **3**: 395 (1877).[82]

Kleinia compositarum (Steetz) Kuntze, Revis. Gen. Pl. **1**: 348 (1891).

Kleinia oliveri Kuntze, Revis. Gen. Pl. **1**: 348 (1891), nom. nud.[83]

Jaumea johnstoni Baker in Bull. Misc. Inform., Kew **1898**: 153 (1898). Types: 'British Central Africa. Nyika plateau, alt. 6000–7000 ft., *Whyte*, 228; Masuku plateau, alt. 6500–7000 ft., and between Mpata and the Nyasa-Tanganyika plateau, alt. 2000–3000 ft., *Whyte*.' 'Nyika plateau, alt. 6000–7000 ft., *Whyte*, 228' (K001097344 syntype); 'Masuku plateau, alt. 6500–7000 ft.' (K000410634 syntype); 'between Mpata and the Nyasa-Tanganyika plateau, alt. 2000–3000 ft., *Whyte*.' (K000410635 syntype).

Hypericophyllum brevipapposum Gilli in Ann. Naturhist. Mus. Wien **78**: 156 (1974). Type: [Tanzania:] '[*Gilli*] 621. Busch hinter der Gärtnerei der Miaasion Lumbila, (Hyparrhenietum filipendulae) 530 m, 11.VIII.1958, Soc. Aufn. XX, fl., fr.' (W19730001159 holotype).

Erect perennial up to c. 120 cm tall. Roots fleshy. Stem terete, ± ribbed, sparsely pubescent to glabrescent above and pilose to ± scabrously hairy or glabrescent below, with opposite branches above or sometimes simple; branches 4–30 cm long, glabrescent, with opposite sessile bract-like leaves 0.9–5 × 0.3–1.4 cm, lanceolate to narrowly oblong. Leaves simple, opposite, sessile or subpetiolate, ± coriaceous, 3.5–18 × 1.5–6 cm, narrowly oblong to lorate, apex tapering, or lanceolate to narrowly ovate, lowermost leaves small ovate to lanceolate, upper leaves decreasing in size toward stem apex, apex obtuse to acute usually mucronate, margin entire or slightly undulate ± revolute scabrid, base cordate or subauriculate semi-amplexicaul with leaf-bases connate, glabrous or sparsely puberulent, punctate, dark green above paler below, midrib ± impressed above prominent and often scabrid below, reticulation prominent below. Inflorescence cymose; capitula 3–21, terminal on branches, homogamous, 2–3.5 cm diam., peduncles widening toward apex; involucre broadly turbinate; phyllaries 3-seriate, outer phyllaries 4–7 mm long broadly ovate to ovate, inner 10–13 mm long narrowly elliptic to lanceolate, glabrous, apex obtuse to acute mucronate, margins ± ciliate becoming ± serrulate towards base rarely subentire. Florets numerous, corollas bright orange-red, exceeding phyllaries, 10–12 mm long, with a constricted cylindrical basal portion c. 2 mm long widening gradually to a deeply 5-lobed apex, glabrous or ± pubescent at base. Achenes 6–7 mm long, narrowly turbinate, ± densely strigose or hispid; pappus of rigid bristles 6–9(10) mm long, ± paleaceous at base, straight or with uncinate apices, laterally winged in lower $^1/_2$–$^3/_4$ wings membranously fimbriate, ± hispidulous to pubescent becoming glabrous towards apex.

Zambia. N: Mbala Dist., Kawimbe, 1800 m, 2.vi.1957, *Richards* 9976 (K, SRGH). W: 27 km south of Mwinilunga on Kabompo road, 6.vi.1963, *Loveridge* 795 (K, M, SRGH). C: 12 km NE. of Lusaka, 1189 m, 13.vii.1930, *Hutchinson & Gillett* 3593 (K, LISC, SRGH). S: Choma Dist., Mapanza, 1128 m, 19.iv.1954, *Robinson* 689 (K, SRGH). **Zimbabwe**. N: Mtoroshanga, 24-27.iv.1948, *Rodin* 4425 (K, MO, PRE, S, SRGH). C: Umsweswe, 976 m, 13.v.1921, *Borle* 166 (K, PRE, SRGH). E: Umtali, Zimunya Reserve, 1006 m, 15.iv.1956, *Chase* 6075 (K, LISC, PRE, SRGH). S: Chibi, Madzivire Dip, 3.v.1962, *Drummond* 7903 (K, LISC, MO, SRGH). **Malawi**. N: Nyika Plateau,1829–2134 m, *Whyte* 228 (K). S: Mt. Mlanje, Tuchila River valley, 12.vii.1956, *Jackson* 1871 (COI, K, LISC, SRGH). **Mozambique**. ?N: Unangu to Lake Shirwa, 915-1372 m, 1899, *Archdeacon Johnson* 27 (K). T: Tete, entre Casula e Furancungo, a 22.8 km de Casula, 9.vii.1949, *Barbosa & Carvalho* 3514 (K, LMJ, SRGH).

Also in Tanzania and D.R. Congo. A relatively widespread but infrequent plant of miombo woodland, usually on rocky granite outcrops or escarpments; 750–1800 m.

Conservation Status: Widely distributed and considered LC (Least Concern).

---

[82] No such combination was actually made by Bentham & Hooker f., Gen. Pl. **2**: 397 (1873) – it was only inferred, like so many other 'well-documented combinations'–, although cited by Lisowski (1991: 486), and earlier intimated by Eyles (1916: 516), the combination was validated by Oliver & Hiern. Similarly, the reference by Hoffmann (1893), to Klatt's name (a combination based on Steetz's basionym), is also illegitimate.

[83] Although cited as 'Oliveri (Vatke) OK.' by Kuntze (1891), '*Jaumea oliveri* Vatke' upon which it was based was a herbarium name cited by Baker whilst describing *J. johnstonii*.

3. **Hypericophyllum elatum** (O.Hoffm.) N.E.Br. in J. Linn. Soc., Bot. **35**: 122 (1902). —Moore in J. Linn. Soc., Bot. **40**: 117 (1911). —Verdcourt in Kew Bull. **6**: 363 (1952). —Pope in Kirkia **10**(1): 108–110 (1975). —Lisowski in (Asterac. Fl. Afr. Cent. 2) Fragm. Flor. Geobot. **36** Suppl. 1: 488 (1991). —Beentje & Hind in F.T.E.A., Compositae **3**: 720–722 (2005). Type: [Tanzania:] 'Uhehe: Rungembe, hügeliges Plateau, sumpfige Wiesen, um 1600 m (*GOETZE* n. 723. – Blühend im März 1899).' (B† holotype, BR8875440, BR8875891).[84]

 Jaumea elata O.Hoffm. in Bot. Jahrb. Syst. **28**(4): 506 (1900).

 Jaumea helenae Buscal. & Muschl. in Bot. Jahrb. Syst. **49**(3–4): 507 (1913). Type: [Zambia:] 'Baumsteppe zwischen Banguelo und Tanganyika-See, 1100 m ü. M. ([*?Helena von Aosta*] n. 1081. – 6. April 1910).' [In: Beschreibung der von Ihrer Königlichen Hoheit der Herzogin Helena von Aosta in Zentral-Afrika gesammelten neuen Arten.] (?B† holotype).

 Jaumea speciosa Lawalrée in Bull. Jard. Bot. État Bruxelles **19**: 224 (1949). Type: [D.R. Congo:] '[*Van Meel*] 1737, 18.iv.1947. Presqu'ile d'Ubwari, savane herbeuse sur crête. Très abdt.' (BR8875891 holotype, WAG610).

 Hypericophyllum speciosum (Lawalrée) Lawalrée in Explor. Hydrobiol. Lac Tanganika **4**: 74 (1955).

 Hypericophyllum speciosum (Lawalrée) Gilli, Ann. Naturhist. Mus. Wien **78**: 157 (1974), comb. superfl.

 Hypericophyllum nyassicum Gilli, Ann. Naturhist. Mus. Wien **78**: 157 (1975).[85] Type: 'Nyassa-Hochland – Station Kyimbila, 1000–1400 m, 24. VI. 1912, *A. Stolz* 1380, fl., fr.' (W19150008022 holotype, K001097346).

Erect perennial herb up to c. 180 cm tall. Roots somewhat fleshy, from a woody rhizome. Stems terete, ribbed, ± stiffly and sparsely pilose to glabrescent, sometimes sparsely pubescent in addition, usually much branched above. Leaves shortly petiolate or sessile, 4–23.5 × 1.5–9.0 cm, elliptic to broadly elliptic or oblanceolate, upper leaves smaller, usually oblong and narrowed to apex or lanceolate, apex obtuse to rounded, margins ± revolute entire to undulate ± scabridly ciliate, base subcordate to broadly cuneate, petiole to c. 5 mm long, semi-amplexicaul, lamina pubescent or glabrous on both sides sometimes hispid or sparsely pilose, strongly 3–5-veined from near base, midrib and reticulation ± prominent below. Inflorescence very lax, cymose with opposite branches up to 48 cm long; bracts 0.8–7 × 0.4–1.8 cm, narrowly oblong to lanceolate; capitula usually 3–12, terminal on branches, 1.5–2 × 2.5–4 cm; phyllaries 3–4-seriate, gradate, outer very broadly ovate from 4 mm long, inner narrowly elliptic to lanceolate, 10–18 mm long with an acute apex, margins subentire becoming ± ciliate toward apex, ± longitudinally striate, glabrous. Florets numerous; corollas reddish-orange, 9–13(15) mm long, tubular, constricted in lower c. 2 mm and widening gradually to a deeply 5-lobed apex, glabrous or ± pubescent towards base. Achenes c. 7 mm long, very narrowly obpyramidal to narrowly turbinate, somewhat 4–5-angled, ribbed, ± densely strigose or hispid; pappus of numerous setae 2–5 mm long, ± terete, uncinate or sometimes with a few straight setae, usually with narrow membranous ± fimbriate lateral wings in lower 1/3, glabrous.

Zambia. N: Kawimbe Mission, 20.v.1955, *Richards* 5780 (SRGH). E: Luangwa River, 25.iii.1955, *E.M. & W.* 1183 (BM, LISC, SRGH). **Zimbabwe**. E: Chipinga, iii.1958, *Crook* in GHS 84601 (COI, LISC, SRGH). **Malawi**. N: Mzuzu, Mzaba Forest, 27.iv.1967, *Salubeni* 687 (COI, K, M, S, SRGH). S: Mlanje Mountain, above Fort Lister Forestry Depot on path to Sombani, 1220 m, 8.vi.1970, *Brummitt* 11373 (K). **Mozambique**. N: Nampula, 12.iv.1961, *Balsinhas & Marrime* 382 (COI, K, LISC, PRE, SRGH). Z: Mocuba, 23.v.1949, *Barbosa & Carvalho* 2828 (SRGH). T: Moatize 11.iii.1964, *Torre & Paiva* 11132 (LISC). MS: Corone, region of Inhaminga, 50 m, 20.iv.1956, *Gomes e Sousa* 4308 (COI, K, LISC, PRE).

Also in the Transvaal (Zoutpansberg), Tanzania and D.R. Congo. Relatively widespread but infrequent in mixed deciduous woodland; 50–1600 m.

Conservation Status: Widely distributed and considered LC (Least Concern).

Wild (1975: 110) also noted '*H. elatum* specimens from N. Mozambique tend to be very similar in appearance to *H. congoense* (O.Hoffm.) N.E.Br., which differs from *H. elatum* mainly in its wingless pappus setae and narrower capitula.'

---

[84] Several collections were erroneously annotated as 'Type' by Brown – they are none other than cited specimens in his revision.

[85] I have examined the K isotype of *H. nyassicum* and consider it conspecific with *H. compositarum*.

4. **Hypericophyllum tessmannii** (Mattf.) Pope in Kirkia **10**(1): 110 (1975). —Lisowski in (Asterac. Fl. Afr. Cent. 2) Fragm. Flor. Geobot. **36** Suppl. 1: 489–490 (1991). Type: 'Kamerun: Uam-Gebeit: Bosum, 620 N. Br., 1620 E. L. (*TESSMANN* n. 2661, an einem Bach, blühend 18. Juli 1914; [*Tessmann*] n. 2676c, Grassteppe).' *Tessmann* 2261 (B† syntype); *Tessmann* 2676c (B† syntype).[86]

Jaumea tessmannii Mattf. in Bot. Jahrb. Syst. **59**(4, Beibl. 133): 23 (1924).

Erect perennial herb up to c. 100 cm tall. Stems ribbed, glabrous, rarely sparsely pilose, sessile-glandular, becoming branched in upper part: branches 15–28 cm long, glabrous with 1–3 pairs of reduced narrowly ovate leaves 1.5–6 cm long. Leaves opposite, sessile, 7–12(17) × 4–7(10) cm, oblong-elliptic to broadly elliptic, glabrous or sparsely puberulent on both sides, sessile-glandular, apex obtuse or rounded, margins entire ± revolute and sparsely scabridly ciliate with cilia pointing to apex, base subcordate, strongly 3–5-veined from near base, veins prominent below, reticulation ± conspicuous below. Inflorescence cymose, occasionally reduced to a single terminal capitulum; capitula usually 3–5, solitary at ends of branches, 1.5–2 × 2.5–3 cm wide, campanulate to broadly turbinate, homogamous; phyllaries 3–4-seriate, glabrous on both sides or sometimes puberulent inside, margins entire hyaline sometimes purple-tinged and ± finely ciliate on inner phyllaries; outer phyllaries 5–10 × 5–8 mm, broadly ovate with an obtuse to subacute apex; inner bracts 14–16 × 5–7 mm, narrowly oblong with an obtuse apex, innermost phyllaries often c. 2 mm wide, lorate, apex acute. Florets numerous; corollas reddish-orange, 8–10.5 mm long, slightly exceeding phyllaries, narrowly funnel-shaped with lower c. 2 mm constricted, apex deeply 5-lobed, glabrous. Achenes 6–7 mm long, very narrowly obpyramidal to very narrowly turbinate, ± 4–5-angled, ribbed, ribs sparsely finely pubescent, glandular-punctate; pappus usually of 5–8 rigid uncinate bristles 3.5–5 mm long and c. 6 ± paleaceous bristles c. 1 mm long apices straight, all bristles glabrous and without lateral wings, or lateral wings poorly developed.

Zambia. N: Chibutubutu, 50 km S. of Kasama, 1320 m, 24.ii.1960. *Richards* 12566 (K, M, SRGH). W: Kafue National Park, Moshi, 3.iv.1963, *Mataundi in Mitchell* 19/62 (M, SRGH).

Also in Cameroon, Central African Republic, D.R. Congo, ?Tanzania. Uncommon, usually on termitaria in marshy or moist grassland; 650–1320 m.

I suspect this species is under-collected as it superficially resembles *H. elatum*. I have been unable to find any material studied by Pope from Tanzania.

Conservation Status: This species is probably best regarded as DD (Data Deficient) since it has probably been both rarely collected and/or overlooked.

124. **SCHKUHRIA** Roth

Schkuhria Roth, Catal. Bot. **1**: 116 (1797), nom. cons. —Heiser in Ann. Missouri Bot. Gard. **32**(3): 265–278 (1945). —Pope in Kirkia **10**(1): 119–121 (1975). —McVaugh, Fl. Novo-Galiciana **12**: 794–795 (1984). —Turner in Phytologia **79**(5): 364–368 (1995). —Mesfin Tadesse, Fl. Ethiopia & Eritrea **4**(2): 279 (2004). —Beentje & Hind in F.T.E.A., Compositae **3**: 722–723 (2005). —Robinson in Harling & Andersson, Fl. Ecuador **77**(1), 190(6): 91–96 (2006). —Baldwin in Phytoneuron **2015-56**: 1–2 (2015).

Tetracarpum Moench, Suppl. Meth. [Moench]: 240 (1802), nom. illeg. superfl. pro *Schkuhria* Roth.

Chamaestephanum Willd. in Mag. Neuesten Entdeck. Gesammten Naturk. Ges. Naturf. Freunde Berlin **1**: 140 (1807).

Mieria La Llave in La Llave & Lexarza, Nov. Veg. Descr. **2**: 5 (1825).

Achyropappus Link & Otto, Ic. Pl. Rar. Pl. 30 (1829), non *Achyropappus* Kunth (1818).

Hopkirkia DC., Prodr. **5**: 660 (1836), non *Hopkirkia* Spreng. (1819).

Cephalobembix Rydb. in N. Amer. Fl. **34**: 46 (1914).

---

[86] Lisowski (1991) incorrectly indicated *Tessmann* 2661 was the holotype, although this could be considered lectotypification.

Slender, erect or decumbent annual or rarely perennial plants; rootstock a taproot. Stems branched from base or unbranched below, 5- to 6-angled, glabrate or puberulous with small flattened hairs to hirsute, weakly to strongly glanduliferous, narrowly fistulose, sometimes chambered. Leaves opposite below, alternate above, petioles short; lamina pinnately or bipinnately divided into linear or filiform lobes, surfaces pilose, with glandular dots or stipitate glands. Inflorescences diffuse, loosely paniculate or corymbiform, with capitula terminal on most branches; pedicels slender, glandular-punctate or stipitate glandular. Capitula homogamous or heterogamous; involucres narrowly to broadly obconic to obovoid; phyllaries herbaceous, mostly or completely subequal, obovate to oblanceolate, obtuse to short-acute, with whitish to reddish, scarious distal margins, stipitate-glandular or glandular-punctate outside, sometimes with 1 or 2 smaller outer bracts present; receptacle epaleaceous. Ray florets absent or very few, female and fertile when present; corollas yellow or white, tube narrow, short stipitate-glandular, limb oblong, apex minutely 2-lobed, scarcely mamillose with elongate cells above; style arms with 2 stigmatic lines. Disc florets few or many, hermaphrodite, fertile; corollas yellow, sometimes red distally, tube base broad below a constricted tube with short- to long-stalked glands, limb with few or no glands, throat campanulate, twice as long as lobes, lobes 5, broadly triangular, covered with broad papillae inside; anther thecae pale brownish, short-hastate at base; apical anther appendages ovate with few glands outside; style base slightly enlarged, arms with paired separate stigmatic lines, apex short-acute, slightly mamillose outside. Achenes elongate, obpyramidal, mostly quadrangular, with appressed stiff setulae along angles, body blackened with slender pale striations, base narrowed; carpopodium minute; pappus of few scarious squamellae, callosed at base, sometimes with wings and apical awns smooth or scabrid outside.

A genus of about five spp. and one, *S. pinnata* (present in the Flora area) is a widespread and weedy species of both N and S America, and is known now from sporadic occurrences in the Old World. Baldwin (2015) has proposed that other than the generitype, *S. pinnata*, the other species belong in other genera yet to be specified, as well as *Nothoschkuhria* B.G.Baldwin; I do not presently accept this and broadly maintain Heiser's concept of *Schkuhria*.

Schkuhria pinnata (Lam.) Kuntze ex Thell. in Repert. Spec. Nov. Regni Veg. **11**: 308 (1912). —Heiser in Ann. Miss. Bot. Gard. **32**: 265–278 (1945). —Martineau, Rhod. Wild Flowers: 90, t. 32 (1953). —Wild, Common Rhod. Weeds, fig. 73 (1955). —Cabrera, Fl. Prov. Buenos Aires, **6**: 238, Fl.71 (1963). —Henderson & Anderson, Common Weeds in S. Afr.: 386, fig. 192 (1966). —Merxmüller, Prodr. Fl. Südwestafr. **139**: 161 (1967). —Guillarmod, Fl. Lesotho: 285 (1971). —Biegel & Mavi in Wild, Rhod. Bot. Dict. of African & English Plant Names: 238 (1972). —Hilliard in Ross, Fl. Natal: 367 (1972). —Pope in Kirkia **10**(1): 119–121 (1975). —McVaugh, Fl. Novo-Galiciana **12**: 795–798 (1984). —Mesfin Tadesse, Fl. Ethiopia & Eritrea 4(2): 279 (2004). Type: 'Habitat . . . . K s. L. ... On le cultive au jardin Botanique national de Paris, ou il forme des touffes presque diffuses, ... Je n'ai pu savoir quel est son lieu natal; mais je crois qu'il a été envoyé avec le *Sanvitalia*, que je ferai connoître à nos lecteurs dans le numéro prochain; or, comme ce *Sainvitalia* est originaire de l'Amérique méridionale, il est vraisemblable que le *Pectis* pinné croît aussi dans le même pays, ... Je crois même qu'il est du Pérou.' (P-LA – P00342129 holotype).[87] FIGURE 6.5.**17**.

    Pectis pinnata Lam. in J. Hist. Nat. **2**: 150, pl. 31 (1792).

    Schkuhria abrotanoides Roth, Catal. Bot. **1**: 116 (1797). Type: 'Sub nomine *Bellii minuti* mihi communicata sunt semina ab amico perdilecto Ill. *de Voigt*.' Holotype unknown, but probably B† or B-W.

    Hopkirkia anthemoidea DC., Prodr. **5**: 660 (1836). Type: '– in Mexici agro Regiomontano legit cl. *Haenke*. (v.s. in h. Haenk. à cl. de Sternberg comm.)' (PRC453239 holotype, G-DC-G00456682).

    Schkuhria bonariensis Hook. & Arn. in J. Bot. (Hooker) **3**(No. 22): 321 (1841), p.p. —Eyles in Trans. Roy. Soc. S. Afr. **5**: 516 (1916). —Sussenguth & Merxmüller in Proc. & Trans. Rhod. Sci. Ass. **43**: 142 (1951). Types: 'Pampas of Buenos Ayres; *Dr. Gillies*. Buenos Ayres; *Tweedie* [1120].' *Gillies* 74 (E00265075 syntype); *Tweedie* 1120 (K000502245 syntype); *Tweedie* s.n. (E00265076, K000502246 syntypes).

---

[87] There are four other sheets in P-LA (P00309066, P00309067, P00309068, P00342128) corresponding to this species.

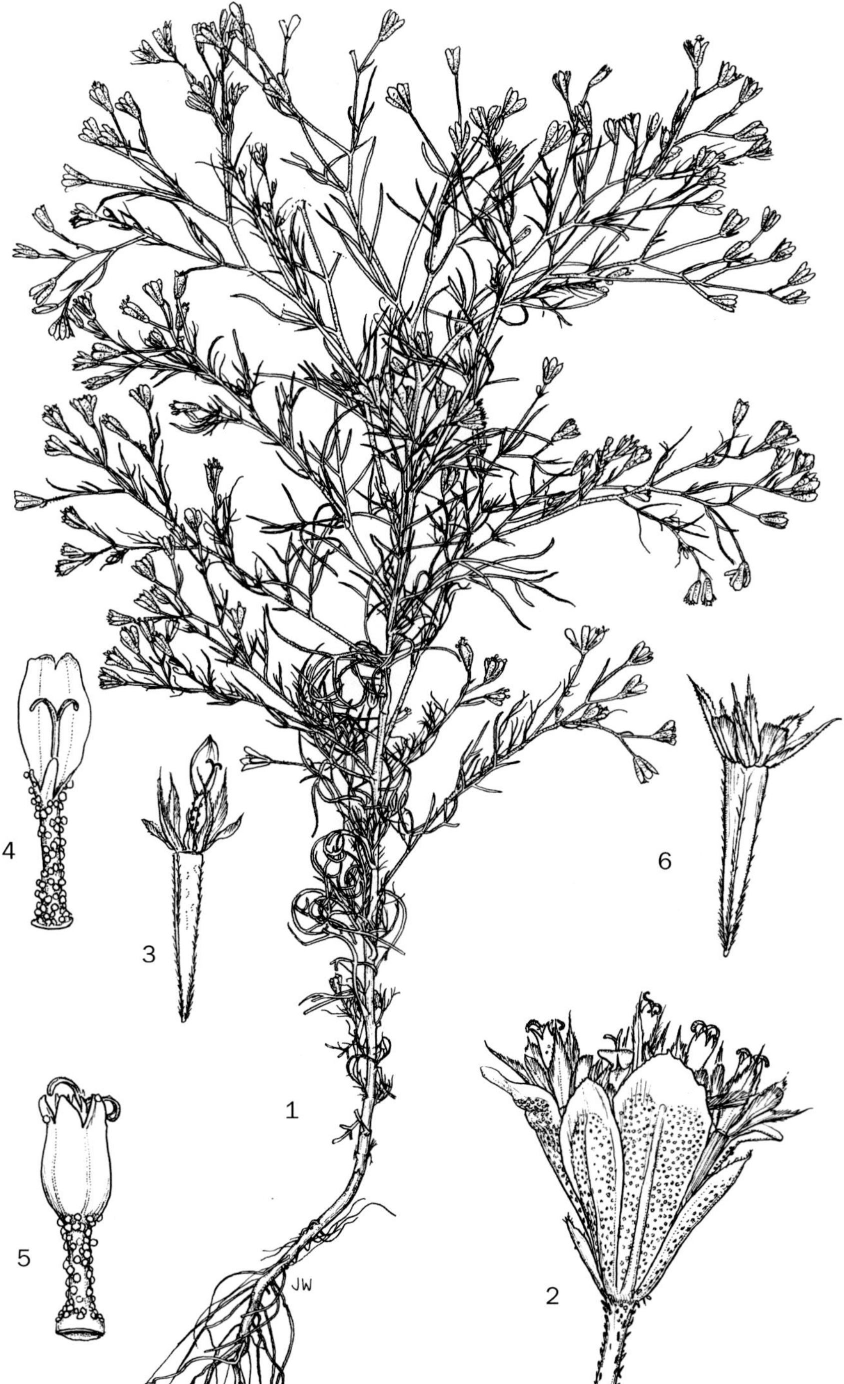

Fig. 6.5.**17**. SCHKUHRIA PINNATA. 1, habit of flowering plant (× 2/3); 2, capitulum(× 6); 3, ray floret (× 6); 4, ray floret corolla (× 16); 5. disc floret corolla (× 16); 6. ray floret achene (× 6). 1 from *Verdcourt* 415; 2–7 from *Hepper & Jaeger* 6692. Drawn by Juliet Williamson. From Flora of Tropical East Africa.

Schkuhria isopappa Benth., Pl. Hartweg.: 205 (1845). Type: [Ecuador:] 'Ad pontem Guapulo prope Quito.' [*Hartweg* 1141] (K000502250 – 1141. 'Annual. "Escoba" (broom)./Near the bridge of Guapulo./alt 8000 ft. nr Quito' holotype, B†, E00433110, LD1217669).

Schkuhria coquimbana Phil. in Anales Univ. Chile **90**: 29 (1895). Type: 'Prope Paihuano in provincia Coquimbo legit *Frid. Philippi* incolae eam *cachanlahuen cimarron* vocant.' ['Cerca de Paihuano, en la provincia de Coquimbo, la recolectó Federico Philippi; la gente la llama "cachanlahuén cimarrón". 60580' – Muñoz Pizarro, 1960: 154]

Rothia pinnata (Lam.) Kuntze, Revis. Gen. Pl. **3**(3): 170 (1898).

Schkuhria pinnata (Lam.) Kuntze, Revis. Gen. Pl. **3**(3): 170 (1898), pro syn.

Rothia pinnata (Lam.) Kuntze α *pallida* Kuntze, Revis. Gen. Pl. **3**(3): 170 (1898), p.p. Type: 'Argentina: Cordoba.' (NY00232813 holotype).

Rothia pinnata (Lam.) Kuntze β *purpuascens* Kuntze, Revis. Gen. Pl. **3**(3): 170 (1898), p.p. Type: 'Argentina Jujuy.' (NY00232814 holotype).

Schkuhria abrotanoides var. *pomasquiensis* Hieron. in Bot. Jahrb. Syst. **29**(1): 53 [May 1900](1901).[88] Type: [Ecuador:] 'Crescit in arvis arenosis prope Pomasqui (*S.*[*odiro*] n. 47).' (B† holotype).

Schkuhria abrotanoides var. *isopappa* (Benth.) Hieron. in Bot. Jahrb. Syst. **29**(1): 53 [May 1900](1901).[89]

Schkuhria advena Thell. in Repert. Spec. Nov. Regni Veg. **11**: 308 (1918). Type: 'Süd-Afrika: Transvaal: Florida, 1906, *H. Hutton* no. 630 (Albany Museum), ab *S. bonariensis* in herb. Univ. Zürich.' (Z000003844 holotype).

Schkuhria pinnata var. *abrotanoides* (Roth) Cabrera in Anal. Soc. Ci. Argent. **114**: 189 (1932). — Merxmüller, Prodr. Fl. Südwestafr. **139**: 161 (1967).[89]

Annual herb 8–60(150) cm high. Stems usually simple at base and moderately to profusely branched in inflorescence, terete, ± ribbed, sparsely glandular-pubescent. Leaves sessile, opposite (becoming alternate in inflorescence, as do branches), 2–6 cm long, 1– or 2–pinnatisect or upper leaves simple and filiform, rachis filiform to linear, lobes 3–7, c. 0.3–3.8 cm long, filiform, lower lobes often divided again in larger leaves, glabrous or with few scattered hairs. Inflorescence a lax panicle or corymb, capitula pedicellate at ends of branches, pedicels 5–35 mm long, filiform, naked, glandular-pubescent. Capitula 5–8 × 2.5–4 mm at apex; involucre turbinate; phyllaries green with a yellow apex, outer phyllaries 1–3, 2–4.5 mm long, inner phyllaries 5–6, imbricate, 4.5–6.2 ×1.5-2.8 mm, oblanceolate to narrowly obovate, margins hyaline and very finely ciliate towards the apex, glabrous, apices obtuse. Ray florets female, 1, rarely absent, ray limb yellow, 1.3–2.0 mm long, narrowly obovate to obovate or narrowly oblong, apex obtuse often notched, glandular-punctate outside; corolla tube green, c. 1.5 mm long, cylindrical, glandular-punctate. Disc florets bisexual, 4–8; corolla yellow, c. 2 mm long, funnel-shaped, corolla lobes 5, 0.5 mm long, glandular-punctate, glabrous, lower part green with protuberant cylindrical glands (at least in fresh material), apices acute. Achenes black, 3.2–4.5 mm long, narrowly obpyramidal, ± longitudinally 4-ribbed, ± appressed setuliferous on angles especially toward base, setulae of twin-hairs, apices bifid; pappus of paleaceous scales, 8, 1–2.5 mm long, obovate or lanceolate, usually conspicuously flecked purple, all apices obtuse and laciniate or scales with obtuse and acuminate, or awned, apices alternating, abaxially ± appressed setuliferous, setulae of twin-hairs with acute, undivided apices.

Botswana. N: Bokalaka, Francistown, 26.xii.1966, *McClintock* K48 (K). SW: Ghanzi & Kgalagadi Dist., Okwa Valley, 5.iii.1978, *Skarpe* S-276 (K). SE: Gaberone, 1972, *Morwe* 2 (SRGH). **Zambia**. C: Lusaka Province, Grasmere Farm, 4.1.1995, *Bingham & Zyambo* 10236 (K). **Zimbabwe**. N: Trelawney, 22.ii.1943, *Jack* 126 (PRE, SRGH). W: Matobo Dist., Besna Kobila Farm, ii.1961, *Miller* 7704 (COI, K, LISC, MO, PRE, SRGH). C: Salisbury, Epworth, xi.1956, *Drummond* 4942 (K, LISC, PRE, S, SRGH). E: Umtali, Maranki Reserve, 22.i.1950, *Chase* 1933 (BM, PRE, SRGH). S: Beitbridge Dist., Nottingham Ranch, 25.iii.1959, *Drummond*

---

[88] The publication dates supplied with this volume indicate categorically that Heft 1 was published in May 1900! This is nearly eight months before Heft 5 of the preceding volume, 28, published in January 1901, q.v.

[89] This only represents a partial synonymy based on names previously referred to this taxon in Africa, using Heiser's concept of *S. pinnata* var. *pinnata* and *S. pinnata* var. *abrotanoides*. Retention of varieties is only possible by recognizing the extremes of variation and it may be more sensible to accept a very variable species.

5999 (K, PRE, SRGH). **Malawi**. C: Ncheu Dist., Dedza Plateau, 1550 m, 29.iii.1978, *Pawek* 14148 (K, MA, MO, UC). **Mozambique**. GI: Vale do Limpopo, 13.ix. 1948, *Myre & Carvalho* 165 (COI, PRE, SRGH). M: Lourenzo Marques, xi.1945, *Pementa* 43772 (LISC, SRGH).

Native of S America, now common in southern and E Africa. Frequent weed of arable fields, disturbed ground (especially by roadsides), very rarely cultivated and has sometimes recorded as a garden escape; 0–3700 m; flowering throughout the year.

Conservation Status: A widespread weed in the tropics; LC (Least Concern).

125. MONTANOA Cerv.

Montanoa Cerv. in La Llave & Lexarza, Nov. Veg. Descr. **2**: 11 (1825). —Robinson & Greenman in Proc. Amer. Acad. Arts **34**(20): 507–534 (1899). —Wild in Kirkia **6**(1): 57 (1967). —Funk, Mem. New York Bot. Gard. **36**: 1–133 (1982). —Lisowski, (Asterac. Fl. Afr. Cent. 2) Fragm. Flor. Geobot. **36** Suppl. 1: 221–222 (1991). —Hind in Fl. Masc., Composées **109**: 183–185 (1993). —Karis & Ryding in Bremer, Asterac. Cladist. Classif.: 583 (1994). —Beentje & Hind in F.T.E.A., Compositae **3**: 756–757 (2005). —Panero in Kubitzki, Fam. Gen. Vasc. Pl. **8**: 469–470 [2006](2007).

Eriocoma Kunth in Humb., Bonpl. & Kunth, Nov. Gen. Sp. Pl. **4** (ed. fol.): 210 (Oct. 1818), nom. illeg. non Nutt. (Jul. 1818)[= Oryzopsis, GRAMINEAE].

Eriocarpha Cass. in Dict. Sci. Nat., ed. 2, **59**: 236 (1829), as nom. nov. pro *Eriocoma* Kunth.

Montagnaea DC. Prodr. **5**: 564 (1836), orth. var.

Priestleya Sessé & Moc. ex DC., Prodr. **5**: 564 (1836), nom. nud. pro syn.

Uhdea Kunth, Ind. Seminum: 13 (1847).

Shrubs or trees, rarely perennial herbs. Leaves opposite, usually ovate, entire or pinnatifid, often lobed, 5–7-veined. Inflorescence many-headed, corymbose paniculate or thyrsoid-cymose. Capitula radiate and heterogamous, rarely discoid; involucre hemispherical; phyllaries 1- or 2-seriate, subequal; receptacle convex, paleaceous, paleae accrescent after anthesis, scarious or membranous, usually spinescent or caudate-tipped, broad, conduplicate and enfolding achenes at maturity. Ray florets neuter, limbs white or creamish. Disc florets hermaphrodite, fertile or innermost sterile, corollas 5-lobed, yellow, greenish-yellow, greyish-white or creamy white; stamens 5, anthers black or yellow, apical anther appendages ovate with glandular hairs; style arms with divided stigmatic surfaces, apices deltoid with a linear appendage. Achenes ± bilaterally compressed, obovoid, black to brownish black or reddish brown, striate, glabrous or sparsely setuliferous; pappus absent.

A genus of c. 25 spp. largely from Mexico and Central America with two spp. widely cultivated and escaped in the tropics of both Old and New Worlds. One species is present in the Flora area having been cultivated naturalized; *Montanoa bipinnatifida* (Kunth) K.Koch is also cultivated but appears not to have naturalized.

Leaves usually deeply 3–5-lobed, sometimes broadly ovate and unlobed; petiole base exauriculate; disc floret corollas moderately to densely pubescent throughout; ray florets 7–8, ray limb 15–17 mm long . . . . **1.** *hibiscifolia*
Leaves pinnatifid to bipinnatifid; petiole base auriculate; disc florets corollas glabrous except for densely pubescent corolla lobes; ray florets 10–12, ray limb 25–35 mm long . . . . **2.** *bipinnatifida*

1. **Montanoa hibiscifolia** Benth. in Oersted, Vidensk. Meddel. Naturhist. Foren Kjøbenhavn **1852**(5–7): 89 (1853), as '*Montagnaea*'. Types: [Nicaragua:] '... fandt jeg i Bjegskovene i Segovia i Naerheden af Matagalpa (4500′) og paa den sydlige Skraauing af Vulkanen Barba i Costa–Rica (6000 ′).' 'In provincia Segovia, *Oersted* 235' (K487561 lectotype, C10007687, C10007688), lectotypified by Morley in J. Adelaide Bot. Gard. **2**(2): 159 (1980); 'Ad Barba in Costarica', *Oersted* 134 (K000487562 syntype). FIGURE 6.5.**18**.

Fig. 6.5.**18**. MONTANOA HIBISCIFOLIA. 1, flowering branch (× 2/3); 2, l.s. capitulum (× 2); 3, phyllary (× 6); 4, immature palea from flowering capitulum (× 10); 5, mature palea from fruiting capitulum (× 2); 6, ray floret corolla (× 3); 7, disc floret corolla (× 8); 8, partial anther cylinder opened out (× 18); 9, disc floret style (× 12); 10, disc floret achene (× 8). 1–4, 6–9 from *Duljeet* s.n.; 5, 10 from *Ogilvie* 15. Drawn by Pat Halliday. From Flore des Mascareignes.

Montanoa hibiscifolia (Benth.) Sch.Bip. ex K.Koch in Wochenschr. Vereines Beförd. Gartenbaues Königl. Preuss. Staaten **7**: 407 (1864), nom. illeg.

Eriocoma hibiscifolia (Benth.) Kuntze, Revis. Gen. Pl. **1**: 336 (1891).

Montanoa samalensis J.M.Coult. in Bot. Gaz. **20**: 49 (1895). Type: [Guatamala:] 'Rio Samalá, Depart. Retalhulen, alt. 1,700 ft, April 1892, *John Donnell Smith* 2,858.' (F0050771 ?holotype, GH00010571, G-BOISS – G00300721, K000487560, US1415704).[90]

Montanoa pittieri B.L.Rob. & Greenm. in Proc. Amer. Acad. Arts **34**(20): 517 (1899). Type: 'Costa Ricam in a hedge on the llanos of Alajuelita, *Pittier*, no. 1455.' (GH10560 holotype, BR5428632, BR5428960, GH00010560, GH 00010592 – s. coll., s.n.).[91]

Montanoa wercklei Berger in Gard. Chron. ser. 3, **50**: 122 (1911). Type: 'This plant was grown from seeds kindly sent to us by *Mr. C. Werckle*, of San José in Costa Rica, in 1905. In its native land it is known as Toona quirita. ... in Horto Mortolensi culta. Floret mensibus, Dec.-Jan.' (K000487550 – flowering branch & K000487551 – leaves and part of stem holotype, G00300722 – there are 2 other unbarcoded duplicates with this number).[92]

Montanoa hibiscifolia (Benth.) D'Arcy in Phytologia **30**(1): 5 (1975), nom. illeg.

Perennial shrub to 6 m. tall; stems cylindrical, pubescent. Leaves petiolate; petioles 1–8 cm, pubescent, lamina 7–40 × 2.5–30 cm, broadly ovate to pentagonal, 3-5-veined from near base, base cordate and abruptly cuneate and with a pair of auricular ± oblong lobes c. 1.5 × 1cm at base, margins serrate-crenate, deeply sinuately and ± palmately lobed, lobes acuminate or obtuse often with smaller secondary lobes, apex acuminate, densely pilose above, shortly grey-tomentose below. Inflorescences terminal, of corymbose panicles c. 25 cm diam.; pedicels 2–6 cm long, densely pilose. Capitula pendulous, 1–1.2 cm diam. in flower, 2–2.5 cm diam. in fruit; involucre subglobose; phyllaries 5–7, essentially uniseriate but appearing biseriate, outer up to 4–5 × 1–2 mm, narrowly ovate, acute or acuminate, pubescent, longitudinally 3-veined; inner somewhat shorter and narrower than outer, glabrescent; receptacle ± conical; paleae 3–3.5 × 2 mm in flower, 9–15 × 4.5–6 mm in fruit, membranous, obovate, stramineous to purplish, margins ciliate, entire, apices truncate-emarginate and caudate, glabrous. Ray florets 7 or 8, ray limb white, 15–17 × 8–9 mm, narrowly elliptic, minutely apex 3-toothed, dorsally puberulent; tube 0.5–0.75 × 0.5 mm, pubescent. Disc florets numerous (85–105), corollas yellow, tube c. 0.5 × 0.5 mm, throat 1.5 × 1–1.25 mm, densely pubescent, lobes 5, 0.5–0.7 × 0.5 mm, apices acute, densely glandular-punctate and pubescent; anthers not fully exserted from corolla. Achenes brownish to reddish-brown, c. 3 mm long, somewhat bilaterally compressed, obovate, 4-ribbed, glabrous, apex narrowed to a very short neck.

Zimbabwe. C: Salisbury, v.1954, *Armitage in* GHS 46807 (K, SRGH). E: Umtali, 6.vi.1967, *Pole Evans* 6804 (K, PRE). (? cult.) Melsetter, 23.vi.1958, *Allott* 1 (SRGH).

A native of Mexico to Costa Rica and once widely cultivated in the tropics and subtropics of both Old and New Worlds. Funk (1982) only reported scattered cultivated material in the Old World including southern India, Myanmar and Sri Lanka, but nothing in Africa. Rarely cultivated in the Flora area nowadays, but sometimes naturalized as a weed of old cultivated areas; 350–2500 m.

Conservation Status: Although relatively rare as a cultivated plant, and even as a naturalized species, with its widespread distribution it is best recorded as LC (Least Concern).

Vernacular name: Daisy Tree.

Very widely, and incorrectly, most material in the continent has previously been determined as *Montanoa bipinnatifida* (Kunth) K.Koch (e.g. Wild in Kirkia **6**(1): 57, 1967), a species with very short pedicels, much larger flowering (and fruiting) capitula, and one of the most diagnostic leaf shapes of the genus, q.v.

---

[90] Funk (1982) stated, and I have no reason to disbelieve the evidence of the material, that the holotype is in F, isotypes in G-Boiss, GH, K, US. Coulter only said that he took 'All of the most critical material' to the Gray herbarium… for final study', but did not state where the holotype was located; if this casts doubt on the presence of a holotype, then all are syntypes.

[91] Material, with the original label, indicates that the collector was *Tonduz*. The unlocalized/unannotated material in GH (ex herb. Klatt) is considered as an isotype.

[92] Morley (1980: 159) selected the 1908 collection as lectotype of this name.

2. **Montanoa** \ (Kunth) K.Koch in Wochenschr. Vereines Beförd. Gartenbaues Königl. Preuss. Staaten **7**: 407 (1864).[93] Type: 'Mexico. Uhde semina misit. Floret Februario.'[94] (P – 'Herb. Schultz Bip. s.n. ex Hort. 25.ii.1864' neotype), neotypified by Morley (1980: 153).

> *Uhdea bipinnatifida* Kunth, Ind. Seminum **1847**: 13 (1847).
>
> *Polymnia grandis* hort. ex Kunth, Ind. Seminum **1847**: 13 (1847), nom. nud. pro syn.
>
> *Montanoa heracleifolia* Brongn. ex Groenl. in Rev. Hort. ser. 4, **5**: 544 (1857), nom. nud.
>
> *Montanoa elegans* K.Koch in Wochenschr. Vereines Beförd. Gartenbaues Königl. Preuss. Staaten **7**: 408 (1864). Type: 'Wie sie nach Europa gekommen, wissen wir nicht; in den Handel kam sie aber von Wien aus durch den Handelsgärtner Abel unter dem Namen Uhdea bipinnatifida vera.' Type material unknown.
>
> *Montanoa pyramidata* Sch.Bip. in K.Koch, Wochenschr. Vereines Beförd. Gartenbaues Königl. Preuss. Staaten **7**: 408 (1864). Type: 'Eine vom *Dr. Oliva* am Guadalajara und ausserdem von *Aschenborn* in Mexiko entdeckte und den beiden letzten Arten im Habitus änliche Art.' (P02441589 – '349' 'Mexico, Jalisco, Guadalajara, 1853, *D. Oliva* s.n.' lectotype), lectotypified by Morley (1980: 155).
>
> *Eriocoma elegans* (K.Koch) Kuntze, Revis. Gen. Pl. **1**: 336 (1891).
>
> *Eriocoma pyramidata* (Sch.Bip.) Kuntze, Revis. Gen. Pl. **1**: 336 (1891).

Shrubs 2–10 m tall. Stems at first quadrangular becoming terete with age, brown to grey, puberulent at first later glabrescent. Leaves petiolate, petiole 4–20 cm long, base auriculate, often lobed beneath lamina, sparsely pubescent, lamina ovate to ovate-lanceolate, 12–30 × 9–40 cm, dark green, margins regularly to irregularly serrate, pinnatifid to bipinnatifid, upper surface densely pubescent, lower surface sparsely to densely glandular and pubescent, apices acute to acuminate. Inflorescences of lax, open, many-headed compound corymbs, branches opposite, capitula pendulous, pedicellate, pedicels 2–7 mm long, sometimes purple, densely glandular and pubescent. Capitula 1.8–2.5 cm diam. in flower, 3.5–4 cm diam. in fruit; phyllaries 8 or 9, biseriate, subequal, reflexed in fruit, 6–8.5 × 1–3 mm, margins ciliate, entire, abaxially densely pubescent, adaxially glabrous, apices acute to acuminate, mucronate; paleae 2–2.5 × 1.5 mm and pale yellow at flowering, apices long-acuminate, growing to 16–21 × 4–5 mm and stramineous and indurate in fruit, margins entire, abaxially sparsely glandular-punctate. Ray florets 10–12, ray limbs white, ovate-lanceolate, 25–35 × 6–11 mm, apex 2-notched, pubescent. Disc florets numerous (95–125), corollas at first green, turning yellow, tube 1.5 × 0.5–1 mm, glabrous to sparsely glandular and pubescent, throat cylindrical 4–5 × 1.5–2 mm, sparsely to moderately glandular and pubescent, lobes 5, 1–1.25 × 0.5–0.75 mm, apices acute and densely pubescent; anthers well-exserted, apical anther appendages yellow, acuminate, glandular-punctate; styles enlarged at base, style arm and appendages yellow. Achenes brownish-black, 3.5 × 2 mm, smooth, glabrous.

A native of Mexico, but previously widely cultivated in both the Old and New Worlds, as well as in Mexico; 450–2000 m; flowering at the turn of the year.

I have seen no herbarium material from the Flora area, nor is the species recorded in any of the checklists available for the Flora area. *Montanoa bipinnatifida* is, however, cultivated in Harare, and appears never to have escaped and naturalized, unlike *M. hibiscifolia*.

Conservation Status: Even though it is cultivated in the Flora area it is best recorded as LC (Least Concern).

---

[93] Koch provided the following note on Uhde and *Uhdea bipinnatifida*: '... wurde vom preussischen Konsul Uhde zu Matameros in Mexiko entdeckt und kam 1845 in den botanischen Garten nach Berlin, von wo aus sie weiter verbreitet wurde. Kunth nannte sie zuerst in einer Versammlung des Vereines zur Beförderung des Gartenbaues des Jahres 1847 *Uhdea pinnatifida*, im Herbste desselben Jahres jedoch *U. bipinnatifida*.'

[94] No original herbarium material has been found. Funk (1982: 120) cited Koch's commentary, and noted the journal referred to by Kunth was clearly published much later than Kunth's description, in 1849.

126. **SYNEDRELLA** Gaertn.

Synedrella Gaertn., Fruct. Sem. Pl. **2**(3): 456, t. 171 (1791), nom. cons. —McVaugh, Fl. Novo-Galiciana **12**: 908–910 (1984). —Lisowski, (Asterac. Fl. Afr. Cent. 1) Fragm. Flor. Geobot. **36** Suppl. 1: 117–119 (1991). —Turner in Phytologia **76**(1): 39–51 (1994). —Hind in Fl. Masc., Composées **109**: 181–183 (1994). —Beentje & Hind in F.T.E.A., Compositae **3**: 744–746 (2005). —Chen Yousheng & Hind in Fl. China **20–21**: 868 (2011). —Hind in Kew Bull. **71**(2)-0632: 1–5 (2016).

Ucacou Adans., Fam. Pl. **2**: 131 (1763), nom. rej.

Erect or procumbent, annual or short-lived perennial herbs, moderately to frequently branched. Stems terete, strigose, pith solid. Leave opposite, petioles narrow, often narrowly winged, usually short; laminas herbaceous, ovate, base short-acute, apex acute, surfaces white-strigose, slightly paler below, without glandular dots, 3-veined from near base. Inflorescence of solitary, or clusters of few (2–10), sessile or short-pedicellate, axillary or terminal capitula. Capitula radiate, heterogamous; involucres campanulate; phyllaries 4 or 5, subequal, outer 2 mostly herbaceous, inner scarious and glabrous outside, enclosing ray floret achenes; receptacle small, short-conical, paleaceous; paleae persistent, scarious, flat to slightly concave, narrowly elliptical to linear, apically slightly erose. Ray florets few, 3–6(8), biseriate, female; limb short, oblong, apex rounded to slightly 3-lobed, mamillose above, pale yellow, tube slender, glabrous; style arms with paired stigmatic lines. Disc florets few, 5–10(12), hermaphrodite; corollas yellow, basal tube slender, glabrous, throat narrowly campanulate, lobes 4, short, with strong marginal papillae inside, strigillose outside; anthers 4, anther thecae blackish; apical anther appendage ovate, apices acute or truncate and crenate, concave outside, glabrous, basal appendages sagittate; style base without enlarged node, style shaft glabrous; style arms slender, tapering with brush of papillae in distal half, stigmatic lines separated at base. Ray achenes broadly obovate to ellipsoid, obcompressed, winged, wings broad, coarsely lobed and continuous across top, pappus of 2 spiniform awns, slightly puberulous above; disc achenes slender, obcompressed, wingless, tuberculate on one surface, distally with 2(–4) stiff tapering awns, minutely puberulous especially on awns.

One species from tropical America that has become a pantropic weed, and is present, although poorly represented, in the Flora area.

Synedrella nodiflora (L.) Gaertn., Fruct. Sem. Pl. **2**(3): 456, t. 171 (1791). —Maquet in Fl. Rwanda **3**: 638, fig. 196/1 (1985). —Lisowski, (Asterac. Fl. Afr. Cent. 1) Fragm. Flor. Geobot. **36** Suppl. 1: 118 (1991). —Hind in Fl. Masc., Composées **109**: 183, t. 61 (1993). —Beentje & Hind in F.T.E.A., Compositae **3**: 744–746 (2005). —Chen Yousheng & Hind in Fl. China **20–21**: 868 (2011). Type: 'habitat in Caribaeis' Dillenius, Hort. Eltham.: t. XLV, fig. 53 (1732), iconotype, lectotypified by Hind (2016: 1).[95] FIGURE 6.5.**19**.

> *Verbesina nodiflora* L., Cent. I. Pl.: 28 (1755).
>
> *Ucacou nodiflorum* (L.) Hitchc. in Rep. (Annual) Missouri Bot. Gard. **4**: 100 (1893).
>
> *Wedelia cryptocephala* Peter in [Wasserpflanzen und Sumfgewächse in Deutsch-Ostafrika] Abh. Konigl. Ges. Wiss. Göttingen, Math.-Phys. Kl. Ser. 2, **13**: 94 (1928). Types: 'Ost Usambara : beim Dorf Amani ca. 850 m ü. M. V¹Z² Mai (*P[eter]* 23449); –Handei : an der Sigibrücke von Longusa ca. 300 m ü. M. Z²⁻³ Februar (*P[eter]* 19602); am Sigi-Fluß bei Longusa V¹ Z³ März (*P[eter]* 19832, 19919); Kwamtilli bei Segoma 300 m, Oktober (*P[eter]* 25220); Sigi →Longusa 500 m, August (*P[eter]* 24655).' (B† syntypes).

Annual herb, erect, 15–150 cm tall. Stems scarcely- to much-branched, scabridulous and glabrescent. Leaves petiolate, petiole winged, to 3 cm long, lamina ovate or elliptic, 2–9(13) × 1–5(6) cm, base cuneate and decurrent, margins crenulate-serrate or rarely subentire, apex acute or obtuse, 3-veined from base, scabridulous on both surfaces. Involucre 7–12 × 3–4 mm; outer phyllaries 10–12.5 × 3–3.5 mm, green, inner phyllaries 6–7 × 1.5–2 mm, scarious; paleae 6–8 × 0.8–1.3 mm, margins laciniate or ciliate. Ray

---

[95] Previous statements as to the lectotypification of this name are erroneous and based on Browne material acquired by Linnaeus three years after publication of the name (viz. LINN 1021.7).

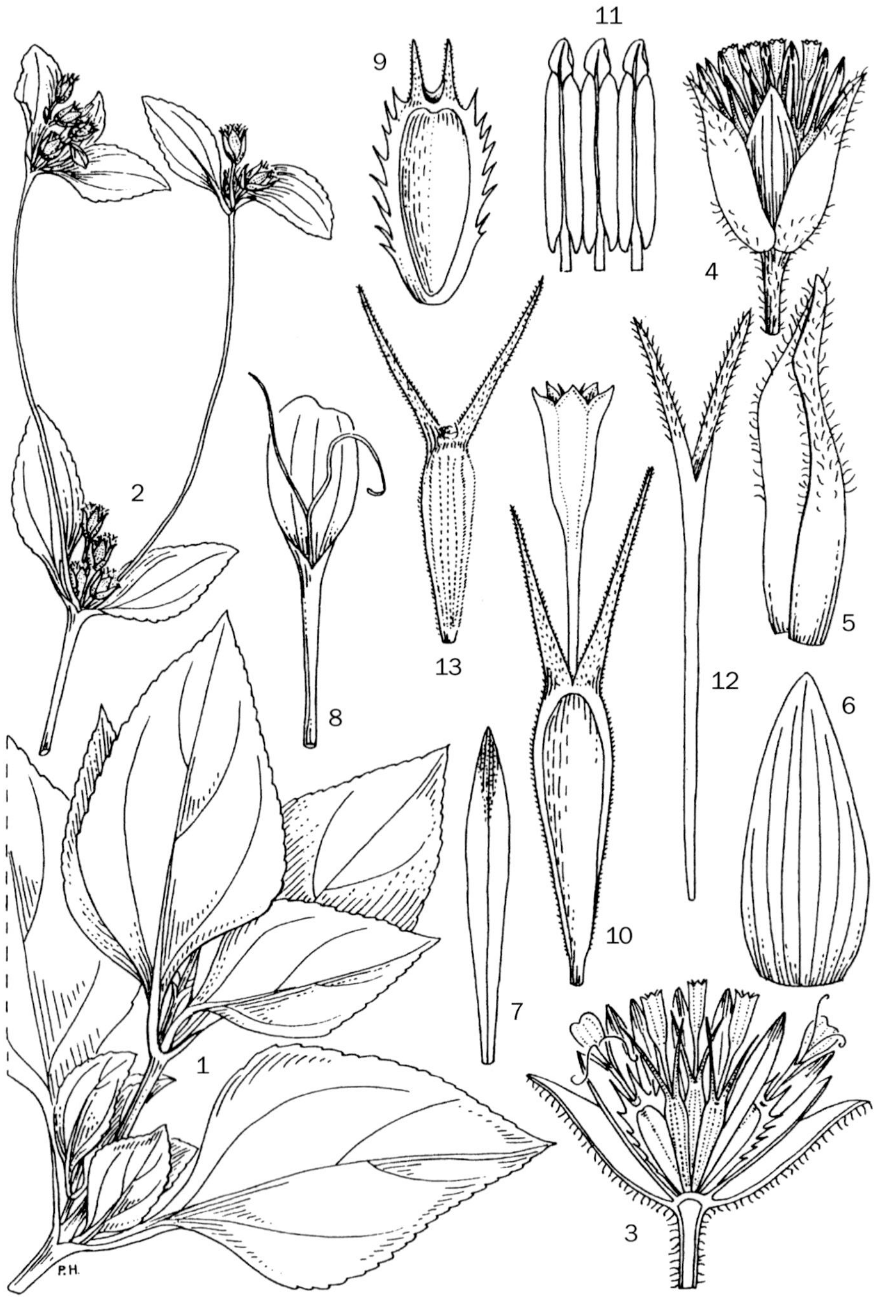

Fig. 6.5.**19**. SYNEDRELLA NODIFLORA. 1, leafy shoot (× 2/3); 2, flowering stem apex (× 2/3); 3, l.s. capitulum (× 4); 4, young flowering capitulum (× 4); 5, outer phyllary (× 6); 6, inner phyllary (× 6); 7, palea (× 6); 8, ray floret corolla (× 12); 9, ray floret achene (× 6); 10, disc floret (× 12); 11, partial anther cylinder opened out (× 32); 12, disc floret style (× 24); 13, disc floret achene (× 6). 1–2 from *Cadet* 1739; 3–13 from *Cadet* 1675. Drawn by Pat Halliday. From Flore des Mascareignes.

florets 3–8, ray limb 1.5–3 mm long, yellow. Disc florets 4–11, corollas yellow, 2–4 mm long; style arms c. 1 mm long. Achenes of ray florets 4–5 mm long, winged, with 2 (rarely 3) short awns; achenes of disc florets 4–6 × 1 mm, outer winged, inner wingless, pappus of 2–3 awns 2–4.5 mm long.

Zambia. N: North Luanga National Park, 11°51'S 32°25'E, 600 m, 22.iv.1994, *Smith* 510 (K). W: Kitwe Dist., Mwekera Forestry Training School, 12°51'S, 28°21'E, 26.iv.1989, *Pope et al.* 2207 (BR, K, LISC, NDU). **Malawi**. N: Nkhata Bay Dist., Nkhata Bay, 840 m, 10.v.1970, *Brummitt* 10541 (K).

Described from the Caribbean and originally from the New World, this species is a pantropical weed. It is also very widespread in surrounding floras. Waste places, cultivated land, secondary vegetation, scrub, roadsides, paths, disturbed ground; a common weed often forming large colonies; 0–1500 m.

Conservation Status: Although not particularly widespread, and probably under-collected, in the Flora area it is not threatened: LC (Least Concern).

127. **BLAINVILLEA** Cass.

Blainvillea Cass. in J. Phys. Chim. Hist. Nat. Arts **96**: 216 (May 1823); —in Dict. Sci. Nat., ed. 2, **29**: 493 (Dec. 1823). —Koster & Philipson in Blumea **6**(2): 349–354 (1950). —Adams in Webbia **12**(1): 217–250 (1956). —Wild in Kirkia **6**(1): 38–39 (1967). —Boulos & Hind in Fl. Egypt **3**: 232–233 (2002). —Mesfin Tadesse, Fl. Ethiopia & Eritrea **4**(2): 299 (2004). —Beentje & Hind in F.T.E.A., Compositae **3**: 735–737 (2005). —Orchard in Austrobaileya **8**(4): 653–669 (2012).

Eisenmannia Sch.Bip. in Flora **24**(1), Intell. 3: 42 (1841), nom. nud. (as *Eisenmannia clandestina* Sch.Bip.)

Galophthalmum Nees & Mart. in Nova Acta Acad. Caes. Leop.-Carol. Nat. Cur. **12**: 7, t. 2 (1824).[96]

Erect annual, possibly short-lived perennial, sometimes aromatic herbs. Rootstock a short taproot, and fibrous. Stems simple to well-branched, branches usually terminating in inflorescences. Leaves opposite, often becoming alternate towards and in inflorescences, simple, petiolate, 3-veined from base, base cuneate, lamina ovate to narrowly ovate, scabrid, margins crenate to serrate, apices acute. Inflorescences leafy, cymose or laxly paniculate, often with solitary capitula in forks of lower inflorescence branches. Capitula heterogamous and radiate, sometimes appearing disciform, or sometimes discoid by abortion; involucre ± campanulate or hemispherical; phyllaries biseriate, subequal, outer usually green with darker green longitudinal striae, inner scarious; receptacle small, flat to slightly convex, paleaceous, paleae scarious, oblong, conduplicate about florets, sometimes glandular-punctate outside, apices truncate and laciniate, sometimes also ciliate. Ray florets female and fertile, 1- or 2-seriate, few, limb short, minutely 2(3)-lobed, yellow or white; style arms linear or slightly clavate pubescent externally. Disc florets hermaphrodite and fertile, few, corollas yellow or white, tube narrow, expanding abruptly into campanulate limb, lobes 5; anther cylinders black; apical anther appendages ovate, basal anther appendages obtuse or sagittate. Ray floret achenes obcompressed or obconical, glabrous, black or grey, sometimes weakly rugose or tuberculate; disc floret achenes often 3-gonous (sometimes 2–4-angled) or compressed, setuliferous above; pappus of 2–3(5) stout barbellate awns or setae arising from an entire narrow cupule of subequal connate scales, or pappus absent.

A genus of probably few spp., possibly four (cf. Orchard, 2012) or less, certainly less than the ten accepted by many authors. The genus is pantropical, although that may be due to the widespread nature of one or two spp.. I consider that only one species is present in the Flora area.

---

[96] Turner in Phytologia **64**(3): 214 (1988) provided a nomenclaturally superfluous name in transferring *Galophthalmum brasiliense* Nees & Mart., the type of *Galophthalmum*, to *Calyptocarpus* Less. as in his concept, *Galophthalmum* and *Calyptocarpus* were congeneric; this was clearly in agreement with Robinson's (1981: 48) concept. Panero's concept of *Blainvillea* (Panero, 2006) clearly included *Galophthalmum* Nees & Mart. Until further work can be done I prefer to keep *Galophthalmum* as a generic synonym of *Blainvillea*. Orchard (2012) apparently agreed with Turner.

Blainvillea dichotoma (Murray) Cass. ex Hemsl., Biol. Centr.-Amer. Bot. **4**(22): 112 (March 1887). Type: not cited, but based on cultivated material in the botanic garden in Göttingen.[97] FIGURE 6.5.**20**.

Verbesina dichotoma Murray in Commentat. Soc. Regiae Sci. Gott. Phys. **2**: 15, t. 4 (1779), non Wall. (1831) (nom. nud.), nec Sieber ex Steud. (1841), nec Reinw. ex Miq. (1856: 75) [= *Quadribractea moluccana* (Blume) Orchard].

Verbesina lanceolata Poir., Encycl. **8**: 460 (1808). Type: 'J'ignore le lieu natal de cette plante. (V. s. in herb. Desfont.)' (FI006469 holotype – the material is localized as 'Brésil – *Commerson*').

Blainvillea rhomboidea Cass., J. Phys. Chim. Hist. Nat. Arts **96**: 216 (May 1823); —Dict. Sci. Nat., ed. 2, **29**: 494 (Dec 1823). Type: ' … cultivés au Jardin du Roi, …' (?P holotype).[98]

Blainvillea gayana Cass., Dict. Sci. Nat., ed. 2, **47**: 90 (1827). —de Candolle, Prodr. **5**: 492 (1996). —Oliver & Hiern in F.T.A. **3**: 375 (1877). —Adams in Webbia **12**(1): 220 (1956); —in Hepper, F.W.T.A., ed. 2, **2**: 237 (1963). Type: 'Nous avons fait cette description sur des échantillons secs, qui nous ont été libéralement donnés par M. Gay: ils provenaient de graines recueillies dans le Sénégal, envoyées à cet habile botaniste sous le nom d'*Ageratum* ou de *Bidens*, et semées par lui dans le Jardin du Luxembourg, où ces échantillons ont fleuri en Septembre 1826.' (P00069616 ?holotype, K000410221).

Blainvillea latifolia (L.f.) DC. ex Wight, Contr. Bot. India: 17 (1834)

Blainvillea rhomboidea var. *lanceolata* (Poir.) DC. Prodr. **5**: 492. (1-10 Oct. 1836).

Oligogyne burchellii Hook.f. in Icon. Pl. **2**: t. 101 (1837). Type: [Brazil:] 'Hab. Rio Janeiro. *Wm. J. Burchell*, Esq. (n. 12.)' (K001092319 holotype – mounted with a *Swainson* s.n. collection).

Eisenmannia clandestina Sch.Bip. in Flora **24**(1), Intell. 3: 42 (1841), nom. nud. (based on Kotschy's *Iter Nubicum* 191).

Blainvillea hispida Edgew. in Trans. Linn. Soc. London **20**: 70 (1846). Type: [India:] 'Han/ Himala, in arvis, alt. ped. 4000–5000. Junio.' [*Edgeworth* 79] (K001065683 holotype).

Blainvillea polycephala Gardner in London J. Bot. **7**: 89 (1848). Type: [Brazil:] '[*Gardner*] 6053. ... Hab. In dry bushy places near the city of Maranham. May, 1841.' (BM, K001092331 – ex Herb. Benthamianum, K001092332 – ex Herb. Hookerianum, US00128749 – a small portion of a flowering branch in a capsule, ex K, together with a photo of the type material, syntypes).

Blainvillea racemosa Gardner in London J. Bot. **7**: 89 (1848). Type: [Brazil:] '[*Gardner*] 1740. ... Hab. In dry, sandy, shady places near Villa do Icó, Proivince of Ceará. Aug. 1838.' (BM, K000054384 – ex Herb. Hookerianum, mounted with a *Gay* s.n. collection of cultivated material K001092329, K000054385 – ex Herb. Benthamianum, mounted with *Guillemin* 177, US00128750 – a photograph of the type material in K, together with 2 capitula and a bract in a capsule, syntypes).

Calyptocarpus burchellii (Hook.) Sch.Bip. in Bot. Zeitung **24**(21): 165 (1866).

Blainvillea rhomboidea sensu Oliv. & Hiern in F.T.A. **3**: 375 (1873).

Blainvillea rhomboidea var. *polycephala* (Gardner) Baker in Mart., Fl. Bras. **6**(3): 176 (1884).

Blainvillea rhomboidea var. *racemosa* (Gardner) Baker in Mart., Fl. Bras. **6**(3): 176 (1884).

Blainvillea gayana Cass. var. *lanceolata* Chiov. in Annuario Reale Ist. Bot. Roma **8**(2): 182 (1904). Type: [Eritrea:] 'Assaorta: Farras, Kankis-Adeita, 200 m., circa, 17. III. 1892 (*P*[*appi*] n. 2943).' (FT003742 holotype).

Wedelia gossweileri S.Moore in J. Bot. **56**(667): 232 (1918). Type: 'Angola, Libob; *Gossweiler*, 6388.' (BM000924359 holotype, COI00005596, LISC).

Wedelia triseta Peter in [Wasserpflanzen und Sumpfgewäsche in Deutsch-Ostafrika] Abh. Ges. Wiss. Göttingen, Math.-Phys. Kl. ser 2, **13**: 94 (1928). Types: [Tanzania:] 'Spare: Buiko →[right arrow] Hedaru 600 m ü. M., Juni (*P*[*eter*] 11066); Graspori am Panganifluß 6 km nördlich von Buiko 600 m, Mai (*P*[*eter*] 10436).' *Peter* 10436 (B100347129, B100347130 syntypes); *Peter* 11066 (B10 0347128 syntype).

---

[97] It is unknown whether a herbarium specimen was prepared. The plate could easily serve as the lectotype in the event no material is found.

[98] The generic and specific description that appeared in the *Dictionnaire* are copied from those that appeared in the *Journal* earlier in the same year, the latter being the place of valid publication.

Fig. 6.5.**20**. BLAINVILLEA DICHOTOMA. 1, flowering shoot (× 2/3); 2, capitulum (× 6); 3, ray floret corolla (× 16); 4, disc floret with enclosing palea (× 5); 5, disc floret corolla (× 4); 6, achene (× 5). 1, 3–5 from *Abdallah & Vollesen* 95/12, 2 from *Gillett* 13045, 6 from *Gilbert* 6274. Drawn by Juliet Williamson. From Flora of Tropical East Africa.

Erect annual herb, 0.3–1.2(2) m tall. Stems poorly- to much-branched, striate, patently scabrid. Leaves 3.0–12 × 1.5–8 cm, elliptic to ovate, ± membranous, base cuneate, 3-veined from base, margins serrate-crenate, rather sparsely pilose, apex tapering, acuminate. Pedicels to c. 1.7 cm long but often shorter, ± patently pilose. Capitula somewhat elongate-campanulate, c. 1.2 cm long and c. 0.5 cm diam.; phyllaries c. 10 mm long, oblong-elliptic to elliptic, pilose and glandular outside; paleae c. 7 mm. long, ± oblong, c. 10-longitudinally veined, apex lacerate-truncate, glandular. Ray florets 3–8, ray limb 1–3 mm long, whitish, glabrous, 2- or 3-dentate; corolla tube c. 2.2 mm long. Disc florets 6–8, throat campanulate, c. 1.4 mm long, tube slender c. 2.2 mm long; corolla lobes and tube base glandular; anthers c. 0.8 mm long, anther thecae dark. Achenes 3–5 mm long (disc achenes angled, to c. 6 mm long), trigonous, narrowly obovoid, brownish to blackish, setuliferous above; pappus of 1–3 weak setae, c. 1.5–2.5 mm long, antrorsely pilose.

Botswana. N: 15 km. N. of Tsau, 11.iii.1965, *Wild & Drummond* 6864 (K, SRGH). **Zambia**. B: Machili, 19.ii.1961, *Fanshawe* 6285 (EA, K, M, SRGH). W: Seshke Dist., Masese Forest Reserve, 21.iii.1984, *Brummitt* 16911 (K). C: Mpika Dist., Luangwa Valley, Mfuwe, 610 m, 13.iii.1969, *Astle* 5611 (K). **Zimbabwe**. N: Urungwe, N. bank of R. Mauore, c. 610 m, 1.iii.1958, *Phipps* 1011 (K, SRGH). W: Wankie Dist., Gwaai-Lutope Junction, 610 m, 27.ii.1963, *Wild* 6024 (K, SRGH). S: 45 km. on rd. to Tuli Breeding Sta. from Koodoovale Motel, 20.iii.1959, *Drummond* 5877 (EA, K, LD, LISC, SRGH). **Malawi**. N: Karonga Dist., Ngala, 460 m, 25.iv.1975, *Pawek* 9547 (K, MAL, MO, SRGH, UC). **Mozambique**. MS: Shiramba dunbe, 14.iv.1860, *Kirk* s.n. (K).

Most probably a native of tropical S America, in the Old World it is widely distributed at the lower altitudes in tropical Africa to the N of our area, through to southern Africa, Madagascar and into Asia where it is common. Shady sites in dry bushland or woodland, especially in disturbed areas, as well as damp marginal areas; 450–1400 m.

Conservation Status: A pantropic weed, often of cultivated areas; not threatened.

Wild (1967) suggested that Indian, and perhaps Madagascan, material of this genus, should be referred to *B. latifolia* (L.f.) DC. He recognized Brazilian material as referable to *B. rhomboidea* Cass. It was distinguished from it in having hemispherical rather than rather elongate-campanulate capitula, but Wild noted that there was 'apparently no sound character distinguishing the floral structures' of the taxa mentioned, and that perhaps recognizing them at subspecific level might be warranted. Wild (1967) only recorded the presence of *Blainvillea gayana* Cass. in the Flora region; the valid name for this material is *Blainvillea dichotoma* (Murr.) Cass. ex Hemsl. However, Boulos & Hind (2002) indicated that '*Spilanthes mauritiana* (Rich. ex Pers.) DC.' of Wild (1967) also represented this taxon but, in a full revision of the synonymy, indicated that the correct name for both taxa is *Blainvillea acmella* (L.) Philipson. The synonymy of the species in the Flora area is based on examination of several of the specimens from Gay's herbarium (largely from cultivated material) that are in K. It is clear that they are conspecific. However, Orchard (2012) has suggested that *B. gayana* is the only taxon present in the Flora area (but is also present in S America and with a narrow range in Australia), *B. acmella* restricted to northern Africa, the Indian subcontinent and Australia.

Further complications are evident with Adams (1956), and later Orchard's (2012) comments concerning the position of *Blainvillea prieureana* DC. Adams (1956: 246) placed the name under *Aspilia helianthoides* (Schumach. & Thonn.) Oliv. & Hiern subsp. *prieuriana* (DC.) C.D.Adams, a position accepted by Wild in Kirkia 5(2): 208 (1966). Wild (1967: 41) modified his concepts, removing *Aspilia helianthoides* subsp. *ciliata* (Schumach..) C.D.Adams from *A. helianthoides* and recognizing it as a separate species but with *Blainvillea prieuriana* in synonymy – a position maintained by Beentje in Beentje & Hind (2005: 748). However, Orchard (2012: 656) placed de Candolle's entity in synonymy under *Blainvillea gayana* Cass. but neglecting to provide a full synonymy or commentary. Apart from strikingly different leaves, capitula, and especially the outer phyllaries I have found no other literature suggesting this synonymy. *Blainvillea prieureana* has usually been placed in the synonymy of *Aspilia ciliata* (Schumach. & Thonn.) Wild in African floras – where it clearly belongs, q.v.

128. **ECLIPTA** L.

Eclipta L., Mant. Pl. Altera: 157, 286 (1771), nom. cons. —Wild in Kirkia **6**(1): 59–60 (1967). —Gibbs Russell in Kirkia **10**(2): 501 (1977). —McVaugh in Fl. Novo-Galiciana **12**: 314–315 (1984). —Hind in Fl. Masc., Composées **109**: 179–181 (1993). —Boulos & Hind in Fl. Egypt **3**: 230 (2002). —Mesfin Tadesse, Fl. Ethiopia & Eritrea **4**(2): 286 (2004). —Beentje & Hind in F.T.E.A., Compositae **3**: 728–730 (2005). —Umemoto & Koyama in Thai Forestry Bull. **35**: 108–118. —Chen Yousheng & Hind in Fl. China **20-21**: 869 (2011). —Orchard & Cross in Nuytsia **23**(1): 43–62 (2013).

Eupatoriophalacron Mill., Gard. Dict. Abr., ed. 4, (1754), nom. rej.

Micrelium Forssk., Fl. Aegypt.-Arab.: 152 (1775), nom. inval.[99]

Abasoloa La Llave & Lex., Nov. Veg. Descr. **1**: 11 (1824).

Paleista Raf., New. Fl. **2**: 43 (1837)[1836].

Clipteria Raf., New. Fl. **2**: 44 (1837)[1836].

Polygyne Phil. in Linnaea **33**(2): 170 (1864/65).[100]

Ecliptica Rumph. ex Kuntze, Revis. Gen. Pl. **1**: 334 (1891), nom. illeg., based on *Eclipta* L.

Annual or perennial herbs; rootstock short, creeping, rooting at nodes, strigose. Stems decumbent to erect, terete, pith narrowly fistulose or solid. Leaves opposite scarcely connate or not at nodes, petiolate or pseudopetiolate, lamina elliptic or narrowly oblong to narrowly ovate, lanceolate or linear, base acute or acuminate, margins entire to serrate, apices narrowly rounded to narrowly acuminate, surfaces strigose, without glandular dots, weakly 3-veined to subpinnate, secondary veins strongly ascending. Inflorescences terminal or axillary, 1–5 capitula per node, pedicels thin. Capitula broadly campanulate, heterogamous, radiate, corollas and achenes rapidly deciduous (mature capitula rarely intact in dry material); phyllaries 8–11, herbaceous, c. 2-seriate, subequal or inner phyllaries somewhat shorter, elliptical to obovate, acute, strigose; receptacle ± convex, paleaceous; paleae linear to filiform, apices slightly broadened. Ray florets 8–c. 50, 1–3-seriate, female and mostly fertile, corollas whitish to yellow, ray limb narrowly elliptical to linear, short-bilobed, smooth to mamillose on upper surface. Disc florets 8–c. 40, hermaphrodite, corollas greenish-white to yellowish, mostly glabrous outside, sometimes with few hairs on lobes, throat narrowly campanulate, usually 4-lobed, lobes broadly oblong-ovate, papillose; anther thecae usually slightly blackened, apical anther appendages broader than wide, with or without a single gland outside; style base not enlarged; style arm apices with tuft of hairs forming acute or truncate tip. Achenes maturing and falling quickly, obpyramidal, tetragonal or outer ray achenes triangular with spiculiform setulae near top, surface usually becoming tuberculate with age, winged or wingless; pappus absent, scarcely coroniform or with 2 or 3 awns.

A genus of c. 5 spp., one a very widespread pantropical, and pansubtropical, weed, *Eclipta prostrata* (L.) L., present in the Flora area.

Eclipta prostrata (L.) L., Mant. Pl. Altera: 286 (1771). —Exell, Cat. Vasc. Pl. S. Tomé: 225 (1944). —Andrews, Fl. Pl. Sudan **3**: 26, t. 4 (1956). —Adams in F.W.T.A., ed. 2, **2**: 241 (1963). —McVaugh in Fl. Novo-Galiciana **12**: 315–317 (1984). —Hind in Fl. Masc., Composées **109**: 179–181 (1993). —Boulos & Hind in Fl. Egypt **3**: 230 (2002). —Mesfin Tadesse, Fl. Ethiopia & Eritrea **4**(2): 286 (2004). —Beentje & Hind in F.T.E.A., Compositae **3**: 728–730 (2005). —Chen Yousheng & Hind in Fl. China **20-21**: 869 (2011). Type: 'Habitat in India' 'Chrysanthemum Maderaspatanum, Menthae arvensis folio & facie, floribus bigemellis, ad foliorum alas, pediculis curtis' in Plukenet, Phytographia: t. 118, f. 5. (1691); Almag. Bot.: 100 (1696), iconotype, lectotypified by Wijnands, Bot. Commelins: 74 (1983); typotype: Herb. Sloane 94: 175 (BM-SL). FIGURE 6.5.**21**.

---

[99] Two species were described, but no generic description was provided.

[100] *Index Kewensis* suggested that *Polygyne* was equated with *Plagiocheilus*. However, the description of *P. inconspicua* suggests it is conspecific with *Eclipta*, albeit a small plant.

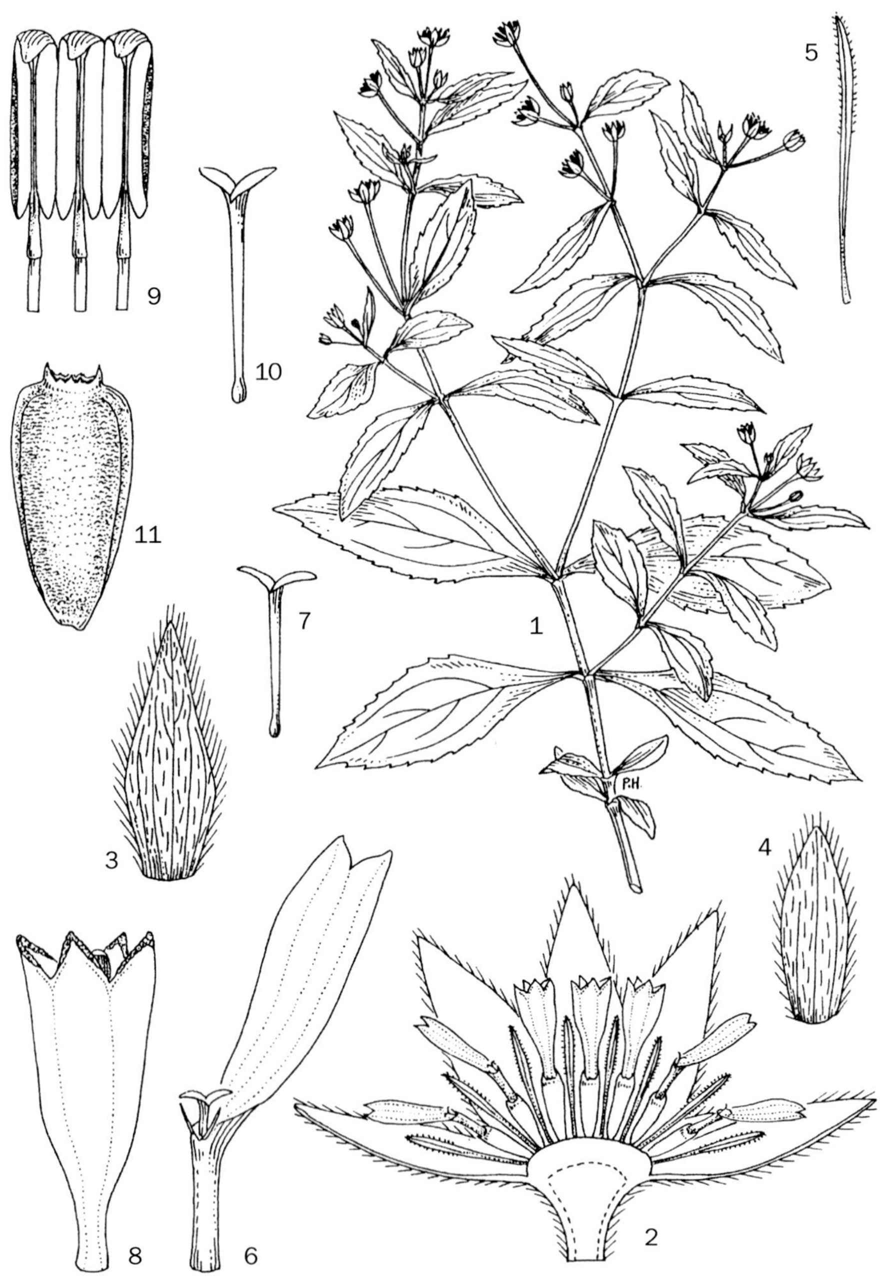

Fig. 6.5.**21**. ECLIPTA PROSTRATA. 1, flowering shoot (× ²/₃); 2, l.s. capitulum (× 12); 3, outer phyllary (× 6); 4, inner phyllary (× 6); 5, palea (× 18); 6, ray floret corolla (× 18); 7, style of ray floret (× 18); 8, disc floret corolla (× 24); 9, partial anther cylinder opened out (× 18); 10, style of disc floret (× 12); 11, achene (× 8). 1–7, 9–11 from *Morin* s.n.; 8 from *Guého* s.n. Drawn by Pat Halliday. From Flore des Mascareignes.

Verbesina pseudoacmella L., Sp. Pl. **2**: 901 (1753). Type: 'Habitat in Zeylona.' (BM000621911 – Herb. Hermann 3: 29, No. 308 lectotype), lectotypified by Koster & Philipson in Blumea **6**: 349, f. 2. (1950).

Verbesina alba L., Sp. Pl. **2**: 902 (1753). Type: 'Habitat in Virginia, Surinamo. F.' (LINN – Herb. Linn. No. 1020.1 lectotype), lectotypified by D'Arcy in Woodson & Schery in Ann. Missouri Bot. Gard. **62**: 1102 (1975).

Verbesina prostrata L., Sp. Pl. **2**: 902 (1753).

Verbesina conyzoides Trew, Pl. Rar.: 8 (1763). Type: 'Loco natalem ignoro, quum in horto reperi.' (?ER holotype).[101]

Bellis ramosa Jacq., Select. Stirp. Amer. Hist.: 216, t. 129 (1763). Type: 'Habitat in Domingo & Martinica, in pratensibus humidis & inundatis aeque maritimis atque aliis.' Herbarium material unknown.

Cotula alba (L.) L., Syst. Nat., ed. 12, **2**: 564 (1767).

Cotula oederi Murray, Prodr. Stirp. Gott.: 228 (1770). Type: 'Cel. OEDERI, Professoris Botanices Hafniensis, de horto nostro eximie meriti, nomine nouam speciem insignio, quoad a. 1769. semina eiusdem mihi misit. Annua. Initio Augusti in vaporario floriut, & semina matura reliquit.' Herbarium material unknown.[102]

Eclipta erecta L., Mant. Pl. Altera: 286 (1771), nom. illeg., based on *Verbesina alba* L.

Eclipta punctata L., Mant. Pl. Altera: 286 (1771), nom. illeg., an illegitimate replacement name for *Bellis ramosa* Jacq. (1760).

Spilanthes pseudoacmella (L.) L., Syst. Veg., ed. 13: 610 (1774).

Micrelium asteroides Forssk., Fl. Aegypt.-Arab.: LXXIV, 152[103](1775), nom. inval.[104] 'Type': [Egypt: Rashid] 'Rosettae' [*Forsskål* 1470] (C10002610 holotype, BM).

Micrelium tolak Forssk., Fl. Aegypt.-Arab.: 153 (1775), nom. inval. 'Type': [Yemen:] 'Hadîe. Arab. Talak.' [*Forsskål* s.n.] (C holotype[105], BM[106]).

Eclipta oederi (Murray) Weigel, Diss. Hort. Gryph.: 33 (1782).

Amellus? *carolinianus* Walter, Fl. Carol.: 213 (1788). Type: not stated.

Eclipta strumosa Salisb., Prodr.: 205 (1796), nom. illeg., based on *Eclipta prostrata* (L.) L.

Eclipta adpressa Moench, Suppl. Meth.: 245 (1802), nom. illeg., based on *Eclipta erecta* L.

Eclipta procumbens Michx., Fl. Bor. Amer. **2**: 129 (1803). Type: 'Hab. in Carolina.' (?P – if not lost, holotype).

Eclipta brachypoda Michx., Fl. Bor. Amer. **2**: 130 (1803), nom superfl., based on *Amellus?* *carolinianus* Walter.

Eclipta undulata Willd., Sp. Pl., ed. 4, **3**(3): 2219 (1803). Type: 'Habitat in India orientalis. 0. (v.s.)' (B-W16373 holotype).[107]

Verbesina pusilla Poir., Encycl. **8**: 459 (1808). Type: 'Cette plante a été recueillie à Porto-Ricco, par M. Ledru. (V. s. in herb. Desf.)' Herbarium material unknown.

Abasoloa taboada La Llave in La Llave & Lexarza, Nov. Veg. Descr. **1**: 11 (1824). Type: [Mexico:] 'Floret februario et martio in S. Jose del Corral ad ripas Huehueyapa. – *Llav*[*e*].' Herbarium material unknown.[108]

Acmella lanceolata Link ex Spreng., Syst. Veg., ed. 16, **3**: 444 (1826), nom. nud. pro syn. sub *Spilanthes pseudacmella* (L.) Spreng. (as '*Acmella lanceolata* Link var.'), non Raf. (1836: 52)[= ?].

---

[101] Trew's herbarium is in Erlangen University, ER.

[102] Following the Bourke Lambert sale in 1863, Murray's herbarium was split up, with material bought by several herbarium. The location of any material referrable to *Cotula oederi* is unknown.

[103] Species without epithet.

[104] At the time of description *Micrelium* contained two species, but no generic description was provided; both 'species' binomials are nom. inval.

[105] Herbarium Forsskålii is in C and plants covered in Forsskål's *Flora Aegyptiaco-Arabica* have been well documented by Hepper & Friis (1994).

[106] Robinson (2006: 136) cited the holotype as in BM – an unnumbered collection, which is counted as an isotype.

[107] This single sheet has '(*Roxburgh*)' written at the bottom of the sheet.

[108] Although La Llave's Mexican material is supposedly in G.

Eclipta palustris G.Forst. ex Spreng., Syst. Veg., ed. 16, **3**: 603 (1826). Type: 'Ins. maris pacifici.' [Habitat in India, *Forster*] (K001065758[109] lectotype, FR0031025), lectotypified by Nicolson & Fobserg, Regnum Veg. **139** (2003: 307).

Eclipta zippeliana Blume, Bijdr. Fl. Ned. Ind. **15**: 914 (Dec. 1826). Type: 'Crescit: in graminosis paludosis Bataviae ab Hortulano Zippelio primo detecta.' (L0001932, L0001933, L0001934 syntypes).

Eclipta linearis Otto ex Sweet, Hort. Brit., ed. 2: 308 (1830), nom. inval.[110]

Wiborgia? oblongifolia Hook., Bot. Misc. **2**: 226 (1831). Type: 'Hab. Lurin, near Lima. [*A. Cruckshanks*, s.n.]' (K000009860 ?holotype, E00433237, US00124481 – fragments).

Eclipta longifolia Schrad., Index Seminum (Göttingen) **1831**: 3 (1831); Linnaea **8**(Lit.): 23 (1833) – a duplication of the Index Seminum account. Type: not cited (G-DC-G00468615 – sent by Schrader, ex 'H. Gotting.' and is considered type material).[111]

Eclipta patula Schrad., Index Seminum (Göttingen) **1831**: 3 (1831); Linnaea **8**(Lit.): 23 (1833) – a duplication of the Index Seminum account. Type: 'Hujus semina haud raro prostratae nomine missa, quae diversissima.' (G-DC-G00468641 – sent by Schrader, ex 'H. Gotting.' and is considered type material).[111]

Eclipta dentata B.Heyne ex Wall., Numer. List (Wallich), n. 3211 (1831), nom. inval., non H.Lév. & Vaniot (1910: 11) [= *Wollastonia dentata* (H.Lév. & Vaniot) Orchard], based on Wall. Cat. No. 3211E.

Eclipta thermalis Bunge, Enum. Pl. China Bor.: 39 (1833) (also published as a separate from Mém. Sav. Étr. Acad. Pertersb. **2**: 113, 1835). Type: 'Hab. ad thermas prope Tan-schan. Floret Majo. ☉.' (?LE holotype, G00468637 – as 'No. 224', US1803835 = US00128630 – 2 portions of leafy flowering stems in a capsule).

Galinsoga? oblongifolia (Hook.) DC., Prodr. **5**: 677 (1836).

Eclipta erecta [var.] β *diffusa* DC., Prodr. **5**: 490 (1836). Type: '① in Senegaliae regione Walo cum var. erecta in argilosis inundatis legit cl. *Perrottet.* (v.s. comm. à cl. inv.)' *Perrottet* 'No 6'(G-DC-G00468611 syntype); *Perrottet* 'No 7' (G-DC-G00468610 syntype).

Eclipta longifolia Schrad. ex DC., Prodr. **5**: 490 (1836). Type: '① in insulis Caribaeis, Guadalupâ; fortè in Porto-Ricco, Domingo et Martinicâ si *Bellis ramosa* Jacq. amer. 216. t. 129 et ideò *E. punctata* Linn. mant. 2. p. 286 huc reitè referencdae? *E. erecta* Linn. mant. 286. (v.s.)' *Schrader* s.n. 1832 (G-DC-G00468615 syntype); *Bertero* (or *Balbis*) *s.n.* (G-DC-G00468614 syntype); s. coll. 151 (G-DC-G00468615 syntype).

Eclipta prostrata [var.] β *undulata* (Willd.) DC., Prodr. **5**: 490 (1836).

Eclipta parviflora Wall. ex DC., Prodr. **5**: 490 (1836). Types: '(Wall.! cat. et herb. n. 323), ... ① in Indiae orientalis humidis legerunt cl. *Heyne* et *Wight*. Hic etiam pertinere videtur E. prostrata D et E Wall.! cat. et herb. comp. n. 319 et Ecl. prostrata à *Chamisso* in Luzoniâ lecta. ... (v.s.)' *Wallich* 3209/319 (G-DC-G00468628, K-W-K001065764, K-W-K0010657660 – with a duplicate Numerical Catalogue label, syntypes).[112]

---

[109] This sheet in K has not been annotated by Nicolson, but is certainly housed amongst the Polynesian material.

[110] The plant was introduced in 1825 but from an unknown locality.

[111] No material has yet been found in GOET.

[112] There appear to be two sheets bearing a numbered label 319 that may well correspond to the 'Wallich' Herbarium material cited by de Candolle, but they do not bear the typical cut-up labels (from the 'Numerical List') that duplicates normally have. However, the supplementary numbers were added by de Candolle to the Compositae from this collection. The corresponding number in the 'Numerical List ...' (= 'Wallich Catalogue') is 3209; there are supplementary entries in the catalogue, A–E, of which the last two correspond to *Heyne* (D) and *Wight* (E), E has two collecting labels at the top indicating differing dates of collection, neither are numbered. Material from the 'Numerical List' corresponding to '323' bears the catalogue number 3213, under which there are three sheets, A, B, and C. An examination of the material in K-W shows A is ex Herb. Heyn, B was collected in 'Martabania' in 1827 (according to the Catalogue) and, C collected in Rampoore and numbered '150/Found in Rampoore [and also illegibly dated]'. Specimens B and C are individually determined as '323' on separate determination slips which appear to originate from de Candolle.

Eclipta procumbens [var.] β *patula* DC., Prodr. **5**: 491 (1836). Type: '① Patr. ign. Ecl. patula Schrad.! in litt. 1832. Buphthalmum diffusum Vahl herb. ex Puer.! herb. (v.s.)' *Schrader* s.n.(G-DCG00468641 holotype).[113]

Eclipta patula Schrad. ex DC., Prodr. **5**: 491 (1836), nom. nud. pro syn.

Buphthalmum diffusum Vahl ex DC., Prodr. **5**: 491 (1836), nom. nud. pro syn.

Eclipta palustris DC., Prodr. **5**: 491 (1836), nom. nud. pro syn.

Eclipta dubia Raf., New Fl. Amer. **2**: 40 (1836). Type: '–Virginia to Florida in gravelly soils, flowers estival: …' Herbarium material unknown.

Eclipta flexuosa Raf., New Fl. Amer. **2**: 41 (1836). Type: '– in Guyana and South America, biennial 2 or 3 feet high, thus totally unlike the two above.' Herbarium material unknown.

Eclipta tinctoria Raf., New Fl. Amer. **2**: 41 (1836). Type: '– in Asia and Egypt, used to die black, certainly different again from all the American sp. but requiring a better description.' Herbarium material unknown.

Eclipta simplex Raf., New Fl. Amer. **2**: 41 (1836). Type: '– sent to me from Alabama and Tennessee as *E. procumbens* although quite erect.' Herbarium material unknown.

Eclipta sulcata Raf., New Fl. Amer. **2**: 41 (1836). Type: '– Louisiana, sent me by *Riddell* as the *E. procumbens*? …' Herbarium material unknown.

Eclipta dichotoma Raf., New Fl. Amer. **2**: 42 (1836). Type: 'Arkanzas, found by *Nuttall*, mistaken also for *E. erecta*, …' Herbarium material unknown.

Eclipta pumila Raf., New Fl. Amer. **2**: 42 (1836). Type: '– Mts. Cumberland of East Kentucky, …' Herbarium material unknown.

Eclipta nutans Raf., New Fl. Amer. **2**: 42 (1836). Type: '– in Kentucky also the banks of the Ohio and Potiowmak, …' Herbarium material unknown.

Paleista procumbens [(Michx.)] Raf., New Fl. Amer. **2**: 43 (1836).[114]

Paleista? brachypoda (Michx.) Raf., New Fl. Amer. **2**: 43 (1836).

Brachypoda prostrata Raf., New Fl. Amer. **2**: 43 (1836), nom. inval., nom. nud. 'perhaps a peculiar G[enus]. of subgenus [of *Paleista*] to be called *Brachypoda prostrata*!', based on *Paleista procumbens*, etc.

Clipteria dichotoma Raf., New Fl. Amer. **2**: 44 (1836). Type: 'Sent to me anonymously from West Tennessee and the Chacta Country, stem bipedal, leaves and peduncles unical, flowers green and small. I had first called it *Eclipta levigata* [sic!], but it appears a peculiar genus by habit, even if there should be short rays, my specimens have none, but the dry Ecliptas seldom show them.' Herbarium material unknown.

Anthemis cotula Blanco, Fl. Filip.: 633 (1837), nom. illeg., non L. (1753: 894). Type: 'Esta planta mui comun, apenas es conocida de los naturales.' Herbarium material unkown.

Eclipta hirsuta Bartl. in Linnaea **13**(1): 95 (1839). Type: [Under the heading 'Index Seminum horti academici Gottingensis 1838. 4 to.] 'Pro *Ecl. latifolia* miss. ex *H. Francof.* 1838.' Type material probably in GOET.

Eclipta arabica Steud., Nomencl. Bot., ed. 2, **1**: 542 (1840), nom. illeg., superfl. pro *E. erecta* L., non Boiss. (1875: 249).

Eclipta marginata Steud., Nomencl. Bot., ed. 2, **1**: 542 (1840), nom. illeg., superfl. pro *E. erecta* L., non Boiss. (1875: 249).

Eclipta patula (Schrad. ex DC.) Endl., Cat. Horti Vindob. **1**: 335 (1842).

Eclipta erecta var. *brachypoda* (Michx.) Torr. & A.Gray, Fl. N. Amer. **2**(2): 269 (1842).

Wedelia psammophila Poepp., Nov. Gen. Sp. Pl. **3**(5–6): 50 (1843). Type: 'Crescit in insulis saepe inundatis arenosis fluminis Amazonum circum Coary. [1832, Poeppig s.n.]' (W0048673 holotype).

Eclipta longifolia var. *linearis* Fisch. & E.Mey., Index Seminum [St Petersburg (Petropolitanus)] **9**: 72 (1843). Types: 'Morison hist. III p. 47. sect. 6 tab. 13 fig 16 (excl. fig. dextr. angustifol.). Habemus ex h. Pragensi sub *E. linearis* nomine; in herbariis vidimis specimina simillima pr. Bahiam a D. Salzmann lecta.'[115]

---

[113] The material from Schrader (s.n., 1832) in G-DC is not separated from the general material of *Eclipta procumbens*.

[114] While based on Elliott's *Eclipta procumbens* (clearly *Eclipta procumbens* Michx.) Rafinesque combined it with *Eclipta brachypoda*, also presumably of Michaux (which was also recognized by Elliott). WCVP contended that this was 'nom. illeg.'

[115] It it unclear who the collector of the material seen in PR might have been, but the Salzmann collection is undoubtedly *Salzmann* 13 (as 'Compos. 13./ Rad. annua. Bahia in paludosis'), a widespread collection – e.g. G-DC-G00468606, K000053329, K000053330, P03824737.

Eclipta erecta [var.] γ *latifolia* Willd. ex Walp., Nov. Actorum Acad. Caes. Leop.-Carol. Nat. Cur. **19** (Suppl. 1): 266 (1843). Type: 'Willd., Hb. l.c. [= Willd. Hb. n. 16371.] fol. 1!. ... Brasilia: Rio de Ianero. (v.s.)' (B-W16371-01 holotype).

Artemisia viridis Blanco, Fl. Filip., ed. 2: 436 (1845). Type: 'Esta planta, muy comum, apenas es conocida de los naturales. ...'.[116]

Eclipta angustifolia C.Presl., Abh. Königl. Böhm. Ges. Wiss. ser. 5, **3**: 535 (1845), nom. nud., pro '*Eclypta erecta* Sieb.'

Eclipta angustifolia C.Presl, Bot. Bemerk.: 105 (1844)[1846], nom. nud., pro '*Eclypta erecta* Sieb.'

Eclipta alba (L.) Hassk., Pl. Rar. Jav.: 528 (1848).

Eclipta alba [forma] α *erecta* (L.) Hassk., Pl. Rar. Jav.: 528 (1848).

Eclipta alba forma β *longifolia* (Schrad. ex DC.) Hassk., Pl. Rar. Jav.: 530 (1848).

Eclipta alba forma *prostrata* (L.) Hassk., Pl. Rar. Jav.: 530 (1848).

Eclipta alba forma γ *zippeliana* (Blume) Hassk., Pl. Rar. Jav.: 530 (1848).

Eclipta procumbens var. *brachypoda* (Michx.) A.Gray, Manual (Gray): 218 (1848).

Eclipta alba var. α *erecta* (L.) Miq., Fl. Ned. Ind. Bat. **2**: 65 (1856).

Eclipta alba var. β *zippeliana* (Blume) Miq., Fl. Ind. Bat. **2**: 66 (1856).

Eclipta alba var. γ *prostrata* (L.) Miq., Fl. Ned. Ind. Bat. **2**: 66 (1856).

Eclipta alba var. δ *parviflora* (Wall. ex DC.) Miq., Fl. Ned. Ind. Bat. **2**: 66 (1856).

Polygyne inconspicua Phil. in Linnaea **33**(2): 171 (1864/65). Type: [Chile:] 'In litore lacus de Aculco dicti occurrit, autumno floret. [Philippi]' (SGO60523 holotype, LP).[117]

Eleutheranthera prostrata (L.) Sch.Bip., Bot. Zeit. (Berlin) **24**: 239 (1866).

Eclipta marginata Boiss., Fl. Orient. **3**: 249 (1875), nom. illeg., non Steudel (1840: 542)[= *Eclipta prostrata* (L.) L.]. Types: 'Hab. in humidis circa Lenkoran ad mare Caspium (Hoh[enacker] !), in maritimis prov. Ghilan Persiae (Auch[er]. 4766!). Valde affinis pracedenti a quâ notis indicatis differre videtur. Ar. Geogr. Insula Java (Zoll!).'[118]

Eclipta erecta var. *prostrata* (L.) Baker, Fl. Maurit. Seych.: 169 (1877).

Ecliptica alba (L.) Kuntze, Revis. Gen. Pl. **1**: 334 (1891), based on the illegitimate generic name *Ecliptica*.

Ecliptica alba β *zippeliana* (Blume) Kuntze, Revis. Gen. Pl. **1**: 334 (1891), based on the illegitimate generic name *Ecliptica*.

Ecliptica alba γ *prostrata* (L.) Kuntze, Revis. Gen. Pl. **1**: 334 (1891), based on the illegitimate generic name *Ecliptica*.

Ecliptica alba δ *parviflora* (Wall. ex DC.) Kuntze, Revis. Gen. Pl. **1**: 334 (1891), based on the illegitimate generic name *Ecliptica*.

Eupatoriophalacron album (L.) Hitchc. in Rep. (Annual) Missouri Bot. Gard. **4**: 99 (1893).

Eclipta alba var. *longifolia* Bettfr., Fl. Argent. **2**: 110, pl. 19(2), Lám.68 (1899), nom. illeg. non (Schrad. ex DC.) Hassk. (1848: 530). Type: 'Florece: Diciembre á Febrero. Palermo, Saavedra, San Isidro.' Herbarium material unknown.[119]

Eclipta philippinensis Gand. in Bull. Soc. Bot. France **65**: 40 (1918). Types: 'HAB.: Oceania, insulae Philippinae (*Cuming* n. 2436!); Nova Caledonia in planitie Dombea (*Debeaux*!).' *Debeaux* s.n. (?P syntype); *Cuming* 2436 (K001065763 syntype[120]).

Eclipta prostrata var. *zippeliana* (Blume) J.Kost. in Backer & Brink, Fl. Java **2**: 402 (1965), comb. inval.[121]

---

[116] The description is identical to *Anthemis cotula* Blanco above, and may well be effectively a new name for his *Anthemis cotula*.

[117] Muñoz Pizarro (1960) cited SGO60523 as the type.

[118] Boissier equated his taxon with Ledebour's *Eclipta prostrata* and 'prob. L. Mant. ex parte', other than it being a later homonym of Steudel's name.

[119] Type material not yet located, but Bettfreund's collections are stated to be in B and SI.

[120] The Kew syntype has had 'Phillip. I.' crossed out and 'Singapore' has been written underneath in pencil, suggesting somebody was doubtful over the collecting locality.

[121] Koster's footnote combination provided only the basionym and no direct reference to place of publication – contrary to Art. 33.4 of the Code.

Eclipta prostrata var. *aureoreticulata* (as '*aureo-reticulata*') Y.T.Chang in Wuyi Sci.J. **5**: 235 (1985). Type: [China:] 'Fujian: Xiamen nec non Tonganxian, ad regionem propre marginem maris, in arvis arenosis et udis, Hortis et ad marginem agrorum vel viarum. alt. 0–100 m', 5.ix.1984, *Tsai* 923 (FJSI holotype).

Eclipta prostrata var. *dixitii* An.Kumar & K.K.Khanna in J. Econ. Taxon. Bot. **23**(3): 713 (1999). Type: [India:] 'Atner, Betul district, Madhya Pradesh; ca. 600 m above msl.', 16.iv.1997, *Kumar* 50015 (CAL - as 50015A holotype, BSA ×2 - as '50015B' and '50015C').

Eclipta angustata Umemoto & H.Koyama in Thai Forest Bull., Bot. **35**: 114 (2007). Type: 'Thailand, Central, Phra Nakhon, Bang Khen, north of Bangkok, 13 Nov. 1965, *Iwatsuki* T-254' (KYO holotype, BKF, TNS).

Erect, ascending or decumbent annual (perhaps sometimes lasting more than one season, rarely a short-lived perennial) herb to 0.05–1 m tall. Stems usually solitary, simple or branched, striate or sulcate, appressed strigose, often rooting at nodes in prostrate to decumbent plants. Leaves opposite, short-petiolate, petiole to c. 3 mm. long, pubescent, lamina narrow, 2–12 × 0.3–3 cm, narrowly ovate, narrowly elliptic or elliptic, apex acute or subacute, base cuneate, weakly 3-veined from base, margins serrate or serrate-crenate, sometimes obscurely so, appressed strigose on both surfaces. Capitula solitary or few (2–3), axillary, heterogamous, radiate, at first 3–4 mm diam. × 2–3 mm tall, expanding to 7–12 mm diam. in fruit, pedicellate, pedicels 1–2.5(6) cm long, slender, densely whitish-pubescent; involucre hemispherical, appearing cupulate in fruit and retaining achenes; phyllaries 8–11, 3–6 × 1.5–2.5 mm, ovate, acuminate or acute, appressed-pubescent; receptacle ± flat at anthesis becoming slightly convex in fruit, paleaceous, paleae persistent, 2-3 mm long, outer linear, inner filiform, apex dilate, laciniate. Ray florets female, fertile, multi-seriate, c. 2.7 mm long; limb 1–2 mm long, linear-oblong, white, apex bidentate. Disc florets hermaphrodite, fertile, numerous, 2 mm long, corolla tubular and campanulate above, falling rapidly after anthesis (exposing green immature achenes), white, lobes 0.5 mm long, apices puberulent; anther thecae black, apical appendages incurved; style arms short and blunt. Achenes blackish, 1.6–2.5 mm long, oblong-cylindrical, trigonous, prominently rugose, ± setuliferous in upper half at least when immature, some achenes appearing smooth; pappus absent or visible as a puberulent rim when immature, setae absent.

Caprivi Strip. Linyanti, 27.xii.1958, *Killick & Leistner* 3143 (K, PRE, SRGH). **Botswana**. N: Lake Ngami, 3.i. 1963, *Smithers* in GHS 140,648 (SRGH). **Zambia**. B: Mongu, 7.i.1960, *Gilges* 861 (SRGH). N: Kawambwa, 10.xi.1957, *Fanshawe* 3890 (K). E: Near Petauke, Mvuvje R., 850 m, 5.xii.1958, *Robson* 844 (K, PRE, SRGH). C: Lusaka, 1220 m, 26.v.1953, *Best* 45 (K). S: Mazabuka, 6.iii.1963, *van Rensburg* 1603 (K, SRGH). **Zimbabwe**. N: Darwin, Mkumburu R., 512 m, 23.i.1960, *Phipps* 2402 (K, SRGH). W: Binga, Lake Kariba, 1.xii. 1960, *Crozier & Mwanza* 7621 (SRGH). C: Salisbury, Makabusi R., 29.vii.1931, *Gilliland* in GHS 5193 (SRGH). E: Chipinga, Sabi Exp. Sta., ix.1959, *Soane* 56 (SRGH). S: Gwanda, Bubye R., v.1955, *Davies* 1269 (SRGH). **Malawi**. N: Nyika, iii.1903, *McClounie* 55 (K). C: Dowa, Lake Nyasa, 1.viii.1951, *Chase* 3885 (BM, SRGH). S: Kasupe Dist., Liwonde National Park, banks of Shire River, 480 m, 20.iii.1977, *Brummitt et al.* 14886 (K). **Mozambique**. N: Nova Freixo, 3.viii.1934, *Torre* 428 (COI). Z: Lugela, Mocuba, Namagoa, 61-122 m, vii.1943, *Faulkner* [Pretoria No.] 139 (BM, COI, EA, K, PRE, SRGH). T: Boruma, Sisitso, R. Zambeze, 274 m, 15.vii.1950, *Chase* 2773 (BM, K, SRGH). MS: Vila Fontes, 1.vii.1946, *Simão* 681 (LM, SRGH). GI: Gaza, Chibuto, 19.vi.1960, *Lemos & Balsinhas* 148 (BM, COI, K, LMJ, PRE, SRGH). M: Goba, *Pimenta s.n.* (LM, SRGH).

A widespread weed of the warmer regions of the world, preferring damp or swampy situations (often a weed of irrigation schemes), or in cultivated areas. Also occurs naturally around pans, pond margins, seepage areas, and flood plains, etc.; 0–2100 m; flowering throughout the year.

Conservation Status: A widespread weed throughout its distribution; LC (Least Concern).

Plants are often conspicuous because of the greenish immature achenes and green leaves that turn black on drying.

129. **LIPOTRICHE** R.Br.

Lipotriche R.Br., Observ. Compositae: 118 (pre Sept. 1817); —in Trans Linn. Soc. London **12**(1): 118 (1817)[Feb. 1818]. —Orchard in Nuytsia **23**: 337–466 (2013). —Hind in Kew Bull. **69**(3)-9528: 1–13 (2014).

Melanthera Rohr in Skr. Naturhist.-Selsk. Kiobenh. **2**(1): 213 (1792). sensu lato. —Wild in Kirkia **5**(1): 1–17 (1965). —Wild in Kirkia **6**(1): 48–54 (1967). —Parks in Rhodora **75**(No. 802): 169–210 (1975). —Lisowski, (Asterac. Fl. Afr. Cent. 1) Fragm. Flor. Geobot. **36** Suppl. 1: 204–218 (1991). —Wagner & Robinson in Brittonia **53**(4): 539–561 (2001). —Mesfin Tadesse, Fl. Ethiopia & Eritrea **4**(2): 297–299 (2004). —Beentje & Hind in F.T.E.A., Compositae **3**: 737–744 (2005). —Panero in Kubitzki, Fam. Gen. Vasc. Pl. **8**: 454 [2006](2007).

Psathurochaeta DC., Prodr. **5**: 609 (1836).

Wuerschmittia Sch.Bip. in Flora **24**(1), Intell. 1: 27 (1841), as '*Würschmittia*', nom. nud.

Wuerschmittia Sch.Bip. ex Walp., Repert. Bot. Syst. **6**: 161 (3–5 Sept. 1846), as '*Würschmittia*'; —Richard, Tent. Fl. Abyss. **1**: 413 (Feb. 1848), as '*Wurschmittia*'.

Annual or perennial, sometimes scandent or climbing, herbs. Rootstock (frequently not collected) sometimes xylopodiaceous or rhizomatous and often with coarse thread-like secondary roots. Stems poorly- to well-branched, terete to quadrangular, variously pubescent, sometimes scabrid, sometimes glabrescent. Leaves opposite, sessile or petiolate, lamina basally cordate, truncate or cuneate, simple, oblong to lanceolate or ovate, sometimes basally lobed, 3-veined, surfaces usually scabrid, variously pubescent, margins entire or serrate, sometimes coarsely so, apices obtuse to acuminate. Inflorescences of solitary terminal capitula or of lax, few-headed, terminal cymes. Capitula heterogamous and radiate; involucre hemispherical to campanulate or obconical; phyllaries 2–3(4)-seriate, hispid-pubescent; receptacle convex, paleaceous, paleae linear to oblanceolate, conduplicate, membranous, often yellowish, midrib distinct, apices lacking apical crest, entire or with small lateral lobes. Ray florets pistillate or neuter, ray limb 3-toothed, pilose beneath on venation, yellow to orangeish-yellow, tube glabrous. Disc florets hermaphrodite, fertile, corollas glabrous, yellow, narrowed below, 5-lobed, lobes often pubescent; anther thecae black. Achenes slightly compressed, 2- or 3-angled, or cylindrical, sometimes obovoid, basally lacking elaiosomes, apices cuneate or truncate, apex setuliferous, setulae with long-acute apical cells; pappus of few to several caducous, antrorsely barbed awns or aristae.

A genus of 12 spp., widespread in Africa. *Melanthera* Rohr had gradually become larger and larger through accretion of many taxa in both the Old and New Worlds. Based on discoid capitula, florets with a white corolla, and black anther thecae as gross characters, for taxa from the Caribbean and S America, the genus had been expanded by Bentham (in Bentham & Hooker f., Gen. Pl., **2**(1), 1873), Wild (1965) and Wagner & Robinson (2001) to include various radiate, yellow rayed taxa from both major continents. Orchard (2013) has discussed the '*Wollastonia/Melanthera/Wedelia* generic complex' in relation to his work on the tribe Heliantheae for the *Flora of Australia*, and in conjunction, I have tackled part of the same complex for the Heliantheae treatment for *Flora Zambesiaca*. The conclusion was reached that the majority of African taxa previously placed in *Melanthera* are more appropriately placed in *Lipotriche* R.Br. I have made the remaining combinations (Hind, 2014) for the African taxa, and these names are used in this account. *Melanthera biflora* (L.) Wild is treated under *Wollastonia* in the present Heliantheae account.

1. Styles present in ray florets 2
– Styles absent in ray florets (i.e. neuter). 4
2. Pappus setae absent; leaf margins regularly serrate-crenate; phyllaries 8–12, ovate to elliptic. . . . . **2.** *richardsiae*
– Pappus setae 1–c. 15; leaf margins often irregularly toothed; phyllaries numerous, usually narrowly ovate 3
3. Pappus setae 1–4(5); leaf venation scarcely prominent; ray florets 10–12, ray limb c. 11 mm long . . . . **3.** *marlothiana*

 – Pappus setae 8–15; leaf venation very conspicuous beneath; ray florets up to c. 20, ray limb c. 17 mm long . . . . **4.** *scandens*

 4. Leaves sessile or very short-petiolate; much-branched herb to 2 m . . . . **1.** *abyssinica*

 – Leaves conspicuously petiolate; few-branched herbs to c. 1 m 5

 5. Ray florets c. 12 . . . .**6.** *robinsonii*

 – Ray florets up to 20 . . . .**5.** *pungens*

1. **Lipotriche abyssinica** (Sch.Bip. ex Walp.) Orchard in Nuytsia **23**: 383 (2013).[122] Types: [Ethiopia:] 'Herb. Schimp. it. Abyss. sect. I. no. 334 et 1533. In declivibus australibus regionis mediae montis Abyssinia Scholoda.' *Schimper* 334 (P0072980 – labelled 'Herb. E. Cosson, 18/Herb. Mus. Paris' lectotype[123], B100024123, BM000924415, BM000924416, BR8362797, GH00274778, GOET001821, HBG504884, JE00019607, K000410453 – ×2 – s.n., LG90024887, M0105208, P00072981, P00072982, S07-16902, S07-16903, TUB005508, WAG0025131), first-step lectotypification by Wagner & Robinson (2001: 556)[124], second-step lectotypification by Hind (2014: 3); *Schimper* 1533 (BR8363121, BR8876430, K000410454 – × 2 – s.n., M0105206, M0105207, MO694330, ?P, TUB005507syntypes).

   Würschmittia abyssinica Sch.Bip. in Flora **24**(1), Intell. 1, 2: 27 (1841), nom. nud.

   Wuerschmittia abyssinica Sch.Bip. ex Walp., Repert. Bot. Syst. **6**: 162 (3–5 Sept. 1846), as '*Würschmittia*'; —Richard, Tent. Fl. Abyss. **1**: 413 (Feb. 1848).

   Melanthera abyssinica (Sch.Bip. ex Walp.) Vatke in Linnaea **39**(6): 496 (1875).[125] —Adams in F.W.T.A., ed. 2, **2**: 240 (1963). —Beentje & Hind in F.T.E.A., Compositae **3**: 738 (2005).

   Amellus abyssinicus (Sch.Bip. ex Walp.) Kuntze, Revis. Gen. Pl. **1**: 306 (1891).

   Melanthera abyssinica var. *angustifolia* Chiov. in Annuario Reale Ist. Bot. Roma **8**(2): 183 (1904). Type: [Eritrea:] 'Mensa: Aba Maitan-Dadà, 1600–1400 m., 8. I. 1893 (*T*[*erracciano &*] *P*[*appi*] n. 1975 bis).' (FT003757 holotype).

   Melanthera djalonensis A.Chev. in Explor. Bot. Afrique Occ. Franc. **1**: 372 (1920), nom. nud. (based on *Chevalier* 18699 & 18829).

   Melanthera sokodensis Muschl. ex Hutch. & Dalziel in F.W.T.A. **2**: 146 (1931). Type: 'Togo: Sokode Farm, 1,000 ft. (Oct.)! Exsicc. *Schröder 73*.' (?K holotype).[126]

---

[122] i) The basioym is often cited as '*Wuerschmittia abyssinica* Sch.Bip. ex A.Rich., Tent. Fl. Abyss. **1**: 413' yet it is quite clear that Schultz Bipontinus' name was used in Walpers' Repertorium at least a year earlier as a descriptio generica-specifica; Walper also used Schultz Bipontinus's manuscripts for both many names and descriptions. To this end, an extra 'syntype', of Quartin Dillon, is mentioned, but is of no nomenclatural value –the Schimper collections are the only syntypes; ii) Orchard (2013: 383) appears to have been unaware of the *Schimper* 1533 syntype, although one such isosyntype was mentioned under *Schimper* 334; no mention was made of the lectotype.

[123] P00072980 – consisting of two sheets clearly originally pinned to one another, one with profuse notes by Schultz Bipontinus and only a small capsule of material referred to [*Schimper*] 332, the second sheet a flowering specimen simply labelled 'Herb. E. Cosson, 18/Herb. Mus. Paris' is the material chosen as the lectotype, not the first!). The barcode system unfortunately fails here, since one barcode applies to two sheets which have been databased as being *Schimper* 334 – mentioning nothing about Schultz Bipontinus's annotations of the two capsules as '332'.

[124] This is considered as first step lectotypification since they neither saw the sheet, nor were apparently aware of the presence of other duplicates.

[125] Attribution is made by Oliver & Hiern (1877) and much later by Wild (1965), and Mesfin Tadesse (2004), that this combination was made by Bentham & Hooker (1873: 377); no such combination was made, but only inferred. Aké Assi in Adansonia nouv. sér., **4**(2): 342 (1964) appears to have been the first to provide the correct authority.

[126] Wild (1965) suggested that the holotype was in K, although repeated searches have failed to find the material suggesting it might not have been at K, or is mounted with some other collection and curated elsewhere.

Perennial herb 1–2 m tall. Stem ± tetragonal, sulcate. Leaves sessile or short-petiolate (to 9 mm long), ovate to ovate-lanceolate, 2–12 ×1–5 cm, base cordate, margins serrate, apex acute or attenuate, 3-veined from base, scabrid on both surfaces. Inflorescences of terminal and solitary capitula or in lax few-headed corymbs, capitula pedicellate, pedicels 2–12 cm long, scabrid. Involucre hemispherical, becoming obconical in fruit, 6–8 mm high at anthesis, up to 15 × 12 mm in fruit; phyllaries 3–4-seriate, green, outer 6–8 mm long and strigose, inner yellowish white, 5–8 mm long, hispid outside; paleae yellowish white, acute at apex, hispid outside. Ray florets neuter, ray limbs orange-yellow, up to 15 × 5 mm, pilose beneath. Disc florets 3–5 mm long, corollas yellow; anthers yellow. Achenes dark brown to black, slightly quadrangular, 3–3.5 mm long, glabrous, surfaces convex; pappus of 2 caducous flat aristae 3.5 mm long and minute bristly scales.

Zambia. W: Kitwe, 29.xi.1967, *Mutimushi* JMM 2373 (K, SRGH). **Zimbabwe**. N: Gokwe, 2.xi.1963, *Bingham* 401 (K, SRGH).

Also known from Sierra Leone, Togo, D.R. Congo, Cameroon, Central African Republic, Guinea, Angola, Ethiopia, Eritrea, Sudan, Kenya, Tanzania, Yemen. In bushland, grassland; 500 m.

Conservation Status: A very widespread species, although poorly represented in the Flora area, and certainly requiring more collections; LC (Least Concern).

2. **Lipotriche richardsiae** (Wild) D.J.N.Hind in Kew Bull. **69**(3)-9528: 6 (2014). Type: [Zambia:] 'N. Rhodesia. –Abercorn, Chilongowelo, [alt. 4900 ft., 21.4.1952] *Richards* 1491.' (K000410446 – Sheet I, BR8877147, EA, K000410445 – Sheet II).[127]

   Melanthera richardsiae Wild in Kirkia 5(1): 9 (1965), as '*richardsae*'. —Beentje & Hind in F.T.E.A., Compositae **3**: 741 (2005).

Scrambling perennial herb to 2 m tall. Stems appressed-pilose, glabrescent, internodes shorter than leaves. Leaves petiolate, petiole to 3 cm long, white-pilose, lamina up to 8(10) × 4(6) cm, ovate, membranous, apex acuminate or acute, base abruptly cuneate, margins regularly serrate-crenate, appressed scabrid-pilose on both surfaces, 3-veined from base, veins scarcely prominent reticulate below. Inflorescences terminal, of lax (1)2–3-headed cymes, capitula pedicellate, pedicels to 10 cm long, appressed-pilose. Capitula 10–12 × 10 mm; phyllaries 2–3-seriate, subequal, c. 10 mm long, ovate to elliptic, apex acute, white-pilose outside; paleae c. 8.5 mm long, oblong, slightly keeled, apex acuminate, veins inconspicuous. Ray florets c. 10, female, limb 1.4 cm long, oblong-elliptic, apex 3-dentate; tube c. 2.7 mm long, glabrous, yellow. Disc floret corollas yellow, glabrous except for lobes, tube c. 7.3 mm. long, base constricted, 2.5 mm long. Achenes grey-black, 3.4 mm long, obovoid, slightly 3-angled, apex setuliferous; setae absent.

Zambia. N: Abercorn, Mbeshi River, 1680 m, 3.v.1957, *Richards* 9527 (BR, K).

Also in Tanzania, Uganda and Rwanda. Mushitu, thicket or forest edges, scrambling up trees and shrubs.

Conservation Status: Although widespread there are very few collections of this species both in our and neighbouring Floras – certainly more fieldwork is required; it is best recorded as DD (Data Deficient), although it may well be LC (Least Concern).

3. **Lipotriche marlothiana** (Klatt) D.J.N.Hind in Kew Bull. **69**(3)-9528: 4 (2014). Type: [Namibia:] 'Hereroland, Okahandja, in fruticetis, alt. 1200 m (*Marloth* n. 1332). –Florif. m. Majo 1886.' (B† holotype). FIGURE 6.5.**22**.

   Melanthera marlothiana O.Hoffm. in Bot. Jahrb. Syst. **10**(3): 277 (Oct. 1888).

   Wedelia triternata Klatt in Bull. Herb. Boissier **4**(12): 839 (Dec. 1896). Type: [Mozambique:] 'Sambesigebiet: Nhaondue; nicht selten an feuchten Stellen, *Menyhart*, Juli 1891, No. 735.' (GH00014085 holotype).

---

[127] The epithet correction, noted by Wagner & Robinson (2001: 557) is allowed under the Code (Art. 60.11, Note 4) as the species is named in honour of Mrs H. M. Richards.

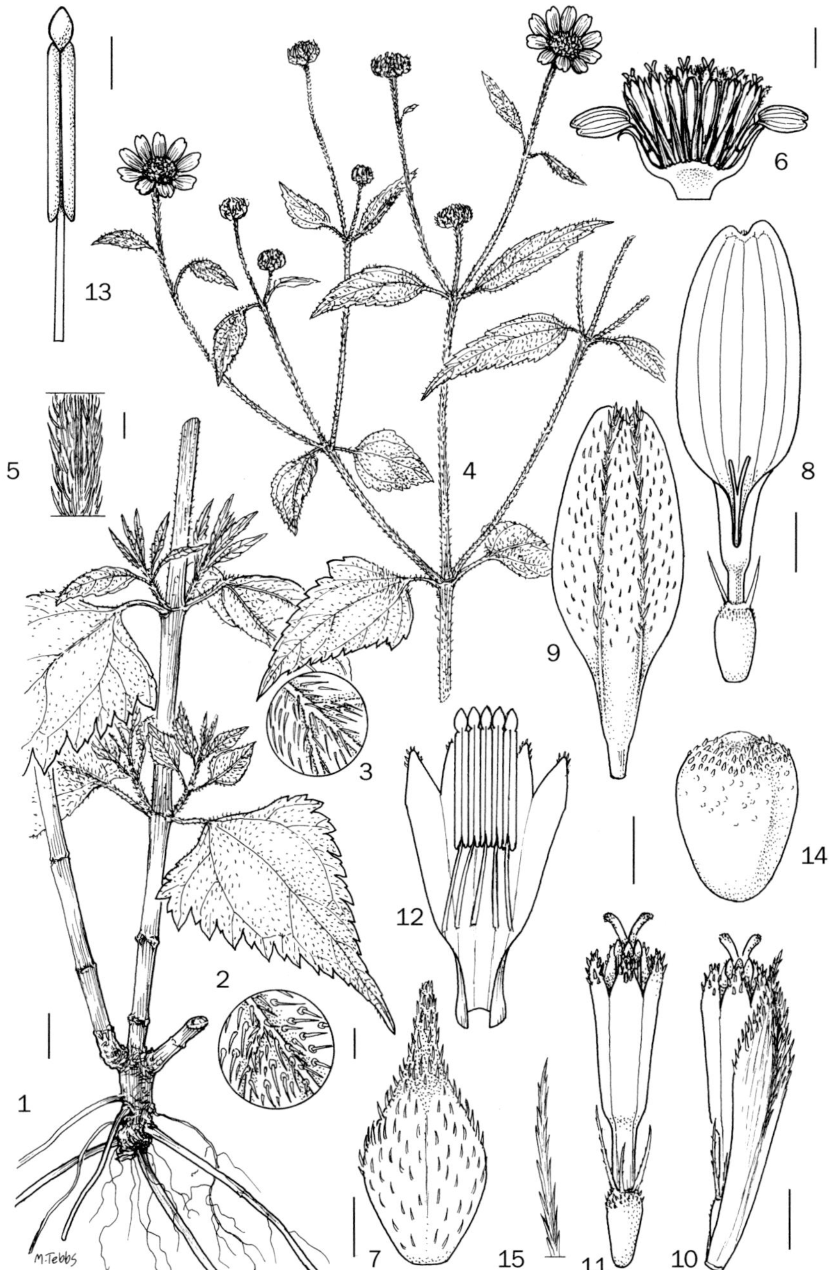

Fig. 6.5.**22**. LIPOTRICHE MARLOTHIANA. 1, basal portion of plant showing rootstock; 2, detail of upper leaf surface; 3, detail of lower leaf surface; 4, portion of flowering branch; 5, detail of pedicel showing indumentum; 6, l.s. capitulum; 7, phyllary; 8, ray floret; 9, lower surface of ray limb; 10, disc floret with accompanying palea; 11, disc floret; 12, disc floret corolla opened out showing attachment of filaments; 13, stamen; 14, achene without caducous pappus setae; 15, detail of pappus seta. 1–3, 5–13 from *Skarpe* 5-111; 4, 14–15 from *Smith* 1247. Scale bars: 13 = 0.5 mm; 2–3, 5, 7–12, 14 = 1 mm; 1, 4, 6 = 10 mm. Drawn by Margaret Tebbs.

Melanthera monochaeta Hiern, Cat. Afr. Pl. **1**(3): 581 (1898). —Mendonça, Contrib. Conhec. Fl. Angola **1** (Compositae): 96 (1943). Types: [Angola:] 'LOANDA. –An annual, subscandent herb. In thickets near Quicuxe; fl. and fr. July 1858. [*Welwitsch*] No. 3547. An apparently annual herb; stem suberect or frequently quasi-scandent among bushes; leaves bright green in the living state. In dry and also in damp thickets between Teba and Quicuxe; fl. and fr. August 1858. [*Welwitsch*] No. 3548. Benguella. –By thickets along the river Cotumbella on the side of the town of Benguella, sporadic; fl. and fr. June 1859. [*Welwitsch*] No. 3550. Mossameded. –On gravelly bushy places along the banks of the river Bero, at Quipola; fl. and fr. June 1860. [*Welwitsch*] No. 3551.'[128] *Welwitsch* 3551 (BM000924424 lectotype, COI00005608, K000410440, LISC219579), effectively lectotypified by Wild (1965: 10); *Welwitsch* 3548 (COI00005609, K000410441, LISU219577 syntypes); *Welwitsch* 3550 (BM000924425, LISU219578 syntypes).[129]

Melanthera varians Hiern, Cat. Afr. Pl. **1**(3): 580 (1898). —Mendonça, Contrib. Conhec. Fl. Angola **1** (Compositae): 97 (1943). Types: [Angola:] 'BARRA DE BENGO. –In swampy places by the river Bengo; fl. and fr. Dec. 1853. [*Welwitsch*] No. 3544. Golungo Alto. –By thickets and in the less dense forests; roadway near Menha-Lula; fl. and fr. July 1855. [*Welwitsch*] No. 3545. In bushy wooded parts of Sobato de Mussengue; fl. and fr. May 1856. [*Welwitsch*] No. 3546. Mossamedes. –Florets golden-yellow. In sandy thickets at the river Bero; fl. July 1859. The ovary of the disk-florets is more rounded than in the three previous Nos. [*Welwitsch*] No. 3543.' *Welwitsch* 3543 (BM000924429, LISU219571 syntypes); *Welwitsch* 3544 (BM000924427, BR8875341, K9000410443, LISU219572, M0105204, M0105205, P00093209 syntypes); *Welwitsch* 3545 (BM000924428, LISU219573 syntypes); *Welwitsch* 3546 (BM000924426, K000410442, LISU219574, LISU219575 syntypes).

Melanthera baumii O.Hoffm. in Warburg, Kunene-Sambesi-Exped.: 418 (1903). —Mendonça, Contrib. Conhec. Fl. Angola **1** (Compositae): 96 (1943). Type: [Angola:] 'Zwischen Goudkopje und Kakele, auf torfigem Boden, 1238 m ü. M. ([*Baum*] Nr. 197, blühend am 3. Oktober 1899.)' (B† holotype, BM000924430, COI00005607, E00239254, K000410444, W19010009468).

Melanthera schinziana S.Moore in Bull. Herb. Boissier, sér. 2, **4**(10): 1018 (1904). Type: [Namibia:] 'Südwest-Afrika: Amboland, Unkuanyama, Omupanda, *Wulfhorst*.' (BM000924423 holotype).

Melanthera rivicola Muschl. ex Dinter in Repert. Spec. Nov. Regni Veg. **19**: 96 (1923), nom. nud. (based on *Dinter* 95).

Melanthera seineri Muschl. ex Dinter in Repert. Spec. Nov. Regni Veg. **19**: 96 (1923), nom. nud. (based on *Dinter* I. 225).

Melanthera triternata (Klatt) Wild in Kirkia **5**(1): 9 (1965).[130]

Lipotriche triternata (Klatt) Orchard in Nuytsia **23**: 385 (2013).

Perennial herb c. 1 m tall. Stems ± 4-angled, sulcate, very sparsely scaberulous. Leaves petiolate, petiole 8–25 mm long, sparsely pilose, lamina up to 8 × 6.5 cm, ovate to broadly ovate, membranous or papyraceous, apex acute to acuminate, base broadly cuneate, or truncate and abruptly cuneate, margins coarsely and irregularly serrate or lobulate, sparsely scabrid-pilose on both sides, 3-veined from base, veins scarcely prominent. Inflorescences of lax terminal cymes, capitula pedicellate, pedicels up to c. 5 cm long but often less, appressed-pilose. Capitula up to 15 × 6 mm; phyllaries 2–3-seriate, c. 5.5 mm long, linear-ovate to ovate, apex acute, appressed-pilose outside; paleae 5–6 mm long, ovate, apex acute to acuminate, keel and upper part of margins ciliate, prominently longitudinally veined. Ray florets

---

[128] Hiern added the following observation: 'The above Nos. from the district of Loanda differ from the others, which must be considered the type of the species, by the rather smaller capitula, also by less truncate achenes (No. 3548), or by the presence of a hairy ring at the top of the otherwise glabrous ovary in addition to the solitary arista (No. 3547); they perhaps belong to distinct species.' pro parte excl. spec. *Welwitsch* 3547 et 3548.

[129] Since Hiern was working on the 'Study set' of the Welwitsch collections in BM it seems unnecessary to provide a second stage lectotypification.

[130] All too often, '*Melanthera triternata* (Klatt) Wild' has been cited as the accepted name for this taxon. Wild (1965: 9) originally made the combination clearly stating the later publication date of Klatt's name (1896), citing Hoffmann's earlier name (from 1889) immediately after in synonymy; I can find no explanation for this lapse, especially because of his comments at the end of this species account – Wild (1967) simply repeated this error, and Orchard (2013: 385) followed suit, accepting Wild, one assumes, in providing the relevant combination in *Lipotriche*.

female, c. 10–12, limb up to 11 mm long, yellow, oblong, apex 3-dentate, tube 1 mm. long, puberulous. Disc florets glabrous except for lobes with 1 or 2, c. 4.2 mm long, constricted base c. 1 mm long. Achenes c. 2 mm long, obovoid, slightly trigonous, apex puberulous; pappus setae 1–4(5).

Botswana. N: 19°02'S, 23°00'E on Jokomotschou Island off Moanachira River, 17.ii.1975, *Smith* 1247 (K, SRGH). SW: 22°23'S, 20°00'E, Okwa Valley, 1 km NE [of] the Namibian Border, 13.xii.1976, *Skarpe* 2-111 (K). **Zambia**. B: Sesheke, i.1922, *Borle* 341 (PRE, SRGH). E: Lusitu R., 19.v.1960, *Fanshawe* 5677 (K, SRGH). S: Victoria Falls, mid-vii.1906, *Kolbe* 3172 (?BOL, K). **Zimbabwe**. N: Sebungwe, 457 m, ix.1955, *Davies* 1477 (K, SRGH). W: Wankie, Victoria Falls Rain Forest, 903 m, 26.iii.1974, *Gonde* 86/74 (K, SRGH). **Mozambique**. T: Nhaondue, *Menyhart* 735 (GH – type).

Also in Angola, Namibia and South Africa (Transvaal). A species of swamp, *Acacia* savanna or in thickets or bushland.

Conservation Status: A widespread species; LC (Least Concern).

4. **Lipotriche scandens** (Schumach. & Thonn.) Orchard in Nuytsia **23**: 383 (2013). Type: not cited, mentioning only flowering and fruiting time: 'Almindelig, blomstrer i den frngtbare [sic!] Aarstid.'[131]

    Buphthalmum scandens Schumach. & Thonn., Beskr. Guin. Pl.: 392 (1827), non Vell. (1831)[= *Tilesia baccata* (L.) Pruski].

    Lipotriche brownii DC., Prodr. **5**: 544 (1836), as '*brownei*'. —Harvey in Harvey & Sonder, Fl. Cap. **3**: 133 (1865). Type: [?Central African Republic:] 'in Africa aequin. ad ripas fluminis Congo legit infeliciss. *Chr. Smith*. Caet. ign.' (G-DC holotype, BM, K000410455, P00073153).

    Melanthera brownii (DC.) Sch.Bip. in Flora **27**(2): 673 (1844). —Moore in J. Linn. Soc., Bot. **40**: 116 (1911). —Eyles in Trans. Roy. Soc. S. Afr. **5**: 515 (1916).

    Amellus scandens (Schumach. & Thonn.) Kuntze, Revis. Gen. Pl. **1**: 306 (1891).

    Melanthera scandens (Schumach. & Thonn.) Roberty in Bull. Inst. Franc. Afr. Noire, Ser. A, **16**: 68 (Jan. 1954). —Brenan in Mem. New York Bot. Gard. **8**(5): 480 (Feb. 1954). —Adams in F.W.T.A., ed. 2, **2**: 240 (1963). —Wild in Kirkia **5**(1): 5 (1965). —Wild in Kirkia **6**(1): 50 (1967).

    Melanthera scandens (Schumach. & Thonn.) Brenan in Mem. New York Bot. Gard. **8**(5): 480 (Feb. 1954), comb. superfl.

Herbs up to 3 m or more tall. Stems branched or ± unbranched, sometimes scandent, branches ± quadrangular, sulcate, scabrous. Leaves petiolate, petioles up to c. 5 cm or more long, scabrous, lamina deltate, broadly ovate, ovate, narrowly ovate or oblong, apex obtuse to acuminate, base truncate cordate or abruptly cuneate, basal angles often ± lobed, margins regularly or irregularly serrate-crenate to almost entire, from sparsely appressed scaberulous-pilose to subtomentose especially below, base strongly 3-veined, midrib and veins often very prominent below, venation strongly reticulate. Inflorescence of solitary terminal capitula or of few-headed lax corymbs, capitula pedicellate, pedicels scabrid, up to c. 18 cm. long. Capitula ovoid to depressed globose, c. 1.5 × 1 cm; phyllaries 2–3-seriate, 4–8 mm long, ovate to broadly ovate, apex subobtuse to acute, appressed pilose outside; paleae 4.5–7 mm long, obovate, abruptly acute, acuminate or slightly acuminate, ± puberulent above, keel and upper margins ciliate, strongly longitudinally veined. Ray florets up to c. 20, pistillate, ray limb up to c. 17 mm long, narrowly oblong, oblong or elliptic, orange yellow or yellow, apex 2 or 3-dentate or emarginate, tube up to 3.5 mm long, puberulous or glabrous. Disc florets glabrous except for lobes, up to 6 mm long, constricted base up to 2.7 mm long. Achenes up to 2.5 mm long, obovoid, somewhat trigonous, apex setuliferous; pappus setae usually about 10–15.

The species is found throughout Tropical Africa and Madagascar, but subsp. *scandens* is confined to the coastal regions of W Africa from Sierra Leone to Angola and is apparently absent from the Flora area.

---

[131] There are 2 *Thonning* 52 sheets in C (C10003431, C10003432) from 'Guinea' (= Ghana), the latter specimen was determined by Adams in 1955. Orchard (2013: 384) was of the opinion that the two sheets were syntypes.

Leaves typically narrowly oblong-ovate, basal angles more or less conspicuously lobed, sometimes lacking lobes, usually cordate . . . . a) subsp. *madagascariensis*
Leaves narrowly oblong-lanceolate, cuneate and basal angles usually lobed, although sometimes inconspicuously . . . . b) subsp. *dregei*

a) Subsp. **madagascariensis** (Baker) D.J.N. Hind in Kew Bull. **69**(3)-9528: 7 (2014). Types: [Madagascar:] '*Baron* 2344! 2534! *Humblot* 410!' *Baron* 2344 (K000410435, P00442917 syntypes); *Baron* 2534 (K, P00442916 syntypes); *Humblot* 410 (K, LD1217175, TAN000160 syntypes).[132]

   Melanthera madagascariensis Baker in J. Linn. Soc., Bot. **21**(137): 418 (1885). —Humbert, Fl. Madagasc. Composées **189**: 658, t. 120 fig. 1–4 (1963).

   Melanthera cuanzensis Hiern, Cat. Afr. Pl. **1**(3): 583 (1898). —Mendonça, Contrib. Conhec. Fl. Angola **1** (Compositae): 97 (1943). Type: [Angola:] 'PUNGO ANDONGO. By wooded thickets along the right bank of the river Cuanza; fl. and fr. Dec. 1856. [*Welwitsch*] No. 3554.' (BM000924434 holotype, K000410447, LISU219581, 219582, P00073176).

   Melanthera cuanzensis var. *altior* Hiern, Cat. Afr. Pl. **1**(3): (1898). —Mendonça, Contrib. Conhec. Fl. Angola **1** (Compositae): 97 (1943). Type: [Angola:] 'PUNGO ANDONGO. By streams in the wooded parts of Sobato Cabanga, growing in masses; fl. and fr. Jan. 1857. [*Welwitsch*] No. 3555.' (BM000924435 holotype, K000410448, LISU219583).[133]

   Amellus madagascariensis (Baker) Kuntze, Revis. Gen. Pl. **1**: 306 (1891).

   Melanthera swynnertonii S.Moore in J. Linn. Soc. Bot. **40**(275): 115 (1911). Type: [Mozambique:] '[Gazaland] Hab. Chibabava, Lower Buzi, 400 ft.; fl. Nov.; [*Swynnerton*] n. 1882.' (BM000924433 holotype).

   Melanthera scaberrima Hiern var. *angustifolia* S.Moore in J. Bot. **65**, Suppl. 2, Gamopet.: 58 (1927). Types: 'Angola: in wet places at Malange, [*Gossweiler*] 1204; among *Glumaceae* in flooded marshes of R. Cuito, [*Gossweiler*] 3194. Angola.' *Gossweiler* 1204 (BM000924431, K, P00073175 syntypes); *Gossweiler* 3194 (BM000924432, BR, COI00005606, K syntypes).

   Melanthera scandens subsp. *madagascariensis* (Baker) Wild in Kirkia **5**(1): 7 (1965). —Wild in Kirkia **6**(1): 51 (1967).

   Melanthera scandens subsp. *subsimplicifolia* Wild in Kirkia **5**(1): 6 (1965). —Wild in Kirkia **6**(1): 51 (1967). Type: 'Cameroons, Likomba, xii.1928, *Milbraed* 10752' (K holotype).

   Melanthera angustifolia Gilli in Ann. Naturhist. Mus. Wien **78**: 158 (1974), nom. illeg. non A.Rich. (1850). Type: [Tanzania:] '[*Gilli*] 626. Lichter Wald von \ nordöstlich von Lupingu, 870 m, 2.VIII.1958, soz. Aufn. XI, fr.' (W19730001155 holotype).

   Lipotriche scandens subsp. *subsimplicifolia* (Wild) Orchard in Nuytsia **23**: 384 (2013).

Also known from D.R. Congo, Angola, Ethiopia, Sudan, Kenya, Uganda, Tanzania, Madagascar, and Namibia.

Leaves typically narrowly oblong-ovate, basal angles more or less conspicuously lobed, sometimes lacking lobes, usually cordate, margins irregularly crenate-dentate, upper surface often somewhat bullate, under surface with prominently reticulate venation.

Botswana. N: Sepopa, Okavango R. 15.iii.1965, *Wild & Drummond* 7103 (SRGH). **Zambia**. N: Abercorn, i.1954, *Nash* 23 (BM). W: Ndola, 30.i.1954, *Fanshawe* 749 (K, NDO). C: Broken Hill, 20.viii.1958, *Seagrief* 3186 (SRGH). S: Katambora, 5–11.xi.1949, *West* 3038 (K, SRGH). **Zimbabwe**. C: Harare, 8.i.1984, *Levine* (K, SRGH). E: Umtali Dist., ravine in centre of Vumba Mts, 1220 m, 26.iv.1959, *Chase* 7111 (BM, K, SRGH). S: Belingwe Dist.,

---

[132] Lisowski (1991: 208) and Mesfin Tadesse (2004: 299) both indicated *Baron* 2344 as the 'holotype', although most subsequent authors have suggested only part of *Baron* 2344 is this entity; in this instance, since there is only one specimen of *Baron* 2344 (marked 'in part' by Baker) at K, this is considered the lectotype.

[133] The mimeographed specimen label on BM(000924435) is marked '*Lipotriche cuanzensis* Welw. β', and '(^{1}/3)' appears after the collector's number; the manuscript label in LISU reads the same.

Mount Buhwa, 1000 m, 2.v.1973, *Biegel* et al. 4253 (K, SRGH). **Malawi**. N: Songue & Kaunga, North Nyorsa, 518–610 m, vii.1896, *Whyte* 69 (K). C: Benga, west shore of Laye Nyasa, Kota-kota Dist., 470 m, 2.ix.1946, *Brass* 17501 (K, NY). S: Upper Shire by river, Fort Johnstone [sic!], 1893-1894, *Scott Elliot* 8428. **Mozambique**. N: Niassa, Vila Cabral, Meponda, 4.ix.1958, *Monteiro* 39 (K, LISM). T: Between Tette and the sea coast [Mazzaro] 21.iii.1860, *Kirk* s.n. (K). MS: Espungabera Mts, 610 m, 21.xi.1960, *Leach & Chase* 10512 (BM, COI, K, SRGH).

Also in the Sierra Leone, Liberia, Ivory Coast, Ghana, Cameroons, Fernando Po, Nigeria, D.R. Congo, Angola, Namibia, Sudan, Ethiopia, Kenya, Uganda, Tanzania, Madagascar. Usually in swamps or near rivers, amongst *Papyrus*, *Cyperus*, etc.

Conservation Status: A widespread taxon; LC (Least Concern).

Both *Swynnerton* 293 and *Chase* 7111 have leaves with slight, or conspicuous, basal lobes, and were cited by Wild (1965) as subsp. *subsimplicifolia* although no different from the bulk of material of subsp. *madagascariensis*. In several areas both lobed and unlobed leaved material can be found.

b) Subsp. **dregei** (DC.) Orchard in Nuytsia **23**: 384 (2013). Type: [South Africa:] 'in Africa australi ad Omsamcoubo, ferè ad maris altitudinem, detexit cl. *Drege*. ... α *latifolia* ... (v. s.) ... β *reticulata* ... (v. s.)'.[134]

   Psathurochaeta dregei DC., Prodr. **5**: 609 (1836).
   Trigonotheca natalensis Sch.Bip. in Flora **27**(2): 672 (1844), nom. nud. pro. syn. (based on *Lipotriche brownii* DC.)
   Melanthera scandens (Schumach. & Thonn.) Roberty subsp. *dregei* (DC.) Wild in Kirkia **5**(1): 8 (1965). —Wild in Kirkia **6**(1): 52 (1967).

Herb 1–2 m tall, sometimes scandent. Leaves narrowly oblong-lanceolate, cuneate and basal angles usually lobed, although sometimes inconspicuously.

Mozambique. GI: Inhambane, entre Morrumbene e Jangamo, 8.xi.1941, *Torre* 3854 (K, LISM). M: Lourenco Marques, 30.5 m, 1.xii.1897, Iter Secundum, *Schlechter* 11563 (BM, COI, K, PRE).

A largely coastal subspecies also in eSwatini, Natal and Cape Province. Often found in swamps with species of *Typha*, mangal woodland adjacent rivers, etc.; 0–600 m.

Conservation Status: A relatively widespread species, and often common in swampy areas, although very poorly represented in the Flora area; LC (Least Concern).

5. **Lipotriche pungens** (Oliv. & Hiern) Orchard in Nuytsia **23**: 383 (2013). Type: [Sudan:] 'Nile Land. Djur-land, *Schweinfurth*!' Material in K and BM give the following handwritten label data: 'Buschwaldung bei Seriba Ghattar. 28 Juni 1869, *Schweinfurth* 1990'; the top and

---

[134] De Candolle included within his new species two 'varieties' making no distinction as to which was typical. There appear to be two collections in G-DC, both numbered *Drège* 5121, each one representing one variety, 'α latifolia − 5121 Omsamcoubo', and 'β reticulata − 5121, [s.loc.]', both dated 1835. There are scattered Drège duplicates named as 'β reticulata' which may be syntypes, e.g. K000410426, K000410428, P00073165 and S07-16911. P00073161 is labelled as both α *latifolia* and β *reticulata* on a slip label, however the main label reads as α *latifolia* and is numbered '5121'. P00073162 is somewhat differently labelled, '14/12 32. Zw. Gebüsch bei Omsamcoubo, 900–1000'/ (V. b).' − but is unnumbered. P00073163 is labelled as α *latifolia*. P00073164 is labelled as β *reticulata* but bears both a date and locality: '25/2 32. Aüf [---] Grasfeldern am Gestade zw. Omtendo & Omsamcoubo 200' (V. c).' − c.f. Drège, Flora **26**(2): 1–230 (1843: 154). Orchard (2013: 384–385) has cited a 'holotype' for the species, and for each of the varieties, which I would disagree with on the basis of what is apparently in G-DC.

bottom of the label are preprinted 'Reise nach Central Africa im Auftrage der Humboldt Stiftung./im Lande der Djur ges. v. G. Schweinfurth.' (K000410451, K000410452 holotype/syntypes[135], BM000924413, M0105203, P00073056, P00073057).

Bushy rhizomatous perennial rarely reaching 1 m tall; branches quadrangular, appressed-scabrous-hairy. Leaves sessile or with petioles occasionally reaching 1 cm long; lamina up to c. 12 x 4 cm, from narrowly oblong to oblong, elliptic, or more typically narrowly ovate to ovate, apex acuminate or acute, base often subcordate or more rarely cuneate, margins usually coarsely serrate, appressed-scabrid pubescent on both surfaces, often paler below, 3-veined from base, veins prominent beneath. Inflorescences of solitary capitula at ends of long pedicels or of lax corymbs; pedicels up to 12 cm long, scabrid-pilose Capitula shallowly hemispherical, up to c. 1.5 × c. 0.7 cm; phyllaries 2–3-seriate, up to 5 mm long, oblong-ovate to broadly ovate, outer often shorter than inner, appressed-pilose and noticeably paler below; paleae 2.6–9 mm long, obovate to ovate, upper margins strongly ciliate, keel serrulate, strongly longitudinally veined and sometimes puberulent above. Ray florets up to c. 20, neuter, ray limb 6–11 mm long, oblong or elliptic, orange or yellow, apex 3-dentate, tube up to 1 mm long, glabrous. Disc floret corollas glabrous except for lobes, up to 5 mm long, constricted base 1 mm long. Achenes c. 2.3 mm long, somewhat trigonous, setuliferous above; setae 1(2) or absent.

A widespread species known from the Congo, Kenya, Uganda, Tanzania, Angola, Zambia, Zimbabwe, Malawi, Mozambique, Namibia (Caprivi Strip), and S Africa.

Leaves usually > 4 cm; palea mucronate or acute, lacking a caudate apex . . . . . a) subsp. *pungens*
Leaves < 4 cm; palea apex caudate-acuminate . . . . .b) subsp. *caudata*

a) Subsp. **pungens**

Melanthera pungens Oliv. & Hiern in F.T.A. **3**: 382 (1877). —Wild in Kirkia **5**(1): 14 (1965). —Lisowski, (Asterac. Fl. Afr. Cent. 1) Fragm. Flor. Geobot. **36** Suppl. 1: 216 (1991). —Beentje & Hind in F.T.E.A., Compositae **3**: 741–742 (2005).

Amellus pungens (Oliv. & Hiern) Kuntze, Revis. Gen. Pl. **1**: 306 (1891).

Melanthera albinervia O.Hoffm. in Bol. Soc. Brot. **13**: 30 (1896). Type: 'Angola, Quidumbo (*Anchieta* [49], [12] dezembro 1887).' (COI00005605 holotype, LISU).

Aspilia zombensis Baker in Bull. Misc. Inform., Kew **1898**(139): 152 (1898). Type: [Malawi:]'BRITISH CENTRAL AFRICA. Mount Zomba, alt. 4000–6000 ft., *Whyte* [s.n.]' (K000410438 holotype, BM, SAM0017630).

Melanthera acuminata S.Moore in J. Linn. Soc., Bot. **35**(245): 344 (1902). Type: [Kenya:] 'Hab. British East Africa, Kavirondo; *G. F. Scott Elliot*, no. 7052. [Ruwenzori Expedition – 1893-94].' (BM000924418 holotype, K000410449 – bearing the original jeweller's tag).

Aspilia zombensis var. *longifolia* S.Moore in J. Linn. Soc., Bot. **35**(245): 345 (1902). Types: [Malawi:] 'Nyassaland, 1895; *J. Buchanan*, no. 24. [Ruwenzori Expedition, 1893–1894.] Shire country; [Dec.] *G. F. Scott Elliot*, no. 8555.' *Buchanan* 24 (BM000924422, MO391642, MO391643, SAM0038658 syntypes); *Scott Elliot* 8555 (BM000924421, K000410439 – bearing the original jeweller's tag, which reads '855/Shire Highlands/Ndirani/Dec', referable to Ndirande, syntypes).

Aspilia eylesii S.Moore in J. Bot. **45**(530): 45 (1907). Type: [Zimbabwe:] 'Hab. Sebakwe; *F. Eyles*, 164.' (BM000924420 holotype, SRGH).

Melanthera ugandensis S.Moore in J. Bot. **54**(649): 256 (1916). Type: 'Hab. Uganda, grassland at Uamananzi; [4000 ft, June 1915] *Dümmer*, 2564.' (BM holotype, K000410450).

Melanthera gossweileri S.Moore in J. Bot. **65**, Suppl. 2, Gamopet.: 58 (1927). Type: 'Hab. ANGOLA: in "Mumua" woods near Forte P[rinceza]. Amelia, Cubango, [December 1906, *Gossweiler*] 4184.' (BM000924419 holotype, COI00005604[136]).

---

[135] Beentje in Beentje & Hind (2005) noted only that the holotype was in K; one of the two sheets bears the full printed and annotated label, the duplicate has the briefest information written on the sheet – both are marked 'bis'; the handwriting on the second sheet (and the 'bis' on the first sheet) is clearly that of Oliver. Orchard (2013: 383) was of the opinion that the duplicates were to be considered as syntypes.

[136] The COI isotype is quite different in the label data: 'Vivas; multicaule; Disseminada pela [?Hiemisilva] na Vila da Ponte, Ganguelas, Huila, Desembro de 1906.'

Melanthera letestui Philipson in J. Bot. **77**(915): 87 (1939). Type: [Central African Republic:] 'French Equatorial Africa: Oubangui-Chari; Upper Kotto Yalinga, 4 Aug. 1921, *Le Testu* 3055 (Type in herb. Le Testu; duplicate in BM000924414, P00073054, P00073055, US00125270 syntypes).[137]

Melanthera albinervia subsp. *acuminata* (S.Moore) Wild in Kirkia **5**(1): 12 (1965).

Melanthera pungens var. *albinervia* (O.Hoffm.) Beentje in Beentje & Hind, F.T.E.A., Compositae **3**: 742 (2005).

Leaves usually > 4 cm; paleae mucronate or acute, not having a caudate apex c. 1.7 mm long.

Caprivi Strip. Katima-Mulilo, *Killick & Leistner* 3062 (K, M, PRE, SRGH). **Zambia**. N: Abercorn, 4.i.1952, *Nash* 72 (BM). W: Ndola, 8.xii.1953, *Fanshawe* 553 (BR, K). E: Fort Jameson, iii.1952, *Benson* 17 (BM). S: Victoria Falls, *Sykes* 199 (SRGH). **Zimbabwe**. N: Darwin, Kandeya, 17.i.1960, *Phipps* 2315 (K, SRGH). W: Victoria Falls, 10.ii.1912, *Rogers* 5581 (BM). C: Marandellas, Grasslands, 29.i.1954, *Corby* 787 (K, SRGH). E: Umtali, 6.i.1955, *Chase* 5436 (K, SRGH). **Malawi**. S: Zomba, *Whyte* (BM, K). **Mozambique**. N: Vila Cabral, ii.1934, *Torre* 40 (COI). Z: Mocuba, iii.1944, *Faulkner* 48 (BM, EA, K, PRE).

Also in Angola, Tanzania and South Africa (Transvaal). Savanna woodlands and woodlands.

Conservation Status: A widespread species, although clearly absent in some of the Flora area, possibly through under-collecting; LC (Least Concern).

Beentje (in Beentje & Hind 2005) was of the opinion that *Melanthera albinervia* was nothing more than a variety of *M. pungens* there being few consistent differences between Wild's subspecies. *M. albinervia* subsp. *caudata* was only mentioned in the note under *M. pungens* var. *albinervia*. I have re-examined the types and other material concerned and agree with this view. However, I do not agree with many of the descriptions of the leaf pubescence. In much material the scabrid hairs are essentially adpressed, the basal portion often conspicuously verrucose but the apical cell (usually short- or long-acute) is smooth. In some material though the hairs are quite obviously erect and much less stock, whilst still possessing the verrucose basal portion and with a much longer, finer smooth apical cell. There appears to be no geographical distribution associated with the difference hair types, nor any apparent link with other characters.

There are three sheets in K from Zimbabwe and S Africa ('TV') that bear the determination, by Burtt Davy, of '*Wedelia holubii*' and are variously marked as 'co-type' or 'type'; the material is undoubtedly *Lipotriche albinervia*. In the public records that exist at Kew, the file on Burtt Davy's 'Flora of the Transvaal' [PRO 1/F/5, cat. no. 112492] indicates that the manuscript for part three of this work (parts I and II were published under the title *A manual of the flowering plants and ferns of the Transvaal ...*) was largely complete; it was never published (although it exists in part as galley proof and in part as type script), hence and new names in the family that were to appear in that part remained unpublished by Burtt Davy – along with the names of many dozens of other prospective new taxa. Burtt Davy's generic concept of *Wedelia* Jacq. was quite broad!

b) Subsp. **caudata** (Wild) D.J.N. Hind in Kew Bull. **69**(3)-9528: 5 (2014). Type: [Zambia:] 'N. Rhodesia. Mufulira, *Cruse* 224' (K ×2 holotype).[138]

---

[137] There seems to be great confusion as to where Le Testu's private herbarium is preserved, although it is most probably the material in P, even though these sheets are marked as isotypes it is likely that P00073054 could serve as lectotype; Philipson clearly worked with the material, and frequently stated that the duplicates were in BM.

[138] *Cruse* 224 at K was mounted as 'Sheet 1' and 'Sheet 2' marked clearly on the labels. However, only 'Sheet 1' (a single flowering specimen complete with rootstock) was sent on loan to Wild (upon which the description was based), and it bears the full collecting label and Wild's det slip indicating 'type'; 'Sheet 2' has a poor transcription onto a 'Herb. Hort. Bot. Reg. Kew.' label, and the material is of four flowering shoots. 'Sheet 2' could be considered part of the holotype, or an isotype sheet.

Melanthera albinervia O.Hoffm. subsp. *caudata* Wild in Kirkia 5(1): 12 (1965).

Leaves less than 4 cm long, 8–15 mm wide. Inflorescences of solitary, long-pedicellate capitula with pedicels 75–165 mm long. Paleae up to 9 mm long, apex caudate-acuminate c. 1.7 mm long.

Zambia. W: Mufulira, 28.xii.1947, *Cruse 224* (K).

Also in the neighbouring Katanga Province of the D.R. Congo. Woodland.

Conservation Status: This taxon is known from a handful of collections from the Copperbelt Province in the Flora area (e.g. *Cruse 63*, *Milne-Redhead* 3042, 3245), and one collection (cited by Wild) from the D.R. Congo and appears to be a relatively local species, although clearly requiring more fieldwork; it is best recorded as DD (Data Deficient), for the time being.

Beentje (in Beentje & Hind 2005) noted 'Subspecies *caudata* ...' but never made any formal combination under *Melanthera pungens* Oliv. & Hiern, nor suggested synonymizing it with a broader concept of *M. pungens*.

6. **Lipotriche robinsonii** (Wild) D.J.N. Hind in Kew Bull. **69**(3)-9528: 7 (2014). Type: [Zambia:] 'N. RHODESIA. Namwala, [1006 m,] 9.i.1957, *Robinson* 2099' (SRGH holotype, K).

    Melanthera robinsonii Wild in Kirkia 5(1): 15 (1965). —Wild in Kirkia 6(1): 53 (1967).

Erect perennial herb to 1 m tall. Stems quadrangular, strongly 4-sulcate, very sparsely appressed-scabrous-pubescent, few-branched. Leaves papyraceous, petiolate, petiole c. 1 cm long sparsely appressed-pilose, lamina up to 8 x 5 cm, ovate, apex acute or tapering to an obtuse apex, mucronate, base cuneate, margins shallowly crenate with mucronate crenations, sparsely appressed-scaberulous, 3-veined from base, veins slightly prominent below, venation reticulate below. Inflorescences terminal, of lax 3–6-headed corymbs; peduncles 1.3–2.3(9) cm long, appressed-pilose. Capitula ± hemispherical, c. 10 × 5–7 mm; phyllaries 2–3-seriate, subequal, ovate, c. 5 mm long, subacute or obtuse, ± pale at base, appressed-pilose outside; paleae c. 6.5 mm long, obovate, apex abruptly acute or mucronate, margins ciliate above, keel serrulate, prominently longitudinally nerved. Ray florets c. 12, neuter, limb c. 8 mm long, elliptic, yellow, apex 2–3-dentate; tube 1 mm long, glabrous. Disc florets glabrous except for the lobes, c. 5 mm long, constricted base 1 mm long. Achenes brown, c. 2.8 mm long, obovoid, slightly 4-angled, apex setuliferous; setae 1–4.

Zambia. N: Luwingu, Chambeshi R., 1189 m, 9.i.1962, *Astle* 1213 (SRGH). S: Kafue, 11.i.1963, *van Rensburg* 1196 (K, SRGH).

Also in D.R. Congo, and possibly Tanzania. [Beentje (Beentje & Hind 2005: 738) determined *Lazarus Thomas* 26 from Tanzania as *L. pungens*.] Swamp grasslands; 1405 m.

Conservation Status: A species with an apparently wide distribution but clearly with few collections throughout its distribution and requiring more fieldwork; DD (Data Deficient).

Wild's protologue description of the habit and longevity of the plant was possibly based upon the label data provided by Robinson. It is quite clear, looking at the type, that the remains of the flowering stems of the previous year are present; the species appears to have short stout rhizomes and wiry roots, and is undoubtedly perennial.

130. **WOLLASTONIA** DC. ex Decne.

Wollastonia DC. ex Decne. in Nouv. Ann. Mus. Hist. Nat. **3**: 414 (1834). —Fosberg & Sachet in Smithson. Contr. Bot. **45**: 30–32 (1980a); in Smithson. Contr. Bot. **46**: 61–62 (1980b). —Chen Yousheng & Hind in Fl. China **20-21**: 871 (2011). —Orchard in Nuytsia **23**: 337–466.

Perennial herbs or weak shrubs, with a resinous odour. Leaves opposite, lamina ovate, 3-veined. Inflorescence of solitary terminal capitula or of open few-headed paniculate-cymes. Capitula heterogamous and radiate; involucre hemispherical to campanulate; phyllaries biseriate, outer series much smaller than inner, inner enfolding ray floret achenes; receptacle convex, paleaceous, paleae acuminate, keeled. Ray florets female, uniseriate, limb yellow. Disc florets hermaphrodite, corollas yellow or greenish yellow, 5-lobed; anthers brown to black. Ray achenes cuneiform, 3-angled, base setuliferous, apex truncate; disc achenes compressed and obscurely 4-angled, base setuliferous; pappus absent or usually a single awn.

A genus of c. 2 spp. predominantly found in the Indo-Pacific area and largely coastal in distribution; one widespread species is present in the Flora area. Fosberg & Sachet (1980 a, b) have discussed the generic problems, and with the taxon represented in the Flora area. Recent studies by Orchard (pers. comm. & 2013) have confirmed my feelings for the 'Melanthera complex': that *Lipochaeta* DC. is best retained as the name for the Hawaiian taxa, *Melanthera* Rohr for the white corolla'd essentially Central and S American taxa, and taking up the available name of *Lipotriche* R.Br. for the African element.

Wollastonia biflora (L.) DC., Prodr. **5**: 546 (1836). —Fosberg & Sachet in Smithson. Contr. Bot. **45**: 32–34 (1980a); in Smithson. Contr. Bot. **46**: 62–67 (1980b). —Orchard in Nuytsia **23**: 360–364 (2013). Type: 'Habitat in India.' (LINN – Herb. Linn. 1021.4 lectotype), lectotypified by Wild (1965: 4). FIGURE 6.5.**23**.

    Verbesina biflora L., Sp. Pl., ed. 2, **2**: 1272 (1763).

    Acmella biflora (L.) Spreng., Syst. Veg., ed. 16, **3**: 591 (1826).

    Verbesina strigulosa Gaudich., Voy. Uranie, Bot.: 463 (1826)[1829]. Types?: 'In insulis Moluccis (Rawak, Waigiou, Pisang, Timor, &c.).' (?P syntypes).

    Verbesina canescens Gaudich., Voy. Uranie, Bot.: 463 (1826)[1829]. Type: 'In insulis Mariannis.' (P00710001 holotype).

    Verbesina argentea Gaudich., Voy. Uranie, Bot.: 463 (1826)[1829]. Type: 'In insulis Mariannis.' (P00123244 holotype).

    Wedelia aristata Less. in Linnaea **6**(1): 160 (1831), nom. illeg., citing *Verbesina strigulosa* Gaudich., etc., in synonymy.

    Wedelia chamissonis Less. in Linnaea **6**(1): 161 (1831), nom. illeg., citing *Verbesina canescens* Gaudich. in synonymy.

    Adenostemma biflorum (L.) Less., Syn. Gen. Compos.: 156 (1832).

    Wollastonia strigulosa (Gaudich.) DC. ex Decne. in Nouv. Ann. Mus. Hist. Nat. **3**: 414 (1834).

    Wollastonia scabriuscula DC. ex Decne. in Nouv. Ann. Mus. Hist. Nat. **3**: 414 (1834), nom. illeg. superfl.

    Wollastonia zanzibarensis DC., Prodr. **5**: 547 (1836). Type: 'in ins. Zanzibar legit cl. *Bojer. Anthemiopsis macrophylla* Bojer.! in litt. 1836. ... (v.s. comm. à cl. inv.)' (G-DC holotype, K000410424).

    Anthemiopsis macrophylla Bojer ex DC., Prodr. **5**: 547 (1836), nom. nud. pro syn.

    Wedelia biflora (L.) DC. in Wight, Contr. Bot. India: 18 (1834).[139]

    Wedelia strigulosa (Gaudich.) K.Schum in Bot. Jahrb. Syst. **9**(3): 223 (1887).[140]

    Wedelia scabriuscula (DC. ex Decne.) Engl. ex K.Schum. in Bot. Jahrb. Syst. **9**(3): 223 (1887), comb. illeg.

    Niebuhria biflora (L.) Britten in J. Bot. **39**: 69 (1901).

    Stemmodontia biflora (L.) W.Wight in Contr. U. S. Natl. Herb. **9**: 377 (1905).

    Wedelia canescens (Gaudich.) Merr. in Phil. J. Sci., C **9**: 155 (1914).

    Wedelia argentea (Gaudich.) Merr. in Phil. J. Sci., C **9**: 155 (1914).

    Wedelia biflora var. *scabriuscula* (DC. ex Decne.) Hochr. in Candollea **5**(2): 321 (1934).

    Wedelia biflora var. *canescens* (Gaudich.) Fosberg in Phytologia **5**(7): 291 (1955).

    Melanthera biflora (L.) Wild in Kirkia **5**(1): 4 (1965).

    Lipochaeta ovata R.C.Gardner in Rhodora **81**(827): 321 (1979). Type: 'Hawaii: Oahu: Honalulu, 1852, *N. J. Anderson* s.n.' (GB holotype).

    Wedelia wallichii Less. var. *megalantha* H. Chuang in Fl. Yunnan. **13**: 833 (2004). Type: [Chine:] 'Yunnan: Xichou ... *S. K. Wu* 61-3780' (KUN holotype).

    Wedelia biflora var. *ryukyuensis* H.Koyama in Acta Phytotax. Geobot. **33**: 245 (1982). Type: 'Japan, RYUKYU: Isl. Irioniote (*Koyama* et al. 177 ...)' (KYO holotype, TNS).

    Melanthera biflora var. *ryujyuensis* (H.Koyama) K.Ohashi & H.Ohashi in J. Jap. Bot. **85**(1): 59 (2010).

---

[139] As '*Wedelia biflora* (L.) Wight' in Contr. Bot. India: 18 (1837). —Oliver & Hiern in F.T.A. **3**: 376 (1877).

[140] Cited as 'DC.' for basionym author.

Fig. 6.5.**23**. WOLLASTONIA BIFLORA. 1, flowering shoot (× ²/₃); 2, capitulum (× 2); 3, ray floret corolla (× 8); 4, disc floret corolla (× 8); 5, achene (× 8). 1, 3–4 from *Drummond & Hemsley* 3987; 2 from *Jex-Blake* 2279; 5 from *Napier* 6233. Drawn by Juliet Williamson. From Flora of Tropical East Africa.

Subshrub or ± woody perennial herb to c. 1 m tall. Branches spreading, decumbent, glabrous or glabrescent, often ± compressed-sulcate. Leaves petiolate, petiole (1.2)2–5 cm long, glabrescent or very sparsely pilose, lamina slightly fleshy, 3–7(14) × 2–5(8) cm, ovate to subcircular, base cuneate to rounded, sparsely scabrid-pubescent above or glabrescent, a little more densely so below, venation slightly prominent below and reticulate, margins crenate or serrate-crenate, apex acute or obtuse. Inflorescence a lax, few-headed ((1)3–6), terminal corymb. Capitula radiate, yellow, 2–3 cm diam., depressed-globose; pedicels 1.5–5.5(8) cm long, slender or thick, glabrous or sparsely pilose; involucre 10–13 × 5–7 mm; phyllaries elliptic to obovate, c. 6 mm long, subequal, appressed-strigose; paleae c. 6 mm long, obovate, apex pubescent, obtuse, strongly longitudinally veined. Ray florets 14 or 15, limb oblong-elliptic 8–13 mm long, apex (2)3-dentate; style present, tube c. 1.3 mm long, glabrous. Disc florets 5–6.2 mm long with a constricted lower portion c. 1.2 mm long, corolla lobes 5. Achenes obovoid, 3-gonous, 3–3.5 × 2–2.5 mm, apex setuliferous, base cuneate; pappus setae 2 or 3, 2–2.5 mm, sometimes absent.

Mozambique. N: Goa Is., 19.v.1961, *Leach & Rutherford-Smith* 10928 (SRGH). M: Catembe, 23.iii.1952, *Barbosa & Balsinhas* 5290 (BM, LMJ).

Widespread from Kenya, Tanzania, along the E coast of Africa (Zanzibar) and the tropical coasts of Asia (India through to Japan) and northern Australia. Plant of coastal dunes, but restricted to coastal islands in Mozambique in the Flora area; flowering almost all year.

Conservation Status: Surprisingly poorly collected in the Flora area; LC (Least Concern).

A variable species throughout its range, but scarcely worth recognizing any of the variability into discreet taxa (cf. Fosberg & Sachet, 1980a, and b, under *Wollastonia*, where two varieties were recognized, and Wagner & Robinson, 2001, albeit under *Melanthera biflora* as one variable species).

131. **ASPILIA** Thouars

Aspilia Thouars, Gen. Nov. Madagasc.: 12(1806). —Adams in Webbia **12**(1): 217–250 (1956). —Wild in Kirkia **5**(2): 197–228 (1966). —Wild in Kirkia **6**(1): 39–48 (1967). — Lisowski, (Asterac. Fl. Afr. Cent. 1) Fragm. Fl. Geobot. **36** Suppl. 1: 184–204 (1991). —Robinson in Phytologia **72**(2): 144–151 (1992). —Turner in Phytologia **72**(5): 389–395 (1992) (New World taxa). —Mesfin Tadesse in Fl. Ethiopia & Eritrea **4**(2): 293 (2004). — Beentje & Hind in F.T.E.A., Compositae **3**: 746–754 (2005). —Orchard in Austrobaileya **8**(4): 653–669 (2012). —Orchard in Nuytsia **23**: 337–466 (2013).

Coronocarpus Schumach. & Thonn. in Beskr. Guin. Pl.: 393 (1827).

Wirtgenia Sch.Bip. in Flora **25**(2): 435 (1842).

Dipterotheca Sch.Bip. in Flora **25**(2): 434 (1842).

Menotriche Steetze in Peters, Naturw. Reise Mossambique **6** (Bot. 2): 475 (1864).

Annual or perennial herbs, rarely suffruticose; rootstock of perennial usually xylopodiaceous. Stems erect or prostrate, simple or branched, variously pubescent, usually hispid. Leaves opposite, petiolate or sessile, lamina ovate, narrowly ovate, linear-oblong, scabrid or hispid. Inflorescences of solitary terminal capitula or of few-headed lax cymes or ± sessile condensed corymbs in upper axils. Capitula heterogamous, radiate; involucre campanulate to shallowly hemispherical; phyllaries 2–3-seriate, outer usually foliaceous and herbaceous, inner scarious at least at base; receptacle convex, paleaceous, paleae narrowly oblong to obovate, conduplicate, keeled, usually indistinctly veined. Ray florets 10–15, uniseriate, neuter and without style or apparently female with style but not producing mature achenes; ray limbs conspicuous yellow, white, lilac or various shades of purple, apex emarginate or 3-toothed. Disc florets hermaphrodite, numerous, corollas tubular, lower part ± abruptly narrower, glabrous or puberulous, corolla lobes 5; apical anther appendages yellow or blackish, anther thecae usually blackish, anther bases obtuse or shortly sagittate; style arms gradually narrowing to an acuminate apex or thickened just below apex, margins pubescent along outside at least in upper part, apex acute. Achenes obovoid and bilaterally compressed when mature (ovaries usually somewhat 4-angled and subcylindrical), setuliferous, setulae usually of twin-hairs with divided apices; pappus a laciniate or ciliate margined cupule often with (1)2(3) barbellate setae.

A problematic genus, relatively recently suffering wholesale transfer to *Wedelia*, initiated by Robinson (1992 – the type), Turner (1992 – S American taxa), and followed by Isawumi in Feddes Repert. **108**(7–8): 549–565 (1997 – taxa predominantly found in W Africa) and Soldano in Compositae Newslett. **32**: 17–22 (1998). Unfortunately, Isawumi's attempt at several new combinations has resulted in a number of invalidly published combinations, etc., for several taxa found in the Flora area. More recently, Orchard (2013) has concurred with Robinson and Turner and considered *Aspilia* to be a synonym of *Wedelia*, having apparently only studied African material based on one collection of *A. thouarsii* DC. The present author's view is that Old World taxa can safely be left in *Aspilia*, New World aspilias belong elsewhere (the necessary combination have yet to be made), and the genus *Wedelia* Jacq. can be restricted to essentially New World taxa separable from the New World aspilias. It is clear that Beentje in Beentje & Hind (2005), Mesfin Tadesse (2004) and Lisowski (1991) had variable concepts of some taxa, notably *A. africana* and *A. mossambicensis*, with differing synonymy or the exclusion of various names; clearly much more detailed work still needs to be done on the species concepts. Orchard's concept of *Blainvillea gayana* (Orchard 2012) also included *Blainvillea prieureana* DC. – considered as a synonym of *Aspilia ciliata* in the account below.

The following key is largely based on Wild (1966, 1967).

1. Ray limbs yellow 3
– Ray limbs purple, purplish-brown, crimson, maroon, rarely pink or rarely white 2
2. Capitula pedicellate; apical anther appendages black; ray limbs white or purple (but outside of the Flora area); disc floret corollas yellow. . . . . **2.** *ciliata*
– Capitula usually clustered and subsessile or sometimes short-pedicellate (if distinctly pedicellate then disc floret corollas purple); apical anther appendages purple; ray limbs purple, purple-brown, crimson, maroon, rarely pink or white; disc floret corollas purple . . . . **1.** *kostchyi*
3. Plants annual . . . . **2.** *ciliata*
– Plants perennial 4
4. Paleae caudate-acuminate (especially in fruit then exceeding phyllaries) 5
– Paleae obtuse or acute, rarely acuminate. 6
5. Phyllaries subequal, biseriate, ovate; leaves narrow-oblong to elliptic. . . . . **3.** *mendoncae*
– Phyllaries gradate, 3-seriate, narrow-oblong or very narrowly triangular; leaves ovate to broadly ovate . . . . **4.** *pluriseta*
6. Phyllaries usually oblong, apices mostly obtuse; capitula solitary on pedicels; anther-appendages yellow. . . . . **5.** *natalensis*
– Phyllaries linear to ovate; apices ± acute; inflorescence usually corymbose; anther-appendages dark or yellow. 7
7. Leaves ± sessile; apical anther appendages dark; paleae keel concolorous with margins . . . . **6.** *angolensis*
– Leaves petiolate or rarely subsessile; apical anther appendages yellow; paleae often with a deep purple keel . . . . **7.** *mossambicensis*

1. **Aspilia kotschyi** (Sch.Bip.) Oliv. in Trans. Linn. Soc., London **29**: 98 (1873). —Oliver & Hiern in F.T.A. **3**: 381 (1877). —Muschler in Bot. Jahrb. Syst. **50** (Suppl.): 342 (1914). —Hutchinson & Dalziel in F.W.T.A. **2**: 146 (1931). —Adams in Webbia **12**(1): 241 (1956); —in Hepper, F.W.T.A., ed. 2, 2:239 (1963). —Wild in Kirkia **5**(2): 202 (1966). —Mesfin Tadesse in Fl. Ethiopia & Eritrea **4**(2): 294 (2004). —Beentje & Hind in F.T.E.A., Compositae **3**: 747 (2005). Type: [Sudan:] 'in Kotschyii it. nubico n. 103. an. 1841.)'[In umbrosis inter rupes montis Cordofani Arasch-Cool d. 1. Oct. 1839.] (P00072735 holotype, BM000798664, BR8873460, HAL0111841, HBG504024, K000410296, K000410298, LD1213345, M0105465, NY00168240, NY00168241, P00072736, P00167569, S07-16795, S07-16797, STU000414, TUB005495, TUB005496, TUB005497, TUB005498, WAG0000510). FIGURE 6.5.**24**.

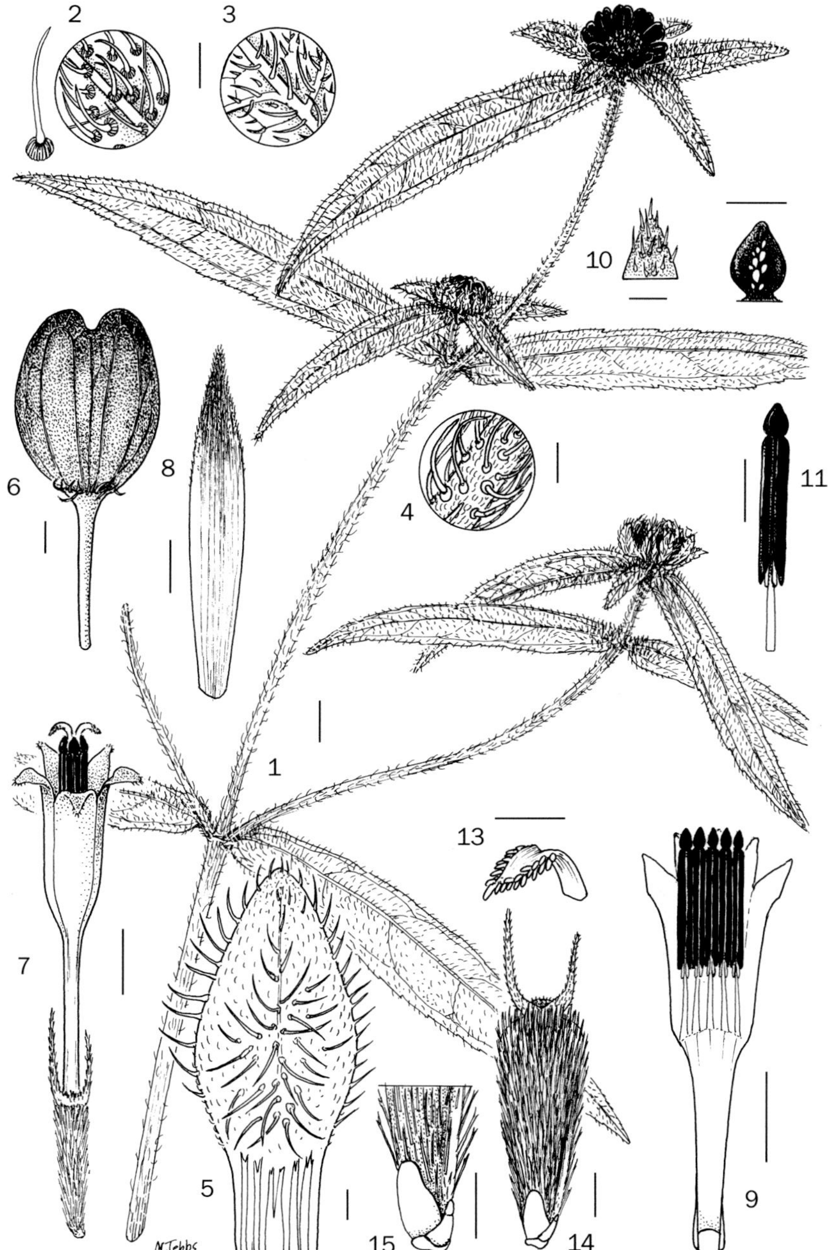

Fig. 6.5.**24**. ASPILIA KOTSCHYI var. KOTSCHYI. 1, flowering branch; 2, detail of upper leaf surface and single hair; 3, detail of lower leaf surface; 4, detail of stem pubescence; 5, phyllary; 6, ray limb; 7, disc floret; 8, palea; 9, disc floret corolla opened out showing attachment of stamen filaments; 10, detail of outer surface of disc floret corolla lobe; 11, stamen, noting black pigmentation; 12, detail of abaxial view of concave apical anther appendage showing glands; 13, detail of disc floret style arm; 14, disc floret achene; 15, detail of base of disc floret achene showing elaiosome. 1, 8 from *Zimba et al.* 931; 2–7, 9–13 from *White* 7012; 14–15 from *Nshingo* 222. Scale bars: 11–13 = 0.5 mm; 2–9 = 1 mm; 1 = 10 mm. Drawn by Margaret Tebbs

Dipterotheca kotschyi Sch.Bip. in Flora **24**(1): Intell. 1: 42 (1841), nom. nud.[141]

Dipterotheca kotschyi Sch.Bip. in Flora **25**(2, Nr. 28): 435 (1842), as genero-specific description.

Wirtgenia kotschyi (Sch.Bip.) A.Rich., Tent. Fl. Abyss. **1**: 412 (1848).

Coronocarpus kotschyi (Sch.Bip.) Benth. in Hooker & Bentham, Niger Fl.: 433 (1849).

Aspilia kotschyi (Sch.Bip.) B.D.Jacks. in Index Kewensis **1**(1): 217 (1893), as '*kotschyana*', comb. superfl.

Aspilia courtetii O.Hoffm. & Muschl., Bull. Soc. Bot. France **57** (Mém. 2 (8c)): 116 (1910). Types: [Chad:] 'Bas-Chari. –Baguirmi entre Madfata et Bousso, 1–10 novembre 1903, nos 10438 et 10439 [*Aug. Chevalier*].' *Chevalier* 10438 (P00072831 syntype); *Chevalier* 10439 (P00072832 syntype).

Wedelia kotschyi (Sch.Bip.) Isawumi, Feddes Repert. **108**(7–8): 559 (1997), comb. illeg.[142]

Wedelia kotschyi (Sch.Bip.) Soldano in Compositae Newslett. **32**: 20 (1998).

Erect annual herb to 0.3–1.8 m tall. Stems simple or few-branched, terete, scabrid or hispid, patently hairy. Leaves sessile or very shortly petiolate, 3.5–20 × 0.6–5 cm, broad-linear, narrow-ovate or narrow-elliptic, rarely lanceolate, strigose or hispid pubescent, hairs longer below especially on venation, base truncate or subamplexicaul, rarely cuneate, margins shallowly serrate-crenate to subentire, apex acute to acuminate. Inflorescence of solitary axillary, usually subsessile to short-pedicellate capitula; pedicels short (up to c. 2.5 cm long), rarely to c. 7 cm, densely hispid. Capitula hemispherical or campanulate, c. 2–4 × 2 cm at anthesis; phyllaries narrowly ovate, foliose and often exceeding disc florets, inner to c. 14 mm long, ovate, acute, long-pilose, hairs spreading, phyllaries diminishing in size towards inside, eventually patent to reflexed when achenes lost, phyllaries and paleae eventually becoming pale brown; receptacle flat to convex (especially at maturity), paleaceous, paleae persistent, c. 10 mm long, obovate, narrowing abruptly into an obtuse or acute often purple-tinged apex, keel tip short-scabrid, margins laciniate. Ray florets 2–5(11), corollas purple, purple-brown, more rarely white or very rarely pink, ray limb ovate to oblong c. 5 mm. long; tube glabrous, narrow, c. 6 mm long, glabrous; styles absent. Disc florets tubular, corollas usually dark purple, rarely pale, c. 6.5 mm long, constricted lower portion c. 4 mm long, glabrous except for lobes, lobes spiculiferous outside; apical anther appendages dark, often purple; usually with several amber-coloured glands in middle of abaxial surface. Achenes 5–7 × 2–2.5 mm, obovoid-cylindrical, densely setuliferous, appearing sericeous, setulae of twin-hairs, apices bifid, subequal or distinctly unequal, pappus setae (0 or 1)2, 1.3–1.7 mm long, appearing antrorsely barbellate or scabrid, cupule scabrid between setae.

Widely distributed, from Senegal, Guinea, Nigeria, Chad, Cameroon, Central African Republic, Sudan, Ethiopia, D.R. Congo, Rwanda, Burundi, Kenya, Uganda, Tanzania, Angola, and throughout the Flora area.

Ray limbs purple, purple-brown, crimson, maroon (or rarely pink) . . . . .a) var. *kotschyi*
Ray limbs white . . . . b) var. *alba*

a) Var. **kotschyi**

Ray floret limbs purplish, reddish purple, crimson, maroon, wine red, blood red or brown, very rarely pink.

Zambia. N: Mpulungu, Lae Tanganyika, 884 m, 19.i.1952, *Richards* 480 (K). W: Ndola, 30.i.1954, *Fanshawe* 744 (K, LISC, PRE). E: Fort Jameson Dist., Nzamane, 900 m, 6.i.1959, *Robson* 1041 (K). S: Livingstone Dist., along Livingstone-Lusaka Road, 18.9 km from the main water tank, 17°40'S, 53°12'E, 1020 m, 23.ii.1997, *Zimba* et al. 931 (K, MO). **Zimbabwe**. N:

---

[141] Crediting Hochstetter with authorship (as has happened in many references) is incorrect; Hochstetter merely presented Schultz Bipontinus's original list in Flora **24**(1). It is quite clear Schultz Bipontinus validated both his names in the later paper on Kotschy's Sudanese expedition, as sole author.

[142] This name appeared as '*Wedelia kotschyi* (Sch.Bip.) Isawumi var. *kotschyi* comb. nova', as neither the volume nor page number on which Schultz Bipontinus' protologue appeared is provided, and the valid combination of the species was not provided either; Isawumi also cited the 'basionym' as Oliver's combination in *Aspilia*.

Urungwe National Park, 302.5 km from Salisbury on Chirundu Road, 875 m, 20.ii.1981, *Philcox* et al. 8784 (K). W: Victoria Falls, *Brain* 9227 (LD, SRGH). **Malawi**. N: Mzimba/ Rumphi Dist., Rumphi Gorge, 1067 m, 4.v.1974, *Pawek* 8579 (K, MO, SRGH, UC). C: Lilongwe Dist., 2 km NE of Malingunde, 20 km SW of Lilongwe, 1160 m, 25.iii.1977, *Brummitt* et al. 14915 (K). S: Machinga Dist., north of Mposa Harbour, 7.iii.1985, *Kwatha & Patel* 8 (K, MAL). **Mozambique**. N: Mutuali, 19.iii.1953, *Gomes e Sousa* 4082 (COI, K, LISC, LMJ, PRE). Z: Mocuba Dist., Namagoa Plantations, 61–122 m, 2.xii. 1943, *Faulkner* Pretoria No. 166 in G.H. 15427 (BM, COI, EA, LMJ, PRE, SRGH).

Brachystegia and *Acacia* woodland, sometimes on rock outcrops, on rocky soil, in bogs or wet grassland, bamboo thickets, often a weed in cotton fields; 0–1800 m.

Conservation Status: A widespread, often weedy species; LC (Least Concern).

Uses: Meller's unnumbered collection, made on the Livingstone Expedition has an additional label 'Called by natives M'tandua segete (an antidote of mosquito bites).'

I have been unable to find any descriptions of populations in the Flora area (unlike in Zanzibar) where var. *alba* is present with var. *kotschyi*, although *Nshingo* 222 from Zambia is pink suggesting some mix; likewise label information the var. *alba* does not mention the presence of the other variety.

b) Var. **alba** Berhaut in Bull. Soc. Bot. France **101**(7–9): 375 (1954). —Adams in Webbia **12**(1): 242 (1956); in Hepper, F.W.T.A., ed. 2, **2**: 239 (1963). —Wild in Kirkia **5**(2): 204 (1966). —Mesfin Tadesse in Fl. Ethiopia & Eritrea **4**(2): 294 (2004). —Beentje & Hind in F.T.E.A., Compositae **3**: 747–748 (2005). Type: 'Bargny (Sénégal) in «Terre argileuses». *Berhaut* 1726.' (P00072769 holotype, BR8874139).

   Aspilia polycephala S.Moore in J. Bot. **45**(530): 45 (1907). Type: [Uganda:] 'Hab. Fort Portal, Toro; [5000 ft, April 20th 1906] *Bagshawe*, 993.' (BM000924391 holotype).

   Wedelia kotschyi (Sch.Bip.) Isawumi var. *alba* (Berhault) Isawumi in Feddes Repert. **108**(7–8): 559 (1997), comb. illeg., since the species combination was not validly published.

Paleae margins laciniate, apices purplish, short-scabrid pubescent adaxially, short-scabrid pubescent abaxially towards apex. Ray floret limbs white. Disc floret corollas purple, sometimes almost lilac-purple; anther thecae black, apical anther appendages black; style arms purple. Achenes densely long-setuliferous, appearing sericeous; setae often 2, 1.4–1.7 mm long.

Zambia. S: Livingstone to Kazengula mile 10, near Kapanda Bridge Camp, 11.iii.1960, *White* 7715 (FHO, K, SRGH). C: Lusaka Province, Kafue Fisheries, Kafue town 10 km W of Lusaka, 980 m, 4.iv.1993, *Bingham* 9010 (K). **Zimbabwe**. W: Victoria Falls, i.1960, *Allen* 254 (K, SRGH). **Malawi**. N: Nyika Plateau, ii & iii.1903, *McClounie* 165 (K, P). **Mozambique**. N: Nampula, 4.iv.1937, *Torre* 1249 (COI, LISC). Z: Entre Nangaze e as cascatas do Rio Mugoé, 26.v.1949, *Barbosa & Carvalho* 2886 (K, LMJ).

Also known from Senegal, Mali, Nigeria, Ethiopia, Sudan, D.R. Congo, Rwanda, Burundi, Uganda, Tanzania. *Mopane-Terminalia* woodland, clay dambo, moist grassland; 90–1500(2400) m.

Conservation Status: A widespread variety found throughout the distribution of the species, but conspicuously less common, especially in the Flora area, sometimes even rare; LC (Least Concern).

Adams (1956) was quite wrong in stating that the 'flower colour … cannot be determined from the dried specimens.' It is quite straightforward to detect the various red/maroon/ crimson-coloured ray limbs as they are all dark in dry material; white limbs are merely a pale fawn colour. I feel Adams (1956: 242, Fig. 2: 10, 11) was merely observing part of the variation of *A. kotschyi* s.l., as I cannot see this distinct difference between the achenes of the two varieties. Lisowski (1991: 186–191) also suggested that, apart from ray limb colour differences, the limbs were a different shape – var. *kotschyi* appearing ovate to oblong-orbicular, var. *alba* appearing elliptic; I have been unable to substantiate this, and certainly not from photographs of the two varieties.

2. **Aspilia ciliata** (Schumach.) Wild in Kirkia **6**(1): 41 (1967). —Maquet in Fl. Rwanda **3**: 624, fig. 192/2 (1985). —Lisowski, (Asterac. Fl. Afr. Cent. 1) Fragm. Flor. Geobot. **36** Suppl. 1: 191, t. 42 (1991). —Mesfin Tadesse in Fl. Ethiopia & Eritrea **4**(2): 294 (2004). —Beentje & Hind in F.T.E.A., Compositae **3**: 748 (2005). Type: 'Hist og her.' [Ghana:] [Guinea, *Thonning* s.n.] (C10004678, C10004679, S-G-6314 syntypes).[143]

   Verbesina ciliata Schumach. in Schumach. & Thonn., Beskr. Guin. Pl.: 391 (1827).

   Blainvillea prieureana DC., Prodr. **5**: 492 (1836). —Oliver in Oliver & Hiern, F.T.A. **3**: 375 (1877). Type: [Senegal:] 'in Senegaliâ superiore ad montes Bakel legit cl. *Leprieur.* ... (v. s. comm. à cl. inv.)' (G00023503 holotype, ?P00069092[144]).

   Wirtgenia abyssinica Sch.Bip. ex Walp., Repert. Bot. Syst. **6**: 146 (1846). Type: 'Crescit in Abyssinia.' [Schimperi iter Abyssinicum. Sectio secunda. 819. Prope Gapdiam in incultis d. 12. Sep. 1838.] (P00072517 holotype[145], BM000924385, BM000924386, BR8362735, BR8873484, E00433071, HAL0111838, K000410291, M0105217, MICH1107921, NY00278073, P00072518, P00418376, S07-16799, STU000305).

   Wirtgenia schimperi Sch.Bip. ex A.Rich., Tent. Fl. Abyss. **1**: 412 (1848). Type: 'in pl. Schimp. Abyss., sect. III, 1441 et 1684. [citing Var. : Robusta. and Var. : Gracilis. – these are printed on the labels themselves] ... Crescit in montosis circa Tchélatchékanné, in convalle fluvii Taccazé, mense Augusto florens (Schimper).' *Schimper* 1441 (K000410294, P00072506, TUB005506 syntypes); *Schimper* 1684 (BM000924382, BR8363510, BR8873491, GOET0010167, HAL0111840, HOH009466, M0105215, M0105216, P00072505, P00072507, P00072520, 00167568, S07-16800, S07-16801, TUB005504, TUB005505 syntypes).

   Aspilia abyssinica (Sch.Bip. ex Walp.) Vatke in Linnaea **39**(6): 495 (1875).

   Aspilia schimperi (Sch.Bip. ex A. Rich.) Oliv. & Hiern in F.T.A. **3**: 379 (1877).

   Aspilia fischeri O.Hoffm. in Planzenw. Ost-Afrikas C: 413 (1895). —Muschler in Bot. Jahrb. Syst. **50** (Suppl.): 340 (1914). Type: [Tanzania:] 'Mit. *A. multiflora* verwandt. – 17 [= Seengebiet] (Unja[mwesi], Kahegi. – *Fischer* n. 370).' (B† holotype).

   Aspilia dewevrei O.Hoffm. in Bull. Soc. Bot. Belg. **39**(3) (Compt. Rend. Soc. Bot. Belg.): 32 (1900).

   Aspilia helianthoides (Schumach. & Thonn.) Oliv. & Hiern var. *papposa* O.Hoffm. & Muschl. in Bot. Jahrb. Syst. **50** (Suppl.): 341 (post March 1914). Types: [Cameroon:] 'Kamerun-Hinterland: (*PASSARGE* n. 57). Togo: Bassari-Station (*KERSTING* n. 127). Sierra Leone: [On alluvium, Kora River, ... Feb. 13, 1892] (*SCOTT ELLIOT* n. 4593).' *Kersting* 127 (?B† syntype); *Passarge* 57 (?B† syntype); *Scott Elliot* 4593 (BM, K000410308 syntypes).

   Wedelia ringoetii De Wild. in Repert. Spec. Nov. Regni Veg. **13**(359–362): 210 (1914). Types: 'Katanga: Shinsenda, mars 1912 (Ringoet, coll. *Homblé* , no. 530) vallée de Kapiri, février 1913 (*Homblé*, no. 1103).' *Homblé* 530 (BR8873781, BR8874153 syntypes); *Homblé* 1103 (BR8873811, BR8873828 syntypes).

   Aspilia bracteosa C.D.Adams in Webbia **12**(1): 238 (1956). Type: 'GOLD COAST: Northern Territories; In damp cultivated land by road east of Yendi, 31st Dec. 1954 (fl. and fr.), *Morton A* 1473 (GC 5594)' (K holotype, GC).

   Aspilia helianthoides subsp. *ciliata* (Schumach.) C.D.Adams in Webbia **12**(1): 245 (1956); in Hepper, F.W.T.A., ed. 2, **2**: 239 (1963). —Wild in Kirkia **5**(1): 206 (1966).

   Aspilia helianthoides subsp. *papposa* (O.Hoffm. & Muschl.) C.D.Adams in Webbia **12**(1): 247 (1956); in Hepper, F.W.T.A., ed., 2 **2**: 239 (1963).

   Aspilia helianthoides subsp. *prieuriana* (DC.) C.D.Adams in Webbia **12**(1): 246 (1956); in Hepper, F.W.T.A., ed. 2, **2**: 239 (1963). —Wild in Kirkia **5**(1): 208 (1966).

   Aspilia helianthoides subsp. *papposa* (O.Hoffm. & Muschl.) Adams in Webbia **12**(1): 247 (1956).

   Wedelia helianthoides (Schumach. & Thonn.) Isawumi subsp. *ciliata* (Schumach.) Isawumi in Feddes Repert. **108**(7–8): 560 (1997), comb. inval., without citing basionym page.

---

[143] There are two sheets in C(C10004678, C10004679), neither determined as 'holotype', but both *Thonning* duplicates are from 'Guinea'; there is a duplicate, S-G-6314. None of the literature I have seen has attempted to lectotypify the name, most simply state the holotype is in C. Adams determined C10004678 as an *Aspilia* (25.vi.1955), but failed to make any comments later (Adams, 1956).

[144] The P collection is marked as 'Galam' (a kingdom in Senegal), and is dated 1830.

[145] The sheet also has profuse notes by Schultz Bipontinus, even though it is labelled an isotype!

Wedelia helianthoides subsp. *prieureana* (DC.) Isawumi in Feddes Repert. **108**(7–8): 560 (1997), comb. illeg., without citing the basionym volume and page in de Candolle's Prodromus.

Wedelia helianthoides subsp. *papposa* (O.Hoffm. & Muschl.) Isawumi in Feddes Repert. **108**(7–8): 560 (1997).

Wedelia ciliata (Schumach.) Soldano in Compositae Newslett. **32**: 20 (1998).

Annual herb from 0.3(1.3) m tall. Stems simple or branched, branches subcylindrical or quadrangular, sulcate, patent- or appressed-pilose. Leaves sessile or petiolate, petioles to c. 1.8 cm long, pilose, lamina papyraceous, up to 6(8) × 4 cm, ovate or narrowly ovate, apex acute, base cuneate or rounded and abruptly cuneate, margins shallowly serrate to subentire, scabrid-pilose on both surfaces, ± discolorous, 3-veined from base. Inflorescence of in lax leafy corymbs, capitula pedicellate, pedicels c. 1–3 cm long, pilose. Capitula heterogamous and radiate; involucre hemispherical or campanulate, c. 1.3 × 1 cm; phyllaries linear, oblong or elliptic, outer sometimes foliose and exceeding disc florets, subequal, diminishing in size inwards, c. 7–18 mm long, apex obtuse or subacute, appressed-pilose; paleae 7.2–10 mm long, oblong or obovate, apex acute or abruptly acuminate, mucronate, upper margins and upper part of keel ciliolate, apex scaberulous or puberulent. Ray florets c. 15, ray limb oblong to broadly elliptic, 7–10 mm long, yellow (or white or purple but not in our area), apex emarginate, tube c. 2 mm long glabrous; styles absent. Disc florets 3.8–5.3 mm long, glabrous except on lobes or minutely puberulent at base of upper part, lower part 1–2 mm long, constricted; apical anther appendages dark. Achenes c. 3 mm long, ± compressed, obovoid-oblong, setuliferous, setulae of long twin hairs, apices unequal, tips finely-acute; pappus a well-developed lacerate corona, longer setae absent although some lacerations of corona may be longer than others (setae sometimes present outside our area).

Zambia. N: Mweru Wantipa, 17.iv.1961, *Phipps & Vesey FitzGerald* 3254 (M, SRGH). W: Mwinilunga, 18.iv.1960, *Robinson* 3567 (K, SRGH). S: Mazabuka, 6.ii.1963, *van Rensburg* 1332 (K, SRGH).

Widespread through tropical Africa from Ethiopia and Senegal southwards but only as far as the Zambezi Valley. Weedy and a colonizer of bare ground.

Conservation Status: A widespread species, but clearly infrequently/under collected in the Flora area; LC (Least Concern).

Orchard's inclusion of *Blainvillea prieureana* DC. in the synonymy of *Blainvillea gayana* Cass. (Orchard, 2012: 656) is in error; an examination of the types concerned shows no similarity at all, hence the expanded synonymy of *Aspilia ciliata* above.

3. **Aspilia mendoncae** Wild in Kirkia 5(2): 212 (1966). Type: [Malawi:] '*Exell, Mendonça & Wild* 945' [Nyasaland: Southern Prov.: Ncheu Dist., Lower Kirk Range, Chipusiri, 1460 m., 17.iii.55, *Exell, Mendonça & Wild* 945.] (SRGH holotype, BM000924379, LISC002771).

Erect perennial herb, xylopodiaceous, often appearing rhizomatous from leathery secondary roots, up to 0.6 m tall. Stems sparsely-branched, branches pilose, hairs ascending. Leaves sessile, 6–13 × 0.6–1.8 cm, linear-oblong to narrowly elliptic, base cuneate, margins entire or shallowly and sparsely serrate, appressed scabrous-pilose above, hairs of 2 types, both scabrid, of shorter hairs and longer arcuate scabrid hairs, scabrous-pilose below, 3-veined from base, midrib prominent below, veins inconspicuous, apex tapering, acuminate. Inflorescences at ends of branches in lax 2- or 3-headed corymbs; pedicels 1.5–5 cm long, slender, densely pilose, hairs ascending. Capitula heterogamous and radiate; involucre campanulate, c. 0.8 × 1.2 cm; phyllaries ± 2-seriate, subequal, c. 1 cm long, narrowly ovate to ovate, apex acute, appressed-pilose outside; paleae c. 8 mm. long, oblong, keeled, scabrous-puberulous towards apex, apex caudate-acuminate. Ray florets c. 10, neuter, ray limb c. 8 mm long, oblong-elliptic, yellow, apex 2-lobed, lobes c. 2 mm long, tube 3.5 mm long, puberulous. Disc florets c. 6.3 mm long, puberulous near middle of corolla and on outside of lobes, otherwise glabrous, base constricted 2 mm long; apical anther-appendages yellow. Achenes dark brown, c. 4.5 mm long, narrowly obovoid, laterally somewhat compressed, setuliferous, setulae of twin hairs, apices unequal, tips acute; pappus a lacerate to almost ciliate cupule and 1 or 2 setae slightly exceeding cupule.

Zambia. N: Abercorn, 20.ii.1964, *Richards* 19056 (K, SRGH). **Malawi**. C: Lilongwe Dist., Bunda Agricultural College, 23 km south of Lilongwe, 1175 m, 19.ii.1970, *Brummitt* 8628 (K). S: Zomba, North side of Makwapala Expt. Station, 716 m, 17.i.1937, *Lawrence* 247 (BR, K). **Mozambique**. T: Furancungo–Pandalanjala Mts., 15.v.1948, *Mendonça* 4242 (LISC).

Brachystegia woodland or grassland; 700-1230 m; flowering from January–May.

Conservation Status: Clearly a fairly widely distributed taxon but still represented by relatively few specimens in herbaria; LC (Least Concern).

Wild (1966) originally cited this taxon for Tanzania based on a *Milne-Redhead & Taylor* collection; this has proven to be a specimen of *A. mossambicensis* (see Beentje in Beentje & Hind, 2005: 749).

4. **Aspilia pluriseta** Schweinf. in Höhnel, Zum Rudolph-See und Stephanie-See: 862 (1892). —Engler, Hochgebirgsfl.: 433 (1892). —Wild in Kirkia **5**(2): 210 (1966); in Kirkia **6**(1): 44 (1967). —Beentje & Hind in F.T.E.A., Compositae **3**: 749 (2005). Type: [Kenya:] 'Ndoro, Rifuju-Gebiet, 1939 Meter.' [Massaihochland, Ndoro, *v. Hoehnel* 66] (B† holotype); 'Kenya, Ukamba, *Hildebrandt* 2712' (K000410287 neotype, B†, BM000798762, M0105210, P00069850, P00069851), neotypified by Wild, (1966: 210).

   Aspilia asperifolia O.Hoffm. in Pflanzenw. Ost-Afr. C: 413 (1895). —Muschler in Bot. Jahrb. Syst. **50** (Suppl.): 335 (1914). Types: [Tanzania:] '13 [= Usagara-Usambara] (Kwa Mshusa, Msinga. –*Holst* n. 9127) 15 [=Kilimandscharo] (Kl., Marangu. –*Volk*[*ens*] n. 367, 541). –Auf trockenen Graslächen im Gebirgsbusch.' *Holst* 9127 (COI, HBG504022, K000410286, M0105209 syntypes); *Volkens* 367 (BM000798674 syntype); *Volkens* 541 (E00433384 syntype).

   Aspilia gondensis O.Hoffm. in Pflanzenw. Ost-Afr. C: 413 (1895). —Muschler in Bot. Jahrb. Syst. **50** (Suppl.): 336 (1914). Type: [Tanzania:] '17 [=Seengebiet] [Ug., Gonda. – *Böhm* n. 41).' (B† holotype); 'Tanzania, *Busse* 1366' (EA neotype, B†), neotypified by Wild (1966: 212).

   ?*Aspilia involucrata* O.Hoffm. in Pflanzenw. Ost-Afr. C: 413 (1895). Type: [Kenya:] '14[= Massaisteppe] (Wadiboma – *Fischer* n. 330). – Steppen.' (B† holotype).

   Aspilia vulgaris N.E.Br. in Bull. Misc. Inform., Kew **1906**: 164 (1906). —Eyles in Trans. Roy. Soc. S. Afr. **5**: 515 (1916). Type: 'Rhodesia, Mashonaland: very common between Umtali and Salisbury, *Hon. Mrs. Evelyn Cecil*, 43.' (K000410464 holotype).

   Aspilia brachyphylla S.Moore in J. Linn. Soc., Bot. **40**: 115 (1911). —Eyles in Trans. Roy. Soc. S. Afr. **5**: 514 (1916). Types: [Zimbabwe:] 'Hab. Chirinda, a common weed in cultivated ground; in fl. May; [*Swynnerton*] nos 292, 495.' *Swynnerton* 292 (BM000924381, K000410465, K000410466 syntypes); *Swynnerton* 495 (BM000798763 syntype).

   Wedelia africana sensu Eyles in Trans. Roy. Soc. S. Afr. **5**: 514 (1916).

   Aspilia pluriseta subsp. *gondensis* (O.Hoffm.) Wild in Kirkia **5**(2): 212 (1966); in Kirkia **6**(1): 44 (1967).

Subshrub; rootstock a xylopodium, c. 2–3 cm diam. also with stout secondary roots. Stems usually several, usually simple, c. 0.3–0.6(1.5) m long, prostrate or suberect, appressed scabrid-pilose or hairs ascending. Leaves sessile or subsessile, lamina 2–5(8) × 0.8–2(3.5) cm, ovate to broadly ovate or narrowly oblong-ovate, apex obtuse or acute, base cordate or broadly cuneate, margins serrate, both surfaces appressed scabrous-pilose, sometimes densely so, usually scabrid, 3-veined from base, veins and reticulate venation somewhat prominent below. Inflorescences of solitary capitula at ends of branches or of c. 3-headed corymbs; peduncles densely pilose, 0.5–2(rarely 6) cm long, often ± sessile among upper leaves. Capitula hemispherical to subglobose, up to c. 1.8 × 1.3 cm; phyllaries 3-seriate, up to c. 8 mm long, subequal or gradate, outer often reflexed, very narrowly triangular to narrowly ovate or ovate, densely appressed-pilose outside, apex acute; paleae 7–11(16) mm long, obovate, developing especially in fruit a scaberulous caudate-acuminate apex, upper half sometimes puberulous. Ray florets (9)10–16, neuter (style absent), ray limb 10–15 mm long, oblong, obovate-oblong or elliptic, yellow, apex emarginate or 3-dentate; tube 2.3–3.2 mm long, glabrous or puberulous. Disc florets (4.7) 6–9 mm long, glabrous or sparsely puberulous, base constricted 1.2–2.5 mm long; apical anther appendages yellow. Achenes grey-brown, 2.5–4.5 mm long, obovoid and ± compressed, setuliferous, setulae of relatively stout twin hairs, apices subequal to unequal, tips acute; pappus a lacerate corona, c. 0.7 mm tall, setae absent although sometimes 1 or 2 lacerations longer than others, to 2.5 mm long.

Zambia. N: Abercorn to Tunduma Road, 22.x.1947, *Greenway & Brenan* 8256 (EA, K, PRE). W: Ndola Dist., 20 miles N. of Kapiri Mposhi on the Great North Road, 1.xii.1952, *Angus* 909 (BM, K). C: Nr. Mt. Makulu Research Stn. 12 mls. S. Lusaka, 20.vii.1956, *Angus* 1377 (K, LISC, SRGH). S: Choma Dist., along Sinazongwe/Choma Road, 11.6 km from junction, 16°52'08"S, 27°16'23"E, 1200 m, 10.iii.1997, *Luwika* et al. 596 (K, MO). **Zimbabwe**. N: Nr. Zwipani Camp, Urungwe Nature Reserve, 20.xi.1957, *Goodier* 389 (K, PRE, SRGH). C: Marandellas, 5.xii.1940, *Dehn* 210 (K, M, SRGH). E: Chirinda, xii.1937, *Obermeyer* 2252 (PRE, SRGH). **Malawi**. N: Vipya, Mpampala Lookout above Nyungwe Forest ca. 25 mi. SW of Mzuzu, 1920 m, 10.iv.1971, *Pawek* 4647 (K, MAL). C: Btwn. Chamama & Bua Drift, Kasungu Dist., c. 1000 m, 15.i.1959, *Robson & Jackson* 1202 (BM, K, LISC, PRE, SRGH). **Mozambique**. N: Nampula, 27.vii.1937, *Torre* 769 (COI). Z: Zambézia entre Pebane e Mocubela, 25.x.1942, *Torre* 4694 (K, LISC). MS: Manica, Dombe, entre Dombe e a povoação de Mapira, a 18 kms do Posto, baixas do R. Mucutuco, pousios, 27.x.1953, *Gomez Pedro* 4459 (K, LMJ, PRE).

Also in Kenya, Tanganyika, Uganda, Rwanda, Burundi, D.R. Congo and the Transvaal. *Brachystegia* woodland or savanna woodland.

Conservation Status: A widespread species and common throughout; LC (Least Concern).

Beentje (2005) rightly combined both subspecies; indeed, there are many long-pedicelled, small leaved taxa so Wild's distinction is somewhat artificial applying to few specimens.

5. **Aspilia natalensis** (Sond.) Wild in Kirkia **5**(2): 213 (1966). Types: 'Port Natal. *Gueinzius*, No. 9, 351, 609.' '*Gueinzius* 9.351' (S07-16792[146] syntype); *Gueinzius* [s.n.] (TCD0003154 syntype); *Gueinzius* 9 (P00072839 syntype); *Gueinzius* 351 (P00072840 syntype).

   Wedelia natalensis Sond. in Linnaea **23**(1): 63 (1850).

   Aspilia welwitschii O. Hoffm. in Bol. Soc. Brot. **13**: 29 (1896). —Hiern, Cat. Afr. Pl. **1**(3): 578 (1898). —Moore in J. Bot. **65**, Suppl. 2, Gamopet.: 58 (1927). —Mendonça, Contrib. Conhec. Fl. Angola **1** (Compositae): 92 (1943). Type: 'Angola (*Welwitsch*, n.º 3559)' [In pastures amidst tall bushes on the right bank of the Lopollo stream, fl. and fr. Oct. and Nov. 1859 and again in Jan. 1860; also in hilly places on a rich soil covered with acacia shrubs, on the right bank of the Lopollo stream, plentiful, fl. and fr. April 1860. [*Welwitsch*] No. 3559.' – citation from Hiern (1898: 578) (LISU219555 holotype, BM000924374, BM000924375, BR8873774, K000410282, LD1226908, LISU219554, M0105211, P00072833).

   Aspilia welwitschii var. *serrata* Hiern, Cat. Afr. Pl. **1**(3): 578 (1898). Type: [Angola:] 'Huilla. – In thickets near Mumpulla; fl. Oct. 1859. [*Welwitsch*] No. 3560.' (BM000924377 holotype, LISU219556).

   Aspilia fontinalis Hiern, Cat. Afr. Pl. **1**(3): 578 (1898). —Moore in J. Bot. **65**, Suppl. 2, Gamopet.: 57 (1927). —Mendonça, Contrib. Conhec. Fl. Angola **1** (Compositae): 93 (1943). Type: [Angola:] 'Pungo Andongo. –In the Panda forest near Condo, close to the cataract of the river Cuanza; fl. and fr. 15 March 1857. [*Welwitsch*] No. 3558.' (BM holotype, K000410278, LISU219562).

   Aspilia monocephala Baker in Bull. Misc. Inform., Kew **1898**(139): 152 (1898). Type: [Malawi:] 'British Central Africa. Zomba, *Whyte and McClounie* [s.n.]' (K000410283 holotype).

   Wedelia katangensis De Wild in Repert. Spec. Nov. Reg. Veg. **13**(359–362): 209 (1914). Types: [D.R. Congo:] 'Katanga: Katentania, novembre 1912 (*Homblé*, no. 794); Eschen 1912 (*Homblé*, no. 887 et 894)' *Homblé* 794 (BR6421595 syntype); *Homblé* 887 (BR8873408, BR8873439 syntypes); *Homblé* 894 (BR6422998, BR6423209 syntypes).

   Wedelia affinis De Wild in Repert. Spec. Nov. Reg. Veg. **13**(359–362): 209 (1914), nom. illeg. non DC. (1836). Types: [D.R. Congo:] 'Katanga: Wegelegen, 1912 (Corbisier, coll. *Homblé*, no. 606 et 612).' *Homblé* 606 (BR6422660, BR6422875 syntypes); *Homblé* 612 (BR8873446, BR8873798 syntypes).

---

[146] The number on the label appears to be a later addition in pencil.

Aspilia kakondensis S.Moore in J. Linn. Soc., Bot. **47**(314): 277 (1925). —Mendonça in Contrib. Conhec. Fl. Angola **1** (Compositae): 93 (1943). Type: 'Angola, Kakonda, open woods near the barrage of the river Kunene; *Gossweiler*, 2822 [sic!].' (BM000924376 holotype, COI100005602, LISC014893).[147]

Xylopodiaceous perennial (although rootstock very rarely collected). Stems few to numerous, suberect or erect, 4–60 cm tall, simple or few-branched (mostly in flowering portion), stems and branches appressed or patently scabrous-pilose, basal part with several persistent, scale-like leaves. Leaves sessile or subsessile, lamina up to 12 × 3 cm, elliptic, narrowly ovate or oblong, rarely narrowly lanceolate, apex acute or obtuse, base cuneate to rounded, margins serrate to subentire, appressed scabrid-pilose above, more densely pilose and less scabrid below, 3-veined from base, veins ± prominent beneath, veins ± reticulate. Inflorescences of solitary capitula, usually on long pedicels (1.5)5–15 cm long, densely pilose with ascending, patent or ± reflexed hairs. Capitula heterogamous and radiate; involucre campanulate, c. 1.5 × c. 1.7 cm; phyllaries typically oblong but varying to broadly elliptic and broadly ovate, usually 6–8, 2-seriate and usually exceeding disc florets, to 1.8 × 1.0 cm, apex obtuse or rounded, densely pilose; paleae 7–9(12) mm long, narrowly obovate, apex subacuminate but not elongating noticeably in fruit, upper margins serrulate, keel serrulate and puberulent above. Ray florets neuter, c. 10, ray limb 5–14.5 mm long, oblong, broadly elliptic or broadly obovate, apex emarginate or 3-dentate, yellow; corolla tube 3.3–5 mm long, puberulous or glabrescent; styles absent. Disc florets 6.8 mm long, puberulous or glabrescent, sometimes hairs confined to base of expanded portion, base constricted 1.7–3 mm long; apical anther-appendages yellow. Achenes 3.5–4.5 mm long, pilose, obovoid-cylindrical, setuliferous, setulae of very short scabrid hairs; pappus a finely laciniate or ciliate cupule, setae absent.

Zambia. N: Abercorn, 27.xii.1951, *Nash 9* (BM). W: Mwinil unga, 30.xi.1952, *White* 3420 (BM, BR, COI, K, PRE). E: Katete, 14.xii.1955, *Wright* 57 (K). S: Mazabuka, 14.i.1960, *White* 6269 (K). **Zimbabwe**. E: Chipinga, E. Sabi, 22.i.1957, *Phipps* 108A (SRGH). **Malawi**. N: Mzimba, Mt. Hora, 11.i.1950, *Krippner* 24 (BM). S: Zomba, *White & McClounie* (K). **Mozambique**. N: Massangulo, Fl. 1938, *Gomes e Sousa* 1203 (COI). Z: Mocuba, Fl. 27.xi.1949, *Faulkner* 475 (COI, K, SRGH).

Also in D.R. Congo, Angola, Tanganyika and KwaZulu Natal. Woodland or savanna woodland; 1050–1900 m.

Conservation Status: A widespread, very variable species; LC (Least Concern).

Some of the shortest flowering stems can be seen in *Pawek* 10591 (from Malawi) noting that there was a 'thick mass of stems from perennial root'.

Beentje in Beentje & Hind (2005: 753) clearly considered that Wild's concept of *A. natalensis* consisted of two entities, one from S Africa, the other more northerly and corresponding to *A. mossambicensis*. If, as Beentje suggested, that the synonymy (of *A. natalensis*) belonged under his (Beentje's) concept of *A. mossambicensis* then the oldest name for anything in this complex is *Wedelia natalensis*. Wild's combination would indeed be the name for everything – and Wild excluded nothing from his concept which also had a distribution well outside of S Africa.

6. **Aspilia angolensis** (Klatt) Muschl. in Bot. Jahrb. Syst. **50** (Suppl.): 337 (1914). — Mendonça in Contrib. Conhec. Fl. Angola **1** (Compositae): 93 (1943). —Wild in Kirkia **5**(2): 220–221 (1966). Type: 'Hab.: Angola, Pongo-Andongo, leg. Jan.-Apr. 1879, [*Mechow*] No. 35.' (GH00065094 holotype, S07-16803).

   Wedelia angolensis Klatt in Annal. Naturh. Hofmus. Wien **7**: 102 (1892).

   Wedelia huillensis Hiern, Cat. Afr. Pl. **1**(3): 576 (1898). Type: [Angola:] 'HUILLA. –In herbaceous places at the outskirts of forests, from Lopollo towards Jãu; fl. March 1860. [*Welwitsch*] No. 3566.' (BM000924369 holotype, K000410281, LISU219567).

   Aspilia huillensis (Hiern) S.Moore in J. Bot. **65**, Suppl. 2, Gamopet.: 57 (1927). —Mendonça in Contrib. Conhec. Fl. Angola **1** (Compositae): 92 (1943).

---

[147] The BM holotype has a handwritten label with '2882' clearly written on it, agreeing with the isotypes in COI and LISC, suggesting a typographical error in the protologue.

Wedelia albiflora Hiern, Cat. Afr. Pl. **1**(3): 577 (1898). —Mendonça in Contrib. Conhec. Fl. Angola **1** (Compositae): 91 (1943). Type: [Angola:] 'PUNGO ANDONGO.–In thickets very frequently flooded, near Candumba; fl. Jan. 1857. [*Welwitsch*] No. 3557.' (BM000924367 holotype, BM000924366, K000410280, P00069976).

Aspilia engleriana Muschl. in Bot. Jahrb. Syst. **50** (Suppl.): [339, fig. 1] 340 (1914). Type: [Angola:] 'Humpata: Halbstrauch an felsen (*Bertha Fritzsche* n. 116. –Mai 1903).' (B† holotype).

Aspilia culuensis S.Moore in J. Bot. **65**, Suppl. 2, Gamopet.: 57 (1927). Type: 'Hab. Angola: in open "Mumua" woods at Calui, near Forte Dom Affonso, [*Gossweiler*] 2900.' (BM000924368 holotype, K000410279, LISC014892).

Erect perennial to 1.7 m tall; rootstock a small woody xylopodium (infrequently collected!); stems few to several, subcylindrical, scabrous-puberulent (or patently hispid outside our area), striate. Leaves sessile or subsessile; lamina 2.5–9 × 0.6–4 cm, from very narrowly ovate (outside our area) to ovate, base slightly cordate, rounded or cuneate, margins serrate to subentire, appressed scabrous-pilose and very rough to touch on both sides (or with patent pilose hairs outside our area), 3-veined from above base, venation pale and prominent beneath, veins ± reticulate, apex acute to obtuse. Capitula campanulate, 1.4 × c. 1.2 cm, in terminal few-(c. 3) headed corymbs, pedicels 0.5 (in lateral capitula) to c. 4 cm long, slender, with ascending or patent eglandular hairs; phyllaries c. 8 mm long, oblong or oblong-elliptic, apices obtuse or subobtuse, pilose outside; paleae 6.8–8 mm long, oblong to ovate, keel serrulate above, margins serrulate above, apex with an oblong ± blunt puberulous appendage c. 1.5 mm long. Ray florets neuter, 10–12, limb 5–9.5 mm long, oblong to narrowly oblong, yellow to pale yellow, apex emarginate; tube 2.7–3.3 mm long, slightly puberulous or glabrous; styles absent. Disc florets 5–6 mm long, glabrous except for the lobes or puberulous near middle, constricted lower part c. 2 mm long; apical anther-appendages dark. Achenes c. 3.6 mm long, narrowly obovoid, ± compressed, ± conspicuously setuliferous in upper part of achene, setulae of fine twin hairs, apices equally forked, tips finely-acute, achene apex with lacerate cupule; setae absent but some lacerations sometimes exceed remainder.

Zambia. W: Solwezi, 20.iii.1961, *Drummond & Rutherford-Smith* 7104 (K, SRGH).
Also in the D.R. Congo (Katanga) and Angola. *Brachystegia* woodland.

Wild (1966: 221) noted the rather variable nature of this species and the types of the synonyms alone support this. However, I was concerned to note that one of the cited Zambian collections (*Richards* 10792) was hardly the same taxon as the *Drummond & Rutherford-Smith* 7104 material and certainly not the same as the type of *Wedelia angolensis* (*Mechow* 35); I think that the *Richard's* collection belongs in *Aspilia natalensis* which has much larger phyllaries and much broader leaves than most material of *A. angolensis*. This may well account for Wild's comment on the phyllaries exceeding the disc florets in his original description (and that of other species too).

7. **Aspilia mossambicensis** (Oliv.) Wild in Kirkia 5(2): 221 (1966). —Lisowski, (Asterac. Fl. Afr. Cent. 1) Fragm. Fl. Geobot. **36** Suppl. 1: 203–204 (1991). —Mesfin Tadesse in Fl. Ethiopia & Eritrea **4**(2): 295–296 (2004). —Beentje & Hind in F.T.E.A., Compositae **3**: 751–753 (2005). Type: [Tanzania:] 'Hab. 6°S. lat., alt. 3800 feet (*Wedelia* no. 1, App. Speke's Journ. 638), *Col. Grant!*' (K000410288 holotype).

Menotriche strigosa Steetz in Peters, Naturw. Reise Mossambique **6** (Bot. 2): 475 (1864), non *Aspilia strigosa* (Hook. & Arn.) Hemsl. (1881). Type: [Mozambique:] 'Standort: Das einzige Exemplar dieser interessanten Gattung, welche bis jetzt nur durch eine Art repräsentirt ist, wurde von *Dr. Peters* in Rios de Sena entdeckt. Unicum specimen, quod suppetit, ...' (B† holotype).

Wedelia mossambicensis Oliv. in Trans. Linn. Soc. London **29**: 97 (1873). —Oliver & Hiern in F.T.A. **3**: 377 (1877).

Wedelia menotriche Oliv. & Hiern in F.T.A. **3**: 377 (1877), nom. nov. pro *Menotriche strigosa* Steetz. —Moore in J. Linn. Soc., Bot. **40**: 114 (1911). —Eyles in Trans. Roy. Soc. S. Afr. **5**: 514 (1916).

Aspilia wedeliiformis Vatke in Osterr. Bot. Zeit. 27: 197 (1877), as '*wedeliaeformis*'. —Oliver & Hiern in F.T.A. **3**: 461 (1877). Type: [Kenya:] 'Lamu in Sansibariae ore in pratis udis rara dec. 1875. [*Hildebrandt* 1908]' (B† holotype).

Aspilia holstii O.Hoffm. in Pflanzenw. Ost-Afr. C: 413 (1895). Types: '13 [= Usagara-Usambara] (Usb., an Bachufern und in Krantlichtungen. – *Holst* n. 148, 4334). 15 [= Kilimandscharo] (Teita,

N'dara. *–Hildebr.*[*andt*] n. 2380). –Im Gebirgsbusch, um 1500 m.' *Hildebrandt* 2380 (B† syntype); *Holst* 148 (B† syntype); *Holst* 4334 (BM000924363, COI, EA, HBG504023, K000410289, M0105212, P00069852 syntypes).

Coreopsis aspilioides Baker in Bull. Misc. Inform., Kew. **1898**: 153 (1898). Type: [Malawi:] 'British Central Africa. Zomba, alt. 2500-3500 ft., [Dec. 1896] *Whyte* [s.n.]' (K000410284 holotype).

Wedelia diversipapposa S.Moore in J. Bot. **37**(442): 401 (1899). —Eyles in Trans. Roy. Soc. S. Afr. **5**: 514 (1916). Type: [Zimbabwe:] 'Hab. Buluwayo, Rhodesia; *Dr. R. Frank Rand* [111]' (BM000924361 holotype).

Wedelia instar S.Moore in J. Linn. Soc., Bot. **35**(245): 343 (1902). Type: [Malawi:] 'Hab. Nyassaland; *John Buchanan*, 1895, no. 67.' (BM000924362 holotype).

Aspilia aspilioides (Baker) S.Moore in J. Linn. Soc., Bot. **40**(1): 115 (1911). —Lisowski, (Asterac. Fl. Afr. Cent. 1) Fragm. Fl. Geobot. **36** Suppl. 1: 202–203 (1991).

Aspilia chrysops S.Moore in J. Bot. **38**(456): 459 (1900). Type: [Somalia:] 'Hab. Laskarato, Somaliland, 1899; *Dr. Donaldson Smith*' (BM000924365 holotype).

Aspilia tanganyikensis Lawalrée in Bull. Jard. Bot. État Bruxelles **19**(3): 222 (1949). Type: [D.R. Congo:] 'Congo Belge: Haut-Katanga: baie de Mtoto, au pied de la montagne, mars 1947, *Van Meel* 1169' (BR8874092 holotype, YBI159937844).[148]

Aspilia vernayi Brenan in Mem. New York Bot. Gard. **8**(5): 479 (1954). Type: [Malawi:] '[Vernay Nyasaland Expedition – 1946]. Zomba District: Zomba Plateau, common in Brachystegia woodlands, woody herb 1–1.5 m., high, much branched, erect, leaves scabrous, flowers yellow, 1500 m., June 4, 1946, [*Brass*] 16215' (K000410285 holotype, BR8873767, PRE0192259, SRGH0106881).

Aspilia gillettii Wild in Kirkia **5**(1): 215 (1965). —Mesfin Tadesse in Fl. Ethiopia & Eritrea **4**(2): 296 (2004). Type: 'ETHIOPIA: Mega Mt., *Gillett* 14355' (K000410302 – 'Sheet 1', K000410303 – 'Sheet 2' syntypes).

Wedelia abyssinica sensu Eyles in Trans. Roy. Soc. S. Afr. **5**: 514 (1916).

Wedelia sp. —Brenan in Mem. New York Bot. Gard. **8**(5): 478 (1954).

Erect perennial herb (sometimes appearing shrub-like) 0.1–2.5 m tall, branches rather stiff. Stems and branches often reddish, appressed scabrous-pubescent or with longer patent hispid hairs. Leaves sessile, subsessile or short-petiolate, petioles up to c. 1 cm long, pubescent or patently hispid, lamina 13(20) × 2.5–5.5 cm, usually ovate, sometimes narrowly oblong, narrowly elliptic or oblong, apex gradually tapering, acuminate or more rarely subobtuse, base rounded or cuneate, margins serrate or subentire, appressed scabrous-pubescent above and scabrid, more densely pubescent beneath but less scabrous sometimes subtomentose, 3-veined from base, venation insculpate above, prominent and reticulate beneath. Inflorescences terminal, usually of few- (3-)headed corymbs, capitula pedicellate, pedicels up to c. 5 cm long, with ascending or patent hairs. Capitula heterogamous and radiate; involucre c. 1 × 1 cm, ± hemispherical to campanulate; phyllaries 2- or 3-seriate, 6–12.5 mm long (sometimes slightly longer than disc florets), ovate, elliptic or oblong, apices subacute or obtuse, pilose and sometimes glandular outside, glands sometimes purplish; paleae 6.7–9.3 mm, narrowly elliptic to oblong, apex obtuse to acute, sometimes mucronate, rarely acuminate, margins usually puberulent above, keel spiculate and often with a strong purple coloration. Ray florets female or neuter, c. 10–12, limb 7–18.5 mm long, oblong or elliptic, orange to yellow (rarely pale yellow), apex 3-dentate or emarginate, tube 1.5–3.5 mm long, glabrous or pubescent above; styles present or absent. Disc florets hermaphrodite, 5.6–7.3 mm, glabrous except for lobes or pubescent near middle, lower constricted part 1.8–2.2 mm long, sometimes with a strong purple line extending longitudinally downwards from between lobes; anther appendages yellow, thecae blackish to purplish. Achenes c. 2.5 mm long, narrowly obovoid, setuliferous, setulae of ascending twin hairs, apices unequal, apical cells finely acute, lacerate cupule often with lacerations varying in length; setae absent or 1 or 2.

Zambia. B: Machili, 10.i.1961, *Fanshawe* 6109 (K). N: Mpika, 29.i.1955, *Fanshawe* 1884 (K). C: Chilanga, ix.1929, *Sandwith* 164 (K). S: Victoria Falls, *Fanshawe* 5693 (K, SRGH). **Zimbabwe**. N: Darwin, Umsengedzi R., 14.v.1955, *Watmough* 121 (SRGH). W: Bulawayo, i.1944, *Martineau* 825 (SRGH). C: Marandellas, Delta, 5.iv.1950, *Wild* 3303 (K, M, SRGH). E: Inyanga, Stapleford, 6.iv.1962, *Wild* 5715 (M, PRE, SRGH). S: 24 km. S. of Fort Victoria,

---

[148] Wild (1966, 1967) was incorrect in citing the type collection as a *Humbert* collection from Tanzania.

Fl. 17.iv.1948, *Rodin* 4255 (K, PRE). **Malawi**. N: *Whyte* s.n. (K). C: Dedza Mt., 20.iii.1955, *Exell, Mendonça & Wild* 1073 (BM, LISC, SRGH). S: Cholo, 5.x.1960, *Chapman* 960 (SRGH). **Mozambique**. N: Nampula, 12.ii.1937, *Torre* 1149 (COI). Z: Quelimane, *Torre* 1518 (COI). T: Zobué, *Mendonça* 595 (LISC). MS: Chimoio, 2.iv.1948, *Barbosa* 1347 (LISC). M: Macia, 17.vii. 1947, *Pedro & Pedrógão* 1475 (K, PRE, SRGH).

Also in E Africa from Ethiopia and Somaliland through Kenya, Uganda, Zanzibar and Tanganyika to KwaZulu Natal, Lesotho and S Africa (Transvaal) as well as in the D.R. Congo and SW Africa. Woodland and savanna woodland.

Conservation Status: A widespread and very variable species; LC (Least Concern).

Beentje in Beentje & Hind (2005) was clearly of the opinion that much of Wild's concept of *Aspilia natalensis* belonged within *A. mossambicensis*, only the S African material belonging to the former. I have commented on the implications of this above, in that *Aspilia natalensis* (Sond.) Wild would be the name for the whole complex as it is based on Sonder's earlier name. Wild excluded no material in making the combination, however, the two taxa can be separated.

132. **VERBESINA** L.

Verbesina L., Sp. Pl. **2**: 901 (1753), nom. cons.; Gen. Pl., ed. 5: 384 (1754). —Robinson & Greenman in Proc. Amer. Acad. Arts **34**: 534–566 (1899). —Wild in Kirkia **6**(1): 61–62 (1967). —Hind in Fl. Masc., Composées **109**: 185–187 (1993). —Boulos & Hind in Fl. Egypt **3**: 236–237 (2002). —Mesfin Tadesse, Fl. Ethiopia & Eritrea **4**(2): 297–298 (2004).

Ridan Adans., Fam. Pl. **2**: 120, 598 (1763), nom. rej. [= *Actinomeris* Nutt. (1818), nom. cons.]

Phaethusa Gaertn., Fruct. Sem. Pl. **2**: 425, pl .169 (1791).

Ximenesia Cav., Icon. **2**: 60, pl. 178 (1793–1794).

Hingstonia Raf. in Med. Repos. **5**: 352 (1808).

Ditrichum Cass. in Bull. Sci. Soc. Philom. Paris **1817**: 33 (1817), nom. rej., non Hampe.

Actinomeris Nutt., Gen. N. Amer. Pl. **2**: 181 (1818), nom. cons. vs. *Ridan* Adans.

Platypteris Kunth in Humb., Bonpl. & Kunth, Nov. Gen. Sp. Pl. **4** (ed. folio): 156 (1818).

Actimeris Raf. in Amer. Monthly Mag. & Crit. Rev. **4**: 195 (1819) ≡ *Actinomeris* Nutt.

Hamulium Cass. in Bull. Sci. Soc. Philom. Paris **1820**: 173 (1820).

Ochronelis Raf. in Atlantic J.: 153 (1832).

Saubinetia J.Rémy in Gay, Fl. Chil. **4**: 282, pl. 49 (1849).

Ancistrophora A.Gray in Mem. Amer. Acad. ser. 2, **6**: 457 (1859).

Chaenocephalus Griseb., Fl. Brit. W. I.: 374 (1861).

Achaenopodium K.Brandegee in Zoe **5**: 239 (1906).

Annual or perennial herbs, shrubs (sometimes vine-like or scrambling) or trees. Stems terete or winged. Leaves sometimes opposite, sometimes most or all alternate, sometimes rosettiform, often rough, sessile, pseudopetiolate or distinctly petiolate, petiole sometimes auriculate, lamina linear, lanceolate, deltoid, ovate or suborbicular, base decurrent, or sagittate, sometimes perfoliate, surfaces strigose, or glabrous, sometimes tomentose, 3-veined or rarely pinnately-veined, margins serrate, dentate, deeply dissected or variously lobed, apices acute, obtuse or rounded. Inflorescence corymbose, capitula few to many, sometimes scapiform and capitula solitary on long terminal pedicels. Capitula heterogamous and radiate or homogamous and discoid, small to large; involucre hemispherical or campanulate; phyllaries 2–6-seriate, imbricate; receptacle usually conical, sometimes flat, rarely globose, paleaceous, paleae persistent, conduplicate about achenes, usually hooded. Ray florets when present, female and fertile, sometimes female and sterile or sometimes neutral, ray limb usually long and showy, sometimes short and ± exserted, orange-yellow, yellow or cream to white. Disc florets hermaphrodite, fertile, 5-lobed (rarely 4-lobed), yellow, orange, red, green white or purple; anther cylinders brown or black, rarely reddish, pinkish or yellowish, apical anther appendages glandular or eglandular; style-arms acute or attenuate at apex. Achenes monomorphic and achenes strongly obcompressed and winged (symmetrically or asymmetrically), or sometimes dimorphic and ray achenes smaller than disc achenes, wings reduced often trigonous, disc achenes strongly obcompressed and conspicuously winged, wings usually white, cartilaginous and contiguous with base of persistent awns when present, often ciliate, body glabrous or setuliferous towards apex; pappus usually of 2 (sometimes 3 in triquetrous ray achenes) deciduous or persistent awns (very rarely uncinate), rarely absent or rudimentary.

A large genus of 200–300 spp. mostly from Mexico and Andean tropical America; the genus is in need of a modern revisionary treatment. One species is both widely cultivated and weedy, and is present in the Flora area, but oddly absent in neighbouring floras.

Verbesina encelioides (Cav.) A.Gray in Bot. California **1**: 350 (1876). —Oliver & Hiern in F.T.A. **3**: 383 (1877). —Andrews, Fl. Pl. Sudan **3**: 55 (1956). —Cronquist in Gleason, New Illustr. Fl. NE U.S. & Adj. Canada **3**: 676 (1963). —Adams in F.W.T.A., ed. 2, **2**: 241 (1963). Type: 'Habitat in Mexico unde introducta in Regium hortum Matritensem. Floruit prima vice mense Novembri: videtur perennis.' (MA syntypes[149]). FIGURE 6.5.**25**.

 Ximenesia encelioides Cav., Icon. **2**: 60, t. 178 (1793).

 Ximenesia encelioides var. α *hortensis* DC., Prodr. **5**: 627 (1836). Type: 'Frequens in hortis botanicis sed nunquàm specimen certè spontaneum vidi. (v.v.c.)'.[150]

 Ximenesia encelioides var. β *pachyptera* DC., Prodr. **5**: 627 (1836). Types: 'in Mexico legit cl. *Berlandier* (pl. exs. n. 2068 et 2286!). (v.s.sp.)'.[151]

 Ximenesia encelioides var. γ *oblongifolia* DC., Prodr. **5**: 627 (1836). Types: 'in Senegaliâ ad ripas flum. regionis Walo legit *Perrottet*, in insulis Mauritianis *Sieber* (fl. maur. 2. n. 118!) et *Belanger*, et in hort. Calcuttensi *Wallich* (v.s.)' *Bélanger* 72 (G-DC-G00456623 syntype); *Perrotet* 32 (G-DC-G00456609 syntype); *Sieber* 118 (G-DC-G00456699 syntype); *Wallich* s.n. (G-DC-G00456612 syntype).

 Ximenesia encelioides Cav. var. δ? *cana* DC., Prodr. **5**: 627 (1836). Types: 'in Mexico circa Laredo legit cl. *Berlandier* (pl. exs. n. 2068 et 2074). ...v.s.)' *Berlandier* 2068 (G-DC-G00456627 syntype); *Berlandier* 2074 (G-DC-G00456622 syntype).[152]

 Ximenesia microptera DC., Prodr. **5**: 627 (1836). Type: 'circa Buenos-Ayres legit cl. *Bacle* [32]. ... (v.s.)' (G-DC-G00456615 – mounted with *Wight* s.n. G-DC-G00456615, holotype).

 Ximenesia australis Hook. & Arn. ex DC., Prodr. **7**: 291 (1838), nom. nud. pro syn.[153]

 Verbesina scabra Phil. in Anales Univ. Chile **36**: 186 (1870),[154] non Benth. (1840). Type: 'Mendoza'.[155]

 Encelia (Geraea) albescens A.Gray in Proc. Amer. Acad. Arts **8**: 658 (1873). Type: 'In the western Mexican province of Sonora, *Dr. Edward Palmer*, coll. 1869, no. 21.' 'Sonora, Mexico. *Palmer* 21' (GH6545 lectotype, US00125360), lectotypified by Coleman (1966: 477).

---

[149] There are four sheets in MA, MA476510 (– Fiche 80/C8) with a handwritten label 'Ximenesia encelioides Cav./ex Bonariensi planitie/Née Iter.', MA476508 (– Fiche 81/A1) has three handwritten labels, the upper 'Ximenesia encelioides/Cav. Ic./ta. 178/Willd. Syng. pa. 2886/ex Hort. Reg. Matr./anno 1808', the lower left 'Ximenesia encelioides/Cav. Ic. 2 ta. 178/X. pachyptera DC. Prod./Pallasia serratifolia Smith/Ex H. M. Aug. 1844' and the middle label 'Genus Encelia affine/ Ximenesia encelioides/Icon. tab. 178/Mexico Culta in R. h. M.1792-93.', MA476508 (– sheet 2 – Fiche 81/A2) merely has a printed label, and MA476509 (– Fiche 81/A3) has two handwritten labels, the upper 'Ximenesia encelioides/Cav. Ic./Ex H. M. Aug. 1848' the lower 'Ximenesia enceli/ odes [sic!] Cav. Ic. ta. 178/Willd. Syng. pa. 2886/ex Hort. Reg. Matr./anno 1808'. This suggests material was continually grown in the botanic garden and specimens prepared and incorporated in the herbarium. Two of these sheets might be considered as type material.

[150] It is not clear from de Candolle's 'v.v.c.' that any herbarium material was prepared from the living material.

[151] I do not consider Coleman's citation of a single *Berlandier* specimen 'P?' as effective lectotypification (cf. Coleman in Amer. Midl. Naturalist **76**(2): 475–481, 1966).

[152] I do not consider Coleman's selection of a single *Berlandier* specimen 'P?' as effective lectotypification.

[153] Coleman (1966: 477) both cited erroneous authorities ('Benth. & Arn. in DC.') and assumed there was a type for this name.

[154] In a separately paginated reprint/preprint of 'Sertum Mendocinum alterum', pp. 1–54, this appears on p. 28.

[155] Muñoz Pizarro (1960: 165) cited SGO44007 and SGO65406.

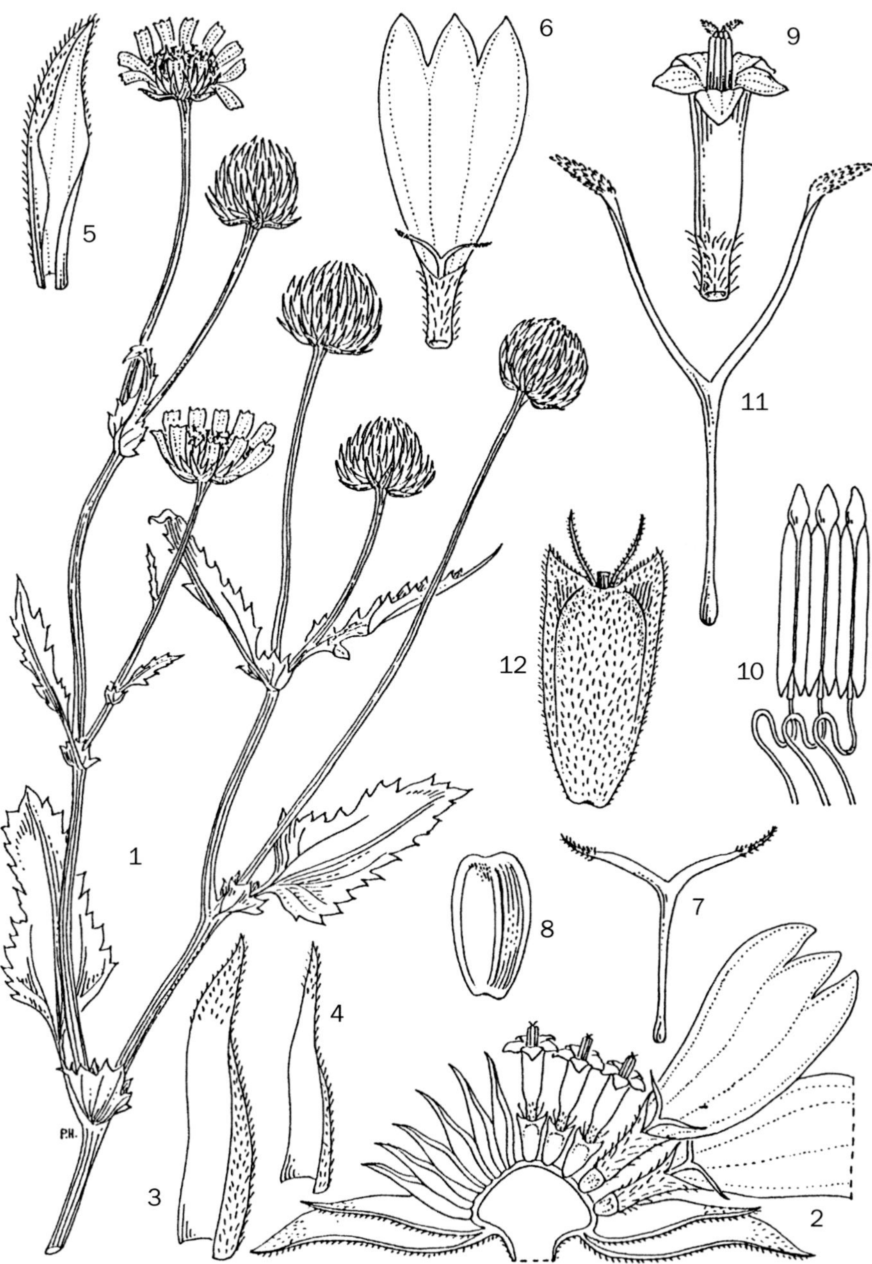

Fig. 6.5.**25**. VERBESINA ENCELIOIDES. 1, flowering branch ($\times$ ²/₃); 2, l.s. capitulum ($\times$ 3); 3, outer phyllary ($\times$ 4); 4, inner phyllary ($\times$ 4); 5, palea ($\times$ 6); 6, ray floret corolla ($\times$ 3); 7, ray floret style ($\times$ 8); 8, ray floret achene ($\times$ 6); 9, disc floret corolla ($\times$ 8); 10, partial anther cylinder opened out ($\times$ 12); 11, disc floret style ($\times$ 12); 12, disc floret achene ($\times$ 6). 1 from *Cadet* 2768; 2–12 from *Morin* s.n. Drawn by Pat Halliday. From Flore des Mascareignes.

Verbesina australis Baker in Mart., Fl. Bras. **6**: 215 (1884), nom. nov. pro *Ximenesia microptera* DC.

Verbesina encelioides var. *cana* (DC.) B.L.Rob. & Greenm. in Proc. Amer. Acad. Arts **34**: 544 (1899), p.p.

Verbesina microptera (DC.) Herter in Revista Sudamer. Bot. **7**: 235 (1843), comb. illeg. non DC. (1836: 616).

Annual herb 0.2–1 m tall. Stems pilose to tomentose, striate. Leaves mostly alternate but sometimes lower opposite, petiolate, petiole 1–4(5) cm long narrowly winged auriculate at base, wings entire or denticulate, auricles up to c. 1 cm long irregularly serrate suborbicular, lamina 3–8(15) × 1–6(7.5) cm, ovate-deltate, apex acute, base abruptly cuneate, margins coarsely serrate, appressed grey-pubescent, more densely so to tomentose below, discolorous, 3-veined from base. Capitula c. 2.5 cm diam., solitary on terminal branches, hemispherical to subglobose; pedicels up to 13(20) cm, pilose, striate; phyllaries biseriate, 1.0–1.3 cm long, very narrowly ovate to linear, apex acute, appressed grey-pilose; receptacle convex; paleae 6–8 × 0.5 mm, narrowly oblong, convex, margins pilose at least towards acute apex. Ray florets limbs bright yellow, 10–20 × 5–8 mm, elliptic or elliptic-obovate, apex 3-dentate, tube c. 1.6 mm long puberulent. Disc florets numerous (c. 100), c. 5.7 mm long, cylindrical but widening slightly upwards, narrower and puberulent at base for c. 1 mm. Achenes c. 7 mm long, obovoid, black with irregular paler patterning, surrounded by a whitish c. 1–1.5 mm broad irregular margined puberulent wing; pappus of 2 slightly incurved setae c. 1.2 mm long.

Botswana. N: Boteti River, 2.xii.1978, *Smith* 2539 (K, SRGH). SW: Tshabong, 25.ii.1960, *Wild* 5147 (K, LD, SRGH). SE: Artesia, 19.i.1960, *Leach & Noel* 241 (K, SRGH). **Zimbabwe**. S: Beitbridge. Limpopo R., Sentinel Ranch, 25.iii.1959, *Drummond* 5991 (K, SRGH). **Mozambique**. GI: Gaza, Dumela, 30.iv.1961, *Drummond & Rutherford-Smith* 7607 (K, PRE, SRGH).

Native to southwestern United States and Mexico, introduced into the W Indies, tropical Africa and the Transvaal; frequently cultivated, even in Temperate regions when plants can get extremely large. Occurs as a weed in low rainfall areas, often on riverbanks and, in the Flora area, along roadsides; 0–2200 m.

Conservation Status: Although naturalized it is probably under-collected in the Flora area; considered as LC (Least Concern).

133. **PARTHENIUM** L.

Parthenium L., Sp. Pl. **2**: 988 (1753); Gen. Pl., ed. 5: 426 (1754). —Rollins in Contrib. Gray Herb. **172**: 1–72 (1950). —Wild in Kirkia **6**(1): 8–9 (1967). —Stuessy in Ann. Missouri Bot. Gard. **62**(4): 1094 (1975). —Hind in Fl. Masc., Composées **109**: 214–216 (1993). —Karis & Rydberg in Bremer, Asterac. Cladist. Classif.: 622–623 (1994). —Mesfin Tadesse, Fl. Ethiopia & Eritrea **4**(2): 339 (2004). —Beentje & Hind in F.T.E.A., Compositae **3**: 815–816 (2005). —Panero in Kubitzki, Fam. Gen. Vasc. Pl. **8**: 445 [2006](2007). —Chen Yousheng & Hind in Fl. China **20-21**: 877 (2011).

Hysterophorus Adans., Fam. Pl. **2**: 128 (1763).

Villanova Ortega, Nov. Rar. Pl. Descr. Dec. **4**: 47, t. 6 (1797), nom. rej., non Lag. (1816).

Argyrochaeta Cav., Icon. **4**: 54, t. 378 (1798).

Bolophyta Nutt. in Trans. Amer. Philos. Soc. ser. 2, **7**: 347 (1840).

Echetrosis Phil. in Anales Univ. Chil. **43**: 504 (1873).[156]

Herbs or shrubs. Leaves alternate, entire to deeply pinnately lobed. Inflorescences paniculate, terminal. Capitula small, radiate (usually inconspicuously so), heterogamous; involucre campanulate or hemispherical; phyllaries 2- or 3-seriate, imbricate, ± gradate to subequal, outer somewhat smaller than inner; receptacle convex or conic, disc florets subtended by paleae, paleae with inrolled margins and densely pubescent apices. Ray florets 5, fertile, female, corollas white or greenish-white,

---

[156] In a separately paginated reprint/preprint in Kew this appeared on p. 28.

limb small, suborbicular, 2- or 3-lobed; style arm apices bifid, with paired stigmatic lines. Disc florets hermaphrodite, sterile, usually numerous, corollas whitish to yellowish, puberulous and gland-dotted outside, narrowly funnel-shaped, 4- or 5-lobed; anthers connate, anther thecae pale; apical anther appendages broadly triangular, apices obtuse, anther bases obtuse; styles entire and obtuse. Achenes of ray florets obcompressed, keeled, crowned by persistent rays and clasped by a phyllary and two lateral concave paleae; carpopodium inconspicuous; pappus of 2(3) recurved awns forming an apparent margin to achene and enclosed by lateral paleae, or absent.

A genus of 16 spp. from N and Central America with one species a widespread pantropical weed which is present in the Flora area. One herbarium record suggests that *P. argentatum* A.Gray has been cultivated in Zimbabwe at the Experimental Station in Harare [Salisbury] (*Arnold* s.n. [SRGH13197], 24.i.1945, K, SRGH). *Parthenium argentatum* is a shrub with densely leafy stems, a characteristically long naked peduncle beneath a dense corymbose cluster of capitula, and yellow pollen.

Parthenium hysterophorus L., Sp. Pl. **2**: 988 (1753). —S. Moore in Fawcett & Rendle, Fl. Jamaica **7**: 214, fig. 67 (1936). —Humbert, Fl. Madagasc., Composées **189**: 636, t. cxv fig. 13–21 (1963). —Wild in Kirkia **6**(1): 8–9 (1967). —Stuessy in Ann. Missouri Bot. Gard. **62**(4): 1094–1096 (1975). —Hind in Fl. Masc., Composées **109**: 214–216 (1993). —Mesfin Tadesse, Fl. Ethiopia & Eritrea **4**(2): 339–340 (2004). —Beentje & Hind in F.T.E.A., Compositae **3**: 815–816 (2005). —Chen Yousheng & Hind in Fl. China **20-21**: 877 (2011). Type: 'Habitat in Jamaicae glareosis'. 'ex H.U. [Hortus Upsaliensis]' (LINN – Herb. Linn. No. 1115.1. lectotype), lectotypified by Stuessy (1975: 1094, albeit as holotype). FIGURE 6.5.**26**.

 Argyrochaeta bipinnatifida Cav., Icon. **4**: 54, tab. 378 (1791). Type: 'Habitat prope Ixmiquilpan oppidum mexicanum, ubi eam reperit *D. Ludovicus Née*. Vidi vidam in Regio horto Matiritense florentem mense Iulio anni 1797. ... Obs. Descriptiones huius plantae et sequentium numeris 417, 418, 432, 435, 437 legi in Regia Medicinae Academia die 12 Octobris. Eas postea die scilicet 16 Novembris publici iuris fecit Matriti D. Casimirus Ortega in opusculo cui titulus *Novarum aut rariorum plantarum hortii Regii botanici Matritensis descriptionum decades.*' (?MA syntypes).

 Argyrochaeta parviflora Cav., Descr. Pl.: 233 (1802). Type: 'Se cria en Cumaná: nació de semillas enviadas por el cuidadano *Bompland*; florecio desde Julio hasta Octubre en el Jardim Botanico.' (?MA holotype).

 Parthenium pinnatifidum Stokes, Bot. Mat. Med. **4**: 278 (1812), nom. illeg. superfl. based on *Parthenium hysterophorus* L.

 Parthenium lobatum Buckley in Proc. Acad. Nat. Sci. Philadelphia **13**: 457 (1861)[1862] Type: [USA:] 'Western Texas. June.[*S. B. Buckley*]' (?PH holotype; ?P-DU).

 Echetrosis pentasperma Phil. in Anales Univ. Chile **43**: 504 (1873).[157] Type: 'Se halla en la provincia de Mendoza de la República Arjentina.'[158]

Annual herb c. 0.6 m tall. Stems branched towards inflorescence, striate, puberulous, scabrid. Leaves sessile, up to (6)10–20(30) × 3–10 cm, ± elliptic or ovate in outline, profoundly bipinnatisect, ultimate segments obtuse, segments 1.5–4 mm wide, slightly puberulous above, somewhat more densely so below. Inflorescence a lax panicle, to 10 cm diam., branches of inflorescence slender, scabrid, pedicels 3–10 mm long. Capitula hemispherical, 3–5 mm diam.; phyllaries 2-seriate, 2–3 mm long, outer oblong-ovate, 1–3 mm long, apex obtuse, entire, abaxially pubescent, inner oblong, 1–3 × 2–3 mm, glandular-punctate above, glabrous below; lateral paleae of ray florets conduplicate and clasping achenes (forming an 'achene-palea complex' for dispersal), apex truncate and tapering downwards, c. 3.3 mm long, apex glandular-puberulous; disc paleae c. 2.3 mm long, membranous, apex ciliate. Ray limbs whitish to creamish, ovate, 0.5–0.7 × 0.5–1 mm wide. Disc florets yellowish, 25–50, c. 2.7 × 0.3 mm, narrowly funnel-shaped, throat c. 1 mm long, limb 4–5-dentate; anthers dark grey or whitish; pollen white; achene filiform, abortive. Achenes broadly obovoid, ± compressed, black, c. 2 ×1.2–1.5 mm, apex setuliferous; pappus of 2 recurved awns forming an apparent rim to achene, c. 5 mm long.

---

[157] In a separately paginated reprint/preprint in K this appeared on p. 28.

[158] Muñoz Pizarro (1960: 140) cited no material in SGO against this name.

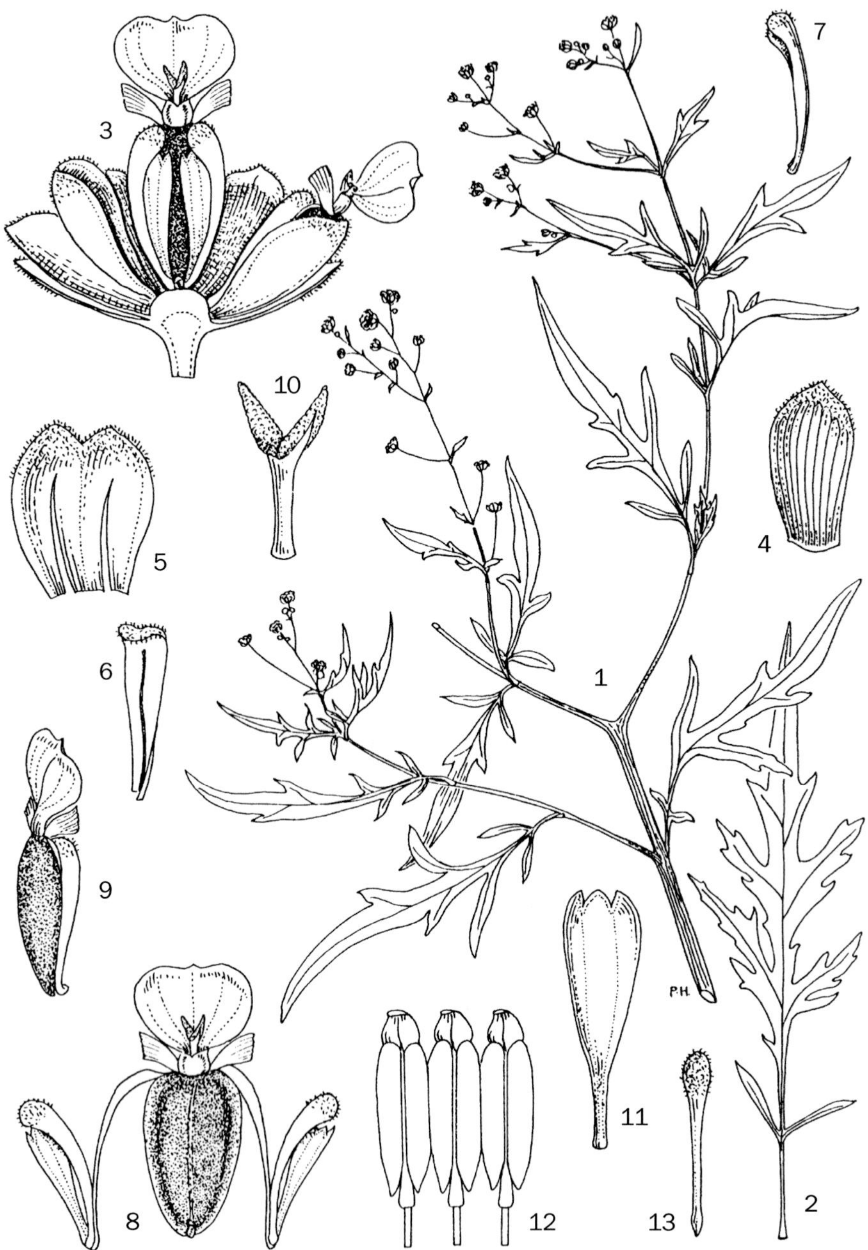

Fig. 6.5.**26**. PARTHENIUM HYSTEROPHORUS. 1, flowering branch ($\times 2/3$); 2, basal leaf ($\times 2/3$); 3, l.s. capitulum ($\times 12$); 4, outer phyllary ($\times 12$); 5, inner phyllary ($\times 12$); 6, palea, abaxial view ($\times 12$); 7, palea, side view ($\times 12$); 8, adaxial view of ray floret complex, showing central ray floret and associated disc florets and adnate paleae ($\times 12$); 9, ray floret complex in profile ($\times 12$); 10, ray floret style ($\times 36$); 11, disc floret corolla ($\times 24$); 12, partial anther cylinder opened out ($\times 36$); 13, disc floret style ($\times 36$). All from *Cadet* 2522. Drawn by Pat Halliday. From Flore des Mascareignes.

Mozambique. M: Polana, 61 m, viii.1961, *Miller* 8002 (K, SRGH).

Native of S and Central America from Chile and the Argentine to Mexico, also in Texas and Florida. An introduced weed in the Lourenço Marques region and Madagascar. Common at roadsides and in disturbed ground. Most certainly under-collected (all from Maputo) rather than particularly rare, and only rarely recorded in the neighbouring FTEA area.

Conservation Status: Probably best recorded as DD (Data Deficient), but likely to be LC (Least Concern) once more fieldwork has been carried out.

134. **AMBROSIA** L.

Ambrosia L., Sp. Pl. **2**: 987 (1753); Gen. Pl., ed. 5: 425 (1754). —Wild in Kirkia **6**(1): 1–3 (1967). —Hind in Fl. Masc, Composées **109**: 212–214 (1993). —Mesfin Tadesse, Fl. Ethiopia & Eritrea **4**(2): 340 (2004). —Beentje & Hind in F.T.E.A., Compositae **3**: 813–815 (2005).

Gaertneria Medik., Philos. Bot. **1**: 45 (1789), non *Gaertneria* Schreb. (1789), nec *Gaertneria* Retz. (1791), nec *Gaertneria* Lam. (1791)

Franseria Cav., Icon. **2**: 78 (1793), nom. cons.

Hemiambrosia Delpino, Studi Lign. Anemof.: 16 (1871).

Hemixanthidium Delpino, Studi Lign. Anemof.: 17 (1871).

Xanthidium Delpino, Studi Lign. Anemof.: 17 (1871).

Acanthambrosia Rydb. in N. Amer. Fl. **33**: 22 (1922).

Annual or perennial, often aromatic, herbs or shrubs. Stems erect or ascending to decumbent or prostrate, poorly branched or well branched near base. Leaves petiolate or sessile, alternate or sometimes opposite, simple and ovate to lanceolate or variously dissected to bipinnatisect, sparsely to densely pubescent, usually glandular-punctate on both surfaces. Capitula small, unisexual, few- to many-flowered. Female capitula sessile or clustered in upper axils of upper leaves and bracts and below male capitula on inflorescence branches, 1–7-flowered; involucre fused into a hard bur, usually with few to many rows of spines, surface pubescent, glandular-punctate, beak smooth, truncate, or with 2–5 teeth. Female florets with only styles, style arms elongated, linear, and separate marginal stigmatic lines. Male capitula spicate or racemose, sessile or short-pedicellate, often pendent, usually maturing simultaneously; florets 6–65 per capitulum; phyllaries connate into a funnel-shaped involucre; receptacle paleaceous, paleae filiform or absent. Male corollas yellowish to whitish, 5-lobed; anthers usually 5, ± free, anther-thecae entire at base, filaments often fused at base; style without swollen base, style arms connate, penicillate, papillae spreading. Achenes ovoid or globose, enclosed in tightly clasping female involucre which develops a circle of 4–6 tubercles or spines and is narrowed above into a beak, achene body black; pappus absent.

A genus of c. 40 spp., widespread in N America south to S America and in the eastern hemisphere; widely introduced. One somewhat weedy species occurs in the Flora area.

Ambrosia maritima L., Sp. Pl. **2**: 988 (1753). —Oliver & Hiern in F.T.A. **3**: 370 (1877). —Adams in F.W.T.A., ed. 2, **2**: 268 (1963). —Humbert, Fl. Madagasc. Composées **189**: 628, t. 115 fig. 1–4 (1963). —Mesfin Tadesse, Fl. Ethiopia & Eritrea **4**(2): 341–342 (2004). —Beentje & Hind in F.T.E.A., Compositae **3**: 813–815 (2005).Type: 'Habitat in Hetruriae, Cappadociae maritimis arenosis.' (BM000647392 – Herb. Clifford: 403, Ambrosia 1, lectotype), lectotypified by Alavi in Jafri & El-Gadi, Fl. Libya **107**: 120 (1983). FIGURE 6.5.**27**.

Ambrosia senegalensis DC., Prodr. **5**: 525 (1836). Type: 'secus ripas fl. Senegal locis humidis argillosis leger. cl. *Bacle* et *Perrottet*. ... (v. s.)'. 'Senegal. M. *Bacle* 1820' (G-DC-G00469964 syntype); 'Senegal. *Perrottet* [s.n.]' (G-DC-G00469965 syntype – two duplicates, one without a printed barcode), 'Sur les bords humides de Sénégal dans les terrain argileux le 15 9bre 1828, *Perrottet* [?]401' (G-DC-G00469966 syntype – two duplicates, one without a printed barcode); 'Places de [?] sur le bord de fleuve au Senegal le 16 Septem. 1824 [?] 1825, *Perrottet* [s.n.]' (G-DC-G00469965 syntype); *Perrotet* 437 (P00064061 syntype); *Perrotet* s.n. (P00064062 syntype); *Perrotet* s.n. (P00365720 syntype).[159]

---

[159] P syntypes all determined by Beentje.

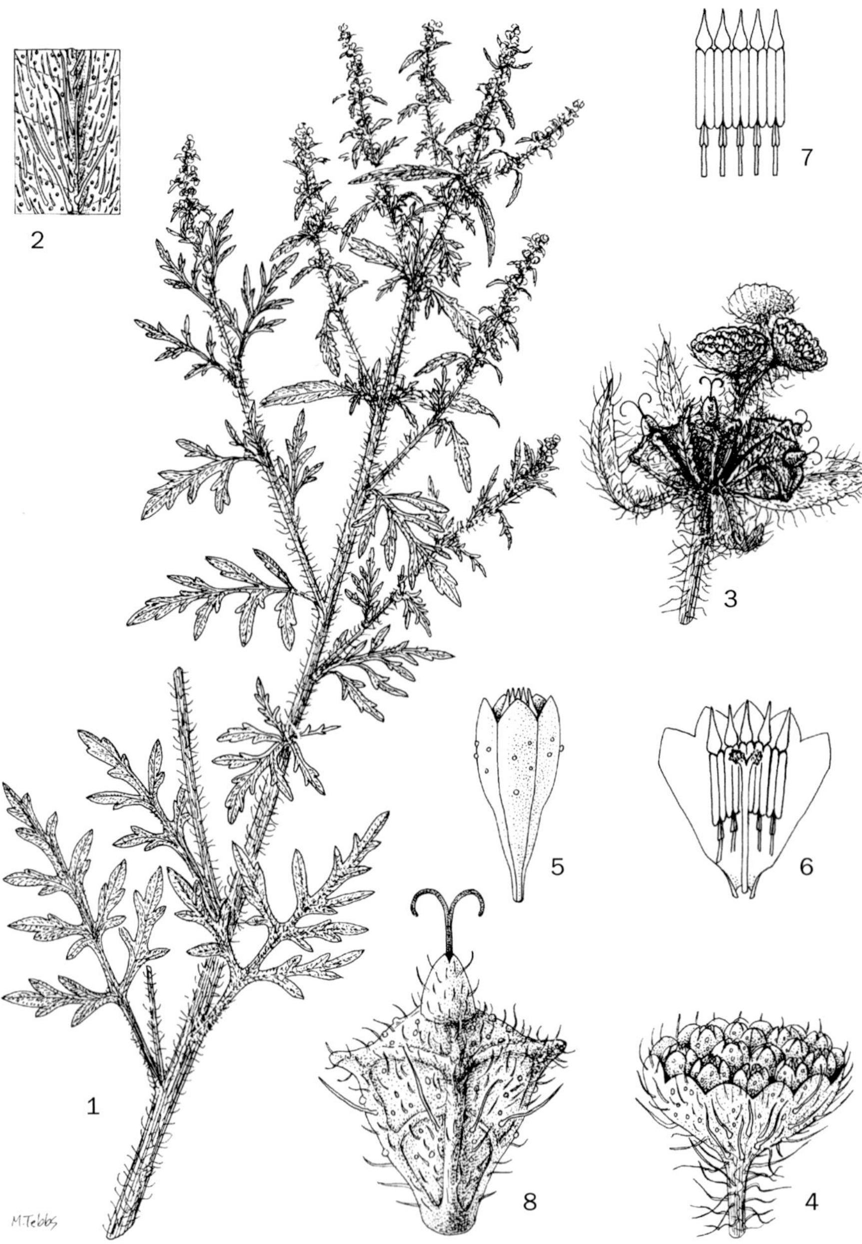

Fig. 6.5.**27**. AMBROSIA MARITIMA. 1, flowering shoot (× 2/3); 2, detail of abaxial leaf surface (× 9); 3, male (upper) and female (lower) capitula (× 4); 4, male capitulum (× 10); 5, male corolla (× 14); 6, male corolla opened out showing attachment point of filaments (× 14); 7, anther cylinder opened out (× 14); 8, female floret (× 14). All from *Bautista et al.* 3457. Drawn by Margaret Tebbs. From Flora of Tropical East Africa.

Annual poorly branched herb, up to 1 m tall, aromatic. Stems subcylindrical, striate, pilose. Leaves alternate, petiolate; petioles up to 5 cm long, pilose, lamina up to c. 8 cm long, ovate, bipinnatipartite, divided almost to rachis, lobes ± obtuse, sometimes serrate-crenate or lobulate, pubescent above, more densely so and ± discolorous below. Male capitula in long racemes 2–13 cm long, subsessile or on short pubescent pedicels up to 2 mm long; involucre 2–5 mm diam., funnel-shaped or hemispherical, margins dentate, appressed-pubescent; florets up to 20 per capitulum, corollas white, c. 2.5 mm long, campanulate, narrowed into a short tube below; anthers exserted, apical anther appendages linear. Female capitula few together below base of male inflorescences; involucre up to 6 mm long, turbinate, neck c. 2 mm long, apex minutely bifid, with 4 or 5 spines pointing slightly upwards c. 2 mm long straight or slightly upcurved placed in a ring about up from base of body of involucre, surface ± reticulate, minutely glandular-puberulent and sparsely pilose; style arms exserted, appendages subulate. Achene dark greyish, c. 1.2 mm long, ovoid, smooth.

Botswana. SE: Gaborone Dam (E. side), s.dat. *Turton*, s.n. (K). **Zambia**. B: Barotseland, i.1924, *Borle* 7 (PRE, SRGH). S: Mazabuka, Kafue Polder, ll.i.1963, *van Rensburg* 1191 (K, SRGH). **Zimbabwe**. E: Umtali, 7.xii.1955, *Chase* 5903 (BM, K, SRGH). **Malawi**. S: Shire R., 15.i.1963, *Kirk* (K). **Mozambique**. Z: Quelimane, 1908, *Sim* (PRE). MS: Gorongoza, 27.ix.1953, *Chase* 5095 (BM, SRGH). GI: Guijá, 9.vi.1947, *Pedrógão* 268 (K, PRE).

Widespread in the Mediterranean Region, E & W tropical Africa, also in NE Africa, the Mascareignes and S Africa. Floodplains, on disturbed ground, alluvial basins, near rivers and in coastal areas; 0–1200 m.

Conservation Status: a relatively widespread weed in the Flora area, albeit relatively infrequently collected; LC (Least Concern).

135. **XANTHIUM** L.

Xanthium L., Sp. Pl. **2**: 987 (1753); Gen. Pl., ed. 5: 424 (1754). —Widder in Repert. Spec. Nov. Regni Veg., Beih. **20**: 1–222 (1923). —Widder in Repert. Spec. No. Regni Veg. **41**(Nr. 1059/1070): 272–284. —Löve & Danserau in Canad. J. Bot. **37**: 173–208 (1959). —Wild in Kirkia **6**(1): 3–5 (1967). —Widder in Phyton (Horn) **12**(1–4): 182–190 (1967). —Löve in Lagascalia **5**: 55–71 (1975). —McVaugh, Fl. Novo-Galiciana **12**: 1092–1095. —Hind in Fl. Masc., Composées **109**: 216–218 (1993). —Boulos & Hind in Fl. Egypt **3**: 227–229 (2002). —Mesfin Tadesse, Fl. Ethiopia & Eritrea **4**(2): 341 (2004). —Beentje & Hind in F.T.E.A., Compositae **3**: 817–818 (2005). —Chen Yousheng & Hind in Fl. China **20–21**: 875–876 (2011).

Xanthium sect. *Acanthoxanthium* DC., Prodr. **5**: 523 (1836).

Xanthium sect. II. *Akanthoplium* Wallr. in Beitr. Bot. **1**(2): 228/241 (1844), nom. illeg. superfl., based on *Xanthium* sect. *Acanthoxanthium* DC.

Acanthoxanthium (DC.) Fourr. in Ann. Soc. Linn. Lyon, sér. 2, **17**: 110 (1869).

Erect and unbranched, or poorly-branched and decumbent, annual herbs. Stems simple to moderately-branched, axils unarmed/spineless or with 1–3-furcate spines. Leaves alternate, petiolate, petioles very short to markedly long, lamina 3-veined, entire, 3-lobed or pinnatifid, ovate-triangular to broadly lanceolate, apices rounded to acute, strigose-pubescent above and beneath, or just above, or sparsely pubescent above and white-tomentose beneath, eglandular or glandular-punctate, glands yellowish. Inflorescences of axillary male and female capitula. Male capitula globose or subglobose, pedicellate, solitary or few at apices of inflorescences, erect, many-flowered; phyllaries ± uniseriate, few; receptacle convex, paleaceous, paleae linear to linear-spathulate or oblanceolate, margins laciniate, surface sometimes pubescent, apices acute to acuminate; corollas funnelform, glabrous or pubescent, 5-lobed, lobes smooth; filaments fused throughout length into tube, anther thecae free, apical anther appendages minute, markedly inflexed; style undivided; ovary rudimentary. Female capitula sessile, produced laterally and axillary beneath male capitula, 2-flowered; phyllaries connate fused into hard bilocular bur; corollas absent; style bifid, style arms linear protruding adaxially to beak of bur, stigmatic lines marginal and separate, stigmatic papillae dense. Fruits (burs) ellipsoid, surface usually densely spiny, spines uncinate, body pubescent between spines, apex with pair of beaks, usually prominent, sometimes obscure; pappus absent.

A genus of three spp. Two spp. are widespread cosmopolitan weeds, particularly in the tropics and subtropics. Early treatments of the genus suggested that there were as many as 25 spp., although the work of Löve & Dansereau (1959) and Cronquist in Cronquist *et al.*, Intermountain Fl. Vasc. Pl. Intermountain W **5**: 64–65 (1994) promoted a much broader species concept and only two or three species, *X. strumarium* being considered very variable throughout its range. Only a minimal synonymy, based largely on that suggested by Widder (1967), for the two taxa in E Africa has been provided; the synonymy could extend to many dozen names especially following Löve & Danserau's comments!

The term 'conceptacle' has been applied to the bur in both *Xanthium* and *Ambrosia* by Robinson (2006) and Panero (2006), but 'perigynium' has also been used (alongside bur) in the *Flora North America* accounts of these genera (Strother in Fl. N. Amer. editorial comm., Fl. N. Amer. **21**: 19–20, 2006); Cronquist (1994) used bur, the preferred term in this account. 'Conceptacle' is an unusual term, often applied to the fruit case of a sporocarp in ferns – it is not used here.

Leaves broadly deltate, base usually mostly cordate, concolorous; plant spineless . . . .
. . . . **1.** *strumarium*
Leaves narrowly ovate, base cuneate, discolorous; plants with 3-partite axillary spines . . .
. . . . **2.** *spinosum*

1. **Xanthium strumarium** L., Sp. Pl. **2**: 987 (1753). —Oliver & Hiern in F.T.A. **3**: 371 (1877). —Löve & Dansereau in Canad. J. Bot. **37**: 174 (1959). —McVaugh, Fl. Novo-Galiciana **12**: 1093–1095 (1984). —Hind in Fl. Masc., Composées **109**: 217–218 (1993). —Boulos & Hind in Boulos, Fl. Egypt **3**: 228–229 (2002). —Mesfin Tadesse, Fl. Ethiopia & Eritrea 4(2): 341 (2004). —Beentje & Hind in F.T.E.A., Compositae **3**: 817 (2005). —Chen Yousheng & Hind in Fl. China **20–21**: 876 (2011). Type: 'Habitat in Europa, Canada, Virginia, Jamaica, Zeylona, Japonia.' (LINN – Herb. Linn. No. 1113.1 lectotype), lectotypified by Rechinger, Fl. Iranica **164**: 39 (1989). FIGURE 6.5.**28B**.
   Xanthium orientale L., Sp. Pl., ed. 2, **2**: 1400 (1763). Type: 'Habitat in China, Japonia, Zeylona.' (LINN – Herb. Linn. No. 1113.2 lectotype), lectotypified by Jeanmonod in Gamisans & Jeanmonod, Compl. Prodr. Fl. Corse, Asteraceae I: 190 (1998);[160] France: Perigord, rechtes Ufer der Dordogne bei Bezenac, 14.ii.1987, *Wisskirchen* 230 (BM000576318 epitype, BOCH) epitypified by Wisskirchen in Jarvis & Turland (1998: 369).[161]
   Xanthium canadense Mill., Gard. Dict., ed. 8: No. 2 (1768). Type: 'The second sort grows naturally in North America.'
   Xanthium chinense Mill., Gard. Dict., ed. 8: No. 4 (1768). Type: 'The fourth sort grows naturally in China, from whence I have often received the seeds; …'
   Xanthium monoicum Gilib., Fl. Lituan. Inch. **1**: 170 (1782), nom. inval. – opera utique oppressa.
   Xanthium americanum Walter, Fl. Carol.: 231 (1788). Type: not cited.
   Xanthium cuneatum Moench, Suppl. Meth.: 300 (1802), nom. illeg., superfl. pro *X. orientale* L.
   Xanthium cordifolium Stokes, Bot. Mat. Med. **4**: 380 (1812). Type: 'Specimen gathered in Sole's Garden.'
   Xanthium macrocarpum DC., Fl. France, Suppl. [5], **6**: 356 (1815). Type: 'Elle a été trouvée dans les vignes du bas Languedoc, par Mademoiselle Lucie Dunal.' (?G holotype).
   Xanthium homothalamum Spreng., Neue Entdeck. Pflanzenk. **1**: 259 (1819 or before Aug. 1820). Type: 'Habitat in Brasilia, unde Mertensius adlatam largitus est.'[162] Herbarium material unknown.
   Xanthium maculatum Raf., Amer. Monthly Mag. & Crit. Rev. 2(5): 344 (1818); —Amer. J. Sci., Ser. 1(1): 151 (1819), nom. nud.

---

[160] Not by Wisskirchen in Jarvis & Turland in Taxon **47**: 369 (1998).

[161] Because the material Herb. Linn. No. 1113.2 is immature.

[162] Sprengel noted 'Licet haut perfecta huius plantae exemplaria investigare potuerim' suggesting that he had poor material available.

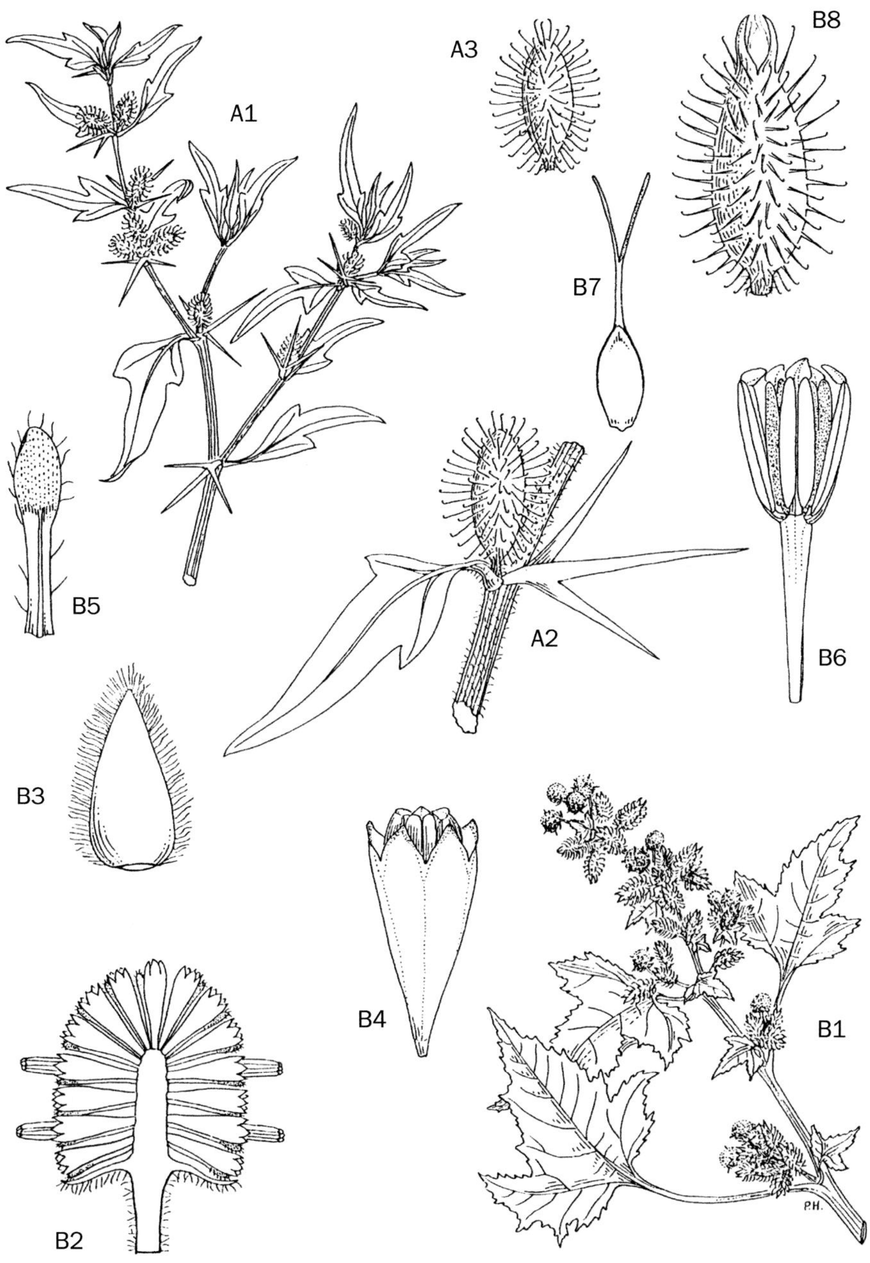

Fig. 6.5.**28**. A. —XANTHIUM SPINOSUM. A1, flowering branch (× 2/3); A2, node showing mature bur (× 2); A3, bur (× 2). B. —XANTHIUM STRUMARIUM. B1, flowering branch (× 2/3); B2, l.s. male capitulum (× 6); B3, phyllary of male capitulum (× 12); B4, male floret corolla (× 16); B5, palea of male capitulum (× 16); B6, anther cylinder of male floret (× 16); B7, female floret (× 6); B8, bur (× 2). A from *Guého* s.n.; B from *Guého* s.n. Drawn by Pat Halliday. From Flore des Mascareignes.

Xanthium italicum Moretti, Diar. Phys. Chem. Hist. Nat. **5**: 8 (1822); —De quibusdam plantis Italiae, Decas Quinta : 8 (1822). Types: 'Habitat in multis Italiae locis: illam inveni prope mare in regione Porto di Fermo, eamque abuntissimam secus decursum fluminis Padi a viciniis Augustae Taurinorum Ticinum usque perspexi.'(L0002841 syntype).[163]

Xanthium occidentale Bertol., Lucubr. Re Herb.: 38 (1822), nom. illeg., superfl. pro *X. orientale* L.

Xanthium brasilicum Vell., Fl. Flumin. Icon. **10**: tab. 23 (1827)[29 Oct. 1831]. Type: not cited.[164, 165]

Xanthium macrocarpum DC. [var.] β *glabratum* DC., Prodr. **5**: 523 (1836), nom. illeg., citing *X. americanum* Walter and *X. canadense* Mill. in synonymy.

Xanthium indicum DC., Prodr. **5**: 523 (1836). Types: 'X. indicum (Roxb.! cat. calc. 67) ... in Ægypto (Coqueb.!), in ruderatis Indiae orientalis frequens, verisim. etiam in Chinâ. X. orientale Linn. sp. 1400 excl. syn. X. Chinense Mill. dict. n. 5? X. Indicum Wall.! cat. et herb. n. 291. Wight herb. ... (v. s.)' *Coquebert* s.n. (G-DC-G00469885 – mounted with [Wallich 3181/]291B syntype); *Roxburgh* s.n. (G-DC-G00469892 – mounted with [Wallich 3181/]291B syntype); *Wallich* [3181/]291 [A Sillet FD.] (G-DC-G00469890[166] syntype); *Wallich* [3181/]291[B Xanth. Strumarium Hb. Wight e Trichinopoly] (G-DC-G00469891 – mounted with *Roxburgh* s.n., G-DC-G00469886 – mounted with *Coquebert* s.n. syntypes); *Wallich* [3181/]291[C var. cordatum, Irawaddi 1826] (G-DC-G00469889 – 'C. var. cordata ripa Irrawaddy inter Yandabo et Pagamen. (X. cordatum Burm. car. 575)' syntype).[167]

Xanthium inaequilaterum DC., Prodr. **5**: 523 (1836). Type: '... in ruderatis ins. Javae prope Buitenzorg legit cl. Blume. ... (v.s. comm. à cl. invent.)' (G-DC-G00469911[168] holotype).[169]

Xanthium strumarium var. *canadense* (Mill.) Torr. & A. Gray, Fl. N. Amer. **2**: 294 (1841–43).

Xanthium priscorum Wallr., Beitr. Bot. **1**(2): 227 (1844). Type: not cited, but the same as in *Xanthium antiquorum* Wallr., which was simultaneously published in the same work.[170]

Xanthium abyssinicum Wallr., Beitr. Bot. **1**(2): 227/230 (1844). Type: '*Xanthium strumarium* Schimp. it. abyss. II. n. 1343./Auf den Sorgho-Feldern Abyssiniens von *Schimper* d. 19. Juli. 1838 entdeckt. (Herb. amic. Lucae.)' (KEIL† holotype, BR8362438, K000410236, M0105425, NY00278110, S07-16469).

Xanthium pungens Wallr., Beitr. Bot. **1**(2): 231 (1844). —Wild, Common Rhod. Weeds: t. 78 (1955). Type: not cited.

Xanthium laevigatum Muhl. ex Wallr., Beitr. Bot. **1**(2): 228/231 (1844). Type: 'Willd. herb. n. 17469. fol. 7. ... In Nordamerika von Mühlenberg gesammelt und an Willdenow eingeschickt.' (B-W holotype).

Xanthium discolor Wallr., Beitr. Bot. **1**(2): 228/232 (1844). Type: '*Xanthium indicum* Wallich. herb. indic. (1824.) n. 3181. A. in herbar. gener. berolin., ...' (B† holotype).

Xanthium roxburghii Wallr., Beitr. Bot. **1**(2): 228/233 (1844). ?Type/s: '*Xanthium strumarium* Willd. herb. n. 17469. fol 6. ... In Ost-Indien. *Roxburgh*.' (?B-W holotype).[171]

Xanthium brevirostre Wallr., Beitr. Bot. **1**(2): 228/235 (1844). Types?: '*Xanthium indicum* herb. Wight. n. 1446 [In Ostindien. Wight. (Herbar. gener. berol.)], ... var. *cordata* pl. *Wallich.* n. 3181.' (?B†, K-W syntypes).

---

[163] L0002841 is marked by Widder as an 'Originalexemplar' and is certainly a iso/syntype, apparently regarded as the lectotype according to annotations, although this is unlikely.

[164] The text provided in Arch. Mus. Nac. Rio de Janeiro **5**: 399 (1881) gave 'Habitat maritimis aeque, ac mediterraneis ad loca strumaria.'

[165] The name is valid as it is accompanied by a diagnostic dissection.

[166] To which is attached an unbarcoded duplicate with a label in pencil with the barcode number written on it, the plant labelled simply as '291a'

[167] An unbarcoded duplicate of 'G00469886' is mounted with a '*Royle* 170 Cachemire' collection, G-DC-G00469887) suggesting that this is also a further syntype.

[168] A det. slip, by Rolf Wisskirchen in 1996, is attached to the sheet noting that this is 'A part of the original specimen collected by Blume', which may be why the material is considered an isotype.

[169] It is probable that the 'original' material is in L.

[170] In Wallroth's account this later appeared as *Xanthium antiquorum* Wallr. (1844: 229), q.v.

[171] This would also appear to be a 'nom. nov.' for '*Xanthium indicum* Roxb.', a nom. nud.

Xanthium pensylvanicum Wallr., Beitr. Bot. **1**(2): 228, 236 (1844), non Gand. (1918). Types: 'α glandulosum *Xanthium strumarium* Beyrich in Herb. amic. Sporleder. … von Beyrich auf unfruchtbaren Plätzen bei Ashville (1833) häufig'; 'β *Xanthium occidentale* Poepp. fl. pensylv. … von *Poeppig* auf feuchten Wiesen in Pensylvanien im Monat September 1824 entdeckt.'[172]

Xanthium longirostre Wallr., Beitr. Bot. **1**(2): 228/237 (1844). Types: '*Xanthium macrocarpum* C. Ehrenb. n. 195. … In Westindien auf St. Thomas und auch auf Haiti von C. Ehrenberg gesammelt (herb. gener. berolin.), und vermuthlich früher schon auf St. Domingo von Bertero entdeckt.' (B† syntype).

Xanthium saccharatum Wallr., Beitr. Bot. **1**(2): 228/238 (1844). Type: '*Xanthium macrocarpum* Berlandier n. 1865.' Type: (BM001009565, GH00014130 ?syntypes).[173]

Xanthium oviforme Wallr., Beitr. Bot. **1**(2): 228/240 (1844). Type: '*Xanthium canadense* Hook. in litt. (herb. general berol.), nec Herm., Mill. et Linn./Angeblich in Nordamerica und vermutlich in Canada.' (B† holotype, K001065884 − ex Herb. Benthamianum).

Xanthium antiquorum Wallr., Beitr. Bot. **1**(2): 229 (1844). Types: '*Kotsch.* iter. nubic. n. 319., *Ehrenb.* fl. dalmat. cent. VIII. n. 83./*Xanthium dioscoridis* Gundelsh. in herb. general. berol. …/An den Ufern des Nils, z. B. bei Chartun in der Provinz Sennaaroder des Tigris. Im Monat Mai mit reifen Früchten; auch in Dalmatien, z. B. bei Castel nuovo *von Ehrenberg* gefunden und vermuthlich im Orient ziemlich verbreitet.'

Xanthium cavanillesii Schouw in Ann. Sci. Nat. Bot., ser. 3, **12**: 357 (1849). Type: 'Buenos-Ayres. Didrichsen.' [ex 'Index seminum horti academici Hauniensis'.]

Xanthium riparium Itzigs. & Hertsch in Bot. Zeitung **12**(2): 34 (1854), nom. illeg., superfl. pro *X. macrocarpum* DC.

Xanthium arenarium Lasch in Bot. Zeitung **14**(24): 411 (1856). Type: not cited, nor for the four unranked named infraspecific taxa. Herbarium material unknown.

Xanthium riparium Lasch in Bot. Zeitung **14**(24): 412 (1856). Type: not cited, nor for the four unranked named infraspecific taxa. Herbarium material unknown.

Xanthium macrocarpum var. *laciniatum* Pouzolz, Fl. Dép. Gard **2**(1): 2, tab. 6 (1862). Type: 'Le var. B, rare dans les vignes, à Manduel.' Herbarium material unknown.

Xanthium strumarium var. *arenarium* (Lasch) R.Uechtr. in Verh. Bot. Vereins Prov. Brandenburg **3–4**: 210 (1862).

Xanthium fuscescens Jord. & Fourr., Brev. Pl. Nov. fasc. **1**: 36 (1866). Type: 'Hab. in arenosis subhumidis Corsicae; Biguglia prope Bastia, ex *dom. E. Revelière*.' Herbarium material unknown.
Xanthium nigri Ces., Pass. & Gibelli, Comp. Fl. Ital. **2**(19): 437 (1877). Type: 'Raccolto alle paludi delle Apertole nel Vercellese dal *Sig. Avv. F. Negri*. Settembre 1869.' Herbarium material unknown.

Xanthium sphaerocephalum Salzm. ex Ball in J. Linn. Soc., Bot. **16**: 503 (1878), nom. nud. pro syn.

Xanthium speciosum Kearney in Bull. Torrey Bot. Club **24**(12): 574 (1897). Type: [USA: Eastern Tennessee:] 'Collected by the writer September 16 [1897], near Wolf Creek Station ([T.H. Kearney, Jr.] no. 785) where it grows on the sandy bottom-lands near the French Broad River and is almost certainly indigenous.' (OS0000379 holotype, US313095 = US00128551).

Xanthium varians Greene, Pittonia **4**(21): 59 (1899). Type: 'Shady banks of the Columbia River, Klickitat Co., Washington, Oct. 1893, *W. N. Suksdorf*, n. 1583, distributed as *X. strumarium*.' (?WIS holotype,[174] F0051946, GH00014131, MICH1107924, NDG59621, NY00278108, NY00278109, OSC0000299, US228570 = US00128552).

Xanthium affine Greene, Pittonia **4**(21): 60 (1899). Type: 'Habitat of the preceding species, and by the same collector, distributed without a specific name, under [*Suksdorf*] n. 1584.' (?WIS holotype,[164] holotype, F0051933, GH00014152, MICH1107923, NDG59598, NY00278099, OSC0000298, US228571 = US00128540).

Xanthium silphiifolium Greene, Pittonia **4**(21): 60 (1899), as '*silphifolium*'. Type: 'The type of this strongly marked species is of *Mr. Suksdorf's* collecting from the banks of the Columbia, Sept., 1883, the specimen preserved in the U.S. Herbarium.' (US46671 = US00128550 holotype, F ×2, NY00278106).

---

[172] The two varieties were described at the same time without reference as to which was typical.

[173] Both duplicates are considered to be isotypes.

[174] Originally in LCU but probably now in WIS, q.v. Tucker *et al.*, Taxon **38**(2): 202 (1989).

Xanthium glanduliferum Greene, Pittonia **4**(21): 61 (1899). Type: 'Collected at Walsh, Assiniboia, 15 Aug., 1895, by *Mr. John Macoun*, and distributed for *X. Canadense*, but the species is evidently new, and thoroughly distinct. The ticket accompanying my specimen bears the Canadian Survey number 10,910.' (?WIS holotype[174], NDG59611).

Xanthium campestre Greene, Pittonia **4**(21): 61 (1899). Types: 'Fertile plains of the Sacramento River, in middle and northern California; the best specimens collected by myself, near Chico, June, 1890; but there exists in the U.S. Herbarium a good one from the Wilkes Expedition obtained near Sacramento.' Although unnumbered, there is a *Pickering* 1361 which may represent the latter collection – but is was also used by Widder as the type of his *Xanthium decalvatum*, q.v. (?WIS syntype).[174]

Xanthium californicum Greene, Pittonia **4**(21): 62 (1899). Type: 'Common in middle California, especially about San Francisco Bay, being the *X. Canadense* of my Manual and of the Flora Franciscana in large part.' (?WIS holotype).[174]

Xanthium acutum Greene, Pittonia **4**(21): 62 (1899). Type: 'Known by a single specimen obtained at Stockton, California, by *Mr. J. A. Sanford*, in 1888.' (?WIS holotype).[174]

Xanthium palustre Greene, Pittonia **4**(21): 63 (1899). Type: 'Known only from the brackish marshes of Suisun Bay, middle California. An exceedingly well marked species, referred to by me as an indiginous state of *X. Canadense* in the Flora Franciscana.' (NDG59612, NDG59613 syntypes).

Xanthium acerosum Greene, Pittonia **4**(21): 63 (1899). Type: 'Known only from the valley of the Red River of the North, where it was collected by the writer [*E.L. Greene*], near Fargo, North Dakota, 4 Sept., 1893.' (?WIS holotype).[174]

Xanthium glabratum (DC.) Britton, Man. Fl. N. States: 912 (1901).

Xanthium commune Britton, Man. Fl. N. States: 912 (1901). Type: 'Type collected by *N. L. Britton* at Westport, N.Y.' (NY00278101 holotype).

Xanthium macounii Britton, Man. Fl. N. States: 913. (1901). Type: 'Goose Island, Lake Winnipeg, Manitoba, *J. M. Macoun*, Aug. 16, 1884. Specimen in the herbarium of the Geological and Natural History Survey of Canada.'

Xanthium commune forma *wootonii* Cockerell in Proc. Biol. Soc. Washington **16**: 9 (1903). Type: [USA:] 'New Mexico: Las Vagas. *Cockerell*.'(US404186 holotype, COLO00412916 – marked as 'COTYPE', GH00014155 – Cockerell's label indicating only '1902', NY00278102 – both Cockerell's label and the main label are numbered '15', the material apparently collected 'Oct.4.1901', RM0001508 – Cockerell's label indicating '1902' and is numbered '61').[175]

Xanthium inflexum Mack. & Bush in Rep. (Annual) Missouri Bot. Gard. **16**: 106 (1905). Type: 'Sandy bottoms along the Missouri River in western Misosuri. … MISSOURI: Courtney, *Bush* 869, September 13, 1900, 1916, October 5, 1903, type, 1804, 1806, October 21, 1902.' (MO714922 holotype).

Xanthium oligacanthum Piper, Contr. U.S. Natl. Herb. **11**: 551 (1906). Type: 'Bolles, Walla Walla County, *Piper*, September 18, 1893; also found at Waitsburg by *Horner* (no. B272). The type is in the National Herbarium.' (US528824 = US00128548, F0051944, GZU000267239, NY00278105 syntypes).[176]

Xanthium bubalocarpon Bush in Rep. (Annual) Missouri Bot. Gard. **17**: 123 (1906). Type: 'TEXAS. Dallas County, [Common on prairie.] *Bush* 1185, September 29, 1900.' (?MO holotype, F0051934 – 2 burs and a photograph of the type in MO, US386828 = US00128543).

Xanthium wootoni Cockerell ex De Vries, Sp. Var., ed. 2: 140 (1905). Type: 'Last year a very curious instance of a partial loss of prickles was discovered by Mr. Cockerell of East Las Vegas in New Mexico. It is a variety of the American cocklebur, often called sea-burdock, or the hedgehog-burweed, a stout and common weed of the western States. Its latin name is *Xanthium canadense* or *X. commune* and the form referred to is named by Mr. Cockerell *X. Wootoni*, in honour of Professor E. O. Wooton who described the first collected specimens.'[177]

---

[175] From Cockerell's small labels on the GH, NY, and RM material the suggestion is that they are duplicates, most probably collected in 'San Miguel County' – the suggestion of the databased material from the Rocky Mountain Herbarium.

[176] There is a specimen marked as 'holotype' in US(528824 = 00128548), although clearly this can only be a syntype.

[177] Although appearing to describe a new species it is quite likely that this is to be treated as a new combination based on Cockerell's f. described in 1903; the types are listed above.

Xanthium barcinonense Sennen in Bull. Géogr. Bot. **24**(Nos. 295–297): 224 (1914). Type: 'Hab.–Catalogne: Barcelone à Can Tunis et à la Farola; Prat del Llobregat.' Herbarium material unknown, but probably in BCN.

Xanthium globosum C.Shull in Bot. Gaz. **59**: 482 (1915), nom. inval. as a provisional name. Type: not cited.[178, 179]

Xanthium pensylvanicum Gand. in Bull. Soc. Bot. France **65**: 54 (1918), nom. illeg., non Wallr. (1844). Type: 'Hab.: America bor., Pensylvania Delaware ad Darby Creek (*Mac Elwee* n. 1316!). Herbarium material unknown.

Xanthium leptocarpum Millsp. & Sherff, Publ. Field Mus. Nat. Hist., Bot. Ser. **4**(1): 3 (1918). Type: '*L.R. Jones*, Burlington, Vermont, September 12, 1896 (type in Herb. Field Museum, cat. no. 430860).' (F430860 = F0051942 holotype, GH00014128 – a fragment of the holotype in a capsule, although the sheet only have one barcode).

Xanthium arcuatum Millsp. & Sherff, Publ. Field Mus. Nat. Hist., Bot. Ser. **4**(1): 4 (1918). Type: '*T.F. Lucy* [14270], river shores and low places, Chemung County, New York, October 11, 1896 (type in Herb. Field Museum, cat. no. 4953)' (F4953 = F0051932 holotype).

Xanthium cylindricum Millsp. & Sherff, Publ. Field Mus. Nat. Hist., Bot. Ser. **4**(1): 4 (1918). Type: '*J.K. Small and A.M. Huger*, Chimney Rock to Hendersonville, North Carolina, October 3, 1901 (type in Herb. Field Museum, cat. no. 401312).' (F401312 = F0051938 holotype).

Xanthium acutilobum Millsp. & Sherff, Publ. Field Mus. Nat. Hist., Bot. Ser. **4**(1): 6 (1918). Type: '*J. Reverchon*, Oak Cliff, Texas, September 2 (type in Herb. Missouri Botanical Garden no. 85603; duplicate sheets in the same herbarium bear the numbers 85470 and 85485).' (MO holotype, MO ×2).

Xanthium crassifolium Millsp. & Sherff, Publ. Field Mus. Nat. Hist., Bot. Ser. **4**(1): 5 (1918). Type: '*B. Mackensen* 123, San Antonio, Texas, October 8 and November 15, 1911 (type in Herb. Field Museum, cat. no. 324122)'. (F324122 = F0051937 holotype).

Xanthium curvescens Millsp. & Sherff, Publ. Field Mus. Nat. Hist., Bot. Ser. **4**(2): 25 (1919). Type: 'VERMONT: Orwell, *Willard W. Eggleston* 1420 [Sept. 23 1899] (type in Hb. Gray).' (GH00014158 holotype, US364400 = US00128546).

Xanthium cenchroides Millsp. & Sherff., Publ. Field Mus. Nat. Hist., Bot. Ser. **4**(2) : 30 (1919). Type: 'TEXAS: near Ferris, *J. Reverchon* 2332 (type in Hb. Mo. 85563; additional material, ibid., on sheet no. 85564).' (MO714853 holotype, MO2140349).

Xanthium calvum Millsp. & Sherff, Publ. Field Mus. Nat. Hist., Bot. Ser., **4**(2): 35 (1919). Type: 'CALIFORNIA: vicin. of Palo Alto, foothills, [near Stanford Univ. Col.] *C.F. Baker* 1760 [2.3.13] (Hb. Calif. 131236, type: Hb. Field 226601; Hb. Gray; Hb. Mo. 85385; Hb. N.Y.)'. (UC131236 holotype, F22601 = F0051935, GH00014154, K001065883, ?MO, NY00278100, US444155 = US00128545).

Xanthium australe Millsp. & Sherff, Publ. Field Mus. Nat. Hist., Bot. Ser., **4**(2): 37 (1919). Type: [Mexico:] 'TAMAULIPAS: vicin. of La Barra, 8 km. east of Tampico, at sea-level, *Dr. Edward Palmer* 275.' (US463216 = US00128542 holotype).

Xanthium japonicum Widder in Repert. Spec. Nov. Regni Veg., Beih. **20**: 31 (1923). Types: 'From the Herb. of the Royal Gardens, Kew, *Oldham*, No. 409. … Nagasaki (*Oldham*, Hb. Leyd., Berl., Hofm., Gött.) – Japonia (*Buerger*, Hb. Leyd.) – Japonia (T.?, Hb. Leyd.).' *Oldham* 409 (GOET00005275 syntype).

Xanthium sibiricum Patrin ex Widder in Repert. Spec. Nov. Regni Veg., Beih. **20**: 32 (1923). Types: Under the heading of 'Exsiccaten Widder listed 'Herbarium of the late East India Company, Nr. 3183. – Herbarium Schlagintweit from India and High Asia, Nr. 720, 4360. Henry's Collection from Central China 1885–88, Nr. 51.– Schindler, Plantae sinenses, Nr. 205, 212.– Bohnhof, Voyage au lac Hanka et en Mandchourie, Nr. 200.– Cavalerie, Plantes de Chine, Nr. 3814. – Erbario Biondi, Nr. 299.' Widder also listed many more specimens under 'Gesehene Pflanzen'.

Xanthium sibiricum var. *subinerme* (C.Winkl.) Widder in Repert. Spec. Nov. Regni Veg., Beih. **20**: 34, 36 (1923).

Xanthium strumarium var. *hausmanni* Widder in Repert. Spec. Nov. Regni Veg., Beih. **20**: 49 (1923). Type: 'Von Bedeutung ist eine Form, die von Hausmann mehrmals zwischen Bozen und Salurn in Südtirol gefunden wurde.'

---

[178] Shull noted earlier in the discussion that 'this variety was first seen on the Campus of the State University of Kentucky several years ago. … It has been found to breed true …'

[179] There is a duplicate of material cultivated from the 'type' donated to K by Sherff.

Xanthium pungens var. *denudatum* Widder in Repert. Spec. Nov. Regni Veg., Beih. **20**: 69 (1923). Type: '… ich nur in einem Exemplar von Kansas, Atchison Co. im Hb. U. W. (leg. *Hitchcock*). Die Exemplare derselben Nummer (727) desselben Exsiccatenwerkes, die im Hb. Wash. liegen, sin normales *X. pungens*.' Herbarium material unknown.

Xanthium decalvatum Widder in Repert. Spec. Nov. Regni Veg., Beih. **20**: 72 (1923). Type: 'Herbarium of the U.S. Exploring Expedition under the command of *Capt. Wilkes*, Nr. 1361 (als *X. strumarium* var. *canadense*" bezeichnet!).' (US46694 = US00128547 holotype).[180]

Xanthium orientale f. *laciniatum* (Pouzolz) Thell. ex Widder in Repert. Spec. Nov. Regni Veg., Beih. **20**: 80 (1923).

Xanthium italicum var. *albinum* Widder in Repert. Spec. Nov. Regni Veg. Beih. **20**: 105 (1923). Type: 'Im Elbetal und dessen näherer Umgebung kommen jedoch fast ausschließlich Formen mit etwas dicklichen, eifömigen, …' Herbarium material unknown.

Xanthium californicum var. *oligacanthum* (Piper) Widder in Repert. Spec. Nov. Regni Veg., Beih. **20**: 112 (1923).

Xanthium pungens var. *globosum* (C.Shull) Widder in Repert. Spec. Nov. Regni Veg., Beih. **20**: 166 (1923).

Xanthium pungens var. *cylindricum* (Millsp. & Sherff) Widder in Repert. Spec. Nov. Regni Veg., Beih. **20**: 167 (1923).

Xanthium saccharatum subsp. *commune* (Britton) Widder in Repert. Spec. Nov. Regni Veg. **21**(8–20): 286 (1925).

Xanthium saccharatum subsp. *aciculare* Widder in Repert. Spec. Nov. Regni Veg. **21**(8–20): 286 (1925). Type: not cited, but many, many syntypes appear to have been cited.

Xanthium chinense var. *globuliforme* C.Shull in Bot. Gaz. **83**(4): 385 (1927). Type: Cult. material from 'Mr. F. F. Crevecoeur near Onaga, Kansas in summer of 1909.'

Xanthium aridum H.St.John in Northwest Sci. **2**(3): 93, fig. 4 (1928). Type: [U.S.A.: Washington. Gravel benches in Black Canyon, Rattlesnake Hills, 2400 ft. alt. *St. John et al.* 8100, April 8, 1927] (LL00208262 – 1 leaf and 1 burr, US1520630 = US00128541 syntypes).

Xanthium pensylvanicum var. *laciniatum* C.Shull & Sherff in Bot. Gaz. **92**: 208 (1931). Type: '*Earl E. Sherff* 5012, cultivated in experimental garden of University of Chicago, Chicago, Illinois, Oct. 11, 1930'. (F636820 holotype, K, US1184216 = US00288942).[181]

Xanthium mongolicum Kitag. in Rep. First Sc. Exped. Manchoukuo, Sect. 4, **4**: (Index Fl. Jehol.): 97 (1936). Type: 'Hab. Manshuria: Prov. Hsing-an occid. [——]: In arenosis circa collis Szu-l ng-tzu-shan [——] prope O-nyž-to [——] (N.H.K. Oct. 2. 1933 – Typus). Dist. Manshuria.' Herbarium material unknown.

Xanthium natalense Widder in Repert. Spec. Nov. Regni Veg. **41**(Nr. 1059/1070): 274 (1937). Type: 'Natal, near Durban, alt. 100'; I. 1902 (*J. Medley Wood*, NG n. 8943 = Urbeleg = Typus).'(NH0008943-0 – 'NG' Natal Government Herbarium holotype).

Xanthium strumarium var. *glabratum* (DC.) Cronquist in Rhodora **47**: 403 (1945).

Xanthium chasei Fernald in Rhodora **48**: 66 (1946). Type: [USA:] 'ILLINOIS: bottomlands of Illinois River near Peoria …, Sept. 15, 1945, *Chase*, no. 8205.' (GH holotype, 'in herb. Chase').[182]

Xanthium strumarium var. *japonicum* (Widder) H.Hara, Enum. Sperm. Jap. **2**: 279 (1952).

Xanthium strumarium subsp. cavanillesii (Schouw) D.Löve & Dans. in Canad. J. Bot. **37**(2): 205 (1959).

Xanthium albinum (Widder) Scholz & Sukopp in Verh. Bot. Vereins Brandenb. **98-100**: 47 (1960)

Xanthium ripicolum Holub in Folia Geobot. Phytotax. **11**(1): 83 (1976), as '*ripicola*', nom. nov. pro *X. riparium* Lasch, non *X. riparium* Itzigs. & Hertsch (1854), nom. illeg.

Xanthium orientale var. *albinum* (Widder) Adema & M.T.Jansen in Gorteria **9**(9): 302 (1979).

Xanthium orientale var. *riparium* (Itzigs. & Hertsch) Adema & M.T.Jansen in Gorteria **9**(9): 303 (1979).

Xanthium strumarium var. *wootonii* (Cockerell) W.C.Martin & C.R.Hutchins, Fl. New Mexico **2**: 2041 (1981), nom. inval. (without basionym reference).

---

[180] See also comments under *X. campestre* Greene, which indicated this is the type of that name also.

[181] This material originated 'in a clump on loam soil at one end of a corn field belonging to Mr. CREVECOEUR, near Onaga, Kansas, July 19, 1928. … Burs from the CREVECOEUR material were planted by Dr Shell in the summer of 1930 …'

[182] There are two sheets in GH (GH00014156, GH00014157) both apparently marked as 'syntypes'.

Xanthium albinum (Widder) Scholz subsp. *ripicolum* (Holub) Dostál in Folia Mus. Rer. Nat. Bohem. Occid., Bot. **21**: 12 (1984), as '*ripicola*'.

Xanthium strumarium f. *purpurascens* Priszter, Magyar Fl. Veg. **7**: 58 (1985). Type: 'Praematricum: *Kunbaracs*, 1972. HHBp.'

Xanthium strumarium subsp. *brasilicum* (Vell.) O.Bolòs & Vigo in Collect. Bot. **17**(1): 90 (1987) [1988].

Xanthium echinatum subsp. *italicum* (Moretti) O.Bolòs & Vigo in Collect. Bot. **17**(1): 90 (1987)[1988].

Xanthium echinatum var. *italicum* (Moretti) O.Bolòs & Vigo in Collect. Bot. **17**(1): 90 (1987)[1988].

Xanthium echinatum var. *cavanillesii* (Schouw) O.Bolòs & Vigo in Collect. Bot. **17**(1): 90 (1987) [1988].

Xanthium sibiricum var. *jingyuanense* H.G.Hou & Y.T.Lu in Bull. Bot. Res., Harbin **20**(3): 249 (2000). Type: 'China. Gansu, Jimgyan, Hongzuizi, alt. 1450 m, riverside, 1998-08-26, *H.G. Hou and Y. T. Lu* 98018.' (LZU holotype, PE).

Xanthium strumarium subsp. *sibiricum* (Patrin ex Widder) Greuter in Willdenowia **33**(2): 249 (2003).

Coarse annual to c. 1(2) m tall; stems poorly-branched, striate, often reddish, sparsely strigose. Leaves petiolate, petiole up to c. 10 cm long scabrid-pubescent, lamina 5–15 × 4–18 cm, transversely or broadly deltate or broadly ovate, simple or 3–5-lobed, 3-veined from base, ± cordate but usually broadly cuneate near petiole, scabrid and green on both surfaces, margins coarsely crenate or irregularly dentate, lobe apices subacute or obtuse. Male capitula uppermost on stems, many-flowered, 5–8 mm diam, terminal on short axillary branches; phyllaries 2–3 mm long, narrowly oblong, pubescent; paleae spathulate or narrowly ovate, c. 2.5 mm long, acute; corollas 2.5–3 mm long, sparsely puberulent; anther-thecae exserted, c. 1.7 mm long. Female capitulum 2–5 in axils, c. 4 mm long, with a few minute bracts at base; phyllaries connate around 2 achenes. Fruit a bur, 2–3 cm long, ellipsoid, central body 6–8 mm diam., pilose and sparsely glandular-puberulous either throughout or towards base; apical beaks 5–7 mm long, 1.5–2 mm thick at base, nearly straight or very slightly recurved, uncinate, hispid except near tip, spines on body somewhat shorter than apical horns, numerous, straight or slightly curved, uncinate, sparsely puberulent below.

Botswana. N: Gomare, 15.iii.1965, *Wild & Drummond* 7121 (SRGH). SE: Mahalapye, vii.1960, *Yalala 111* (SRGH). **Zambia**. S: Mazabuka, 28.ii.1963, *van Rensburg* 1509 (K, SRGH). **Zimbabwe**. N: Mazoe, Concession, 5.iii.1953, *Wild* 4031 (K, LD. PRE. SRGH). W: Bulawayo, 1372 m, 30.iv.1958, *Drummond* 5518 (K, SRGH). C: Salisbury, 1.x.1952, Phipps 2181 (SRGH). S: Beitbridge, Shashi R., 22.iii.1959, *Drummond* 5922 (K, SRGH). **Malawi**. S: Nsanje Dist., Ndindi Marsh neark Marka, 23.iii.1995, *Kutsaira & Kaunda* 102 (K). **Mozambique**. GI: Gaza, R. Limpopo, Dumela, 30.iv.1961, *Thompson* 12 (K, PRE. SRGH). M: Lourenço Marques. Namacha [sic!], estrada para Matianine andados c. 1.5 km, 5.vii.1974, *Marques* 2493 (K, ?LMU).

Weed of disturbed ground and by streams; 0–2500 m. Potentially flowering throughout the year.

A very variable species that is a widespread weed in the tropics, subtropics and even the temperate zones in both the Old and New Worlds. I have made no attempt at recognizing infraspecific ranks for the material cited as it is unwarranted.

Conservation Status: A widespread weed in the Flora area; LC (Least Concern).

The inspiration for George de Mestral, the inventor of Velcro.

2. **Xanthium spinosum** L., Sp. Pl. **2**: 987 (1753). —Eyles in Trans. Roy. Soc. S. Afr. **5**: 514 (1916). —Wild, Comm. Rhod. Weeds: t. 79 (1955); in Kirkia **6**(1): 4–5 (1967). — McVaugh, Fl. Novo-Galiciana **12**: 1093 (1984). —Hind in Fl. Masc., Composées **109**: 216–217 (1993). —Boulos & Hind in Fl. Egypt **3**: 227–228 (2002). —Mesfin Tadesse, Fl. Ethiopia & Eritrea **4**(2): 341 (2004). —Chen Yousheng & Hind in Fl. China **20–21**: 876 (2011). Type: 'Habitat in Lusitania. ☉.' (LINN – Herb. Linn. No. 1113.3 lectotype), lectotypified by Wijnands (1983: 87). FIGURE 6.5.**28A**.

Xanthium catharticum Kunth in Humb., Bonpl. & Kunth, Nov. Gen. Sp. Pl. **4** (ed. folio): 216 (1818). Type: [Ecuador:] 'Crescit in Regno Quitensi, prope Chillo et Quito, alt. 1315 hex. ... Floret Martio.' (P-Bonpl.-P00320266 – 'Bonpl. mss. n. 3006. Quito' holotype).

Xanthium spinosum [var.] β *brachyacanthum* DC., Prodr. **5**: 523 (1836). Type: 'In specimine Brasiliano midi obvio ex comp. Bras. 867, spinae multò breviores quàm in Europaeis. ... (v.s. in h. Mus. reg. Par.)' (P holotype, G-DC).

Xanthium parvifolium DC., Prodr. **5**: 524 (1836). Type: '... Patr. ign. verisim. Amer. austr. ... (v.s. comm. ab am. Delessert.)' (G-DC-G00469926 holotype).

Xanthium armatum Humb. ex Wallr., Beitr. Bot. **1**(2): 228/242 (1844), nom. illeg., citing *Xanthium catharticum* Kunth in synonymy.

Xanthium xanthocarpon Wallr., Beitr. Bot. **1**(2): 229/241 (1844). Type: '*X. spinosum* Beyrich in herb. amiciss. Sporleder./In den vereinigten Staaten und namentlich in Virginien auf freien Feldern zwischen Staunton und Charlotteville von *Beyrich* im Monat September fruchttragend entdeckt.'[183] Herbarium material unknown.

Xanthium eriocarpon Wallr., Beitr. Bot. **1**(2): 242 (1844). Type: '*Xanthium ambrosioides* Hook. et Lindl. in lit./ Das von W. Arnott in dem Herbar. general. berol. niedergelegte Exemplar scheint aus Süd-Amerika zu stammen.' (B† holotype).

Xanthium spinosum var. *inerme* Bel in Rev. Bot. Bull. Mens. **11**: 481 (1893). Type: 'Au mois d'août dernier [1892], nous avons trouvé sur les bords du Tarn une variété tout à fait inerme qui a attiré notre attention. Nous en avons envoyé des graines à plusieurs jardins botanique de France et, nous-même, nous avons fait un semis considérable. [*Bel*, s.n.]' Herbarium material unknown.

Xanthium spinosum var. *canescens* Costa, Introd. Fl. Cataluña: 160 (1864). Type: 'Hab. ad oras fl Besós pr. Badalona.' (BC or possibly LE – Herbarium Catalonicum holotype).

Acanthoxanthium spinosum (L.) Fourr. in Ann. Soc. Linn. Lyon, ser. 2, **17**: 110 (1869).

Xanthium medium Nossotovsky in Izv. Imp. Bot. Sada Petra Velikago [Bull. Jard. Imp. Bot. Pierre le Grand] **14**(4–6): 454 (1914). Type: 'L'auteur [*A. Nossotovsky*] a decouvert dans le région du Don (Stanitza Gnilovskaia près de Rostov) une remarquable espèce du genre *Xanthium* ...' (LE holotype).

Xanthium multifidum Larrañaga, Escritos Damaso Antonio Larrañaga **1**: 28, pl. 117 (1922) [Pub. Inst. Hist. Geog. Urug.], nom. inval.[184]

Xanthium canescens (Costa) Widder, Repert. Spec. Nov. Regni Veg., Beih. **20**: 121 (1923).

Xanthium spinosum f. *laciniatum* Scheuerm. & Thell. ex Widder, Repert. Spec. Nov. Regni Veg., Beih. **20**: 135 (1923). Type: 'Hannover, auf den Kartoffelfeldern bei Döhren in inzelnen Exemplaren, mit Wolle vorübergehend eingeschleppt, 10.9.1913, *R. SCHEUERMANN*' (GZU holotype).

Xanthium spinosum var. *pseudinerme* Widder ex Parodi in Physis **8**(No. 31): 480 (1927). Type: [Argentina:] 'Buenos Aires: Pueyrredón (F. C. C. A.), leg. *A. BURKART* no 393, 12-IV-26 (Herb. L. R. PARODI no 7188) ejemplar tipo, det. F. J. WIDDER.' (BAA holotype, SI020519 – oddly det. as 'No es un tipo').[185]

Xanthium spinosum f. *praecocius* Bitter ex Widder in Phyton (Horn) **11**(1 & 2): 75 (1964). Type: 'Da mir das Original selbst nicht zugänglich war, kann wohl dessen photographische Wiedergabe in BITTER 1908: tab. 9, fig. 1 vorläufig als Lectotypus anerkannt werden.'

Xanthium spinosum var. synacanthum Widder in Phyton (Horn) **11**(1 & 2): 77 (1964). Type: [Britain:] 'Cabbage field, Charlton, Worcs, V.-c. 37, 15.8.1955, *C. W. BANNISTER*, wool advent. fl. of Brit. distrib. by J. E. LOUSLEY No. W/92'. (RDG – originally Herb. LOUSLEY, holotype).

Xanthium spinosum var. *laciniatum* (Scheuerm. & Thell. ex Widder) Widder in Phyton (Horn) **11**(1 & 2): 76 (1964).

Acanthoxanthium spinosum subsp. *catharticum* (Kunth) D.Löve in Lagascalia **5**(1): 64 (1975).

---

[183] In Wallroth's earlier key this species was numbered 20 in his listing, not 17 as it appeared later in the fuller description and discussion.

[184] This reference is provided in the major databases, yet there is no p. 28 in vol. 1 of Larrañaga's work. Lám. CXVII is a poor sketch (unnamed – the name only appearing in the Explicación at the end of the first volume) of what is undoubtedly of *Xanthium spinosum*. Since it has no dissections, this can only be considered 'nom. inval.' cf. Gray Herb. Card Cat., Issue 123.

[185] This variety is recognized by Ariza Espinar & Freire in Zuloaga *et al.*, Fl. Vasc. Argent. **7**: 266 (2015).

Xanthium cloessplateanum D.Z.Ma in Acta Bot. Bor.-Occid. Sin. **11**(4): 346 (1991), as '*cloessplateaum*'.[186] Type: 'Ningxia: Tongxing, IX.1984. leg. *X. Z. Dong*. No. 424.' ('Ningxia Agr. Col. Conserv.' holotype).

Annual herb 10–60(120) cm tall; stems striate, yellowish, puberulent. Leaves with a 3-partite yellow spine up to 1.5–3 cm long in or adjacent to axil, petiolate, petiole 1–15(25) mm long puberulous or tomentellous, lamina up to 8–10 × 2–3 cm, very narrowly ovate to narrowly ovate, base cuneate or rounded to subtruncate, margins undulate with a few coarse teeth or lamina 2–4(7)-lobed, lobes to 2.5 cm long, green, pubescent (especially on veins) or glabrescent above, white-tomentellous below, apex acute to acuminate. Male capitula terminal, globose, florets numerous; phyllaries c. 1.3 mm long, elliptic, pubescent; paleae c. 1.5 mm long, spathulate-cucullate, densely hairy abaxially in upper half; corollas c. 1.5 mm long, broadly funnel-shaped, pubescent above; anther-thecae c. 1.7 mm long, exserted, spreading. Female capitulum c. 2.5 mm long; phyllaries connate around 2 achenes. Fruit an ellipsoid or oblong-cylindrical bur, 9–14 × 6–8 mm, central body yellowish-green, whitish-puberulent, body spiny, spines numerous, 2–3.5 mm long, yellowish, apex uncinate, apical beaks short c. 1–3 mm long or beaks absent.

Botswana. SE: Mahalapye, vii.1960, *Yalala* 114 (SRGH). **Zimbabwe**. C: Marandellas ix.1917, *Walters* 2390 (K). W: Bulawayo, 13.vi.1941, *McAllister* [Government Herb. No.] 8071 (K, SRGH).

Widespread in both the Old and New Worlds as a weed. In our area occurs in drier habitats as a rule than *X. strumarium*, sometimes as a pasture weed but either rarer, or poorly collected; 0–3000(3500) m. Throughout its range it can potentially flower at any time of the year.

Conservation Status: Although this species would appear to have been rarely collected in the flora area I have recorded it as LC (Least Concern) globally.

136. **ACMELLA** Rich.

Acmella Rich. in Pers., Syn. Pl. **2**: 472 (1807). —Moore in Proc. Amer. Acad. Arts **42**(No. 20): 521–569. (1907). —Moore in Bot. Jahrb. Syst. **45**(4): 426–427 (1911). —Adams in Webbia **12**(1): 325–330 (1956). —Jansen in Syst. Bot. Monogr. **8**: 1–115 (1985). —Diaz-Piedrahita in Revista Acad. Colomb. Ci. Exact. **17**: 645–648 (1990). —Mesfin Tadesse, Fl. Ethiopia & Eritrea **4**(2): 288 (2004). —Robinson in Harling & Andersson, Fl. Ecuador, **77**(1), 190(6): 17–28. (2006).[187]

Ceratocephalus Burm. ex Kuntze, Revis. Gen. Pl. **1**: 326 (1891), nom. illeg. superfl. pro *Spilanthes* Jacq. pp. & *Acmella* Rich. p.p.

Athronia Neck., Elem. Bot. **1**: 32 (1790), nom. inval. rej.

Colobogyne Gagnep., Notul. Syst. **4**: 15 (1920).

Erect, decumbent or repent annual or perennial herbs. Roots fibrous, rarely taprooted. Stems poorly- or well-branched, sometimes rooting at lower nodes, glabrous to pilose or tomentose. Leaves opposite, on elongate leafy stems or basal, petiolate, winged or wingless; lamina filiform to broadly ovate, with 3 main veins from base, base attenuate to cordate, surfaces glabrous to hirsute, strigose or tomentose, margins entire to coarsely dentate, apices obtuse to long-acuminate. Inflorescences terminal or axillary, with solitary long-pedicellate capitula. Capitula heterogamous and radiate or homogamous and discoid, ovoid; phyllaries herbaceous, 4–24, 1–3-seriate, subequal or outer series longer and spreading, linear to broadly ovate or elliptic, margins entire, apices usually obtuse; receptacle becoming columnar with age,

---

[186] I have reason to believe, following contact with my Chinese co-author in the Heliantheae account for the Flora of China', that the specific epithet is missing an 'n' and should be '*cloessplateanum*'.

[187] Lectotype (original lectotypification superseded by the selection of Robinson, 2006: 17): *Anthemis repens* Walter = *Acmella repens* (Walter) Rich. in Pers. In the original lectotypification (Jansen, 1985: 19): *Acmella oppositifolia* (Lam.) R.K.Jansen was chosen, but this was not based on one of the elements in the original publication and proved to be a species of *Heliopsis* Pers. – *Heliopsis buphthalmoides* (Jacq.) Dunal (a New World plant).

paleaceous, paleae persistent, conduplicate, scarious, glabrous. Ray florets 0–3–22, female; corollas white or yellow, sometimes greenish white, violet-purple or orange, with or without conspicuous limbs. Disc florets many (23–620), hermaphrodite; corollas white or greenish-white to yellow or orange, glabrous, lobes 4 or 5, triangular, papillose on inner surfaces; anther thecae usually blackish, apical anther appendages broadly ovate, glandular or eglandular; style base with distinct enlarged node; style arms blunt, with stigmatic papillae continuous across inner surface. Achenes laterally compressed, 2-ribbed, elliptic, body black at maturity, margins glabrous, pubescent or with cork-like margins, shoulder sometimes extended upward beyond base of corolla, sides glabrous to densely setuliferous or tuberculate; pappus absent or of 1–10 soft setae on upper angles of achene.

A pantropical genus of c. 30 spp. with many spp. in Mexico, Central America and S America; two spp. are presently recorded in the Flora area.

Capitula radiate; corollas yellow or orange-yellow . . . . **1.** *caulirhiza*
Capitula discoid; corollas white or greenish white. . . . . **2.** *radicans* var. *radicans*

1. **Acmella caulirhiza** Delile in Cailliaud, Centurie Pl. Afr. Voy. Méroé: 45, Pl. 64, fig. 7 [1826]; —Voy. Méroé **4**: 335, t. 64, fig. 7 (1827).[188] —Jansen in Syst. Bot. Monogr. **8**: 37, map (1985). —Lisowski, (Asterac. Fl. Afr. Cent. 1) Fragm. Fl. Geobot. **36** Suppl. 1: 228, t. 50 (1991). —Hind in Fl. Masc., Composées **109**: 207, t. 71 (1993). —Agnew & Agnew, Upland KenyaWild Fl., ed. 2: 216, t. 88 (1994), as '*calirhiza*'. —Mesfin Tadesse, Fl. Ethiopia & Eritrea **4**(2): 288–289 (2004). Type: [Sudan:] 'Gash El-Ganem. Plante de Sennâr [1815–1820, *Cailliaud* s.n.], usitée contre les maux de tête. (Notes manuscr. de M. Cailliaud.)' (MPU007033 holotype).

  Eclipta filicaulis Schumach. & Thonn., Beskr. Guin. Pl.: 390 (1827). Type: 'I og ved Aquapim.' *Thonning* 277 (C10003806 lectotype, C10003807), first-step lectotypification by Adams (1956: 328),[189] second-step lectotypification by Jansen (1985: 37).

  Spilanthes africana DC., Prodr. **5**: 623 (1836). Type: 'in Africà austr.-orient. ad Zw. Omtata et Omsamwubo legit cl. *Drege* n. 5086! ... (v. s.)' (G-DC-G00456683 holotype, E00239253 – s.n., HAL0111741 – s.n., HBG504082 – s.n., K000410415 – s.n.), P00073399 – s.n., P00073400, P00073401 – s.n., P00073402).

  Spilanthes caulirhiza (Delile) DC., Prodr **5**: 623 (1836).

  Spilanthes caulirhiza β *madagascariensis* DC., Prodr. **5**: 623 (1836). Type: 'in insulà Madagascar legit cl. *Bojer*. ... (v. s. comm. à cl. Bojer.)' (G-DC-G00456998 holotype).

  Spilanthes mauritiana f. *madagascariensis* (DC.) A.H.Moore, Proc. Amer. Acad. Arts **42**: 542 (1907).

  Spilanthes abyssinica Sch.Bip. ex A.Rich., Tent. Fl. Abyss. **1**: 415 (1848). Types: 'SPILANTHES ABYSSINCA. C. H. Schultz, in pl. Schimp. Abyss., sect. I, nº 134. ... Crescit in campis, pratis et ad margines stagnorum circa Adoua et Mariam-Chauvista (*Quartin Dillon, Schimper*), mensibus Junio et Julio florens et in provincia Choa (*Ant. Petit*).' *Schimper* 134 (P lectotype, BM000924393, BR8363244, BR8871626, G00015642, GH00012655, GH00012656, GOET000991, HAL0111740, HBG504083, HOH009337, HOH009338, K000410419, K000410420, K000410421, L0001786, L0001787, L0001788, M0105201, M0105202, MPU012906, MPU012907, NY00260013, OXF, P00073373, P00073374, P00167601, S-G-5740, TUB005512, TUB005513, W), lectotypified by Jansen (1985: 38).

  Spilanthes filicaulis (Schumach. & Thonn.) C.D.Adams, Webbia **12**(1): 326 (1956).

Perennial or sometimes annual herbs. Stems decumbent or scrambling, often rooting at nodes, to 15 cm high or 60 cm long, sappy, often reddish or purplish, glabrous or pilose. Leaves petiolate, petiole 1–15 mm long, narrowly winged, glabrous to sparsely pilose, lamina ovate, 1–7 × 0.8–4 cm, base

---

[188] The *Centurie* (originally in volume 4 of the *Voyage à Méroê*) was published in 1826 and is apparently a preprint, from which the citation has priority.

[189] Adams (1956: 328) cited '*Thonning* 227' as 'typus', but did not specify to which of the two sheets in C it referred.

attenuate, 3-veined from base, margins dentate, sparsely ciliate, apex acute or obtuse, glabrous or pilose. Capitula solitary and hemispherical, becoming conical, heterogamous and radiate, 4–14 × 4–10 mm, pedicels 1.8–7 × 0.9–2 mm; involucre 3.5–6 mm tall; phyllaries biseriate, 9–13, outer usually spreading and longer than inner, 5–6, 3.5–5.9 × 0.8–2.2 mm, margins entire or subentire, apices rounded to acute, inner 4–7, 3.5–6 × 0.9–2 mm, apices rounded to acuminate. Ray florets 10–15, corollas yellow to orange, 2.2–3.3 mm long, ray limb 1.4–2.3 × 0.6–1.8 mm wide. Disc florets numerous (68–250), hermaphrodite, corollas yellow or ± orange, 1.1–1.9 mm long, corolla lobes 4 or 5, 0.2–0.4 × 0.2–0.5 mm; anthers black. Achenes 1.3–2.4 mm long, lacking a cork-like margin, glabrous, or moderately ciliate, cilia with recurved-tipped hairs; pappus usually absent but sometimes some setae present on achenes of outer florets.

Zambia. N: Mpulungu, Lake Tanganyika, 29.xii.1951, *Richards* 177 (K). W: Ndola, 18.vi.1954, *Fanshawe* F1282 (K, SRGH). C: Chilanga Fish Farm, 1248 m, 11.xi.1962, *Lusaka Nat. Hist. Club* 190 (K, SRGH). S: Mapanza E., 1067 m, 6.ix.1953, *Robinson* 311 (K). **Zimbabwe**. N: Urungwe National Park, Chimutsi Dam, NE of Zambezi Escarpment, 16.ii.1981, *Philcox et al.* 8630 (K). W: Matobo Dist., Matopos, 19.xii.1956, *Garley* 102 (K, SRGH). C: Salisbury, 1524 m, i.1921, *Mainwaring* 2892 (K, PRE). E: Inyanga, Pungwe, below the waterfalls in the river valley, 18.xii.1930, *T. C. E. Fries et al.* 3916 (K, S). S: Belingwe Dist., streamside below Murongwe Dam NW of Mt. Buhwa, c. 1100 m, 4.v.1972, *Biegel et al.* 4283 (K, SRGH). **Malawi**. N: Nyika, Apoka village, Rukuru River,1600 m, 29.x.1958, *Robson & Angus* 469 (BM, K, SRGH). C: Kota Kota, Benga, west shore of Lake Nyasa, 470 m, 2.ix.1946, *Brass* 17489 (K, NY). S: Zomba Dist., Mpita Tobacco Estate, Thondwe, 24.vii.1984, *Kaunda & Tawakali* 84 (K, MAL). **Mozambique**. Z: N of Namuli Peaks, 1158 m, 27.vii.1962, *Leach & Schelpe* 11486 (K, SRGH). T: Tete, north of Mlangeni, 1385 m, 25.v.1970, *Brummitt* 11112 (K). MS: Beira, 15 m, 25.xii.1906, *Swynnerton* 1865 (BM, K). GI: Vila de João Belo, 8.x.1945, *Pedro* 271 (K, LMJ. PRE).

Widespread in tropical and subtropical Africa and Madagascar and Mauritius.

Swampy or seasonally wet sites, river-banks, cultivated areas, forest margins, roadsides, grassland; 200–3100 m; flowering throughout the year.

Conservation Status: LC (Least Concern).

2. **Acmella radicans** (Jacq.) R.K.Jansen, Syst. Bot. Monogr. **8**: 69 (1985). Type: 'In Venezuela sponte crescit; in caldariis nostris floret a Julio ad Januarium, quando annua perit.' Herbarium material unknown.

    Spilanthes radicans Jacq., Collectanea **3**: 229 (1791), as '*Spilanthus*'.

    Spilanthes exasperata Jacq., Icon. Pl. Rar. **3**(9): 15, t. 584 (1792), as '*exasperatus*', nom. superfl. but also citing *S. radicans* Jacq.

    Ceratocephalus exasperatus (Jacq.) Kuntze, Revis. Gen. Pl. **1**: 326 (1891).

Var. **radicans** FIGURE 6.5.**29**.

    Spilanthes exasperata [var.] β *cayennensis* DC., Prodr. **5**: 626 (1836). Type: 'in Cayennâ legit cl. *Patris*. Pars inf. plantae mihi ignota. (v.s.)' (G-DC holotype).

    Spilanthes leucophaea Sch.Bip. ex Klatt, Leopoldina **23**: 144 (1887). Type: 'Sp. leucophaea Hort. Berol.? (Schultz Bip.) ... Mexico, Culiacan, leg. *Schaffner*. Herb. Klatt.' (GH12618 holotype).

    Spilanthes botterii S.Watson, Proc. Amer. Acad. Arts **26**: 141 (1891). Types: 'Near Guadalajara; November, 1889 ([*Pringle*] n. 2946); also collected in Orizaba by *Botteri* (n. 825 in Herb. Gray).' *Pringle* 2946 (GH00012612 lectotype,[190] NY00259993), lectotypified by Jansen (1985: 70); *Botteri* 825 (F0051430, GH00012611).

    Spilanthes ocymifolia var. *acutiserrata* A.H.Moore, Proc. Amer. Acad. Arts **42**: 533 (1907). Type: 'Costa Rica: Cartago: [1500 m] *J. J. Cooper*, 5807, Cartago (in Herb. J. D. Sm.) (specimen cum eodem lectum in Herb. A. H. Moore).' (US1403345= US00125332 holotype, F0051436, F0051437, GH00012622, GH00012623, K000497314, US).

---

[190] F0051429 – s.n., but dated '2, Oct.' is probably not an isolectotype, contrary to Jansen's annotation.

Sanvitalia longepedunculata M.E.Jones, Contr. W. Bot. **18**: 78 (1933). Types: '[*Jones*] Nos. 27761 and 27729, La Barranca, Guadalajara Nov. 19 1930. ... A weed in fields.' *Jones* 27761 (POM193601 lectotype, POM0001521, POM, US00128567),[191] lectotypified by Blake in Contr. U.S. Natl. Herb. **29**: 132 (1945).

Annual herbs, 6–30 cm tall. Stems erect to ascending, rarely rooting at nodes, green to purplish, glabrous to pilose. Leaves petiolate, petiole 4–37 mm, lamina ovate to narrowly ovate, 1–8 × 1–5 cm, base attenuate, lamina glabrous to sparsely pilose, margins denticulate to dentate, hispid, apices acute. Capitula homogamous and discoid, usually solitary or in axillary cluster of 2 or 3, pedicellate, pedicels 1.8–6.5 × 0.6–2 mm, sparsely to moderately pilose. Capitula 6–11 × 4.7–9.4 mm; phyllaries 6–9, biseriate, herbaceous, margins entire, sparsely to moderately ciliate, outer 3–5, 3.7–6.1 × 1.2–3.9 mm, lanceolate to broadly ovate, apices acute to acuminate, inner 2–4, 4–6.6 × 0.9–2.2 mm, lanceolate to ovate, apices acute to acuminate; receptacle 3.8–7.5 × 0.5–1.9 mm; paleae 4–5.2 × 0.5–1 mm, stramineous or purple-tinged. Florets 60–150; corollas 1.6–21 mm long, greenish-white to white, tube 0.5–0.6 × 0.2–0.5 mm, throat 1.1–1.5 × 0.2–0.7 mm, lobes 4, 0.2–0.4 × 0.2–0.3 mm; anthers brownish-black. Achenes 2–2.7 × 0.6–1 mm, margins lacking corky-margins, densely ciliate, hairs straight-tipped; pappus of 2 (on inner achenes) or 3 (usually on outer achenes), subequal setae, shorter setae 0.5–1.4 mm, longer setae 1–1.6 mm.

Zambia. N: Isoka Dist., 30 km from Mukombe School on Nakonde road, c. 1500 m, 22.iv.1986, *Philcox et al.* 10067 (K). W: Kitwe Dist., Mwekera Forestry Training School, 12°51'S, 28°21'E, 26.iv.1989, *Pope* et al. 2208 (BR, K, LISC, MO, NDO, SRGH).

A weedy plant widespread throughout Mexico and Central America but also present in northern S America and widely introduced in Cuba, India and Tanzania. Still poorly collected in the Flora area (as well as in neighbouring Floras), with plants wrongly identified as '*Spilanthes costata* Benth.' in several countries in Africa – a taxon absent in our Flora area. Stream banks, roadside ditches, cultivate areas; 60–1800 m; flowering throughout the year.

Conservation Status: Apparently very infrequently collected, but a widespread introduction in a number of areas; LC (Least Concern).

137. **ZINNIA** L.

Zinnia L., Syst. Nat., ed. 10, **2**: 1189, 1221, 1377 (1759), nom. cons. —Robinson & Greenman in Contrib. Gray Herb. n.s. No. X, Proc. Amer. Acad. Arts **32**: 14–20 (1897). —Torres in Brittonia **15**(1): 1–25 (1963). —Wild in Kirkia **6**(1): 54–55 (1967). —McVaugh, Fl. Novo-Galiciana **12**: 1108–1125 (1984). —Hind in Fl. Masc., Composées **109**: 176–177 (1993). —Mesfin Tadesse, Fl. Ethiopia & Eritrea **4**(2): 290 (2004). —Beentje & Hind in F.T.E.A., Compositae **3**: 708 (2005).
Crassina Scepin, Schediasma de Acido Veget.: 42 (1759)[1758], nom. rej. vs. *Zinnia* L.
Tragoceros Kunth in Humb., Bonpl. & Kunth, Nov. Gen. Sp. Pl. **4** (ed. folio): 195 (1818).
Mendezia DC., Prodr. **5**: 532 (1836).
Lejica Hill ex DC., Prodr. **5**: 534 (1836), nom. nud. pro syn. under *Zinnia pauciflora* L. (= *Zinnia peruviana* L.)
Diplothrix DC., Prodr. **5**: 611 (1836).
Anaitis DC., Prodr. **5**: 628 (1836).
Zinnia sect. *Diplothrix* (DC.) A.Gray in Smithsonian Contr. Knowl. **3**: 105 (1850).
Sanvitaliopsis Sch.Bip. ex Greenm. in Proc. Amer. Acad. Arts **4**: 260 (1905).
Zinnia subgen. *Diplothrix* (DC.) Torres in Brittonia **15**(1): 5 (1963).

Annual or perennial, erect or procumbent herbs or low subshrubs or shrubs. Rootstock a slender or thickened taproot. Stems pubescent, sometimes sparsely so, or glabrescent, striate or smooth, angular to terete, green but sometimes yellowish to purplish. Leaves opposite, entire, sessile to short-petiolate, base connate and sheathing stem, pubescent, linear, filiform or acerose, narrowly-lanceolate, narrowly-ovate, broadly ovate to elliptic, glabrous, hispid or glabrescent, margins entire, apices obtuse, subacute or acute.

---

[191] POM specimens now integrated with RSA (post 1966), as RSA-POM – RSA0001522, RSA0001523.

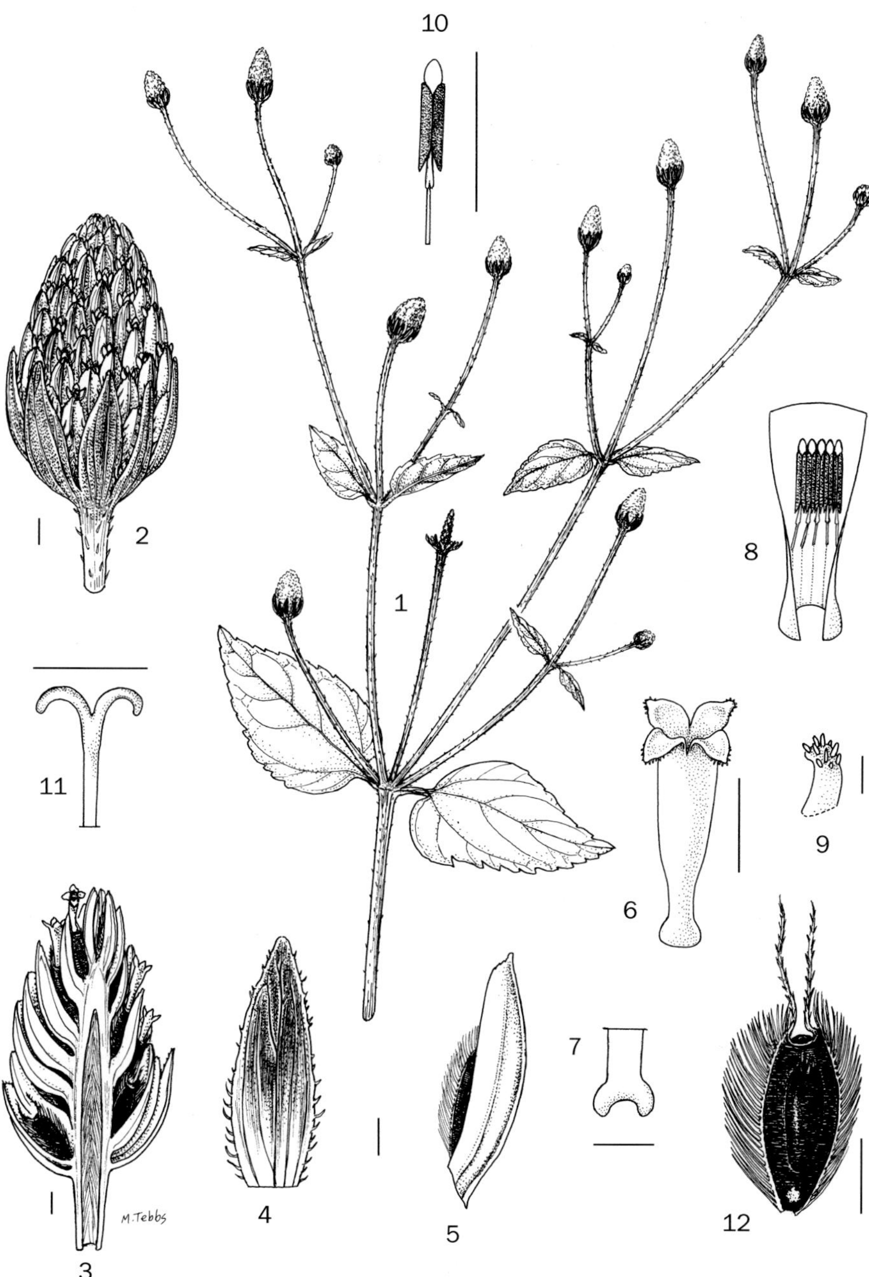

Fig. 6.5.**29**. ACMELLA RADICANS var. RADICANS. 1, flowering branch; 2, capitulum; 3, l.s. capitulum; 4, phyllary; 5, achene with enclosing palea; 6, floret corolla; 7, detail of base of corolla; 8, corolla opened out showing attachment of filaments; 9, detail of corolla lobe; 10, stamen; 11, style; 12, mature achene. 1–4, 6–11 from *Kalawo et al.* 611; 5, 12 from *Phillipson* 4894. Scale bars: 9 = 0.2 mm; 7 = 0.4 mm; 11 = 0.5 mm; 2–6, 8, 10, 12 = 1 mm; 1 = 10 mm. Drawn by Margaret Tebbs.

Inflorescence of solitary, long-pedicellate terminal capitula, pedicels usually inflated beneath involucre. Capitula radiate, heterogamous; involucre cylindrical to hemispherical; phyllaries imbricate, gradate, 3-seriate; receptacle slightly concave to narrowly conical, paleaceous; paleae persistent, conduplicate and enveloping achene, apex rounded to acute, erose, fimbriately lobed or entire and cuspidate, bicolorous or concolourous, hyaline to membranaceous becoming scarious. Ray florets female, fertile, ray limb usually present, persistent, white, yellow, orange, red, purple or lilac; style bifid, pubescent or glabrous. Disc florets hermaphrodite, corollas tubular, 5-lobed, usually zygomorphic with one longer lobe, glabrous outside, smooth or densely velutinous inside; apical anther appendages acute, basal anther appendages truncate to sagittate; style arms truncate and apically penicillate or acute and densely velutinous. Ray achenes ± triquetrous or flattened, body smooth, striate or tuberculate, glabrous or setuliferous; disc achenes compressed or angular, glabrous or setuliferous; pappus of awns or bearing horns, or epappose.

A genus of 23 spp. Two spp. are cultivated in the Flora area. The widely cultivated *Zinnia violacea* Cav., of Mexican origin, is well-known as a garden annual in our area. It sometimes establishes itself as an escape near habitation and along roadsides, although perhaps not as much as *Z. peruviana* L.

Capitula campanulate; paleae erose, rounded; disc achenes with single long awn . . . .**1.** *peruviana*
Capitula hemispherical; paleae fimbriately lobed; disc achenes awnless. . . . . **2.** *violacea*

1. **Zinnia peruviana** L., Nat. Syst., ed. 10, **2**: 1221 (1759), nom. cons.[192] —Miller, Fig. Pl. Gard. Dict. **1**: 43, t. 64 (1756) (right). —Exell, Cat. Vasc. Pl. S. Tomé: 225 (1944). — Torres in Brittonia **15**: 12 (1963). —Hind in Fl. Masc. Composées **109**: 178–179 (1993). —Mesfin Tadesse, Fl. Ethiopia & Eritrea **4**(2): 291 (2004). —Beentje & Hind in F.T.E.A., Compositae **3**: 708 (2005). —Maroyi in Kirkia **18**(2): 202 (2006). —Chen Yousheng & Hind in Fl. China **20–21**: 863 (2011). Type: 'Habitat in Peru.' 'Bidens calyce oblongo squamoso, seminibus radii corolla non decidua coronatis' in Miller, Fig. Pl. Gard. Dict. **1**: 43, t. 64 (right) (1756), iconotype, lectotypified by Jeffrey in Jarvis *et al.* (1993: 100). FIGURE 6.5.**30**.

   Zinnia pauciflora L., Sp. Pl., ed. 2, **2**: 1269 (1763), nom. illeg., replaced synonym of *Z. peruviana*.
   Zinnia multiflora L., Sp. Pl., ed. 2, **2**: 1269 (1763). —Harvey in Harvey & Sonder, Fl. Cap. **3**: 609 (1865). —Moore in Fawcett & Rendle, Fl. Jamaica **7**: 219 (1936). —Martineau, Rhod. Wild Flowers: 89 (1953). Type: 'Habitat - - - - N. L. Burmannus.' (LINN 1019.2?, S –Herb. Linn. No. 353.5? syntypes?).[193]
   Zinnia tenuiflora Jacq., Icon. Pl. Rar. **3**: 15, t. 590 (1793). Type: not cited.
   Zinnia revoluta Cav., Icon. **3**: 26 (1794)[1795]. Type: 'Hab. in Imperio Mexicano. 0 Floret August in Regio horto Matritense.' (?MA holotype).
   Zinnia verticillata Andrews in Bot. Repos. **3**: t. 189 (1801). Type: 'It is a native of Mexico, South America; and was introduced to our gardens about the year 1789, by Monsr Richard, from the Paris gardens, at the same time with the Virgilia; ... Our figure was taken, this year, at the Hammersmith Nursery, where, it was grown first in this kingdom.' t. 189 in Andrews (1801), iconotype, lectotypified by Robinson (2006: 225).[194]
   Zinnia hybrida Ruiz & Pav. ex Sims in Bot. Mag. **47** (n.s. 5): t. 2123 (1820), nom. illeg. non Roem. & Usteri (1787), nec hort. Paris ex Poir. (1808). Types: 'Communicated by Messrs. WHITLEY, BRAME and MILNE, in August last, under the name of *grandiflora*; but as we find the same species in the herbarium of A. B. LAMBERT, Esq. collected in South America by RUIZ and PAVON, under that of *hybrida*, we

---

[192] Although *Chrysogonum peruvianum* L. is often referred to this name the original element is a plate of a wedelioid plant, not a *Zinnia*; the name is definitely not the basionym of *Zinnia peruviana* L.

[193] See comment in Jarvis (2007: 932), who noted that both Torres (1963: 12) and Hind (1993: 179) had expressed doubt over whether LINN 1019.2 should be considered as the type.

[194] Robinson (2006: 225) cited the 'holotype' as 'Botanist's Repository **3**: pl. 189'. This should be the 'lectotype is the original of the plate 189', surely?

Fig. 6.5.**30**. ZINNIA PERUVIANA. 1, flowering shoot (× ²/₃); 2, l.s. capitulum (× 2); 3, phyllary (× 3); 4, palea (× 3); 5, ray floret (× 2); 6, ray floret style (× 14); 7, disc floret corolla (× 8); 8, partial anther cylinder opened out (× 36); 9, disc floret style (× 14); 10, disc floret achene (× 4). All from *Richard* 482. Drawn by Pat Halliday. From Flore des Mascareignes.

have thought it right to adopt this. Mr WHITLEY received the seeds of this plant from the East Indies, by favour of Mrs STUART, but it was most probably introduced there from Brazil.' (?BM holotype).[195]

Zinnia leptopoda DC., Prodr. **5**: 535 (1836). Type: '① in Mexico, ad Tlacolola in Oaxacanâ ditione (*Andr.* fl. exs. n. 314), et in montibus circa urb. Mex. (*Berl.*! pl. exs. n. 686). Z. pauciflora in fl. mex. ined. non Linn. ... (v.s.)' *Berlandier* 686 (G-DC00455343 lectotype, BM001009611, GH00014252, HAL0111103, LD1244060, MO191409), lectotypified by Torres (1963: 12);[196] *Andrieux* 314 G-DC-G00455337, (GH0014253 syntypes).

Zinnia intermedia Engelm. in Wislizenius, Mem. Tour. N. Mex.: 107 (1848). Type: 'Common about Cosihuiriachi, flowers in September. [*Wislizenius*]' (GH0014246 holotype, ?GH0014247).

Zinnia mendocina Phil. in Anales Univ. Chile **36**: 185 (1870).[197] Type: 'Mendoza.' (GOET, SGO syntypes).[198]

Crassina intermedia (Engelm.) Kuntze, Revis. Gen. Pl. **1**: 331 (1891).

Crassina leptopoda (DC.) Kuntze, Revis. Gen. Pl. **1**: 331 (1891).

Crassina multiflora (L.) Kuntze, Revis. Gen. Pl. **1**: 331 (1891).

Crassina peruviana (L.) Kuntze, Revis. Gen. Pl. **1**: 331 (1891).

Crassina tenuiflora (Jacq.) Kuntze, Revis. Gen. Pl. **1**: 331 (1891).

Crassina verticillata (Andrews) Kuntze, Revis. Gen. Pl. **1**: 331 (1891).

Crassina peruviana var. *flava* Kuntze, Revis. Gen. Pl. **3**(3): 143 (1898). Type: 'Bolivia: Rio Tapacari.' [BOLIVIA. Rio Tapacari, 3000 m, 19 Mar 1892, *Kuntze* s.n.' (NY ×2[199]).

Annual herb to 0.6 m tall; stems thinly hairy or glabrescent. Leaves sessile or subpetiolate, up to c. 4 × 1.2 cm, narrowly oblong-ovate, apex obtuse or subacute, rounded or broadly cuneate at 3-veined base, scaberulous-pubescent on both sides, glandular-punctate. Capitula up to 4 cm in diam., with peduncles gradually thickening upwards and 2–7 cm long; involucre cylindrical or slightly funnel-shaped; phyllaries 3-seriate, broadly oblong, rounded at the apex, striate, dark-tinged near apex, glabrous except for ciliolate margins near apex; outer and lowermost bracts c. 5 mm long, inner and uppermost c. 10 mm long, paleae c. 1.4 cm long, obtuse and serrulate at apex, dark-tinged towards apex, pubescent outside towards apex. Ray floret limb reddish or brownish-red (white or yellowish-cream outside of Flora area), drying ± purplish, exserted and spreading, c. 1.2 × 0.8 cm, apex rounded or shallowly emarginate, margin puberulous; tube c. 1 mm long, villous. Corolla of disc florets c. 4 mm long, tubular; lobes pubescent. Achenes c. 1 cm long, linear, triquetrous, pubescent and without pappus awns in those of ray, linear, compressed, pubescent on margins and with 1 setulose awn c. 8 mm long in those of disk.

Botswana. SE: South East Dist., at turn off towards Kanye from Lobatse-Pitsane Road, Stony Hill, 21.x.1978, *Hansen* 3494 (BM, C, GAB, K, PRE, SRGH, WAG). **Zimbabwe**. W: Bulawayo, Hillside Dam, xii.1957, *Miller* 4879 (K, PRE, SRGH). C: Que Que, 1360 m, 4.ii.1972, *Biegel* 3822 (K, SRGH).

Widespread as a roadside weed in Matabeleland, the Transvaal and Natal, and recorded for Kenya and Tanzania. Known also as a weed in the islands of the Gulf of Guinea and Ethiopia. A native of Central America and from Brazil and Bolivia N to Mexico, occurs also in Texas and Florida; 0–3500 m; Probably flowering throughout the year.

Conservation Status: LC (Least Concern).

Interestingly, Maroyi (2006) referred to this taxon as a 'perennial'; all species in *Zinnia* sect. *Zinnia* are considered annuals.

---

[195] Robinson (2006: 225) cited the holotype as ?BM.

[196] Torres actually wrote '(T: *Berlandier* 686, G-DC, Isotype: GH!)' which, in keeping with the remainder of the paper, appears to indicated the 'holotype'. This, in the case of *Z. leptopoda*, can be considered lectotypification. This contrasts with Robinson (2006: 225) who appears to have lectotypified the name based on the *Andrieux* collection, as 's.n.' in G-DC. The *Andrieux* collection, '314', was of course specified in the protologue; Torres selection still takes precedence.

[197] In a separately published reprint/preprint in K this appeared on p. 27.

[198] Muñoz Pizarro (1960: 166) cited two collections in SGO, 44581 and 60589. Duplicate type material appears to be in GOET.

[199] According to Wetter & Zanoni (1985: 330).

2. **Zinnia violacea** Cav., Icon. **1**: 57, t. 81 (1791). Type: 'Habitat in Mexico. ... Vidi floridam in Regio horto Matritense mense Iulio.' (MA246537 holotype).[200]

Zinnia elegans Jacq., Ic. Pl. Rar. **3**: t. 589 (late 1792 or early 1793), nom cons.[201] —Hind in Fl. Masc., Composées **109**: 177 (1993). —Mesfin Tadesse, Fl. Ethiopia & Eritrea **4**(2): 292 (2004). Type: not cited.[202]

Zinnia violacea var. *coccinea* Lindl., Bot. Reg. **15**: t. 1294 (1829). Type: 'Scarlet Zinnia. ... This splendid plant came up among some Mexican seeds presented to the Horticultural Society by J. S. Mill, Esq.'[203]

Zinnia elegans [unranked] α *violacea* (Cav.) DC., Prodr. **5**: 536 (1836).

Zinnia elegans [unranked] β *alba* DC., Prodr. **5**: 536 (1836). Type: '(v. v. c.)'.[204]

Zinnia elegans [unranked] γ *purpurascens* DC., Prodr. **5**: 536 (1836). Type: '(v. v. c.)'.[204]

Zinnia elegans [unranked] δ *coccinea* (Lindl.) DC., Prodr. **5**: 536 (1836).

Zinnia australis F.M.Bailey in Queensland Dept. Agr. Bull. **9**, Bot. [Bull.] **3**: 14 (1891). Type: [Australia:] 'Walsh River, *T. Barclay-Millar*.' (BRI-AQ0354876 holotype).

Annual, to 1 m, usually unbranched (often only few-branched in flowering portion) and flowering when much shorter. Stems markedly hispid, appressed-pilose towards inflorescence. Leaves opposite, sessile, lamina ovate to lanceolate, to 10 × 5 cm, resin-dotted, strigose above and beneath, margins coarsely strigose, apex acute, base truncate or subcordate, rounded or auriculate. Inflorescence usually of solitary terminal capitula, sometimes few-franched, capitula pedicellate, pedicels (2)5–15 × to 3 mm, multi-striate. Capitula heterogamous and radiate, to 6 cm diam. across rays (considerably more when flattened, or in cultivated material); involucre hemispherical, 1–2 × 0.7–1 cm; phyllaries 3–4-seriate, to 10 × 4–7 mm, straw-coloured, rigid, obovate, apices rounded, erose, often pale green or blackish banded; receptacle elongate, 10–23 mm tall, paleaceous, paleae equalling disc florets at anthesis, 15 mm long in fruit, ovate-lanceolate, apex narrowed beneath expanded pectinately lobed appendage, apex reddish or orange-red. Ray florets 5–21 (depending upon size of plants), female, ray limb obovate to spathulate, 20–25(29) × 8–15 mm, scarlet or orange above, greenish yellow to orange-yellow beneath; styles long-exserted, glabrous or pilose. Disc florets hermaphrodite, numerous (100–200+), corollas 7–8 mm long, tube yellow, inflated at base; corolla lobes 1.5–2.5 mm long, recurved, black outside, densely golden-velutinous inside; anther cylinder black, included within corolla, 2.5–2.8 mm long, apical appendages narrowly-triangular, coriaceous; style arms 1.5–2 mm long. Ray florets achenes obovate, abruptly contracted at apex beneath neck of ray limb, prominently angled adaxially, 6–7 × 3–3.5 mm, appressed setuliferous, surface roughened; disc floret achenes flat, smooth or sometimes tuberculate, apically truncate or emarginate, awnless.

Zimbabwe. C: Rhodesville, nr. Salisbury, 4.v.1934, *Gilliland* 25 (K). **Malawi**. N: Mzimba Dist., 2 mi. N of Mzambazi, 13 mi. S of Mperembe, 1250 m, 10.iii.1978, *Pawek* 13953 (K, MA, MO). S: Blantyre Dist., roadside 2 km from Chileka Airport, 775 m, 30.vi.1970, *Brummitt* 11729 (K).

A native of Mexico but widely cultivated, especially in the tropics, and sometimes escaping, but usually persisting only near habitation and on some roadsides; 0–2000 m; probably flowering throughout the year.

Conservation Status: LC (Least Concern).

---

[200] This sheet has a handwritten label 'Zinnia violacea/Mexico. H. R. M.' with a typewritten label 'Zinnia elegans Jacq.'

[201] Many authors have mis-stated the publication date of Cavanilles *Icones* – it pre-dates the publication of *Z. elegans* by at least a couple of years. The proposal to conserve *Z. elegans* against *Z. violacea* is misguided, failing to follow McVaugh's observations (McVaugh, 1984) and would rather accept the superficial treatments in other floras and compendia.

[202] In Collectanea, Suppl. **5**: 152 (1797) Jacquin stated 'Sub hoc titulo semina accepi. Patria mihi ignota est.' Holotype possibly at BM.

[203] It is unclear if Lindley prepared any herbarium material.

[204] De Candolle noted only that he had seen cultivated material; it is unknown if herbarium material was prepared.

138. **SCLEROCARPUS** J.Jacq. ex Murr.

Sclerocarpus Jacq. ex Murr., Syst. Veg., ed. 14: 783 (1784). —Jacquin in Nov. Act. Helvet. **1**: 34 (1787). —Wild in Kirkia **6**(1): 37–38 (1967). —Dyer, Gen. S. Afr. Fl. Pl. **1**: 697 (1975). —Lisowski, (Asterac. Fl. Afr. Cent. 1) Fragm. Flor. Geobot. **36** Suppl. 1: 181 (1991). —Mesfin Tadesse, Fl. Ethiopia & Eritrea **4**(2): 302 (2004). —Beentje & Hind in F.T.E.A., Compositae **3**: 757–759 (2005).

Scabrid or hispid-pubescent erect annual or perennial herbs. Leaves alternate, or occasionally subopposite, simple, lobed or unlobed, 3-veined, glabrous or hispid. Inflorescences of solitary terminal capitula or rarely of lax cymose panicles, capitula sessile or pedunculate. Capitula radiate, heterogamous; involucres campanulate or hemispherical; phyllaries few, leafy, uniseriate, equalling or exceeding florets; receptacle small; paleae cymbiform, indurate and clasping florets, becoming contracted above achene into a ± tubular neck. Ray florets neuter, few (5–8) or absent, corolla tubes long, limb minute, emarginate, ovate to orbicular, yellow to orange. Disc florets 10–50, tubular, hermaphrodite, fertile, corollas tubular, 4-5- lobed, yellow or orange; anther thecae yellow or blackish, stamens usually retained on corolla; basal anther appendages rounded or minutely apiculate; style long, style arms terete, apices obtuse. Achenes asymmetrically obovoid, black or brown, glabrous; pappus a minute lacerate annulus or absent.

A genus of eight spp., most from Mexico and southern N America, with only one species in the Flora area, common in the Old World.

Sclerocarpus africanus Jacq. [Icon. Pl. Rar. **1**: t. 176 (1781), nom. nud.] ex Murr., Syst. Veg., ed. 14: 783 (1784). —Oliver & Hiern in F.T.A. **3**: 374 (1877). —Eyles in Trans. Roy. Soc. S. Afr. **5**: 514 (1916). —Mendonça, Contrib. Conhec. Fl. Angola **1** (Compositae): 90 (1943). —Adams in F.W.T.A., ed. 2, **2**: 235 (1963). —Lisowski, (Asterac. Fl. Afr. Cent. 1) Fragm. Flor. Geobot. **36** Suppl. 1: 181–182 (1991). —Mesfin Tadesse, Fl. Ethiopia & Eritrea **4**(2): 302–302 (2004). —Beentje & Hind in F.T.E.A., Compositae **3**: 757–759 (2005). Type: either a cultivated specimen or the original of Ic. Pl. Rar. **1**: t. 176. FIGURE 6.5.**31**.

Erect annual herb 0.2–1.2(2) m tall. Stems striate, hispid-pubescent, well-branched in inflorescence. Leaves opposite below, alternate above within flowering portion, petiolate, petioles to 4 cm long, hispid-pubescent, hairs eglandular, lamina membranous, 3–12 × 1–6.5 cm, exceeding internodes, ovate, apex acute, base cuneate and tapering gradually into petiole, margin serrate or crenate with apex of teeth minutely mucronate, appressed-hispid pubescent or scabrid-pubescent on both surfaces, hairs predominantly short but with scattered longer hairs on surfaces, young leaves and stems sometimes densely golden-pubescent. Capitula 1–1.5 cm diam., terminal on branches, hemispherical; phyllaries 5, 7–30 × 10 mm, leafy, narrowly ovate, tapering to both base and apex, appressed-pubescent; receptacles small, conical, paleaceous, paleae 9.5–12 mm long, indurate about achene, cymbiform, pubescent above. Ray florets few or absent, c. 5 mm long, tubular, widening into a short emarginate slightly spreading limb, yellow, puberulent. Disc florets 10–14, corollas yellow, c. 7.5 mm long, tubular, widening gradually above, puberulent above, hairs eglandular; apical anther appendages narrow, subacute; anther thecae dark coloured. Achenes black, 5–8 mm long, multistriate; pappus absent.

Botswana. N: [Chobe Dist.] Duma Tau Riverine along edge of lagoon, [Linyanti Wildlife Reserve, bordering Chobe National Park] 18°31.819'S, 23°34.314'E, 956 m, 22.ii.2003, *Heath & Heath* 379 (K). **Zambia**. B: Machili, 31.xii.1960, *Fanshawe* 6060 (SRGH). C: 5 km. W. of Luangwa R., Gt. East Rd., 26.iii.1955, *E. M. & W.* 1201 (BM, SRGH). S: Siavonga Dist., Zambezi Escarpment, along Malenga stream, 730 m, 15.iii.1997, *Harder* et al 4053 (K, MO). **Zimbabwe**. N: Urungwe, Rifa R., 518 m, 24.ii.1953, *Wild* 4083 (K, LD, SRGH). W: Wankie, 19.vi.1934, *Eyles* 8019 (K, SRGH). C: Salisbury, 3.v.1906, *Flanagan* 3037 (PRE). **Malawi**. N: Lake Nyasa, Likoma I., 28.vi.1900, *Johnson* 6 (K). **Mozambique**. T: Tete, 1.ii.1860, *Kirk* (K).

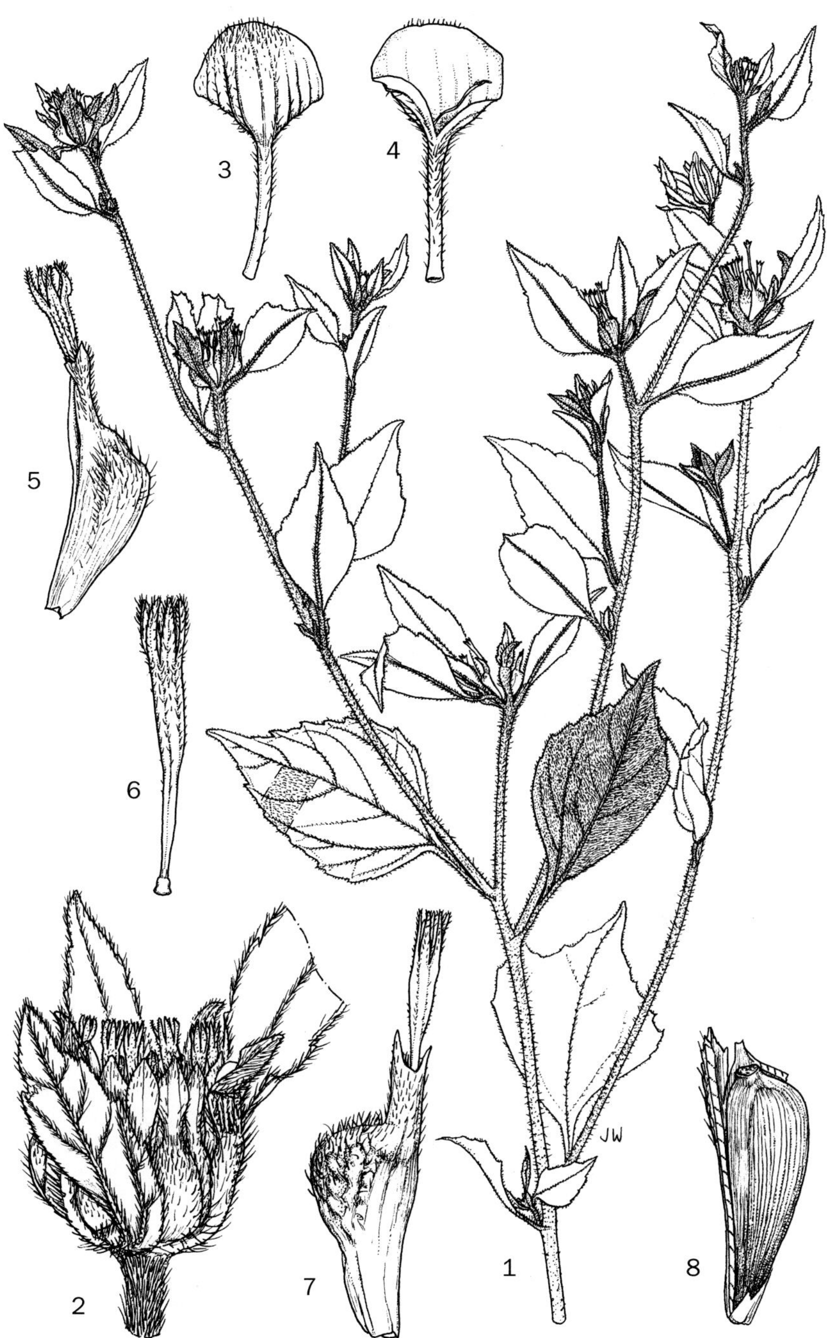

Fig. 6.5.**31**. SCLEROCARPUS AFRICANUS. 1, flowering branch (× 2/3); 2, capitulum (× 3); 3, abaxial view of ray floret corolla (× 6); 4, adaxial view of ray floret corolla (× 6); 5, flowering disc floret with enclosing palea (× 4); 6, disc floret corolla (× 6); 7, desiccated disc floret with enclosing palea (× 5); 8, disc floret achene with palea partially removed (× 5). 1, 6, 8 from *Lye* 4770b; 2 from *Bally* 4714; 3–5, 7 from *Pielou* 169. Drawn by Juliet Williamson. From Flora of Tropical East Africa.

Throughout tropical Africa from the Sudan and Senegal to SW Africa; also in Oman, Yemen, India and Cape Verde Islands. In *Baikiaea mutemwa* forest in low altitude riverine fringes and along streams, also on Kalahari Sand; 150–2200 m.

Conservation Status: Widespread throughout tropical Africa; not threatened, LC (Least Concern).

139. **TITHONIA** Desf. ex Juss.

Tithonia Desf. ex Juss., Gen. Pl.: 189 (1789). —Wild in Kirkia **6**(1): 55–56 (1967). —La Duke in Rhodora **84**(840): 453–522 (1982). —Lisowski, (Asterac. Fl. Afr. Cent. 1) Fragm. Flor. Geobot. **36** Suppl. 1: 218–221 (1991). —Hind in Fl. Masc., Composées **109**: 190–192 (1993). —Isawumi in Compositae Newsl. **29**: 31–39 (1996). —Beentje & Hind in F.T.E.A., Compositae **3**: 759–761 (2005). —Panero in Kubitzki Fam. Gen. Vasc. Pl. **8**: 469 [2006](2007).

Tithonia subgen. *Mirasolia* Sch.Bip. in Seeman, Bot. Voy. Herald: 305 (1856).

Mirasolia (Sch.Bip.) Benth. & Hook.f., Gen. Pl. **2**(1): 367 (1873).

Urbanisol Kuntze, Revis. Gen. Pl. **1**: 370 (1891), nom. illeg. superfl. pro *Tithonia* Desf. ex Juss.

Annual herbs, or perhaps short-lived perennials. Leaves usually alternate, occasionally opposite, entire or lobed, 3-veined from base. Capitula showy, hemispherical, solitary on long lateral pedicels, fistulose towards apex beneath involucre, heterogamous, radiate; phyllaries 2–5-seriate, ribbed and leathery below, herbaceous or papyraceous above; receptacle convex, paleaceous, paleae clasping disc florets, persistent. Ray florets neuter, limb 2–3-dentate. Disc florets hermaphrodite, corolla tubular, narrowed below, 5-lobed; anthers with an apical appendage, ± sagittate at the base; style bifid, style arm apices acute, hairy outside. Achenes 4-sided, oblong-cylindrical; carpopodium present; pappus a crown of short free or connate small scales with usually 1–2 additional awns.

A genus of 11–13 spp. from Mexico and Central America; two spp. are widely cultivated and often escape in the tropics of the New and Old World and both are present in the Flora area as both cultivated and naturalized species.

Leaves entire, or lower 3-lobed, base usually ± cordate; phyllaries biseriate, inner series herbaceous acute; ray florets usually 9 or 10, ray limbs c. 33 mm long, bright orange-red; disc florets >115; stamen filaments pubescent. . . . . **1.** *rotundifolia*
Leaves usually 3–5-lobed, base cuneate not cordate; phyllaries c. 4-seriate, inner series with rounded or obtuse papery apices; ray florets usually 13, ray limbs c. 48–69 mm long, bright yellow or orange-yellow; disc florets 80–100; stamen filaments glabrous . . . . . **2.** *diversifolia*

1. **Tithonia rotundifolia** (Mill.) S.F.Blake in Contrib. Gray Herb., n.s., **52**: 41 (1917); in Contr. U.S. Natl. Herb. **20**: 426 (1921). —Moore in Fawcett & Rendle, Fl. Jamaica **7**: 231 (1936). —Lisowski, (Asterac. Fl. Afr. Cent. 1) Fragm. Flor. Geobot. **36** Suppl. 1: 218–219 (1991). Type: 'This plant was discovered by the late Dr. Houston growing naturally at La Vera Cruz in New Spain, from whence he sent the seeds to England.' [cult. material by ?] (BM001009724 holotype).

 Tagetes rotundifolia Mill., Gard. Dict., ed. 8., *Tagetes* No. 4 (1768).

 Tithonia uniflora Desf. ex J.F.Gmel., Syst. Nat., ed. 13, **2**(2): 1259 (1792), nom. illeg. – no validy published genus at this time.[205]

---

[205] Reference in the protologue, after merely a citation of the name was to 'Des Fontaines. act. Paris. extram. vol. 12'. This, as Stafleu & Cowan, Regnum. Veg. **94**: 629 (1976) pointed out, was never published but it appeared later as 'Description du genre *Tithonia*' in Ann. Mus. Nat. Hist. Nat. **1**: 49–51 (1802).

Tithonia tagetiflora Desf. in Ann. Mus. Natl. Hist. **1**: 49, pl. 4 (1802). Type: '*Thiery*, voyageur connu par des projets utiles à son pays, en envoya, en 1778, au jardin des plantes, des graines qu'il avoit cueillies dans les environs de la Vera Cruz. Elle a été cultivée pendant deux ou trois ans; ...' (P00309495 holotype).

Tithonia tagetiflora Lam., Tab. Encycl. **3**: 284 (1823); **2**: t. 708 (1797), nom. illeg., later homonym. Type: not specified.

Helianthus speciosus Hook. in Curtis's Bot. Mag., ser. 2, 3 [vol **61** of whole]: t. 3295 (1834). Type: 'Along with the very beautiful drawing here figured, my obliging correspondent, THOMAS GLOVER, Esq. of Manchester, sent me the following account of this charming HELIANTHUS. "Mr. EDWARD LEEDS of this place, who has lately commenced business as a Nurseryman and Florist, from among a packet of seeds from the Botanic Garden, Mexico, sent to him by W. HIGSON, Esq. of Manchester, has raised several plants that are not known in this neighbourhood. Only one, the subject of my present communication, has flowered, ... Only one seed vegetated; and ... it is said to have come from Jorullo. ..." (K000487735 – Herb. Hook. holotype, K000373021).[206]

Leighia speciosa (Hook.) DC., Prodr. **5**: 583 (1836).

Tithonia aristata Oersted in Vidensk. Meddel. Dansk Naturhist. Foren. Kjøbenhavn **1852**(5–7): 114 (1852)[1853]. Type: 'Jeg fandt den paa Bjerget Aguacate i Costa-Rica.' (C holotype).

Tithonia heterophylla Griseb. in Bonplandia **6**: 9 (1858). Type: 'Ins. Taboga pr. Panamam: *Duch.* [*assaing* s.n.]' (GOET002121 holotype, GOET002122).

Tithonia speciosa (Hook.) Griseb., Cat. Pl. Cub.: 155 (1866).

Tithonia macrophylla S.Watson in Proc. Amer. Acad. Arts **26**: 140 (1891). Type: [Mexico:] 'Barranca, near Guadalajara; Sept., 1889 ([*Pringle*] n. 2798).' (US00277334 holotype, F106758, GH0013122, MICH, NY00273846, US00048053).

Urbanisol tagetifolius (Desf.) Kuntze, Revis. Gen. Pl. **1**: 370 (1891).

Urbanisol tagetifolius var. *diversifolius* Kuntze f. *grandiflorus* Kuntze, Revis. Gen. Pl. **1**: 370 (1891), nom. illeg.

Urbanisol tagetifolius [unranked] α *normalis* Kuntze, Revis. Gen. Pl. **1**: 370 (1891). Type: 'Singapur, verwildert.'[207]

Urbanisol tagetifolius [unranked] β *speciosus* (Hook.) Kuntze, Revis. Gen. Pl. **1**: 370 (1891).

Urbanisol aristatus (Oerst.) Kuntze, Revis. Gen. Pl. **1**: 371 (1891).

Urbanisol heterophyllus (Griseb.) Kuntze, Revis. Gen. Pl. **1**: 371 (1891).

Tithonia speciosa (Hook.) Klatt in Bull. Soc. Roy. Bot. Belgique **31**(2): 203 (1893), comb. illeg. superfl.

Tithonia vilmoriniana Pamp. in Bull. Soc. Bot. Ital. **1908**: 133 (1908). Type: 'Habitat: Culta in Horto Botanico Florentino e seminibus mexicanis, pr. Jacona (Michoacan) lectis et a cl. *Vilmorin-Andrieux* missis.' (FI holotype).

Tagetes diversifolia sensu Martineau, Rhod. Wild Flowers: 89, t. 32 fig. 3 (1953).

Annual herb up to c. 2 m tall, unbranched; stem often purplish, erect, terete, striate, villous when young, later glabrescent. Leaves alternate, petiolate, petiole 2–7(15) cm long pubescent; lamina up to 20(27) × 10(19.5) cm, deltate-ovate to broadly ovate, apex acuminate, base cordate or rounded at base and then abruptly cuneate and tapering gradually into petiole, margins crenate or serrate, lower leaves often ± deeply 3-lobed with rounded sinuses, scabrid-pubescent on both surfaces but especially below. Peduncle to 30 cm long, stout, c. 1.5 cm diam., villous at first later glabrescent except near apex; phyllaries up to 2.3 cm long, subequal, narrowly ovate, acute, pubescent, herbaceous and green above; paleae c. 1.2(1.5) cm long, narrowly oblong-ovate, acuminate, glabrous. Ray florets orange-red; limb up to 40 mm long, elliptic or narrowly ovate, 2–3-dentate at apex; tube c. 2 mm long, puberulent. Corolla of disc florets c. 8.5 mm long with narrower lower part c. 1 mm long, puberulent in lower half. Achenes c. 5.5 mm long, 4-angled, ± compressed, setuliferous; pappus of short free scales c. 1.5(2.5) mm long, and 2 somewhat longer deciduous setae.

---

[206] La Duke (1982) annotated the sheet in K used for t. 3295 (ex Herb. Hookerianum) as the 'Holotype'. Interestingly there is one sheet in the 'Cultivated' material covers that is of one capitulum on part of a pedicel with a printed label. This label notes that the specimen is from the original plant grown by Mr Edward Leeds and could therefore be counted as an isotype.

[207] This taxon was not mentioned by Wetter & Zanoni (1985).

Botswana. N: Thamalakane River, Maun Camp No. 9, 19°29'S, 23°29'E, 14.i.1974, *Smith* 777 (K, SRGH). **Zambia**. W: Ndola, 13.iii.1954, *Fanshawe* F958 (EA, K). C: Lusaka, Chilanga-Kafu, 26.i.1963, *van Rensburg* 1259 (K, SRGH). **Zimbabwe**. W: Bulawayo, i.1954, *Martineau* 132 (SRGH). C: Salisbury, 15.ii.1950, *Wild* 3224 (K, SRGH). E: Umtali, 5.iv.1951, *Chase* 3734 (BM, K, SRGH). **Malawi**. C: Lilongwe Dist., Lilongwe Nature Sanctuary Forest, 5.v.1984, *Patel et al.* (K, MAL).

Native to Central America from Mexico to Panama, introduced into the W Indies, Venezuela and Brazil; also into southern Africa. A common weed of roadsides and disturbed ground, sometimes still cultivated, reported as a hedge-plant in a neighbouring region; 800–1250 m; flowering from November to June, but probably flowering throughout the year.

Conservation Status: LC (Least Concern).

2. **Tithonia diversifolia** (Hemsl.) A.Gray in Proc. Amer. Acad. Arts **19**: 5 (1883). —Blake in Contrib. U.S. Nat. Herb. **20**: 426 (1921). —Moore in Fawcett & Rendle, Fl. Jamaica **7**: 232 (1936). —Lisowski in (Asterac. Fl. Afr. Cent. 1) Fragm. Flor. Geobot. **36** Suppl. 1: 219–221 (1991). —Hind in Fl. Masc., Composées **109**: 191–192 (1993). —Beentje & Hind in F.T.E.A., Compositae **3**: 761 (2005). Types: 'South Mexico, abundant in the valley of Orizaba, also in the valley of Codova (*Bourgeau*, 2319, 1562); Guatemala, Dueñas (*Fraser, Salvin*)' *Bourgeau* 2319 (K000487726 lectotype, BR5522910, FI, GH00010519, K000487727, S-G-4147, US000128796), lectotypified by La Duke (1982: 498), as per determined sheet; *Fraser* s.n. (K000487730 syntype); *Salvin* s.n. (K000487729 syntype); *Bourgeau* 1562 (K000487728 syntype). FIGURE 6.5.**32**.

Mirasolia diversifolia Hemsl., Biol. Cent.-Amer. Bot. **2**: 168, t. 47 (1881).

Tall, often branching, herb up to c. 3 m., annual or perhaps sometimes a short-lived perennial; stems sometimes woody, ± 4-angled, striate, softly hairy at first, later glabrescent. Leaves alternate, petiolate, petiole to 7(10) cm long pubescent ± biauriculate; lamina to 12(20) × 7 (18.5) cm, narrowly ovate, ovate, obovate-oblong or obovate, ± deeply 3–5-lobed, rarely entire, apex acuminate, cuneate at base and tapering gradually into petiole, margins crenate or crenate-serrate or subentire, scaberulous-pubescent above, densely pubescent below, sometimes almost grey-tomentose below. Capitula solitary on ends of side-branches; peduncles up to c. 10 cm long, broadening up to c. 1 cm diam. near apex, striate, glabrescent; phyllaries unequal, outer 5–10 mm long oblong-ovate acute leathery, inner up to 17 mm. long acute papyraceous towards apex; paleae c. 12 mm long, obovate, abruptly acute or acuminate, pubescent or glabrescent above, striate. Ray florets yellow or orange-yellow, ray limb to 6 cm long, narrowly oblong, oblong or narrowly oblong-obovate, apex 3-dentate, tube to 3 mm long, puberulous. Corolla of disc florets to 8 mm long, cylindrical, base narrowed, densely puberulent in lower half. Achenes c. 6 mm long, ± 4-angled, densely adpressed-setuliferous; pappus of short scales c. 2 mm long and connate below, with 2 somewhat longer setae.

Zambia. N: Abercorn, 29.iii.1960, *Angus* 2196 (FHO, K, PRE, SRGH). W: Mwinilunga Dist., West Lunga River, 16.v.1986, *Philcox* et al. 10324 (K). E: Fort Jameson, cult., 25.iv.1952, *White* 2455 (FHO, K). **Zimbabwe**. C: Hunyani, Prince Edward Dam, 16.v.l934, *Gilliland* 143 (BM, K). E: Umtali, 2.vi.1949, *Chase* 1449 (BM, K, SRGH). **Malawi**. C: Lilongwe, 25 mi S of Lilongwe, 15.viii.1976, *Pawek* 11537 (K, MO, SRGH, UC). S: Mt. Mulanje foor, Esperanza Estate, 800 m, *J. D. & E. G. Chapman* 7484 (K, MO). **Mozambique**. N: Vila Cabral, 15.v.1948, *Pedro & Pedrógão* 3599 (EA, LMJ). M: Namaacha, estrada para Matianine, 5.vii.1974, *Marques* 2489 (K, LMU).

Native to Central America from Mexico to Costa Rica, introduced into the W Indies, Ceylon, India, Malaysia and tropical Africa. A weed of roadsides and waste places, sometimes troublesome in higher rainfall areas, sometimes still cultivated; 800–1600 m; flowering from December to August, but probably flowering throughout the year.

Conservation Status: A troublesome weed in some parts of the Flora area; LC (Least Concern).

Fig. 6.5.**32**. TITHONIA DIVERSIFOLIA. 1, flowering branch (× ²/₃); 2, ½ l.s. capitulum (× 2); 3, inner phyllary (× 2); 4, palea (× 3); 5, ray floret corolla (× 1); 6, ray floret achene (× 4); 7, disc floret corolla (× 6); 8, partial anther cylinder opened out (× 6); 9, disc floret style (× 4); 10, disc floret achene (× 4). All from *Duljeet* 10420. Drawn by Pat Halliday. From Flore des Mascareignes.

140. **HELIANTHUS** L.

Helianthus L., Sp. Pl. **2**: 904 (1753); Gen. Pl., ed. 5: 386 (1754). —Wild in Kirkia **6**(1): 57–59 (1967). —Heiser *et al.* in Mem. Torrey Bot. Club **22**(3): 1–218 (1969). —Schilling & Heiser in Taxon **30**(2): 393–403 (1981). —Hind in Fl. Masc., Composées **109**: 189–190 (1993). —Beentje & Hind in F.T.E.A., Compositae **3**: 706 (2005).

Annual or perennial herbs. Rootstock fibrous, a thick taproot, rhizomatous, or sometimes of round or elliptic tubers. Stems erect, simple or poorly- to well-branched, moderately to densely pilose, villous or tomentose or glabrescent. Leaves simple, alternate, usually opposite, sometimes opposite below and alternate or subopposite towards and within inflorescence, petiolate, bases acuminate, cuneate or cordate, lamina linear to linear-lanceolate, ovate, ovate-lanceolate or oblong, ± concolorous or obviously discolorous, mid-green and scabrid pubescent or pilose above, densely white- or tawn-lanate villous or tomentose beneath, or densely silvery-villous on both surfaces, margins entire, serrulate or crenate-serrate, flat or revolute, apices acute, acuminate or obtuse. Inflorescences of solitary terminal capitula or capitula in paniculate-cymes or lax corymbs. Capitula radiate and heterogamous or discoid and homogamous, moderate-sized to extremely large (especially in some cultivated material); involucres campanulate, sometimes broadly so, or hemispherical; phyllaries 2–5-seriate, imbricate, subequal, narrow-lanceolate, oblong or oblanceolate, base indurate or not, apices acute, acuminate (sometimes abruptly so) or obtuse, pilose to villous or silvery-tomentose outside, sometimes apices darkened; receptacle flat to shallow-convex, paleaceous; paleae obtuse, keeled, straw-coloured or black or deep purplish, sometimes just with black or deep purplish apices, glabrous or pubescent outside towards apex and on keel, conduplicate about and sometimes considerably longer that accompanying achene, apices 3-lobed, middle lobe often markedly longer, acute. Ray florets, when present, neuter, corollas pale yellow to golden yellow, orange yellow, red, burgundy coloured, chestnut, or plumb coloured, limb sparsely pilose beneath, apices 3-lobed. Disc florets numerous, hermaphrodite, fertile, corollas yellow to greenish yellow, or more usually reddish or burgundy coloured to purple (sometimes with both colour variants in species), tube glabrous, usually inflated at base, 5-lobed, lobes sometimes darker in colour than tube, pubescent; anther thecae brown, black or deep purple; style arms flattened, sometimes with cylindrical appendages. Achenes biconvex, obovate to oblong, black, dark speckled or greyish, glabrous or sparsely to densely setuliferous, sometimes only towards apex, often glabrescent, setulae of twin-hairs; elaiosome absent; pappus of 2 deciduous or caducous awns, usually with few deciduous or caducous squamellae in between, whitish or off-white to straw-coloured.

A genus of about 50 spp., predominantly from N America; many spp. are cultivated throughout the temperate regions and one extensively grown for oil-seed – *Helianthus annuus* L. in both temperate and tropical regions. Two spp. are cultivated in the Flora area (*H. annuus* and *H. angustifolius* L. – neither represented by herbarium material) and a further two well-naturalized in Mozambique. The following key accounts for all four, but only the two naturalized species are described.

Anashchenko in Bot. Zhurn. **59**: 1472–1481 (1974) considered several species as no more than varieties of *Helianthus annuus*, taking a very conservative approach to the genus indeed; I have maintained Heiser *et al.*'s view.

1. Plants perennial (from rhizomes, tubers or crown buds); leaves lanceolate to linear-lanceolate (in ours) . . . .*angustifolius*
 – Plants annual; leaves usually long-petiolate 2
2. Leaves and stems densely white-tomentose, silky . . . . **1.** *argophyllus*
 – Leaves and stems glabrous to densely hispid 3
3. Plants usually < 1 m tall, stems usually much-branched; leaf bases cuneate, truncate (rarely cordate); phyllaries c. 2 mm wide, gradually attenuate . . . . **2.** *debilis* subsp. *cucumerifolius*
 – Plants usually >1.5 m tall, stems usually simple (in many cultivars) or much-branched (then stems stout); leaf bases cordate or subcordate; phyllaries (3)5 mm or more wide, abruptly attenuate . . . . *annuus*

1. **Helianthus argophyllus** Torr. & A.Gray, Fl. N. Amer. **2**: 318 (1842). —Watson in Pap. Michigan Acad. Sci. **9**: 337 (1929). Type: 'Texas [III], *Drummond* [196]!' (?NY holotype, E00433086, GH00008798 – s.n., NY00179073 – s.n., 'upper leaves' fragment of holotype, P00710159).[208]

Annual herb 1–3 m tall. Stems densely white-tomentose to floccose, usually much-branched. Leaves sometimes opposite below, upper alternate, petiolate, petiole 3 or more cm long, white-villous; lamina (10)15–25 cm long, ovate to ovate-lanceolate, lowermost broadly ovate, apex acuminate, base cuneate, subcordate below, margins crenate-serrate or subentire, silvery villous on both surfaces. Capitula solitary and terminal on villous peduncles c. 5 cm long; involucre 2–3 cm diam.; phyllaries c. 3-seriate, c. 1.4 × 5–7 mm, ovate to ovate-lanceolate, silvery villous dorsally, apices abruptly long-attenuate, glabrescent; paleae c. 1.2 cm long, conduplicate and keeled, 3-lobed, middle lobe longest and caudate-acuminate, glabrous or villous towards apex, prominently longitudinally veined. Ray florets 15+, limb yellow or orange-yellow, c. 3 cm long, narrowly obovate-elliptic, apex minutely 3-denticulate, dorsally pilose, tube c. 2 mm long sparsely pilose. Disc floret corollas deep purple, lobes pubescent, base of limb villous, tube glabrous. Achenes blackish, 4–6(7) mm long, obcompressed, obovate, apex ± truncate, setuliferous, more densely so towards the apex; pappus of 2 rapidly caducous, very narrowly ovate acuminate awns.

Mozambique. GI: Bilene, 11.iii.1952, *Barbosa & Balsinhas* 4880 (BM, LMJ). M: Marracuene, 6.ix.1945, *Pedro* 45 (K, LMJ, PRE).

Native of the southeastern United States (Texas). Introduced and forming dense colonies on the sand-dunes of the Lourenço Marques district.

Conservation Status: A well-established introduced plant; LC (Least Concern).

Closely related to the common sunflower, *H. annuus* L., but distinguished by its silvery-tomentose appearance of stem, leaves and phyllaries, otherwise it is remarkably similar.

2. **Helianthus debilis** Nutt. in Trans. Amer. Phil. Soc., ser. 2, **7**: 367 (1841). —Watson in Pap. Michigan Acad. Sci. **9**: 355 (1929). —Heiser *et al.*, Mem. Torrey Bot. Club **22**(3): 45–51 (1969). Type: 'Hab. The sea-coast of East Florida. (*Dr. Baldwin.*)'(BM000839315 holotype).

    Helianthus annuus L. var. *debilis* (Nutt.) Anashch. in Bot. Zhurn. **59**: 1476 (1974).

Subsp. **cucumerifolius** (Torr. & A.Gray) Heiser in Madroño **13**: 160 (1956). Type: [USA:] 'Texas. *Drummond* [171]! July-Sept.' (?NY00179083 holotype, GH00263108). FIGURE 6.5.**33**.

    Helianthus cucumerifolius Torr. & A.Gray, Fl. N. Amer. **2**: 319 (1842). —Watson, in Pap. Michigan Acad. Sci. **9**: 356 (1929). —Heiser *et al.*, Mem. Torrey Bot. Club **22**(3): 49–51 (1969).[209]

    Helianthus debilis f. *cucumerifolius* (Torr. & A.Gray) Voss in Vilmorin, Blumengärtn., ed. 3, **1**: 482 (1894).

    Helianthus debilis var. *cucumerifolius* (Torr. & A.Gray) A.Gray, Synopt. Fl. N. Amer. **1**: 273 (1884). —Wild in Kirki **6**(1): 58–59 (1967).

Annual 0.55–0.65(1) m tall, often decumbent, or erect. Stems slender, much-branched, hispid or scabrid, purplish-spotted (although appearing dark with pale blotches in herbarium material). Leaves alternate (at least above), petiolate, petiole to 7 cm or more long, hispid-pilose, lamina 4–9(10) × 3–8 cm, broadly ovate, acute, mucronate, base cordate and abruptly cuneate, margins coarsely and irregularly serrate, scabrid-pubescent on both sides. Capitula solitary, terminal on upper branches; peduncles (16)25–50 cm long, scabrous-pubescent. Involucre 1.6–2 cm diam., phyllaries 2–3-seriate, c. 1 cm long, very narrowly ovate-acuminate, scabrous-pubescent; paleae c. 7 mm long, 3-cuspidate, ± conduplicate and keeled, middle lobe longest and caudate-acuminate, puberulent, prominently longitudinally veined. Ray florets 11–17, limbs yellow, 1.2–2+ × 0.5–1 cm, narrowly elliptic, apex acute

---

[208] It seems odd that only a fragment of the holotype exists in NY (where Torrey's collections are supposed to be).

[209] Data provided in the NY type database is inconsistent with that suggested by the protologue.

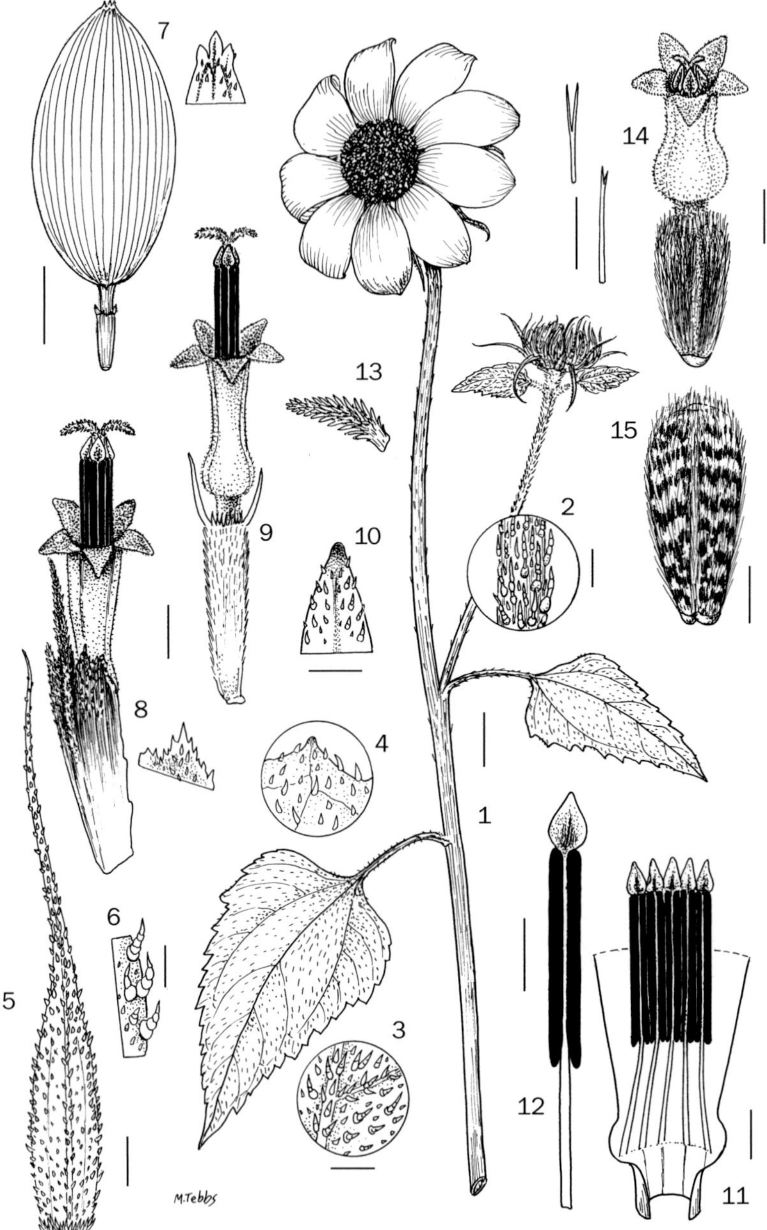

Fig. 6.5.**33**. HELIANTHUS DEBILIS subsp. CUCUMERIFOLIUS. 1, flowering branch; 2, detail of pedicel indumentum; 3, detail of adaxial leaf surface indumentum; 4, detail of adaxial leaf surface indumentum of leaf tooth; 5, phyllary; 6, detail of indumentum of phyllary margin; 7, ray floret with detail of lobed apex; 8, disc floret with accompanying palea and detail of upper palea margin; 9, disc floret with anther cylinder conspicuously exserted from corolla; 10, detail of abaxial surface of corolla lobe; 11, disc floret corolla opened out showing attachment of filaments; 12, stamen; 13, style arm apex; 14, disc floret with maturing achene with anther cylinder included within corolla, and detail of two achene setulae; 15, mature achene lacking apical awns. 1–14 from *Matthew* 53137; 15 from *Amico* 1129. Scale bars: 10 = 0.5 mm; 2–6, 8, 9, 11–15 = 1 mm; 7 = 5 mm; 1 = 10 mm. Drawn by Margaret Tebbs.

and minutely 3-denticulate, dorsally puberulent, tube c. 1.3 mm long puberulent. Disc floret corollas puberulent, lobes deep purple, yellowish below. Achenes 2.5–3 mm long, glabrous or setuliferous; pappus of 2 short, slender rapidly caducous awns.

Mozambique. Z: Chinde, viale della città, 7.i.1970, *Amico & Bavazzano* ?1129 (FI, K). M: Lourenco Marques, Ponta Vermelha, 25.ix.1945, *Pedro* 136 (LMJ, PRE).

Native of southeastern Texas in the USA. An introduced weed on the red loamy sands of Ponta Vermelha; the *Amico & Bazzano* collection may well be cultivated or an escape on the roadside in Chinde.

Conservation Status: An introduction from the USA, albeit rarely escaping and apparently naturalizing; LC (Least Concern).

Anashchenko's annotations of several sheets in K suggested he initially wanted to recognize this taxon as a variety of another subspecies, '*petiolaris*' – of *Helianthus annuus*. Fortunately this combination was not made.

141. **TRIDAX** L.

Tridax L., Sp. Pl. **2**: 900 (1753); Gen. Pl., ed. 5: 382 (1754). —Robinson & Greenman in Proc. Amer. Acad. Arts **32**: 1–10 (1897). —Powell in Brittonia **17**(1): 47–96 (1965). —Wild in Kirkia **6**(1): 10–11 (1967). —Lisowski, (Asterac. Fl. Afr. Cent. 1) Fragm. Flor. Geobot. **36** Suppl. 1: 124 (1991). —Hind in Fl. Masc., Composées **109**: 209–211 (1993). —Mesfin Tadesse, Fl. Ethiopia & Eritrea **4**: 305 (2004). —Beentje & Hind in F.T.E.A., Compositae **3**: 764–765 (2005).

Bartolina Adans., Fam. Pl. **2**: 124 (1763).

Balbisia Willd., Sp. Pl., ed. 4, **3**(3): 2214 (1803), nom. rej., non *Balbisia* Cav., nom. cons.

Sogalgina Cass. in Bull. Sci. Soc. Philom. Paris **1818**: 31 (1818).

Galinsogea Kunth in Humb., Bonpl. & Kunth, Nov. Gen. Sp. Pl. **4** (ed. folio): 198 (1818).

Ptilostephium Kunth in Humb., Bonpl. & Kunth, Nov. Gen. Sp. Pl. **4** (ed. folio): 199 (1818).

Carphostephium Cass. in Dict. Sci. Nat., ed. 2, **44**: 62 (1826).

Mandonia Wedd. in Bull. Soc. Bot. France **11**: 50, pl. 1 (1864).

Annual or perennial herbs. Rootstock a taproot, rarely somewhat woody. Stems erect, procumbent or decumbent, sometimes rooting at nodes. Leaves opposite, rarely alternate above, simple or trilobed to pinnately lobed or divided, margins entire, serrate or dentate, to repand, surfaces scarcely to sparsely or densely pubescent, puberulent, pilose, hirsute, hispid or canescent-tomentose, often stipitate-glandular. Inflorescences of few- to many-headed cymose panicles or of solitary capitula, capitula on slender or stout elongate pedicels. Capitula discoid and homogamous or radiate and heterogamous; involucre campanulate, rarely urceolate; phyllaries ± uniseriate or 2–3-seriate, subequal, or 4–5-seriate and unequal, imbricate, greenish and usually purple to purple-tinged apices, inner phyllaries with scarious, purple margins; receptacle conical to convex or nearly flat, paleaceous, paleae scarious, persistent, rarely deciduous, partially enclosing achenes, yellowish to yellowish-green. Ray florets, when present, female and fertile, corollas white, yellow, pink or purplish, limbs obscurely or obviously bilabiate, external lip (2)3(4)-lobed, inner lobes 1 or 2, obvious, vestigial or absent. Disc florets hermaphrodite, actinomorphic, 5-lobed, corollas yellow, rarely white or greenish-yellow, glabrous or pubescent, throat tubular to narrowly funnelform, lobes oblanceolate, apices acute or obtuse, erect to reflexed or involute; apical anther appendages ovate, basal anther appendages obscurely to obviously sagittate; style arms recurved to revolute, slender, subterete or flattened on inner surface, stigmatic lines weakly defined or obscure. Achenes turbinate or narrowly obconical, terete to ridged, marginal achenes usually recurved and compressed, body glabrous to densely setuliferous, setulae long or short, ascending; pappus of several to many, plumose or fimbriate, linear-lanceolate, rarely oblong, scales and or slender setae, rarely absent.

A genus of c. 25 spp. Mexico, Central America S to southern South America. One species is a widespread pantropic weed and present in the Flora area, *Tridax procumbens*.

Tridax procumbens L., Sp. Pl. **2**: 900 (1753). —Andrews, Fl. Pl. Sudan **3**: 55 (1956). —Adams in F.W.T.A., ed. 2, **2**: 230, fig. 248 (1963). —Humbert, Fl. Madagasc., Composées **189**: 666, t. cxv fig. 2932 (1963). —Powell in Brittonia **17**(1): 80 (1965). —Lisowski, (Asterac. Fl. Afr. Cent. 1) Fragm. Flor. Geobot. **36** Suppl. 1: 124–126 (1991). —Hind in Fl. Masc., Composées **109**: 210–211 (1993). —Mesfin Tadesse, Fl. Ethiopia & Eritrea **4**(2): 306–307 (2004). —Beentje & Hind in F.T.E.A., Compositae, **3**: 764–765 (2005). —Chen Yousheng & Hind in Fl. China **20–21**: 864 (2011). Type: 'Habitat in Vera Cruce.' *Houston* s.n. (BM-000647229 – Herb. Clifford: 418, *Tridax* 1, lectotype), lectotypified by Powell (1965: 80). FIGURE 6.5.**34**.

    Balbisia elongata Willd., Sp. Pl., ed. 4, **3**(3): 2214 (1803). Type: 'Habitat in Mexico. ☉ (v. v.).' (B-W – 16366).[210]

    Amellus pedunculatis Ortega ex Willd., Sp. Pl., ed. 4, **3**(3): 2214 (1803)., nom. nud. pro syn.

    Balbisia canescens Pers., Syn. Pl. **2**: 470 (1807). Type: 'Hab. ad S. Martham Continentalis Amer. australis. *Richard*.'(?P02465330 holotype).

    Tridax procumbens var. *canescens* (Pers.) DC., Prodr. **5**: 679 (1836).

    Balbisia pedunculata Ortega ex Hoffmanns., Verz. Pfl.-Kult.: 228 (1824), nom. illeg., based on *Balbisia elongata* Willd.

    Balbisia divaricata Cass. in Ann. Sci. Nat. **23**: 90 (1831). Type: '*M. Bouton*, qui n'avait point nommé cette plante, dit qu'elle est cultivée dans le jardin des Pamplemousses.' Type material: ?P.

    Tridax procumbens var. *ovatifolia* B.L.Rob. & Greenm. in Proc. Amer. Acad. Arts **32**: 7 (1896). Type: 'Collected by *E. W. Nelson* in the vicinity of Yalalag, Oaxaca, July, 1894, no. 948.' (GH13227 holotype).

Annual, or perennial herb, prostrate to ascending to c. 0.5 m tall. Rootstock a taproot. Stems few, simple or poorly-branched, striate, pilose. Leaves opposite, petiolate, petiole 4–30 mm, pilose, lamina narrowly ovate to ovate, 2–7(12) × 1–4(6) cm, sometimes appearing 3-lobed, base cuneate, margins coarsely dentate or incised-dentate, scabrid to strigose-pubescent on both surfaces, apex acute or acuminate. Capitula solitary, heterogamous and radiate, c. 1–1.5 cm. diam., scapiform, scape 5–25 cm long, ebracteolate or with a pair of reduced leaf-like bracts towards base, otherwise pilose throughout; involucre 7–8 mm long; phyllaries 2–3-seriate, densely pilose, outer 3.5–6.5 (8) × 2–3.5 (4) mm, ovate-acuminate, inner 5.5–8 × 1.5–2 mm, ± oblong, pilose, apices obtuse apiculate, often purplish; receptacle flat to slightly convex, paleaceous; paleae 6–8(9) × 1–1.5 mm, linear, pilose on upper part of keel outside. Ray florets female, (2)3–6, ray limb 2.5–5 × 2–4 (5) mm, ovate, whitish or yellowish-white, apex 3-dentate, pilose, corolla tube c. 3.5 mm long. Disc florets hermaphrodite, corollas 5–7 mm long, cylindrical, yellow, puberulous towards base; corolla lobes lanceolate, apices pubescent inside. Achenes 2–2.5 mm long, narrowly obovoid to obconical or cylindrical, black, densely setuliferous; pappus setae plumose, (2)2.5–3 mm long in marginal florets, and c. 20, of alternately long and short setae, 4–7.5 mm in disc florets.

Zambia. N: North Luangwa National Park, 11°45'S, 32°10'E, 690 m, 20.xi.1993, *Smith* 0017 (K). W: Ndola, 25.ix.1953, *Fanshawe* 370 (EA, K, SRGH). C: Lusaka, 28.i.1963, *van Rensburg* 1278 (K, SRGH). E: Lundazi, 1098 m, 1.vi.1954, *Robinson* 811 (K, SRGH). S: Sinazongwe Dist., Chete Island National Park, 550–650 m, 1.iii.1997, *Zimba* et al. 998 (K, MO). **Zimbabwe**. N: Mazoe, Jumbo, 13.i.1959, *Phipps* 843 (SRGH). W: Wankie, 6.vii.1956, *Clarke* 117 (PRE). C: Hartley, 30.iii.1961, *Whellan* 1824 (SRGH). E: Umtali, 9.ix.1944, *Chase* 1207 (BM, K, LD, SRGH). **Malawi**. N: Northern Province, Rumphi, 1310 m, 27.iii.1976, *Phillips* 1539 (K, MO). C: Mchinji Dist., Namitete, 1130 m, 30.iii.1970, *Brummitt* 9535 (K). S: Port Herald, Makanga, 79 m, 19.iii.1960, *Phipps* 2551 (K, SRGH). **Mozambique**. N: Namapa, Erati, 23.iii.1960, *Lemos & Macacua* 48 (BM, COL, EA, K, LMJ, PRE, SRGH). Z: Mocuba, Namagoa, 61–122 m, 194?, *Faulkner* Pretoria No. 180 (COT, EA, K, PRE, SRGH). MS: S. of Dondo, 12.v.1959, *Williams* 98 (SRGH).

---

[210] There are three sheets under this number in B-W. The first sheet is the most complete material and appears to have the following nom. nud. written beneath the specimen. The second has '(Balbis)' written at the bottom of the sheet, but lacks any capitula, and the third has '(Humboldt)' written at the bottom.

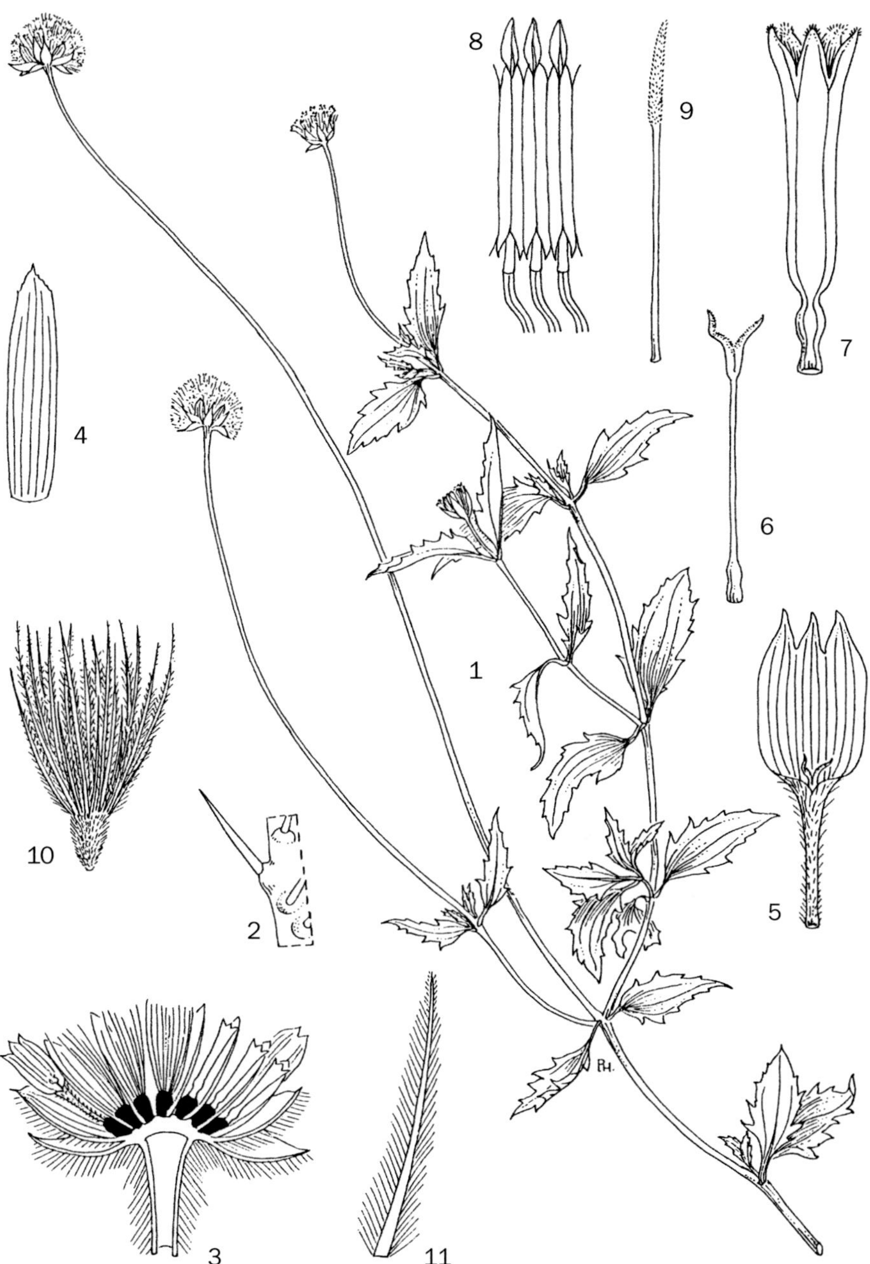

Fig. 6.5.**34**. TRIDAX PROCUMBENS. 1, flowering branch (× 2/3); 2, detail of leaf hair (greatly magnified); 3, l.s. capitulum (× 2); 4, palea (× 4); 5, ray floret corolla (× 4); 6, ray floret style (× 8); 7, disc floret corolla (× 8); 8. partial anther cylinder opened out (× 12); 9, disc floret style (× 12); 10, achene (× 4); 11, detail of pappus seta (greatly magnified). All from *Cadet* 5224. Drawn by Pat Halliday. From Flore des Mascareignes.

A native of Central and S America. Introduced as a weed to Florida, widespread in Africa, Madagascar, Asia, Australia and some Pacific Islands. Roadsides, path edges, waste ground, field margins and cultivated areas; 0–2000(2850) m; flowering throughout the year.

Conservation Status: A widespread weed in the tropics; LC (Least Concern).

142. **GALINSOGA** Ruiz & Pav.

Galinsoga Ruiz & Pav., Fl. Peruv. Prodr.: 110, t. 24 (1794). —St. John & White in Rhodora **22**(258): 97–101 (1920). —Wild in Kirkia **6**(1): 9–10 (1967). —Canne in Rhodora **79**(819): 319–389 (1977); —in Madroño **25**: 81–93 (1978). —Schulz in Feddes Repert. **92**(5–6): 387–395 (1981). —Hind in Fl. Masc., Composées **109**: 204–206 (1993). —Mesfin Tadesse, Fl. Ethiopia & Eritrea **4**: 304–305 (2004). —Beentje & Hind in F.T.E.A., Compositae **3**: 761–764 (2005).

Galinsogea Willd., Sp. Pl., ed. 4, **3**(3): 2228 (1803), orth. var.

Gallinsoga J.St.Hil., Expos. Fam. Nat. **1**: 417 (1805), orth. var.

Wiborgia Roth, Catal. Bot. **2**: 112 (1800), nom. rej., non *Viborgia* Moench (1794), nec *Wiborgia* Thunb. (1800), nom. cons.

Galinsoja Roth, Catal. Bot. **3**: 78 (1806), orth. var.

Vigolina Poir. in Lam., Encycl. **8**: 613 (1808), nom. superfl., based on *Wiborgia* Roth.

Vargasia DC., Prodr. **5**: 676 (1836), nom. illeg., non Bertero ex Spreng. (1825).

Adventina Raf., New Fl. N. Amer. **1**: 67 (1836).

Baziasa Steud., Nomencl. Bot., ed. 2, **1**: 192 (1840), p.p.

Stemmatella Wedd. ex Sch.Bip. in Bull. Soc. Bot. France **12**: 82 (1865), nom. nud.

Stemmatella Wedd. ex Benth. & Hook.f., Gen. Pl. **2**: 193, 359, 360 (1873).

Galinsogaea Himpel, Fl. Elsass-Lothr.: 187 (1891), orth. var.

Stenocarpha S.F.Blake in Bull. Misc. Inform., Kew **1915**: 348 (1915).

Weak annual herbs. Stems simple to well-branched, erect or decumbent, striate, variously pubescent, hairs appressed to spreading, sometimes with long stipitate-glandular hairs. Leaves opposite, sessile or petiolate, petioles narrowly connate around node, lamina elliptic to broad-ovate, base rounded, cuneate or attenuate, glabrous to densely pilose, trinervate, margins entire or serrate, apices acute to acuminate. Inflorescences of solitary capitula or terminal and axillary cymes, capitula pedicellate, pedicels slender, usually pilose. Capitula small, heterogamous, radiate; involucres hemispherical to narrowly-campanulate; phyllaries (1)2(3)-seriate, scarious, ovate, apices usually obtuse, outer 1–4, herbaceous, triangular to elliptic, flattened or convex, glabrous or pilose, margins entire or short-laciniate and sometimes scarious, apices acute to obtuse, inner phyllaries herbaceous or scarious, narrowly to broad-ovate or lanceolate-ovate, glabrous or pilose, margins entire, ciliate or short-laciniate, apices acute or obtuse; receptacle convex or broad- to narrow-conical, paleaceous, paleae membranaceous or scarious, usually dimorphic, outer lanceolate, elliptic or broadly-ovate, flat, apices often 3-fid, sometimes entire, purple or reddish, often joined in groups of 2 or 3 at base to a single phyllary, these enclosing a ray floret; inner paleae linear to narrow-ovate, convex to conduplicate, entire to deeply trifid, apices acute to obtuse, rarely weakly cuspidate. Ray florets (3)5–8(15), rarely absent, female, uniseriate, limb white, pink or purplish to purplish-red, usually 3-lobed, rarely bilabiate, tube pilose; style bifid. Disc florets few to many (5–150), hermaphrodite, corollas yellow, greenish-yellow, rarely purple, throats tubular, rarely campanulate, pilose, rarely glabrous outside, tube minutely pilose, lobes 5, apices acute, papillose, erect or recurved; anther thecae yellow or brown, apical anther appendages ovate to oblong, basal appendages sagittate; style arms recurved, somewhat flattened, apices acute. Achenes dimorphic, obconical to obpyramidal, body black, glabrous or setuliferous, ray achenes slightly larger and more flattened and incurved at base than disc achenes; pappus absent or pappus of lanceolate, laciniate or fimbriate scales with obtuse to aristate apices, or rarely of barbellate setae.

A genus of c. 14 spp. from Mexico, Central and S America; two spp. are widespread pantropical and pansubtropical weeds, sometimes also persisting in temperate regions when introduced. Both the weedy spp. are present in the Flora area.

Pedicel hairs short, appressed-ascending, intermingled with a few short stipitate-glandular hairs; paleae markedly 3-fid; ray limbs white to pink; pappus scales without marked terminal awns . . . . **1.** *parviflora*
Pedicel hairs long and spreading with some long stipitate-glandular hairs; paleae undivided or lobes short and unequal; ray limbs white to dark purplish-red; pappus scales with terminal awns. . . . . **2.** *quadriradiata*

1. **Galinsoga parviflora** Cav., Icon. **3**: 41, t. 281 (1795). —Moore in Fawcett & Rendle, Fl. Jamaica **7**: 257, fig. 88 (1936). —Wild, Common Rhod. Weeds, t. 70 (1955). —Andrews, Fl. Pl. Sudan **3**: 30 (1956). —Ross-Craig, Draw. Brit. Pl. **15**: t. 27 (1960). —Humbert, Fl. Madagasc., Composées **189**: 667 t. cxv fig. 22–28 (1963). —Adams in F.W.T.A., ed. 2, **2**: 230 (1963). Type: 'Verbesina biflora Hort. Reg. Parisiens. ... Habitat in Peruvia.. Floret Septembri. .... Obs. Hanc plantam vidi in Regio horto Pariensiensi anno 1785 nomine Verbesinae biflorae, ibi enata ex seminibus e Peruvia missis a D. Dombeyo; in patriam reduc vidi eam iterim in Regio horto Matritense, cuius nomen tandem mutatum fuit in debitum Galinsogae.' (MA 475684 – Fiche 32/C6 lectotype, MA ×3), lectotypified by Schulz (1981: 389).[211] FIGURE 6.5.**35A**.

    Galinsoga quinqueradiata Ruiz & Pav., Syst. Veg. Fl. Peruv. Chil. **1**: 198 (1798), nom. illeg., superfl. based on *Galinsoga parviflora* Cav.

    Wiborgia acmella Roth, Catal. Bot. **2**: 112 (1800). Type: 'Habitat in Peru. . ... Ab Indis vocatur haec planta Paica-Iullo, quam masticantur in affectibus ovis.' (holotype unknown, but probably B† or B-W).[212]

    Galinsoga laciniata Retx. in Hoffm., Phytogr. Bl.: 46 (1803). Type: not cited.[213]

    Vigolina acmella (Roth) Poir. in Lam., Encycl. **8**: 613 (1808).

    Wiborgia parviflora (Cav.) Kunth in Humb., Bonpl. & Kunth, Nov. Gen. Sp. Pl. **4** (ed. folio): 201 (1818).

    Sabazia microglossa DC., Prodr. **5**: 497 (1836). Type: '① in montanis circa Mexico ad S. Augustinum legit *Berlandier* (pl. exs. n. 733). (v.s.)' (G-DC holotype).

    Sabazia microglossa var. β *puberula* DC., Prodr. **5**: 497 (1836). '① circa Mexico in montanis. (*Berl.* [*andier*]! pl. exs. n. 910). (v.s.)' (G-DC holotype).

    Adventina parviflora Raf., New Fl. N. Amer. **1**: 67 (1836). Type: 'Growing spontaneous for several years in the orchard of Bartram's Garden, come with seeds from the south. ... Probably a Florida plant.' Herbarium material unknown.

    Galinsoga parviflora var. *semicalva* A.Gray in Smithsonian Contr. Knowl. **5**(Art. 6): 98 (1853). Type: 'Side of mountains, at the copper mines; Oct. ([*Wright*, 1851] 1268.)' (GH8241 holotype, GH8239, GH8240).

    Galinsoga hirsuta Baker in Curator's Rep. Thirsk Nat. Hist. Soc. **1861**: 13 (1862), nom. nud.[214]

---

[211] There are four sheets in MA. The lectotype bears three handwritten labels: 'Vulgo Pacoyuyu/R. H. Matritensis/Sept. 1790' on the top label, 'Galinsoga parviflora/Icon. t. 281/Peru. 1794.' on the middle label and 'Galinsoga parviflora Cav./ex Peruvia et Regno Chilensis/Née Iter.' on the bottom label, and one typed label indicating this is the lectotype of Galinsoga parviflora, determined by Dorothea L. Schulz, 19.7.1979 – this sheet was cited as the lectotype by Schulz (1981: 389), but 'p.p.'; it was also determined in 1978 as *G. quadriradiata* by Schulz.

[212] Roth's protologue citation referred to Feuillée's J. Obs. (1714-25) in its German translation, Beschr. Arzen. Pfl. (1756), based on collections from Chile and Peru. Material from Roth's herbarium went in part to B, some into B-W, the former now largely destroyed. It is possible some duplicate material may exist.

[213] Canne (1977: 373) suggested that a specimen in LD represented the holotype.

[214] Canne (1977: 374) accepted valid publication of this name and cited a holotype – which cannot be the case.

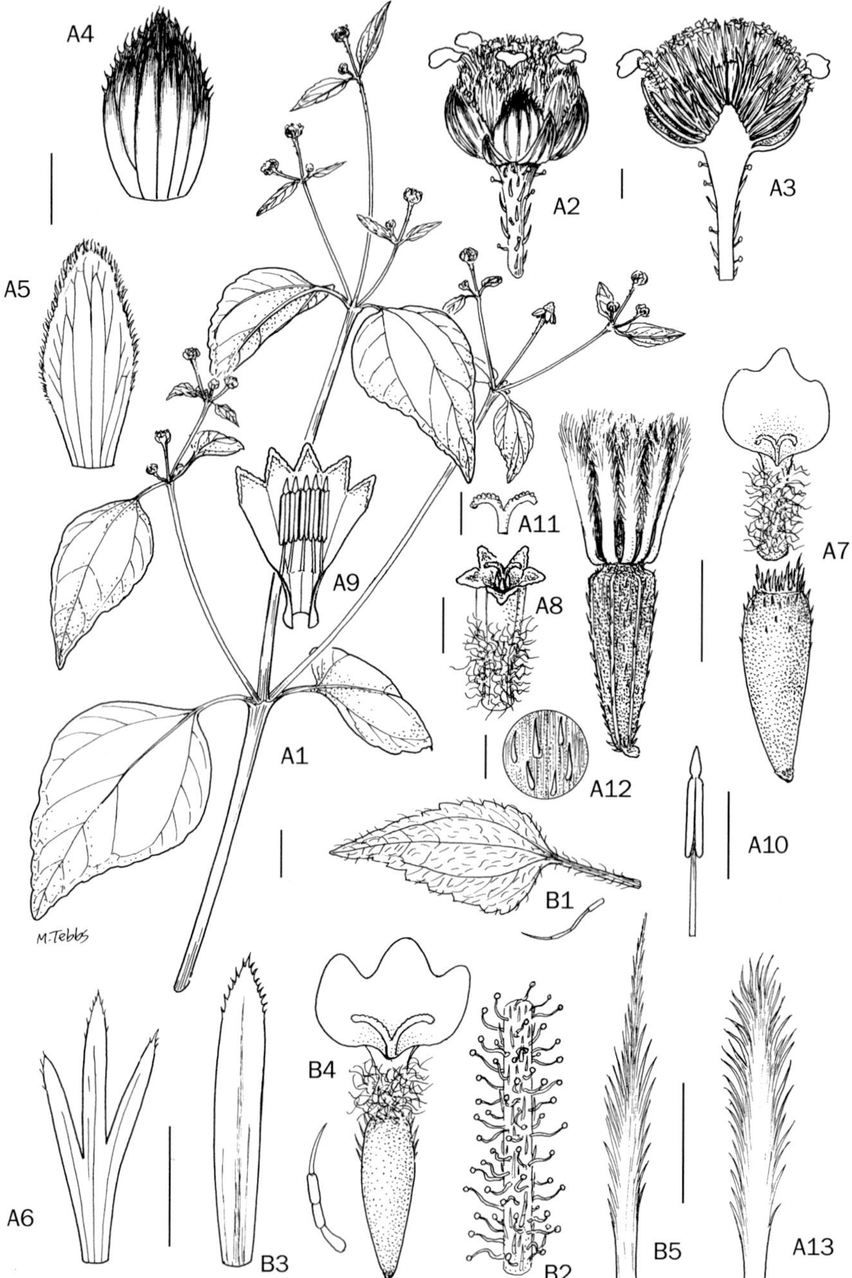

Fig. 6.5.**35**. A. GALINSOGA PARVIFLORA. A1, flowering branch; A2, capitulum; A3, l.s. capitulum; A4, outer phyllary; A5, inner phyllary; A6, palea; A7, ray floret with separated achene; A8, disc floret corolla; A9, disc floret corolla opened out showing attachment of filaments; A10, stamen; A11, disc floret style arms; A12, disc floret achene with detail of surface showing setulae; A13, pappus scale. GALINSOGA QUADRIRADIATA. B1, leaf, with a detail of trichome; B2, detail of pedicel pubescence; B3, palea; B4, ray floret with detail of corolla trichome; B5, disc floret pappus scale. A from *Smith* 1341; B from *Chase* 5953. Scale bars: A8, A10–11, A13, B5 = 0.5 mm; A1–7, A9, A12, B1–4 = 1 mm. Drawn by Margaret Tebbs.

Galinsoga parviflora var. *genuina* f. *parceglandulosa* Thell. in Allg. Bot. Z. Syst. **21**(1–4): 8 (1916). Types: die zweite ist besonders für das südliche Gebiet Mitteleuropas charakteristisch (Tessin, Veltlin, Comersee, Venetien [Vittorio, sonst meist f. 1], auch Zürich!!, Süd-Tirol, Ungarn) und dominert auch in anderen meridionalen Gegenden der Erde (Peru: Lima [*Pavon*!], Mexiko!, Arizona! [auch Rhode Island!]. Süd-Afrika!, Java! Australien!), deutlich drüsig sind jedoch die Exemplare von Markelfingen am Untersee und solche von Frieburg i. Br!!, Stuttgart!!, Berlin und Stolp in Pommern.' 'Australia: New South Wales, Cumberland Co., Erworben, 1913, *T. V. Alkin* s.n.' (Z lectotype), lectotypified by Canne (1977: 374).

Galinsoga parviflora var. β *adenophora* Thell. in Allg. Bot. Z. Syst. **21**(1–4): 9 (1916). Types: 'So in typischer Ausbildung in Mexiko: Puebla, 1908, *F. Arsène*! 1909, *F. Nicolas*! (beide als *G. brachystephana*, Herb. Univ. Zürich), um São Paulo in Brasilien (*A. Usteri*! in Herb. Eidgen. Tech. Hochschule) und bei Buenos Aires (Umgebung des Hafens, 1899, *G. Debeaux* Nr. 92! in Herb. Montpell., comm. J. Daveau). In Europa bisher mit Sicherheit nur in England beobachtet: Kew (Surrey), 1901, *F. H. Davey* n. 812! in Herb. Univ. Zürich; eine angenäherte Form fand sich 1907 im botanischen Garten Zürich (!!) verwildert vor.' *Usteri* s.n. (ZT lectotype, P), lectotypified by Canne (1977: 374).

Galinsoga semicalva (A.Gray) H.St.John & D.White in Rhodora **22**(258): 100 (1920).

Galinsoga semicalva var. *percalva* S.F.Blake in J. Wash. Acad. Sci. **30**: 472 (1940). Type: [USA:] Arizona: Santa Rita Mountains, Pima County, *David Griffiths & J. J. Thornber* 162 (type no. 497226, U. S. Nat. Herb.); ...' (US 497226 holotype, NY00169422).

Galinsoga sphaerocephala Jones ex S.F.Blake, Contr. U.S. Natl. Herb. **29**: 130 (1945), nom. illeg., pro syn.

Galinsoga quadriradiata sensu Merxm. apud Suesseng. & Merxmüller in Proc. & Trans. Rhod. Sci. Assoc. **43**: 65 (1951).

Annual herb to c. 1 m. Stem and branches striate, hispidulous or glabrescent, sometimes sparsely stipitate-glandular above. Leaves opposite, petiolate, petiole up to 2.5 cm long, hispidulous, lamina membranous, 1–11 × 0.5–7 cm, broad- to narrow-ovate, base broadly cuneate, 3-veined from base, margins crenate to subentire, hispidulous or glabrous on both surfaces, apex acute or acuminate. Pedicels 0.1–4 cm long, moderately to densely pilose, hairs appressed and antrorse, sometimes interspersed with stipitate-glands. Capitula 3.5–5 × 2–6 mm; outer (sterile) phyllaries 2–4, 1.2–2.2 × 0.6–1.5 mm, margins broad-scarious, apices obtuse to narrowly-rounded; inner phyllaries 2.5–3.5 × 1.3–2.6 mm, margins scarious; receptacle ± convex, paleaceous, outer paleae broad-elliptic to obovate, 3-veined, 2 or 3 fused at base or greater to form units in groups of 2 or 3 with adjacent phyllaries, complex enclosing ray floret, inner paleae obovate or lanceolate, 2–3.2 × 0.4–1 mm, 3-fid, persistent. Ray florets (3)5(8), ray limb 1.8–2 × c. 1.5 mm, broadly ovate, 3-lobed, whitish, pink or white with pink veins, tube 0.8–1.1 mm long, pilose. Disc florets 8–50, corollas yellow, tube 0.3–0.5 mm long, throat 0.8–1 mm long, pubescent outside; apical anther appendages eglandular. Ray achenes 1.5–2.5 mm long, glabrous or apically setuliferous, pappus absent or of 5–8 narrow laciniate scales c. 1 mm long; disc achenes 1.2–2 mm long, glabrous and epappose or setuliferous and pappus of 15–20 oblong fimbriate scales c. 1.9 mm long, apices usually obtuse, sometimes acute or acuminate.

Caprivi Strip. E. of Kwando R. x.1945, *Curson* 1051 (PRE). **Botswana**. N: Maun, 19°29'S, 23°26'E, 7.iv.1975, Smith 1341 (K, SRGH). **Zambia**. W: Ndola, 20.i.1954, *Fanshawe* 699 (K, PRE). C: Lusaka, 1200 m, 15.v.1955, *Best* 95 (EA, K, SRGH). S: Choma, Muckle Neuk, 1280 m, 11.x.1954, *Robinson* 908 (K, SRGH). **Zimbabwe**. W: Bulawayo, vii. 1959, *Miller* 5959 (SRGH). C: Marandellas, 3.ix.1943, *Dehn* 782 (M, SRGH). E: Umtali, 11.iv.1955, *Chase* 5542 (BM, K, SRGH). S: Victoria, Makoholi Experimental Station, 9.xi.1977, *Benderayi* 106 (K, SRGH). **Malawi**. N: Mzimba, Marymount, Mzuzu, 1372 m, 30.1.1971, *Pawek* 4339 (K, MAL). C: Dedza, 28.v.1963, *Banda* 509 (K, SRGH). S: Mlanje, 25.vii.1956, *Newman & Whitmore* 185 (BM, SRGH).

Central and S America; introduced as a weed into the United States, the Old World tropics, subtropics and even into some temperate countries, i.e. into southern England.

Conservation Status: A widespread weed in the Flora area; LC (Least Concern).

2. **Galinsoga quadriradiata** Ruiz & Pav., Syst. Veg. Fl. Peruv. Chil. **1**: 198 (1798). Type: 'Habitat in Peruviae versuris, ruderatis et segetibus ad Limam et Chancay. Floret toto anno.' 'Peru: H. RUIZ et J. A. PAVÓN s.n.' (MA816405, which is the right hand

specimen on the sheet determined by Schulz – C3 on sheet 290 of microfiche of the Ruiz & Pavón herbarium – lectotype, FI (p.p.), P02140656), lectotypified by Schulz (1981: 392). FIGURE 6.5.**35B**.

  Galinsoga parviflora [var.] * *quadriradiata* (Ruiz & Pav.) Pers., Syn. Pl. **2**: 472 (1807).

  Galinsoga parviflora var. *quadriradiata* (Ruiz & Pav.) Poir. in Lam., Encycl. Suppl. **2**(2): 701 (1812).

  Wiborgia urticifolia Kunth in Humb., Bonpl. & Kunth, Nov. Gen. Sp. Pl. **4** (ed. folio): 201, t. 389 (1818), as '*urticaefolia*'. Type: [Ecuador:] 'Crescit juxta villam Marchionis de Miraflores, inter Mulalo et Pansache, alt. 1700 hex. (Regno Quitensi.) K Floret Junio.' [*Humboldt & Bonpland* 'n. 3055'] (P-Bonpl holotype).

  Jaegeria urticifolia (Kunth) Spreng., Syst. Veg., ed. 16, **3**: 590 (1826).

  Sabazia urticifolia (Kunth) DC., Prodr. **5**: 497 (1836).

  Vargasia caracasana DC., Prodr. **5**: 676 (1836). Type: '① ad Caracas legit c. *Vargas* [267]. (v.s. comm. à cl. inventore.)' (G-DC holotype).

  Galinsoga parviflora var. *hispida* DC., Prodr. **5**: 677 (1836). Types: '① in Mexico circa urbem (*Berl. [andier]*! pl. exs. n. 615), in Chilensibus montibus (h. Haenke!). ... (v.s.)' *Berlandier* 615 (P02140653 lectotype, P02140652), lectotypified by Canne (1977: 356).[215] Haenke sn., PRC(453389), syntype.

  Adventina ciliata Raf., New Fl. N. Amer. **1**: 67 (1836). Type: 'Found with the last, but in a different place and season: ...' [q.v. *Adventina parviflora* – see under *Galinsoga parviflora* above – 'Growing spontaneous for several years in the orchard of Bartram's Garden, come with seeds from the south.'] Location of type material, if extant, unknown.

  Baziasa urticifolia (Kunth) Steud., Nomencl. Bot., ed. 2, **1**: 192 (1840).

  Galinsoga brachystephana Otto, Index Sem. Hort. Berol. (1840), nom. nud.

  Galinsoga hispida Benth., Bot. Voy. Sulphur: 119 (1845). Types: 'Peyta, in Columbia, Guayaquil [*Hinds*]; gathered also by *Cuming* at Lima (n. 1028).' *Hinds* s.n. (K lectotype), lectotypified by Canne (1977: 356). Cuming 1028, K(00037086) 000009853, syntype.

  Wiborgia brachystephana (Otto) Heynh., Nomencl. Bot. Hort.: 707 (1846), comb. illeg. as apparently based on *Galinsoga brachystephana*, published a year later!

  Galinsoga brachystephana Otto ex Heer & Regel, Ind. Sem. Hort. Turic. [≡ p. 2] (1846) [1847]. Type: not cited but based on cultivated material in 'Hort. Berol.'[216]

  Galinsoga hispida var. *purpurascens* Fenzl, Del. Sem. Hort. Vindob. Advers. Bot. Stirp. Sem. **1849/1850**: 2 (1851), based on *Galinsoga brachystephana* Otto ex Heer & Regel

  Galinsoga hispida var. *albiflora* Fenzl, Del. Sem. Hort. Vindob. Advers. Bot. Stirp. Sem. **1849/1850**: 2 (1851), nom. nud.

  Galinsoga urticifolia (Kunth) Benth. in Oerst., Vidensk. Meddel. Dansk Naturhist. Foren. Kjøbenhavn **1852**: 102 (1852) [1853].

  Galinsoga parviflora var. *caracasana* (DC.) A.Gray, Smithsonian Contr. Knowl. **5**: 98 (1853).

  Galinsoga caracasana (DC.) Sch.Bip., Bull. Soc. Bot. France **12**: 80 (1865).

  Stemmatella sodiroi Hieron. in Bot. Jahrb. Syst. **28**(5): 601 (1901). Type: [Ecuador:] 'Crescit in regione interandina (SODIRO n. 31/1).' (B† – F0BN015190 holotype, US – ?fragment of holotype).[217]

  Stemmatella lehmannii Hieron. in Bot. Jahrb. Syst. **28**(5): 602 (1901). Type: 'Columbia: crescit frequenter in vicinitate urbis Popayan, alt. s. m. 1600–2200 m (*L.*[*EHMANN*] n. 5667).' (B† – F0BN015191 holotype, K ×3, US – fragment of holotype).[218]

---

[215] Material of the *Berlandier* collection is in G-DC, but there is no evidence of the *Haenke* material from Chile, this is probably in PR.

[216] Canne (1977: 356) suggested that this material was probably in ZT.

[217] Canne (1977: 374) originally placed this name in synonymy of *G. parviflora*.

[218] There are 3 sheets in K, one of which was sent on loan and annotated as 'lectotype' bears a minimally annotated Lehmann label with merely the collecting number and 'Wedelia'. Of the 2 other sheets one bears the locality data written on the Lehmann label. Schulz (1981: 392) was of the opinion that the fragmentary isotype in US would be better as the lectotype as it is 'more favourable in the interest of an exact determination' (literal translation of Schulz's comment in German). I can find no formal place where Schulz declared the K sheet as the lectotype – but I do not necessarily accept Schulz's comment as to lectotypification based on the US material since the material in K is the same taxon.

Stemmatella urticifolia (Kunth) O.Hoffm. ex Hieron. in Bot. Jahrb. Syst. **28**(5): 603 (1901).

Galinsoga humboldtii Hieron. in Bot. Jahrb. Syst. **28**(5): 618 (1901). Type: 'Exstat inter plantas a cl. HUMBOLDT et BONPLAND collectas loco non indicato, a cl. KUNTHIO nomine »*Wibourgia urticaefolia* var. achaeniis squamulis coronatis« determinata.' Herbarium material unknown.

Stemmatella urticifolia var. *eglandulosa* Hieron. in Bot. Jahrb. Syst. **36**(5): 487 (1905). Type: 'Peruvia: crescit prope Cutervo (*J.[elski]* n. 609, m. Aprili 1879).' (B† holotype).

*Galinsoga hispida (*DC.) Hieron. in Notizbl. Königl. Bot. Gart. Berlin-Dahlem **19**: 15 (1907), nom. illeg., later homonym.

Galinsoga quadriradiata [unranked] *hispida* (DC.) Thell., Allg. Bot. Z. Syst. **21**(1–4): 11 (1916).[219]

Galinsoga quadriradiata [unranked] *quadriradiata* f. *vargasiana* Thell. in Allg. Bot. Z. Syst. **21**(1–4): 14 (1916). Type: 'Die venezuelanische Pflanzen (= *Vargasia caracasana* DC.! sens. strict. ex specim. authent.) weicht, wie schon bemerkt (vgl. S. 3 und Fußn. 2), vom Typus der Rasse durch ...'.[220]

Galinsoga quadriradiata [unranked] *quadriradiata* f. *purpurascens* (Fenzl) Thell. in Allg. Bot. Z. Syst. **21**(1–4): 15 (1916).

Galinsoga quadriradiata [unranked] *quadriradiata* f. *albiflora* (Fenzl) Thell. in Allg. Bot. Z. Syst. **21**(1–4): 15 (1916).

Galinsoga aristulata E.P.Bicknell in Bull. Torrey Bot. Club **43**: 270 (1916), based on *Galinsoga parviflora* Cav. var. *hispida* DC.[221]

Galinsoga bicolorata H.St.John & D. White in Rhodora **22**(258): 99 (1920). Type: 'MEXICO: altitude 4000-5500 feet, Tumbala, Chiapas, Oct. 20, 1895, *E. W. Nelson*, no. 3,356' (GH8242 holotype, US).

Galinsoga ciliata (Raf.) S.F.Blake in Rhodora **24**(278): 35 (1922). —Ross-Craig, Draw. Brit. Pl. **15**: t. 28 (1960). —Adams in F.W.T.A., ed. 2, **2**: 230 (1963).

Sabazia urticifolia var. *venezuelensis* Steyerm. in Fieldiana, Bot. **28**: 672 (1953). Type: [Venezuela:] 'Type in herb. Chi. Nat. Hist. Mus., collected on pastured open slopes of mountain between Santa Domingo and Los Quebraditos, south of Las Sabanetas, above Humocavo Bajo, state of Lara, alt. 2430-2475m., February 8, 1944, *Julian A. Steyermark* 55379, "rays lavender; disc golden."' (F1388807 – although determined as an isotype – holotype, US2046662).

Galinsoga eligulata Cuatrec. in Revista Acad. Colomb. Ci. Exact. **9**: 241 (1954). Type: 'Columbia, Dept. Caldas; Chinchiná, "Centro Nacional de Investigaciones del Café", 1350-1400 m. alt., colect. 22-XI-1946 *J. Cuatrecasas* 23098. "Hierba. Involucro verde, corolas amarillas." ' (F1635243 holotype).

Ageratum perplexans M.F.Johnson in Ann. Missouri Bot. Gard. **58**(1): 80 (1971). Type: 'BOLIVIA. Yungas, 1890, *A. Michael Bang* 235' (MICH1108862 holotype, GH00000752 – a mix of *G. parviflora* and *G. quadriradiata*, MSC – a mix as in GH, MO, MSC – a mix as in GH, NY, US00076498, WIS – a mix as in GH).

Erect or spreading annual herb c. 0.3–1.5 m tall. Branches sulcate or striate and hispid-pilose, often stipitate-glandular distally. Leaves opposite, petiolate, petiole up to 0.2–6 cm long, hispid-pilose, lamina 1.5–9.5 × 0.5–5.5 cm, ovate, base abruptly cuneate, 3-veined from base, margins crenate-serrate, hispid-pilose on both surfaces, apex acute to acuminate. Pedicels 0.2–2.5(5) cm, slender, densely pilose. Capitula 3–8 × 2–10 mm (including rays); outer (sterile) phyllaries 2–4, 0.9–3 × 0.5–2 mm, broad-elliptic to oblong, pubescent and stipitate-glandular or glabrous, longitudinally 3- or more-veined, apices obtuse or outer sometimes acute, inner phyllaries 2.5–4 × 1.5–2.5 mm, margins narrowly scarious, laciniate above; receptacle shallowly convex; outer paleae 2–3 × 0.6–1.8 mm, broad-elliptic to obovate, margins ciliolate above, fused in groups of 2 or 3 at base to adjacent inner phyllaries and enclosing ray floret, inner paleae 1.8–3.3 × 0.4–1 mm, entire or irregularly 3-fid, rapidly deciduous. Ray florets 4 or 5 (rarely to 8), limbs c. 1.6 mm long, broadly ovate-oblong, 3-lobed, whitish (but often reddish in South America!), tube 0.5–1.2 mm long, densely hispidulous. Disc florets 15–65, corollas yellowish-green, tube 0.3–0.5 mm

---

[219] Cited by many authors as 'var.' Thellung actually wrote 'var. (vel subsp.)' and as such must be considered '[unranked]'.

[220] It is unclear what material is left once the apparent single specimen used by de Candolle is excluded! No other material applicable to this form was cited by Thellung, especially since he appears earlier to have ascribed the Vargas type material of de Candolle's name to this forma!

[221] The validating description with its types, citing also the following material: 'A few plants on Easton Street in full flower September.[ber 13, 1907; Fair Street, September 19, 1914; specimen in the herbarium of Miss Grace B. Gardner.'

long, densely pubescent outside, throat 0.8–1.5 mm long, campanulate, narrowed below into tube; apical anther appendages eglandular. Achenes black, c. 1.7 mm long, ray achenes setuliferous or glabrous, epappose or pappus of few setae or of 8–20 fimbriate, sometimes aristate scales, 0.2–1.4 mm long, white, disc floret achenes ± angular, 1.1–1.6 mm long, body setuliferous, epappose or with few fimbriate scales, or to 20 lanceolate or oblanceolate, fimbriate, usually aristate, scales c. 1.5 mm long.

Zambia. W: Kitwe Dist., 6.1.1974, *Fanshawe* F12140 (K, NDO). C: Lusaka airport, 3.ii.1970, *Brummitt* 8401 (K). **Zimbabwe**. C: Salisbury, off Enterprise Road, 1494 m, 21.xii.1968, *Biegel* 2719 (K, SRGH). E: Inyanga, Juliasdale, Punch Rock, c. 1700 m, *Biegel* 3961 (K, SRGH). **Mozambique**. E: Umtali, lmbesa Valley, 25.i.1956, *Chase* 5953 (BM, SRGH).

Central and S America but introduced as a weed into the United States, Canada and Europe. recorded also from W Africa and probably more widespread in Africa than has been thought as it is easily confused with *G. parviflora*; weed of roadsides and waste places.

Conservation Status: Probably under-collected as it is easily confused with *G. parviflora*; LC (Least Concern).

143. **ACANTHOSPERMUM** Schrank

Acanthospermum Schrank, Pl. Rar. Hort. Monac. **2**: t. 53 (1820). —Blake in Contr. U.S. Natl. Herb. **20**(10): 383–392 (1921). —Wild in Kirkia **6**(1): 5–8 (1967). —Stuessy in Rhodora **72**(789): 106–109 (1970). —Lisowski, (Asterac. Fl. Afr. Cent. 1) Fragm. Flor. Geobot. **36** Suppl. 1: 111–117 (1991). —Hind in Fl. Masc., Composées **109**: 170–173 (1993). —Pruski in Taxon **46**(4): 805–806 (1997). —Mesfin Tadesse, Fl. Ethiopia & Eritrea **4**(2): 308 (2004). —Beentje & Hind in F.T.E.A., Compositae **3**: 766–768 (2005). —Chen Yousheng & Hind in Fl. China **20-21**: 865 (2011).

Centrospermum Kunth in Humb., Bonpl. & Kunth, Nov. Gen. Sp. Pl. **4** (ed. folio): 212 (1818).

Orcya Vell., Fl. Flum.: 344 (1825)[7 Sept.–28 Nov. 1829].

Echinodium Poit. ex Cass. in Dict. Sci. Nat., ed. 2, **59**: 235 (1829), nom. nud. pro syn., non Juratzka (1866).

Erect or prostrate annual herbs. Stems unbranched and erect or relatively well-branched and prostrate, pubescent or sometimes glabrous. Leaves opposite, petiolate or pseudopetiolate, lamina elliptic to broadly ovate or spathulate, base usually cuneate, margins crenate or coarsely dentate to subentire, hirsute or glabrous, apex obtuse, rounded or acute. Inflorescences of solitary axillary capitula, capitula pedicellate, pedicels densely pubescent. Capitula heterogamous, radiate. involucre campanulate or hemispherical. phyllaries biseriate, outer herbaceous, flat, ovate, apices acute, inner completely enclosing ray achenes, variously armed with beaks or spiny with uncinate or straight spines. receptacle small, convex, paleaceous, paleae membranous, concave, ± persistent, apices ciliate to laciniate. Ray florets uniseriate, female, few, ray limb elliptic, apically 3-toothed, yellow. Disc florets functionally male, few. corollas tubular, pilose, yellow. anther thecae usually blackened, apical anther appendages ovate, apices obtuse, often glandular-punctate. style clavate, undivided, lacking stigmatic papillae. Fruit cuneiform, flattened, surfaces glandular-punctate, covered in hooked spines and with 2 long divergent apical spines. Ray achenes oblong fusiform, rarely trigonous or compressed; pappus absent. Disc achenes undeveloped.

A small neotropical genus of five spp., containing two widely adventive spp., *A. australe* (Loefl.) Kuntze and *A. hispidum* DC., both present in the Flora area. Three spp. are present.

1. Plants erect; leaves obovate to obovate-oblong, cuneate; capitula ± sessile; fruit ± compressed, wedge-shaped-cuneate in outline, with 2 apical spines. . . . . **1.** *hispidum*
- Plants with prostrate stems; leaves obovate to broadly obovate or subcircular; capitula mostly pedicellate; fruit oblong-fusiform lacking apical spines 2
2. Fruits usually 4 or 5 per capitulum, ± 1 cm long; leaves and stems glabrous or pubescent . . . . **2.** *glabratum*
- Fruits usually (6)8–9 per capitulum, ± 0.7 cm long; leaves and stems usually densely pubescent . . . . **3.** *australe*

1. **Acanthospermum hispidum** DC., Prodr. **5**: 522 (1836). —Blake in Contrib. U.S. Natl. Herb. **20**(10): 386, t. 23c (1921). —Mendonça, Contrib. Conhec. Fl. Angola **1** (Compositae): 89 (1943). —Wild, Common Rhod. Weeds: t. l fig. D, 61 (1955). —Adams in F.W.T.A., ed. **2**, 2: 241 (1963). —Humbert, Fl. Madagasc., Composées **189**: 629, t. CXV fig. 5-12 (1963). —Lisowski, (Asterac. Fl. Afr. Cent. 1) Fragm. Flor. Geobot. **36** Suppl. 1: 112–114 (1991). —Mesfin Tadesse, Fl. Ethiopia & Eritrea **4**(2): 308 (2004). —Beentje & Hind in F.T.E.A., Compositae **3**: 766–768 (2005). —Chen Yousheng & Hind in Fl. China **20-21**: 865 (2011). Type: 'in Brasiliae sabulosis maritimis circa Bahiam legit cl. *Salzman*. [21] .... (v.s. comm. à cl. Salzm.)' (G-DCG(00469866) holotype, H(1023237) HAL(0110829) K000053302, 000053303).

> *Acanthospermum humile* (Sw.) DC. [unranked] β *hispidum* (DC.) Kuntze, Revia. Gen. Pl. **1**:303 (1891).

Coarse bushy erect annual, exceptionally up to 1.3 m tall or more, usually c. 0.3–0.5 m. Stems striate, hispid, with both long hairs and a short puberulous indumentum. Leaves 1–10 × 0.5–4 cm, sessile or lower short-petiolate, oblong or obovate, apex obtuse or subacute, cuneate towards base, margins subentire, repand-dentate or coarsely dentate, both sides hispid and glandular-punctate. Capitula sessile in axils of leaves or in forks of branches, 3–5 mm diam; when in flower, enlarging in fruit to c. 18 mm; outer phyllaries green, 5–6, c. 7 mm long, narrowly ovate, apices acuminate or acute, ciliate outside and at margins; paleae subtending disc florets membranous, 2–2.4 mm long, obovate-truncate, margins and upper part of midrib glandular-dentate. Limb of female florets yellowish, c. 1.5 mm long, elliptic, apex minutely 3-toothed, puberulent. Disc florets hermaphrodite, glandular-puberulent, limb campanulate, 5-lobed, 1 mm long, tube c. 1.5 mm long. achene abortive. Fertile achenes each enclosed in an enlarged inner phyllary forming a wedge-shaped fruit narrowing to base, 5–6 mm long without pair of somewhat divergent 4–5 mm long straight or somewhat hooked terminal spines, shorter lateral hooked spines fairly numerous on body of fruit.

Botswana. N: Nata R., 28.iv.1957. *Drummond* 5263 (K, SRGH). SE: Mahalapye, vii.1960, *Yalala* 113 (SRGH). SW: Ghanzi & Kgalagai Dist., Tshane village, 19.iii.1980, *Skarpe* S-413 (K). **Zambia**. B: Mongu, 10.i.1960, *Gilges* 972 (SRGH). N: Mbereshi, 23.vi.1957, *Robinson* 2379 (K, SRGH). W: Ndola, 23.vi.1955, *Fanshawe* 2340 (K). C: Lusaka, 26.i.1963, *van Rensburg* 1260 (K, SRGH). E: Petauke, Mvuvje R., 5.xii.1958, *Robson* 845 (BM, K, PRE, SRGH). S: Mazabuka, 28.ii.1963, *van Rensburg* 1511 (SRGH). **Zimbabwe**. N: Sinoia, 9.iii.1944, *Hopkins* in GHS 11874 (SRGH). W: Bulalima-Mangwe, Plumtree Station, 3.vii.1962, *Wild* 5836 (SRGH). C: Gatooma, 28.i.l941, *Gatooma Police* in GHS 7901 (SRGH). E: Umtali, 18.ii.1943, *Hopkins* in GHS 9714 (SRGH). S: Gwanda, Bubye Ranch, 9.v.1958, *Drummond* 5721 (K, SRGH). **Mozambique**. N: Nampula, 11.iv.1961, *Balsinhas & Marrime* 380 (BM, K, LMJ, PRE, SRGH). T: Tete Dist., 2 km N of Mlangeni, 1385 m, 25.v.1970, *Brummitt* 11105 (K). MS: Vila Pery, 18.vi.1956, *Myre & Balsinhas* 2464 (LM, SRGH). M: Vila Luiza, 7.iv.1947, *Barbosa* 124 (COL, K, LM, SRGH).

Native to tropical S America now widespread as a weed in tropical and subtropical Africa; 0–2500 m.

Vernacular name: Upright Starbur.

Conservation Status: LC (Least Concern).

2. **Acanthospermum glabratum** (DC.) Wild in Kirkia **6**(1): 6 (1967). —Lisowski, (Asterac. Fl. Afr. Cent. 1) Fragm. Flor. Geobot. **36** Suppl. 1: 114–115 (1991). —Beentje & Hind in F.T.E.A., Compositae **3**: 766 (2005). Type: 'in Brasiliâ (*Sell*[*ow*].!), circa Rio-Janeiro (*Martius! Lund! Gaudichaud!*). *Acanthospermum Brasilum* Mart.! in litt. (v.s.)' 'Brazil, Rio de Janeiro, *Martius* 1823' (G-DCG(00469854) lectotype, SRGH – photo.), lectotypified by Wild (1967: 6). FIGURE 6.5.**36B**.

> *Acanthospermum xanthioides* (Kunth) DC. [var.] γ *glabratum* DC., Prodr. **5**: 522 (1836).
> *Acanthospermum australe* sensu Wild, Common Rhod. Weeds: t. 60 (1955).

Prostrate annual or possibly short-lived perennial herb; branches striate, short whitish-pubescent, hairs ± appressed. Leaves petiolate, petiole to 1 cm long, pubescent; lamina 1–3 × 0.4–2.5 cm, elliptic to broadly

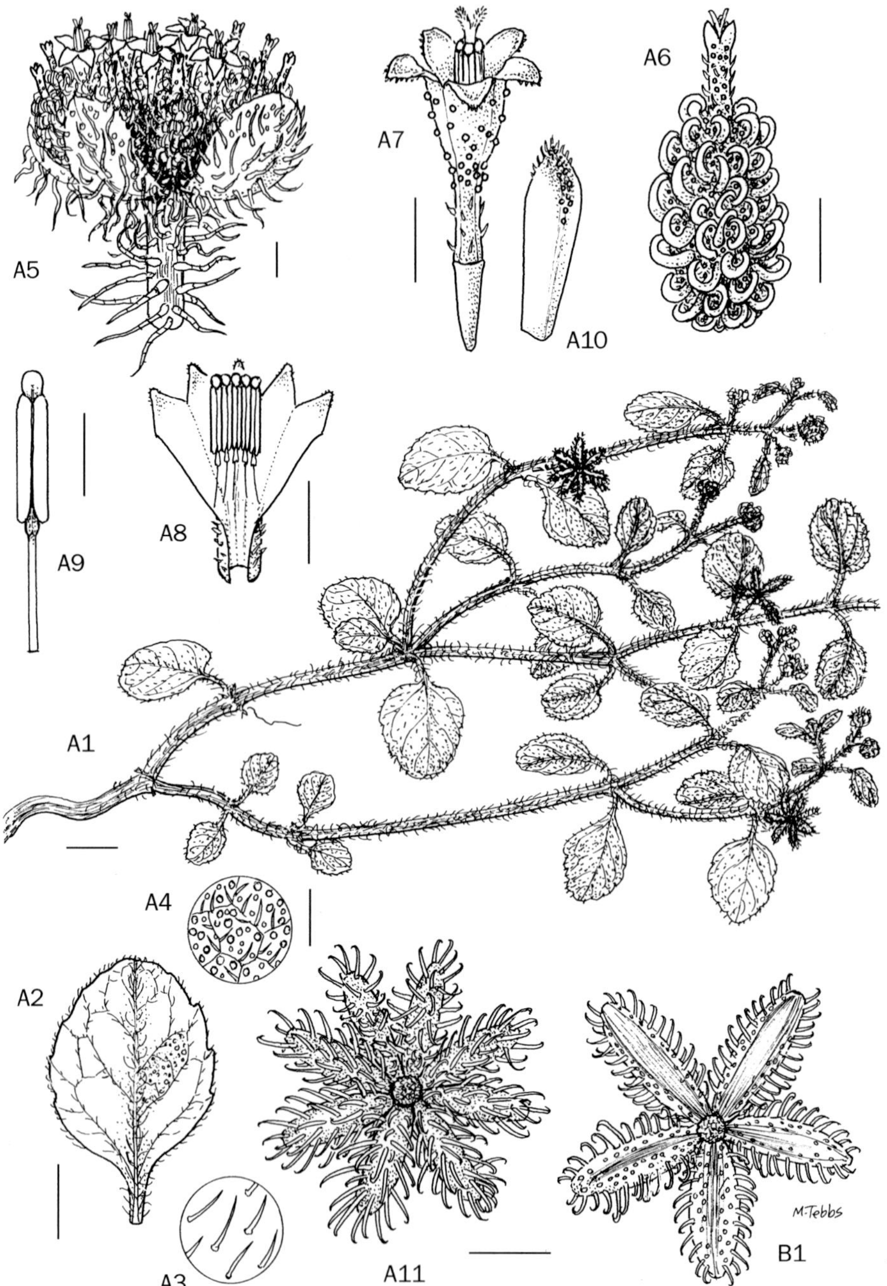

Fig 6.5.**36**. A. ACANTHOSPERMUM AUSTRALE. A1, habit; A2, leaf; A3, detail of adaxial leaf surface; A4, detail of abaxial leaf surface; A5, capitulum; A6, marginal female floret; A7, disc floret; A8, corolla of disc floret opened out showing attachment filaments; A9, stamen; A10, palea; A11, fruiting capitulum. B. ACANTHOSPERMUM GLABRATUM. B1, fruiting capitulum. A1–4 from *Wilberforce* 122; A5–11 from *Fanshawe* 27; B1 from *Strid* 2616. Scale bars: A9 = 0.5 mm; A4–8, A10 = 1 mm; A2, A11, B1 = 5 mm; A1 = 10 mm. Drawn by Margaret Tebbs.

ovate, apex rounded to subacute, base abruptly cuneate into petiole, margins coarsely dentate, glabrous or glabrescent on both surfaces, glandular-punctate. Capitula pedicellate, pedicel to 1 cm long, pubescent, 5–7 mm diam. in flower, enlarging in fruit to 2 cm diam.; outer phyllaries c. 3 mm long, oblong-ovate, apex obtuse, mucronulate, sparsely ciliate, gland-dotted; paleae subtending disc florets c. 2.5 mm long, membranous, obovate, emarginate, margins ciliate towards apex. Ray florets female, limb yellowish, c. 1.4 mm long, ± oblong, apex 3-toothed, gland-dotted. Disc florets hermaphrodite, limb funnel-shaped 5-lobed, c. 1.3 mm long, lobes gland-dotted, tube c. 1 mm long pilose; achenes abortive. Fertile achenes usually 4-5, each enclosed in an enlarged inner phyllary forming a fusiform-oblong fruit 8-10 mm long, body of fruit glandular, c. 6-ribbed, ribs with 1 or 2 rows of uncinate spines about 2 mm long.

Botswana. S: Southern Dist., 20 km SE of Jwaneng, c. 1100 m, 4.iii.1987, *Long & Rae* 79 (K). **Zambia**. W: Mufulira, 21.iii.1949, *Cruse* 509 (K). C: Serenje-Kanona, Great North Rd., 5.iv.1961, *Richards* 14938 (K, SRGH). **Zimbabwe**. N: Gokwe, 24.iv.1962, *Bingham* 245 (K, SRGH). W: Figtree Station, 13.x.1934, *Story* 4830 (K). C: Marandellas, 14.xii.1951, *Corby* 747 (SRGH). E: Umtali, 4.xi.1954, *Chase* 5319 (SRGH). **Mozambique**. GI: Inhambane, iii.1938, *Gomes e Sousa* 2100 (COI, K). M: Maputo, zona verde de Benfica, xii.1989, *Groenendijk* 2243 (K, LMU).

A native of S America recorded from eastern Brazil (as far N as Rio de Janeiro), Paraguay and Uruguay; also in Florida and Georgia, U.S.A., where it is probably introduced; an introduced weed in Kenya, Tanzania and South Africa, and on La Réunion in the Mascarenes. In higher rainfall areas than *A. hispidum*, often at roadsides; 0–1850 m.

Conservation Status: LC (Least Concern).

Often in the past confused with *A. australe*, but easily separated by the fewer ray florets (and fruits), the coarsely dentate leaf margins and the frequently glabrous to sparsely pubescent leaf surfaces.

3. **Acanthospermum australe** (Loefl.) Kuntze, Revis. Gen. Pl. **1**: 303 (1891). Type: '... observatae in itinerere a Cumana ...... ad fluvium Orinoco, per Barcellonam & Las Missiones de Piritu ..... Feb. 1755.'[222] FIGURE 6.5.**36A**.

> *Melampodium australe* Loefl., Iter Hispan.: 268 (1758).
>
> *Centrospermum xanthioides* Kunth in Humb., Bonpl. & Kunth, Nov. Gen. Sp. Pl. **4** (ed. folio): 213 (1818). Type: 'Crescit inter Cumana et Bordones. (Nova Andalusia.). Floret Augusto.', nom. illeg.[223]
>
> *Acanthospermum brasilum* Schrank, Pl. Rar. Hort. Monac. **2**: 53 (1819)[Apr.–May 1820]. Type: 'Patria Brasilia, unde *D. Martius* semina misit' (M holotype).
>
> *Orcya adhaerescens* Vell., Fl. Flumin.: 345 (1825) [7 Sept.–28 Nov. 1829]; Fl. Flumin. Icon. **8**: tab. 83 (1827)[29 Oct. 1831]. Type: [Brazil:] 'Udique habitat praecipue silvis excultis.' Location of extant original material unknown.
>
> *Echinodium prostratum* Poit. ex Cass. in Dict. Sci. Nat., ed. 2, **59**: 245 (1829), nom nud.[224]
>
> *Acanthospermum xanthioides* (Kunth) DC. [var.] α *obtusifolium* DC., Prodr. **5**: 522 (1836). Types: '– in Cumanâ? (*Loefl.*), in Brasiliae prov. Rio-Grande (h. Mus. Bras. n. 1069). Melampodium australe Loefl. itin. 268. Linn. sp. 1303. (v.s. ex Forsyth sine loci design.)' (G-DC-G00469850– 'h. Mus. Bras. 1069', P syntypes).
>
> *Acanthospermum xanthioides* [var.] β *acutifolium* DC., Prodr. **5**: 522 (1836), nom. illeg. pro *Centrospermum xanthioides* Kunth
>
> *Acanthospermum hirsutum* DC., Prodr. **5**: 522. (1836), nom. illeg., superf. pro *Acanthospermum brasilum* Schrank.

---

[222] Not in LINN; the material LINN 1034.1 does not belong to the genus now known as *Acanthospermum*, and is likely to be the reason for Smith's annotation – 'not an original specimen', together with evidence that this material is ex herb. Mutis.

[223] According to the *Index Nominum Genericorum* database, but without supporting comment!

[224] Cassini provided the following comment: 'Nous avons fait cette description générique [of *Centrospermum*!] sur un échantillon sec, en mauvais état, receuilli dans la Guiane françoise par *M. Poiteau*, et qui se trouve dans l'herbier de M. Gay ou il étoit étiqueté *Echinodium prostratum* Poit.' This specimen is in K.

Annual or short-lived perennial herb with prostrate branches; densely ± patently hairy with rather long hairs, often purplish. Leaves petiolate, petiole to 1.4 cm long pilose; leaf-lamina 1–4.5 × 0.8–3.9 cm, elliptic to broadly ovate or subcircular, apex rounded or obtuse, base abruptly cuneate, margins subentire to coarsely dentate, pubescent on both sides, gland-dotted. Capitula pedicellate, pedicel to 1.2 cm long, densely pubescent, 3–5 mm diam. in flower, enlarging in fruit to c. 1.5 cm diam.; outer phyllaries c. 4 mm long, oblong-ovate, apex obtuse, mucronulate, sparsely ciliate, gland-dotted; paleae subtending disc florets c. 2 mm. long, membranous, oblong, emarginate, ciliate towards truncate apex and along midrib. Ray florets female, ray limb pale yellowish, c. 1.3 mm long, ± oblong, apex 3(4) dentate, gland-dotted. Disc florets hermaphrodite, limb funnel-shaped, c. 1.2 mm long, 5-lobed lobes, gland-dotted, tube c. 1 mm. long pilose; achene abortive. Fertile achenes usually 8-9, each enclosed in an enlarged inner phyllary forming a fusiform-oblong fruit c. 7 mm long c. 8-ribbed, ribs with 1 or 2 rows of uncinate spines c. 2 mm long, body of fruit glandular.

Zambia. N: Chilongowelo, *Richards* 2303 (K). W: Ndola, 2.i.1955, *Fanshawe* 1769 (K, SRGH). **Zimbabwe**. N: Mrewa, 2.i.1960, *Leach* 9740 (SRGH). C: Salisbury, 11.xii.1931, *Eyles* 7060 (SRGH). E: Chimanimani Mts., 14.ii.1960, *Goodier* 900 (SRGH). **Malawi**. N: Mzimba, 28.ii.1959, *Robson* 1727 (BM, K, PRE). **Mozambique**. M: Lourenco Marques, *Borle* 574 (PRE).

A native of S and Central America recorded from the W Indies, the Guianas, Brazil (S to about Rio de Janeiro), Venezuela, Colombia, Bolivia and northern Paraguay; commonly occurs as a weed in our area and South Africa, also recorded from Nigeria, but interestingly not yet recorded in the neighbouring F.T.E.A. area; in our area a species of fairly high or high rainfall areas, often on roadsides; 0–2100 m.

Conservation Status: LC (Least Concern).

Acanthospermum australe and *A. glabratum* have been confused in the past, and united by some botanists, either in synonymy (Blake 1921), or as varieties (de Candolle 1836). They are quite separable, the key above summarizing the obvious differences. Wild (1967) has a much expanded commentary.

144. **MICRACTIS** DC.

Micractis DC., Prodr. **5**: 619 (1836). —Schulz in Gleditschia **18**(2): 211–218 (1990). — Beentje & Hind in F.T.E.A., Compositae **3**: 775–776 (2005). —Panero in Kubitzki, Fam. Gen. Vasc. Pl. **8**: 490 [2006](2007).
Limnogenneton Sch.Bip. ex Walp., Repert. Bot. Syst. **6**: 146 (1846).[225]

Annual or possibly short-lived perennial herbs. Leaves opposite, simple, pseudopetiolate or sessile, lamina lanceolate to ovate or deltoid, 3-veined, scabrid pubescent, hairs eglandular. Inflorescences of open paniculiform-cymes. Capitula heterogamous, radiate; involucre hemispherical; phyllaries 1- or 2-seriate, herbaceous; receptacles convex to shallow-conical, paleaceous, paleae enveloping achenes and lower part of florets. Ray florets female, corollas yellow or yellowish-white, ray limbs small, minutely 2-lobed (rarely 3-lobed). Disc florets hermaphrodite, inner sometimes functionally staminate, corollas 4-lobed; style arms with parallel stigmatic lines, apices deltate, papillose. Achenes obcompressed, ovoid-oblong, incurved, glabrous, eglandular, brown to blackish; pappus absent.

A genus of three spp. in eastern Africa and Madagascar.

Micractis bojeri DC., Prodr. **5**: 620 (1836). —Schulz in Gleditschia **18**(2): 214 (1990). Type: 'in ins. Madagascar legit cl., *Bojer* [59] ... (v. s. comm. à cl. Bojer.)' (G-DC-G00456714 holotype). FIGURE 6.5.**37**.
Limnogenneton abyssinicum Sch.Bip. ex Walp., Repert. Bot. Syst. **6**: 147 (1846). Types: 'in Hb. Abyss. *Schimp*[er]. no. 1099 [sic! – 1059 is on the labels of all 'types']. ... Crescit in Abyssinia in paludibus

---

[225] The generic name is usually ascribed solely to Schultz Bipontinus, both the generic name, and single species are more correctly ascribed to 'Sch.Bip. ex Walp.'

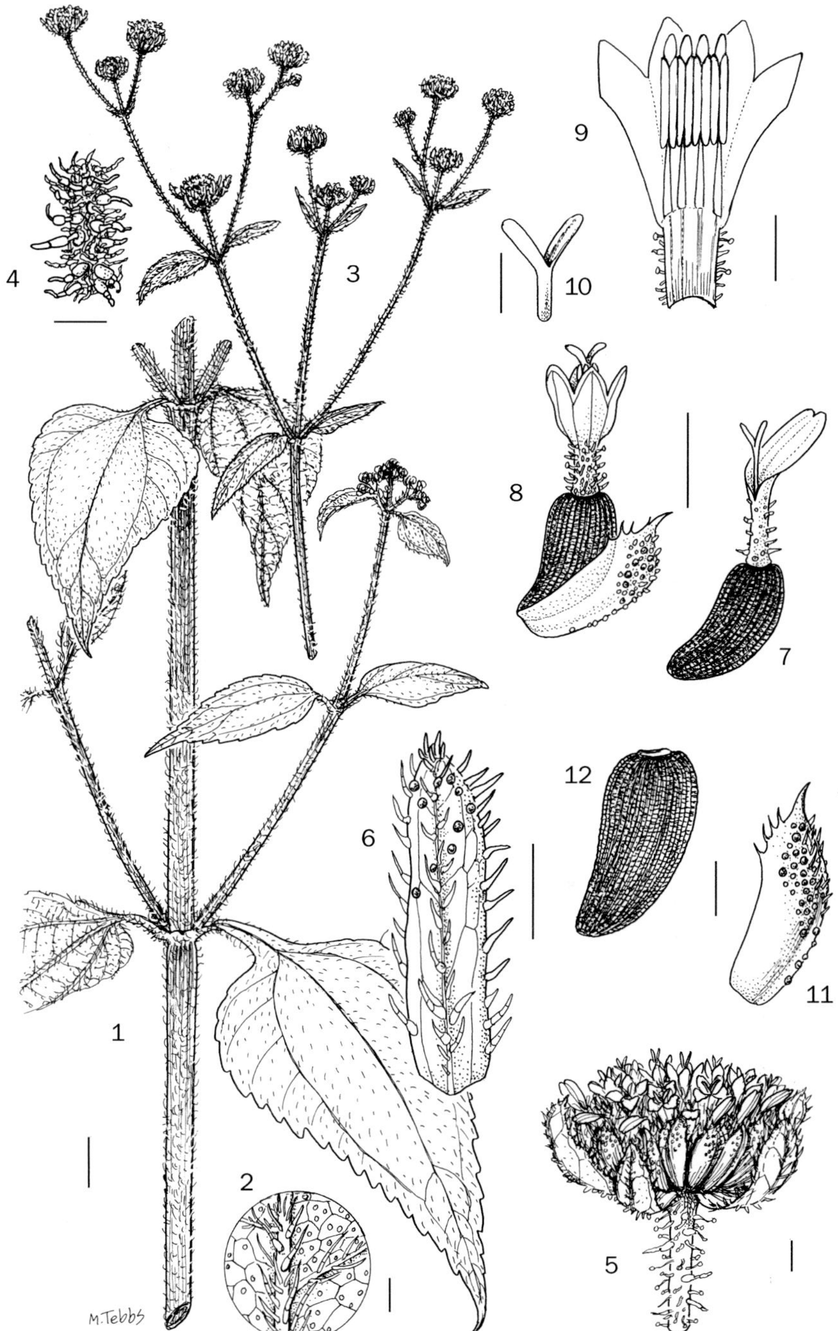

Fig. 6.5.**37**. MICRACTIS BOJERI. 1, leafy stem; 2, detail of abaxial leaf surface; 3, part of inflorescence; 4, detail of pedicel; 5, capitulum; 6, phyllary; 7, ray floret; 8, disc floret and accompanying palea; 9, corolla of disc floret opened out to show attachment of filaments; 10, style arms of disc floret; 11, palea; 12, mature disc floret achene. 1–2 from *Phillips* 1691; 3–11 from *Brummitt* 9990; 12 from *Buchanan* 77. Scale bars: 9–12 = 0.5 mm; 2, 4–8 = 1 mm; 1, 3 = 10 mm. Drawn by Margaret Tebbs.

prope Adoam.' (W0009545 lectotype, BR8360922, BR8879264, C10000363, E00259964, FI1000799, FT, G23942, G23943, GH00012364, GOET001822, GOET1822, HAL0111837, K000410230, K000410231, LG90028359, LZ, M0105422, M0105423, MO, P00069227, P00069228, P00418375, S-G-3672, TUB005470, TUB005471, UPS, US, W009544), lectotypified by Schulz (1990: 213).[226]

Cryphiospermum abyssinicum (Sch.Bip. ex Walp.) Schweinf., Beitr. Fl. Aethiop.: 284 (1867).

Sigesbeckia abyssinica (Sch.Bip. ex Walp.) Oliv. & Hiern in F.T.A. **3**: 372 (1877). —Humbert, Fl. Madagasc., Composées **189**: 638, t. cxv fig. 33–37 (1963). —Adams in F.W.T.A., ed. 2, **2**: 242 (1963). —Wild in Kirkia **6**: 37 (1967). —Maquet in Fl. Rwanda **3**: 620, fig. 191/1 (1985). —Lisowski, (Asterac. Fl. Afr. Cent. 1) Fragm. Flor. Geobot. **36** Suppl. 1: 179, t. 39 (1991). —Agnew & Agnew, Upland Kenya Wild. Fl., ed. 2: 215, t. 88 (1994).

Sigesbeckia emirnensis Baker in J. Linn. Soc. Bot. **20**(127): 188 (1883). Type: 'Central Madagascar, *Baron* 902.' (K000410228 holotype).

Micractis abyssinica (Sch.Bip. ex Walp.) Chiov. in Ann. Bot. Roma **9**: 73 (1911)

Annual or possibly short-lived perennial herb 0.3–2.5 m high; stem often purple, branches scabridulous to glabrous. Leaves ovate or narrowly ovate, 2–22 × 0.7–9.5 cm, base gradually narrowing, base connate with that of opposite leaf, 3-veined from base, margins finely serrate to serrate, apex acuminate or attenuate, scabrid or scabridulous above, thinly pubescent and glandular-punctate beneath. Inflorescences terminal, lax, paniculate, c. 20 cm across, capitula pedicellate, pedicels 1–4 mm long, pubescent, capitula sometimes closely subtended by leafy bracts to 12 mm long. Involucres 3–5.5 × 5–8 mm; phyllaries 3–4 mm long, oblong to ovate, abaxially pilose, margins ciliate, apices subobtuse; paleae c. 3 mm long, glandular. Ray florets many, biseriate but appearing uniseriate, ray limb yellow, 0.8–1 mm long, tube 0.5–0.8 mm long and glandular. Disc florets many, corollas yellow, 1–1.8 mm long, tube glandular. Achenes dark brown, obovoid, 1.8–2.5 mm long, 4-angled, glabrous, finely verrucose, eglandular; pappus absent.

Malawi. N: Zomba Plateau, by Mlunguzi Dam, 1420 m, 19.iv.1970, *Brummitt* 9990 (K). Masuku Plateau, 6500–7000 ft., vii.1896, *Whyte s.n.* (G, K, Z). ?S: s.loc., 1891, *Buchanan* 877 (BM, G, K, W).

Also in Cameroon, D.R. Congo, Ethiopia, Kenya, Uganda and Tanganyika, Nigeria and Madagascar. Only from Malawi in the flora area, and apparently only collected in the Zomba and Nyika Plateaus, from very few collections. certainly under-collected. Forest margins, stream sides, in seasonally moist grasslands, or swamp grassland; c. 1350–2750 m; flowering from March to May.

Conservation Status: A widespread species, albeit poorly collected within a very small area in the Flora region; LC (Least Concern).

145. **GUIZOTIA** Cass.

Guizotia Cass. in Dict. Sci. Nat., ed. 2, **59**: 237, 247 (1829). —Wild in Kirkia **6**(1): 11–12 (1967). —Baagøe in Bot. Tidsskr. **69**(1): 1–39 (1974). —Lisowski, (Asterac. Fl. Afr. Cent. 1) Fragm. Flor. Geobot. **36** Suppl. 1: 126 (1991). —Hind in Fl. Masc., Composées **109**: 202–204 (1993). —Karis & Ryding in Bremer, Asterac. Cladist. Classif.: 610 (1994). —Mesfin Tadesse, Fl. Ethiopia & Eritrea **4**: 309 (2004). —Beentje & Hind in F.T.E.A., Compositae **3**: 770–774 (2005). —Panero in Kubitzki, Fam. Gen. Vasc. Pl. **8**: 489–490 [2006](2007).[227]

---

[226] It is not at all clear why Schulz used the material in W as the lectotype; Walpers herbarium (containing many of the types of his new taxa) was sold after his death, and the location of much material is unknown. The lectotype specimen in W was first determined by Schulz as an isotype – as were all other duplicates. The lectotype citation by Schulz (1990: 214) is somewhat confusing. The isolectotype P69227 is a profusely annotated (by Schultz Bipontinus) sheet, where it is clear that small portions (in small paper capsules) of other *Schimper* collections (559, 659) are pinned to the sheet.

[227] Originally conserved against a vernacular designation of '*Werrinuwia* of Heyn', following Art. 14.13 of the Code, this is no longer necessary.

Ramtilla DC. ex Wight, Contr. Bot. India: 18 (1834).

Veslingia Vis. in Nuovi Saggi Imp. Regia Accad. Sci. Padova **5**: 269 (1840) Index Seminum (PAD, Patavium): 9 (1840), nom. illeg., non Heist. ex Fabr. (1759)[= *Aizoon* L., AIZOACEAE].

Annual or perennial herbs, rarely shrubs. Stems poorly- to well-branched, often scabrid. Leaves opposite or more rarely ternate, simple, sessile or short-petiolate, sometimes connate or perfoliate or with auriculate bases, lamina linear to lanceolate, ovate or deltate, herbaceous or coriaceous, margins entire to serrate. Inflorescences of solitary terminal capitula or of many-headed corymbiform-cymes, capitula sessile or pedicellate. Capitula radiate and heterogamous; involucre campanulate to hemispherical; phyllaries ± biseriate, outer ± foliaceous, 5–6, inner scarious, passing into obovate 3-veined paleae; receptacle hemispherical to conical, paleaceous; paleae persistent, carinate around disc florets, glabrous or pubescent, glandular or eglandular. Ray florets uniseriate, female, ray limb yellow, 3-dentate. Disc florets numerous, hermaphrodite, fertile, caducous, corollas yellow with reddish resin canals, cylindrical, narrowed below, with reflexed pubescence at base, 5-lobed; apical anther appendages ovate, obtuse, basal anther appendages obtuse and short-sagittate; style arms with parallel stigmatic surfaces, apices subulate, appendages papillose. Ray achenes triquetrous, disc achenes obcompressed to ± tetragonous, black or brownish, glabrous; pappus absent.

A genus of six spp. with one species, *G. abyssinica*, widely cultivated.

Annual, glabrous or sparsely pubescent; leaves with sessile glands; outer phyllaries broad-ovate or obovate; ray florets 6–8(15); paleae usually 5-veined . . . . **1.** *abyssinica*
Annual or perennial, usually conspicuously scabrid; leaves with stipitate-glandular hairs; outer phyllaries lanceolate; ray florets 8–16; paleae 3-veined . . . . **2.** *scabra*

1. **Guizotia abyssinica** (L.f.) Cass. in Dict. Sci. Nat., ed. 2, **59**: 237 (1829). —Lisowski, (Asterac. Fl. Afr. Cent. 1) Fragm. Flor. Geobot. **36** Suppl. 1: 127–128 (1991). —Mesfin Tadesse, Fl. Ethiopia & Eritrea **4**(2): 309–310 (2004). —Beentje & Hind in F.T.E.A., Compositae **3**: 770–771 (2005). Type: 'Habitat in Abyssinia.' 'H.U. Polymnia Bidentis (…) abyssinica' (LINN – Herb. Linn. No. 1033.5 lectotype), lectotypified by Baagøe (1974: 20).[228]

   Polymnia abyssinica L.f., Suppl. Pl.: 383 (1781).

   Polymnia frondosa Bruce, Select Spec. Nat. Hist. in Trav. Disc. Source Nile, ed. 2, **7**: 337[–339] (1805). **8**: pl. 52 (1813).Type: specimens unknown.[229]

   Verbesina sativa Roxb. ex Sims in Curtis's Bot. Mag. **26**: t. 1017 (1817). Type: 'We are informed by Dr. Roxburgh, to whom the Botanic Garden at Brompton is indebted for seeds of this hitherto undescribed vegetable, that it is cultivated in the Mysore country and several other parts of India, …'. Herbarium material unknown.

   Parthenium luteum Spreng., Nov. Prov.: 31 (1819), nom. illeg. citing 'Verbesina sativa. Hort. angl. Habitat —' taken to mean reference to Sims' name.[230]

   Heliopsis platyglossa Cass. in Bull. Soc. Philom. Paris **1821**: 187 (1821). Dict. Sci. Nat., ed. 2, **24**: 332 (1822). Type: 'J'ai étudié cette plante en 1821, sur un individu vivant cultivé au Jardin du Roi, où il était innommé, et où il fleurissait au mois de Juillet. On ignore son origine.'

   Tetragonotheca abyssinica (L.f.) Ledeb., Ind. Sem. Hort. Dorpat. Suppl.: 7 (1824).

   Jaegeria abyssinica (L.f.) Spreng., Syst. Veg., ed. 16, **3**: 590 (1826).

   Helianthus oleifer Wall., Numer. List.: 3191 (1831), nom. inval., nom. nud.

---

[228] Material in LINN, LINN 1033.5, is marked as 'H.U. Polymnia Bidentis' alongside of which is pencilled 'abyssinica'. This cultivated sheet is taken to be original material.

[229] Baagøe (1974: 20) indicated the name 'is typified by the table.' However, Bruce made several notes concerning the plant. The complete plate, the original of which would suffice as the type, is pl. 52 in volume 8 of Murray's second edition, but this was doubtless drawn from material grown from seed brought back by Bruce. It is unclear where the original of this plate is, although it is by a French artist – and may well be in P, as might the material upon which it was based also be.

[230] See also Rollins (1950: 70), who clearly followed the suggestion in Index Kewensis.

Ramtilla oleifera DC. in Wight, Contr. Bot. India: 18 (1834), nom. illeg., citing amongst other names, *Verbesina sativa* Roxb. ex Sims.

Ramtilla oleifera var. *angustior* DC. in Wight, Contr. Bot. India: 18 (1834). Type: '*Helianthus oleifer*, Wall! cat. n. 3191.' (K-Wall holotype).

Guizotia oleifera (DC.) DC. in Mém. Soc. Phys. Genève **7**(2): 271 (1836).

Guizotia oleifera var. *sativa* (Roxb. ex Sims) DC. in Mém. Soc. Phys. Genève **7**(2): 272 (1836).

Bidens ramtilla Wall. ex DC., Prodr. **5**: 551 (1836), nom. nud. pro. syn. (of *Guizotia oleifera*).

Anthemis mysorensis [herb. madr. ex] DC., Prodr. **5**: 551 (1836), nom. nud. pro syn. (of *Guizotia oleifera*).

Guizotia abyssinica var. *sativa* (Roxb. ex Sims) Oliv. & Hiern in F.T.A. **3**: 385 (1877).

Guizotia abyssinica var. *angustior* (DC.) Oliv. & Hiern in F.T.A. **3**: 385 (1877).[231]

Annual herb, 1–2 m high, erect. Stems branched, striate, often purplish, scabrous to glabrous. Leaves sessile, subconnate to perfoliate, broadly lanceolate to oblanceolate or ovate, 10–15 × 2–6 cm, base truncate to cordate, scabrid on both surfaces, glandular-punctate, midrib prominent, venation pinnate, margins entire to serrate, ± revolute, apex acute. Inflorescences of terminal few-headed cymes; capitula pedicellate, pedicels 2–12 cm long, densely pilose towards apices, hairs glandular or eglandular. Capitula many-flowered; involucre spherical to campanulate, 7–10(32) × (10)15–20 mm; outer phyllaries 5(6), foliaceous, broad-ovate, 7–10 mm long, pilose, apices acute to obtuse and mucronate, (5)9(12)-veined, (5.3)7–10(33) × (2.3)4–6(11) mm, apices acute to obtuse, pubescent abaxially, inner obovate, 5–8(10)-veined, 5–7(9) × (1)1.5–2(3) mm, scarious, apices acute to obtuse, abaxially puberulous; paleae 5–7(9) × (1)1.5–2.5(3) mm, (3)5(8)-veined, apex margins ciliate to laciniate. Ray floret corollas yellow, 6–8(15), ray 8–14(21) × (5)6–7(12) mm, tube 1–2.8 mm long, distinctly pubescent at base. Disc florets many, corollas yellow, 4–5.5 mm long, distinctly pubescent at base. Achenes 3.5–5.7 × 1.2–1.5 mm, glabrous, black to brownish, shiny; pappus absent.

Malawi. Lisowski (1991: 127) cited the species for Malawi, but no material has been found.

Also known from Ethiopia, Sudan, Cameroon, D.R. Congo, Uganda, Tanzania, South Africa, India, and Europe. Baagøe (1974: 21) considered the species only native to the Ethiopian Highlands and was cultivated elsewhere; 200–2300 m. Widely grown for niger seed, a commercial oil, the achenes have also been a common additive to bird seed feed in Europe.

Conservation Status: Although apparently under-collected in the Flora area the species is otherwise widespread and unlikely to be under threat; LC (Least Concern).

2. **Guizotia scabra** (Vis.) Chiov. in Annuario Reale Ist. Bot. Roma **8**: 184 (1904). —Andrews, Fl. Pl. Sudan **3**: 32 (1956). —Brenan in Mem. New York Bot. Gard. **8**(5): 480 (1954). —Adams in F.W.T.A., ed. 2, **2**: 230 (1963). —Lisowki, (Asterac. Fl. Afr. Cent. 1) Fragm. Flor. Geobot. **36** Suppl. 1: 128–130 (1991). —Mesfin Tadesse, Fl. Ethiopia & Eritrea **4**(2): 310–311 (2004). —Beentje & Hind in F.T.E.A., Compositae **3**: 771–772 (2005). Type: [Ethiopia:] 'Hab. in Africa locis Tumad et Cassan dictis, ubi legit *Th. Kotschy* [501] anno 1837-38. Floret a Majo in aestatem.' (FI000515 lectotype, FI ×2, K000410413, M0105418, P00073666, P00073667, P00418380, S-G-6351, W007323, W007324, W007325, W007326, W18890058307), lectotypified by Baagøe (1974: 25). FIGURE 6.5.**38**.

Veslingia scabra Vis. in Nuovi Saggi Imp. Regia Accad. Sci. Padova **5**: 269 (1840), Index Seminum (PAD, Patavium): 9 (1840)

Guizotia schultzii Hochst. ex A.Rich., Tent. Fl. Abyss. **1**: 407 (1848).[232] —Oliver & Hiern in F.T.A. **3**: 385 (1877). —Hutchinson, Botanist S. Afr: 508 (1946). Types: 'in pl. Schimp. Abyss., sect. 1, 350, III, 1510. ... Crescit circa Memessah (*Quartin Dillon*), et in provincia Ouodgerate (*Ant. Petit*), in regione septentrionali montis Selleuda, mensibus Octobre ad Novembrem florens (*Schimper*).' 'Ethiopia, Scholoda Mts, *Schimper* I, 350' (TUB –according to Baagøe, the specimen with Schimper's 'original label' –lectotype, BM0009244403, BM000924404, BM000924405,

---

[231] Baagøe (1974: 20–21) listed a number of other nomina nuda.

[232] The name '*Guizotia schultzii* Sch.Bip. in Walpers Repert. Bot. Syst. **6**: 158 (1846)' is often cited, presumably as superfluous later homonym. I feel this was merely a reference to Richard's earlier name, not a new name!

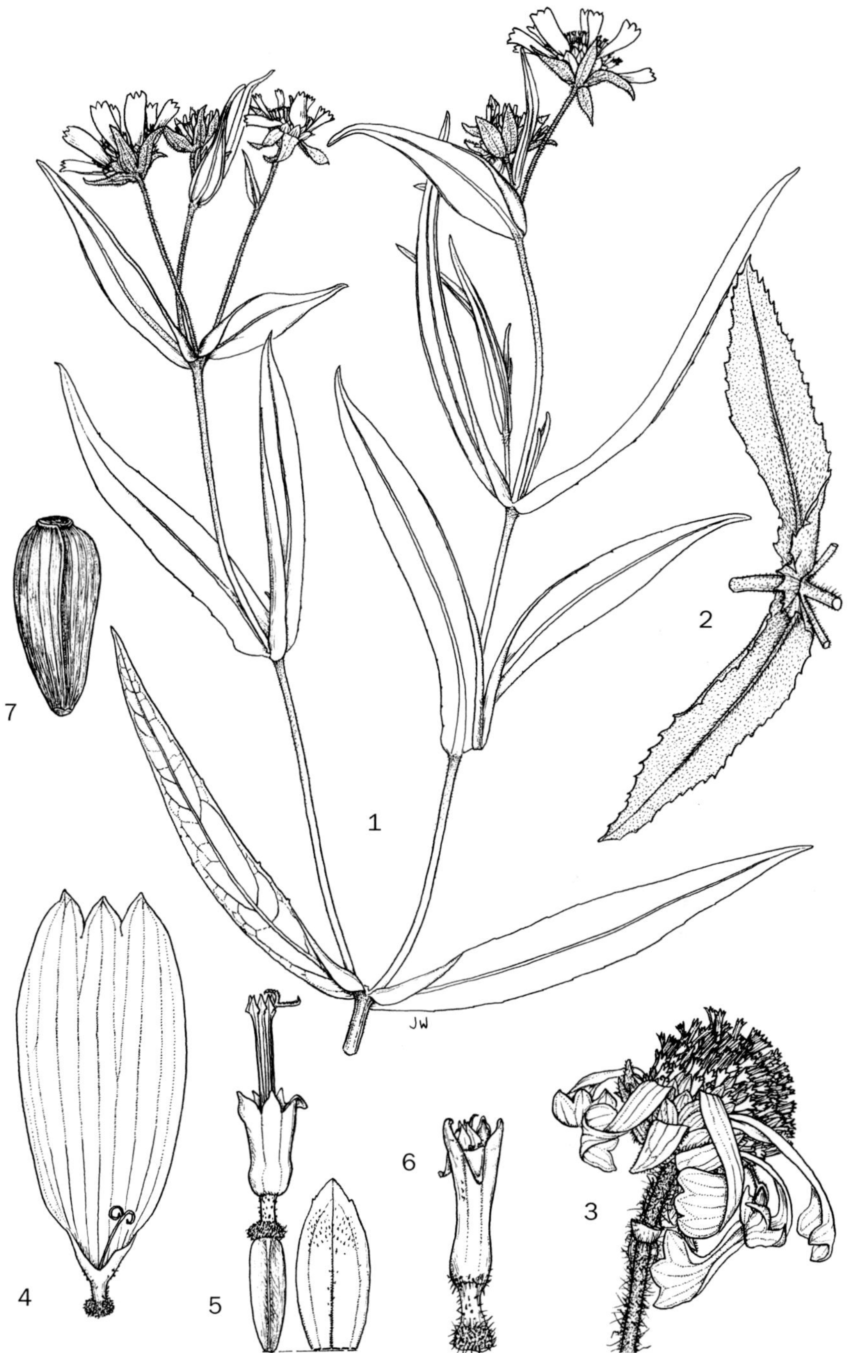

Fig. 6.5.**38**. GUIZOTIA SCABRA. 1, apex of flowering shoot (× ²/₃); 2, detail of leaf pair (× ²/₃); 3, mature capitulum (× 2); 4, ray floret corolla (× 4); 5, disc floret with palea (× 6); 6, disc floret corolla (× 8); 7, disc floret achene (× 10). 1, 4–5, 7 from *Richards* 8453; 2 from *Harley* 9206; 3 from *Jefford* 322a; 6 from *Peter* 44223. Drawn by Juliet Williamson. From Flora of Tropical East Africa.

BR8361653, FI000514, G, GH, GOET001595, K000410412, M, P00073687, P000418381, S07-9668, S07-9669, S-G-2964, TUB005520, TUB005521, WAG0101346), lectotypified by Baagøe (1974: 25); *Quartin Dillon* s.n. (P00073676, P00073677 syntypes).; *Ant. Petit* (? syntype); *Schimper* III, 1510 (BR8361264, BR8679659, K000410411, P0073694, P00418382 syntypes).

Guizotia nyikensis Baker in Bull. Misc. Inform., Kew **1898**: 153 (1898). Types: 'British Central Africa. Nyika plateau, alt. 6000–7000 ft., *Whyte*, 198. Masuku plateau, alt. 6500–7000 ft., *Whyte*.' *Whyte* 198 (K000374812 syntype); *Whyte* s.n. (K000374813 syntype).[233]

Guizotia schultzii var. *sotikensis* S.Moore in J. Linn. Soc. Bot. **35**: 344 (1902). —Fries in Acta Hort. Berg. **9**: 141 (1928). Type: [Kenya, Kericho District Dist.:] 'British East Africa, Sotik, [1889,] *F. J. Jackson* [s.n.]' (BM000924402 holotype).

Guizotia collina S.Moore in J. Linn. Soc. Bot. **38**: 262 (1908). Type: [Uganda:] 'Hab. Ruwenzori E. 5000 ft. March [26.3.1906, *Wollaston* s.n.]' (BM000924401 – mounted with *E. Bruce* 2678-BM000924400 –, holotype).

Guizotia eylesii S.Moore in J. Bot. **46**: 43 (1908). —Eyles in Trans. Roy. Soc. S. Afr. **5**: 515 (1916). Type: [Zimbabwe:] 'Mazoe, in marsh. [4300 ft., April 1906] *F. Eyles*, 349.' (BM000924406 holotype, K00374815, SRGH010674).

Wedelia oblonga Hutch. in Gard. Chron., ser. 3, **45**: 18 (1909). Types: 'The following description was drawn up from a living plant sent to Kew by Messrs. Veitch & Sons ...'[234]

Guizotia ringoetii De Wild. in Feddes Repert. Spec. Nov. Regni Veg. **13**(359–362): 204 (1914). Types: '[D.R. Congo:] 'Katanga: Shinsenda, 1912 (*Ringoet*, no. 6). Uvira (Tanganika), 1908 (*Rouling*).' *Rongoet* 6 (BR8678928, BR8679239 syntypes); *Rouling* s.n. (BR86779383 syntype).

Guizotia kassneri De Wild. in Feddes Repert. Spec. Nov. Regni Veg. **13**(359–362): 205 (1914). Type: '[D.R. Congo:] Katanga: Bords du Tankanika, mai 1908 (*Kassner*, no. 3031[sic! = 3037].' (BR8679055 holotype, K000410552, P0073659).

Guizotia oblonga (Hutch.) Hutch. & Bullock in Bull. Misc. Inform. Kew **1933**: 150 (1933).

Guizotia scabra var. *sotikensis* (S.Moore) Robyns, Fl. Spermatophyt. Parc Nat. Albert **2**: 526 (1947).

Coreopsis galericulata Sherff in Amer. J. Bot. **34**(3): 156 (1947). —Verdcourt in Kew Bull. **17**: 498 (1964). Type: '*Mrs M. V. Webster* 8,858, alt. 6,400 feet, Kitale, Kenya Colony, commun. June, 1942' (EA holotype).

Erect perennial herb c. 1 m. tall, sometimes up to 2 m; stems terete or ± 4-angled, scabrous-pubescent or glabrescent. Leaves opposite or more rarely ternate, up to 15 × 3 cm, variable in shape and size, from very narrowly ovate (almost linear) to oblong or ovate-oblong, apex subobtuse or more often acute, mucronate, base cordate-amplexicaul, rounded or broadly cuneate, sometimes slightly connate, margins subentire or shallowly serrate ± cartilaginous, scabrid-pubescent on both sides especially above but sometimes quite glabrous beneath and very rarely so above. Capitula cymose in c. 3–6-capitulate lax or ± dense inflorescences at ends of branchlets, hemispheric, c. 2 cm diam.; peduncles 3–c. 6 cm long, harshly patent-pilose, rarely glabrous; outer phyllaries up to 1.5 cm long, narrowly ovate, tapering to the acute apex, exceeding the inner ones, scabrid-pubescent or pilose, rarely glabrescent, margins ciliate; paleae 4–4.5 mm long, obovate, pubescent towards apex, margin ciliolate towards apex. Ray florets yellow, limb to c. 2 cm long, oblong, deeply 3-dentate at apex, puberulent beneath. tube c. 2 mm long, puberulent, with hairs at base reflexed. Disc florets yellow, c. 4 mm long, cylindrical, narrower base c. 1 mm long, pubescent, hairs at base very dense and reflexed. Achenes 1.8–3 mm long, narrowly obovoid, ± tetragonous, glabrous.

Zambia. N: 80 km S. of Abercorn, 18.vii.1930, *Hutchinson & Gillett* 3826 (BM, K, SRGH). W: Ndola, 3.iv.1954, *Fanshawe* 1055 (K). C: Walamba, 22.v.1954, *Fanshawe* 1226 (K). **Zimbabwe**. N: Mazoe, iv.1906, *Eyles* 349 (BM, K, SRGH). C: Salisbury, ii.1925, *Eyles* 4764 (SRGH). E: Melsetter, Iona Farm, 15.iii.1953, *Chase* 4855 (BM, SRGH). S: Buhera, Sabi R., iii.1954, *Davies* 703 (SRGH). **Malawi**. N: Nyika, 29.x.1958, *Robson & Angus* 471 (BM, K, SRGH). C: Dedza Dist., Dedza Mountain, 1680 m, *Pawek* 11559 (K, MO, MA). S: Cholo

---

[233] K000374814 was considered as possible syntype material but has none of the protologue detail provided by Baker. It was annotated by Baker.

[234] At the end of the Latin description the following material was cited: 'British East Africa: from seeds collected by Diespecker! Eldoma Ravine, Whyte!', taken by Beentje & Hind (2005: 773) as syntype material. The Diespecker 'seed' was that provided to Veitch and used by Hutchinson to describe the species.

Mt., 26.ix.1946, *Brass* 17822 (BM, PRE, SRGH). **Mozambique**. N: Vila Cabral, 10.v.1934, *Torre* 56 (COI). MS: Mossurize, Jihu, R. Curumadzi, 18.viii.1906, *Swynnerton* 1883 (BM, K).

Also in Nigeria, Cameroun, Sudan, Ethiopia, Eritrea, D.R. Congo, Uganda, Kenya and Tanzania, Ruanda, Burundi. Like *Guizotia abyssinica, G. scabra* is also considered native of the Ethiopian Highlands. Usually in swampy areas or by streamsides, occasionally in grassland in high rainfall areas; 700–2400 m; flowering throughout the year.

Conservation status: A widespread and well-collected species throughout its range, and certainly in the Flora area; LC (Least Concern).

Following Baagøe's revision, only the typical subspecies is present in the Flora area.

146. **SIGESBECKIA** L.

Sigesbeckia L., Sp. Pl. **2**: 900 (1753). Gen. Pl., ed. 5: 383 (1754). —Wild in Kirkia **6**(1): 36–37 (1967). —Humbles in Ciencia y Naturaleza **13**(1–2): 2–19 (1972). —Dyer, Gen. S. Afr. Fl. Pl. **1**: 696 (1975). —Schulz in Haussknechtia **3**: 57–64 (1987). —Hind in Fl. Masc., Composées **109**: 176 (1993). —Beentje & Hind in F.T.E.A., Compositae **3**: 776–778 (2005).[235]

Schkuhria Moench, Methodus: 566 (1794), nom. rej. non *Schkuhria* Roth.

Minyranthes Turcz. in Bull. Soc. Imp. Naturalistes Moscou **24**(1): 181 (1851).

Zandera D.L.Schulz in Haussknechtia **4**: 32 (1988).

Annual or short-lived perennial herbs, usually erect, sparsely- (almost simple) to moderately-branched. roots fibrous. Stems terete, usually viscid with spreading multicellular, stipitate-glandular hairs, rarely glabrous, usually fistulose. Leaves opposite, sessile or petiolate/pseudopetiolate, petioles winged, lamina usually ovate or oblong, base slightly to strongly narrowed, rounded to subtruncate, discolorous, lower surface slightly paler than upper, glandular-punctate or not, 3-veined usually from above base, margins entire, to short-serrate, apex obtuse to acuminate. Inflorescence terminal, cymose, capitula few to many, pedicellate, pedicels usually long, stipitate-glandular, glands usually numerous. Capitula erect or nodding, usually heterogamous and radiate; involucres campanulate or hemispherical; phyllaries biseriate, outer phyllaries spreading or reflexed, linear to spathulate, herbaceous, as long as or longer than inner, often densely long-and/or conspicuously stipitate-glandular or glandular-punctate, inner phyllaries of equal number to, and subtending, ray florets, chartaceous to partly herbaceous, appressed, oblong to ovate, glabrous or stipitate-glandular, obtuse to short-acute; receptacle rounded to conical, paleaceous, paleae similar to innermost phyllaries, membranous, conduplicate about accompanying achene, persistent, glabrous, apices cucullate and somewhat erose. Ray florets few to several (5–15), uniseriate, rarely absent, female, ray limb oblong, 3-lobed or ± 2-lobed, upper surface mamillose, yellow to whitish, sometimes purplish below, tube slender, hispidulous. Disc florets few to many (5–35), hermaphrodite or functionally male, corollas yellow or whitish, sparsely to densely hispidulous below, tube narrow, throat funnelform, lobes 3 or 5, nearly smooth inside; anther thecae yellow or green, apical appendage ovate, eglandular; style base distinctly thickened for some length; style arms with separate, paired, stigmatic lines, apical anther appendages short-acute, penicillate, hairs blunt. Achenes obovate, curved, ± laterally compressed to terete or quadrangular, smooth, wingless, glabrous, body black with pale striations; pappus absent.

A genus of about eight spp. from the New World, but with one species now found as a widespread weed in the Old World tropics and present in the Flora area. The plant often recorded as *Sigesbeckia abyssinica* (Sch.Bip.) Oliv. & Hiern (cf. Wild in Kirkia **6**: 37, 1967) is now recognized as *Micractis bojeri* DC.

Sigesbeckia orientalis L., Sp. Pl. **2**: 900 (1753). —Oliver & Hiern in F.T.A. **3**: 372 (1877). —Harvey in Harvey & Sonder, Fl. Cap. **3**: 132 (1865). —Mendonça, Contrib. Conhec. Fl. Angola **1** (Compositae): 89 (1943). —Humbert, Fl. Madagasc. Composées **189**: 637,

---

[235] Frequently found with the variant generic name spelling of *Siegesbeckia* which reflects the name of Johann Siegesbeck, after whom the genus was named; Linnaeus used the spelling I have used for this account, although the predominant spelling is with the 'e' as shown in the synonymy below.

t. cxv fig. 38–39 (1963). —Adams in F.W.T.A., ed. 2, **2**: 242 (1963). —Mesfin Tadesse in Fl. Ethiopia & Eritrea **4**(2): 316–317 (2004). —Beentje & Hind in F.T.E.A., Compositae **3**: 776–778 (2005). Type: 'Habitat in China, Medioa ad pagos.' '*Sigesbeckia*' in Linnaeus, Hort. Cliff.: 412, t. 23 (1738), iconotype, lecotypified by Stearn, Introd. Linnaeus' Sp. Pl. (Ray Soc. Ed.) (1957: 48).[236] FIGURE 6.5.**39**.

Siegesbeckia triangularis Cav., Icon. **3**: 27, tab. 253 (1795). Type: 'Habitat in Imperio Mexicano. Floret in Regio horto Matritense Augusto et Septembre.' (MA holotype).

Siegesbeckia iberica Willd., Sp. Pl., ed. 4, **3**: 2220 (1803). Type: 'Habitat in Iberia. (v. s.)' (B-W holotype).

Siegesbeckia brachiata Roxb., Fl. Indica **3**: 439 (1832). Type: 'An annual a native of the moist vallies among the Circar mountains. Flowering time the cold season.' (?K holotype, BR5522125, ?K).[237]

Siegesbeckia microcephala DC., Prodr. **5**: 496 (1836). Type: 'in Novâ-Hollandiâ ex herb. Lambert. ... (v.s. comm. à cl. Lambert.)' (G-DCG00468966 holotype).[238]

Siegesbeckia gracilis DC., Prodr. **5**: 496 (1836). Type: 'in Novâ-Hollandiâ ex herb. Lambert. ... (v.s. comm. à cl. Lambert.)' (G-DCG00468967 holotype).

Siegesbeckia caspia Fisch. & C.A.Mey. ex Hohen. in Bull. Soc. Imp. Naturalistes Moscou **11**(3): 284 (1838). Type: '[Provinciae Talysch et regionum adjacentium.] Crescit rarius in pomariis humidis pagorum circa Lenkoram jacentium. Floret Julio m.' (?LE holotype).

Minyranthes heterophylla Turcz. in Bull. Soc. Imp. Naturalistes Moscou **24**(1): 181 (1851). Type: 'In provincia Cagayan insulae Luçon. *Cuming* n. 1351.' (?KW holotype, C10007956, E00414258, GOET002435, ?K).[239]

Siegesbeckia orientalis f. *pubescens* Makino in Bot. Mag. (Tokyo) **18**: 100 (1904). Type: 'Hab. Japan. common.' (?MAK/?TI holotype).

Siegesbeckia orientalis f. *glabrescens* Makino in Bot. Mag. (Tokyo) **18**: 100 (1904). Type: 'Hab. Japan. common.' (?MAK/?TI holotype).

Siegesbeckia orientalis f. *angustifolia* Makino in Bot. Mag. (Tokyo) **18**: 100 (1904). Type: 'Hab. Kiusiu (*T. Kawakami*!). Rarius.' (?MAK/?TI holotype).

Siegesbeckia filarszkyi Pit. in Pit. & Proust, Iles Canaries Fl.: 225 (1908). Type: [Spain, Canary Islands:] 'Palma: de San Andrès à los Sauces (0 à 300 m). Endroits humides, bords des conduites d'eau de la zone maritime où elle est très abondante.' (?P ?holotype/syntypes).

Siegesbeckia pubescens (Makino) Makino in J. Jap. Bot. **1**(7): 24 (1917).

Siegesbeckia orientalis var. *pubescens* Makino ex Makino in J. Jap. Bot. **1**(7): 25 (1917), nom. nud. pro syn. sub. *S. pubescens*.

Siegesbeckia glabrescens (Makino) Makino in J. Jap. Bot. **1**(7): 25 (1917).

Siegesbeckia orientalis var. *glabrescens* Makino ex Makino in J. Jap. Bot. **1**(7): 25 (1917), nom. nud. pro syn. sub. *S. glabrescens*

Siegesbeckia orientalis var. *angustifolia* Makino ex Makino in J. Jap. Bot. **1**(7): 25 (1917), nom. nud. pro syn. sub. *S. orientalis*

Siegesbeckia humilis Koidz. in Bot. Mag. (Tokyo) **39**: 24 (1925). Type: 'Hab. Japonia: Kiushiu, Kumagori, Isshochi, leg. *K. Mayebara*!' (?TI holotype).

Siegesbeckia orientalis var. *tenggerensis* Hochr. in Candollea **5**: 320 (1934). Type: 'Java, Tengger, sur la grande arête circulaire, dans la forêt de *Casuarina* à sous-bois *d'Euphorbia*, alt. 2400 m., 23 janvier 1905, herbe à fleurs jaunes, peu fréquente ([*Hochreutiner*] n. 2742).' (G holotype).

Siegesbeckia formosana Kitam. in Acta Phytotax. Geobot. **6**(2): 87 (1937). Type: [Japan:] 'Hab. Prov. Takao: Kizangun, Naiyeizan ad summum (2 nov. 1931 *T. Hosokawa* n. 478).' (?TI holotype).

Sigesbeckia orientalis subsp. *glabrescens* (Makino) Kitam. in Mem. Coll. Sci. Kyoto Imp. Univ., Ser. B, **16**: 263 (1942), comb. illeg. pro syn.

---

[236] MacVaugh & Anderson's selection (Contr. Univ. Mich. Herb. **9**: 488. 1972) of LINN1018.1 is predated by that of Stearn.

[237] This name is also referred to 'Hort. Bengal. 62'; this dates from 1814 and there the name is a nom. nud. – the name was validated in Flora Indica.

[238] There are specimens with different dates (1816, 1820) in G-DC, both Lambert labelled material, but there is only one barcode on the sheet.

[239] Very oddly, we appear to have a label for a duplicate of this material in K, but it is on the corner of a 'cut up' sheet. I have been unable to trace the reminder of the sheet, and the potential isotype.

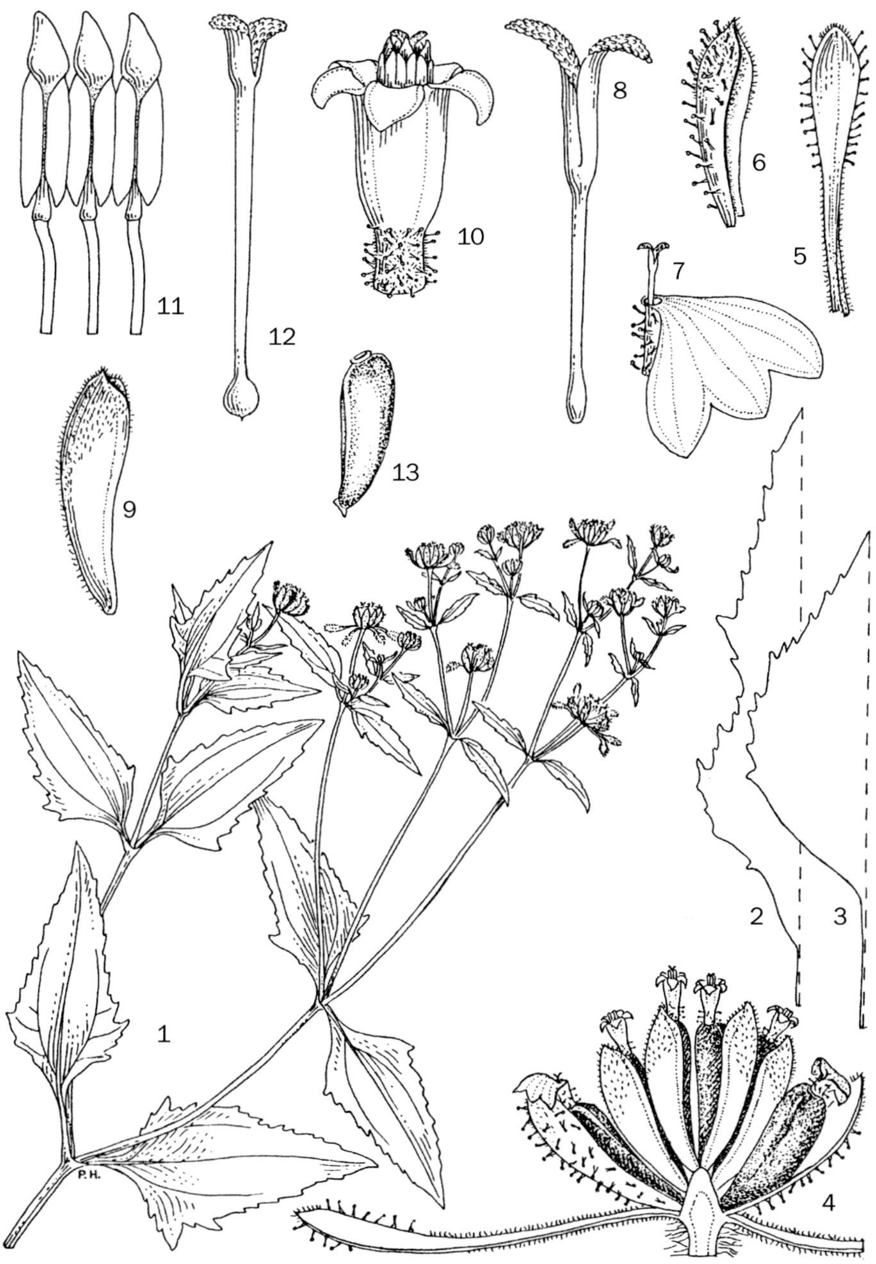

Fig. 6.5.**39**. SIGESBECKIA ORIENTALIS. 1, flowering shoot (× ²/₃); 2–3, outlines of lower leaves (× ²/₃); 4, l.s. capitulum (× 6); 5, outer phyllary (× 4); 6, inner phyllary (× 6); 7, ray floret corolla (× 20); 8, style of ray floret (× 60); 9, palea (× 8); 10, disc floret corolla (× 30); 11, anther cylinder opened out (× 60); 12, style of disc floret (× 60); 13, achene (× 6). 1, 4–13 from *Guého* s.n.; 2 from *Bosser* 11502; 3 from *Bijioux* 889. Drawn by Pat Halliday. From Flore des Mascareignes.

Sigesbeckia orientalis subsp. *pubescens* (Makino) Kitam. in Mem. Coll. Sci. Kyoto Imp. Univ., Ser. B, **16**: 264 (1942), comb. illeg. pro syn.

Sigesbeckia orientalis subsp. *glabrescens* (Makino) H.Koyama in Fl. Jap. (Iwatsuki et al., eds) **3b**: 32 (1995).

Sigesbeckia orientalis subsp. *pubescens* (Makino) H.Koyama in Fl. Jap. (Iwatsuki et al., eds) **3b**: 32 (1995).[240]

Annual herb 0.1–1.5 m tall. Stems simple below, usually moderately-branched in inflorescence, branches erect, striate or sulcate, pilose, 2 or 3 per node. Leaves petiolate/pseudopetiolate, petiole 0.5–5(10) cm long, pubescent, lamina membranous, 5–17 × 3–11 cm, narrowly ovate to broadly ovate or deltate-ovate, discolorous, pubescent above, pubescent and glandular-punctate beneath, base cuneate to truncate and abruptly cuneate near petiole, margins coarsely and irregularly serrate to coarsely serrate-crenate, apex obtuse or acute. Inflorescence a lax, leafy, dichotomous panicle up to c. 30 cm diam., bracts densely pubescent, pubescence of ± dense, simple, uniseriate, multicellular eglandular hairs above, similarly pubescent beneath but denser and moderately glandular-punctate. pedicels slender up to c. 1 cm long, pubescent, hairs uniseriate, multicellular and eglandular. Capitula radiate, c. 1 cm diam.; phyllaries biseriate, at first subequal, outer markedly larger and spreading in fruit, outer usually 5, 3-7 × 0.5–1.5 mm, growing to c. 12 mm in fruit, linear-spathulate, green, puberulent with simple uniseriate, multicellular eglandular hairs towards base, viscid, stipitate-glandular hairs conspicuous throughout, c. 0.6 mm long, inner usually 5 (or equal in number to ray florets), possessing both conspicuous stipitate-glandular hairs and shorter uniseriate, multicellular eglandular hairs throughout, apices cucullate about ray floret achenes; paleae c. 4 mm long, obtuse, puberulent, glandular. Ray florets usually 5, yellow (often changing to purplish post anthesis, although *Pope* 2220 indicates red when in flower), ray limb 3-fid, spreading, 1.3–1.5(2.5) mm long, tube c. 0.7 mm long, with sessile glands. Disc florets 10–30, corollas yellow to yellowish-brown, c. 1 mm long, 5-lobed, campanulate, tube c. 0.8 mm long, glandular-punctate towards base; anther cylinder partially exserted, apical anther appendages acute, yellowish, base ± cordate. Achenes blackish, 2–3.5 mm long, ± quadrangular, obovoid-oblong, curved; carpopodium conspicuous, pale, asymmetrical and present on inner surface.

Zambia. N: Abercorn Dist., Saisi River Marsh, 1500 m, 27.ii.1957, *Richards* 8341 (EA, K). W: Kitwe Dist., Mwerkera Forestry Training School, 27.iv.1989, *Pope* 2220 (BR, K). **Zimbabwe**. E: Chirinda Dist., Chirinda Forest, 1160 m, 23.iv.1947, *Wild* 1923 (K, SRGH). **Malawi**. N: Mugesse For. Res., 13.x.1952, *Chapman* 27 (BM, SRGH). C: Dedza Mountain, 25.iii.1969, *Salubeni* 1287 (K, SRGH). S: Chiradzulu Dist., foot of Chiradzulu Mountain, 1200 m, 11.v.1980, *Brummitt & Patel* 15615 (K). **Mozambique**. Z: Milange (possibly in Malawi!), 14.vii.1949, *Faulkner* 441 (COI, K). T: Tete, entre Vila Mouzinho e Metengo Balama, a 7.7 km de Vila Mouzinho, 17.vii.1949, *Grandvaux Barbosa & Carvalho* 3644 (K, LMJ).

Known throughout tropical Africa and in the Transvaal and Natal, also widespread in the tropics and subtropics, especially in the Old World. Miombo woodland with dambo margins, forest margins and also a weed of cultivation, usually in higher rainfall areas; (0)350–1750 m; flowering from February–July, although possibly flowering throughout the year.

Conservation Status: For such a widespread pantropical weed there are still relatively few collections throughout the Flora area, especially on a province-by-province basis, it is certainly under-collected; best recorded as LC (Least Concern).

Wild (1967: 36) indicates that the achenes were 'glandular-puberulent' a character I haven't seen in material – the achenes appear glabrous and eglandular.

---

[240] '*Siegesbeckia esquirolii* Levl. & Vant. in Feddes Repert. Nov. Spec. Regni Veg. **8**: 59 (1910)' was included in synonymy by Humbles (1972: 5). It is clear from the holotype in E, and the determination made by Lauener & McKean in 1973, that this plant is a species of *Adenostemma* [tribe Eupatorieae], probably *A. lavenia* (L.) Kuntze.

Tribe 14. **EUPATORIEAE** Cass.[241]

Eupatorieae Cass. in J. Phys. Chim. Hist. Nat. Arts **88**: 202–203 (1819). —Robinson & King in Heywood *et al.*, Biol. Chem. Compositae **1**: 437–485 (1977). —King & Robinson, The genera of the Eupatorieae (Asteraceae). Monogr. Syst. Bot. **22**: [i]–xi, 1–581 (1987). —Lisowski, (Asterac. Fl. Afr. Cent. 2) Fragm. Flor. Geobot. **36** Suppl. 1: 449–479 (1991). —Bremer, Anderberg, Karis & Lundberg in Bremer, Asterac. Cladist. Classif.: 625–680 (1994). —Hind & Robinson in Kubitzki, Fam. Gen. Vasc. Pl. **8**: 510–574 [2006](2007). —Robinson, Schilling & Panero in Funk *et al.*, Syst. Evol. Biogeogr. Compositae: 731–744 (2009). —Grossi, Katinas & Nakajima in Phytotaxa **141**(1): 25–39 (2013).
Heliantheae Cass. supersubtribe *Eupatoriodinae* (Cass.) C.Jeffrey in Bot. Zhurn. **87**(11): 11 (2002).

Herbs (rarely aquatic or semi-aquatic), subshrubs, shrubs, climbers, small trees, sometimes epiphytic. Leaves usually opposite, rarely strictly alternate, sometimes rosulate or verticillate, sessile or petiolate, lamina usually simple. Inflorescence usually a corymbose panicle, sometimes spicate. Capitula sessile or distinctly pedicellate, homogamous, discoid, rarely with some zygomorphic outer florets; involucre cylindrical, campanulate or hemispherical, rarely subtended by a subinvolucral bract (*Mikania*); phyllaries in 1 to several series, few or numerous, imbricate, subimbricate or distant, equal, subequal or markedly gradate, persistent or variously deciduous, lanceolate or ovate; receptacle flat to convex, sometimes highly conical, usually naked, glabrous or sometimes pubescent. Florets few, very rarely 1, often 4 or 5 to many, commonly fragrant; corollas funnelform to tubular, never truly yellow, lobes relatively short, commonly 5, very rarely 4; anther cylinders usually included within corolla tube; apical anther appendages obtuse or acute, rarely emarginate or lobed, as long as broad or shorter, sometimes absent, basal appendages short or almost absent, obtuse or rounded; anther collars indistinct, cylindrical or variously pronounced; nectary rarely visible; style base glabrous or pubescent, sometimes with a swollen node; styles usually very conspicuous and much exserted, glabrous or rarely pubescent; style arms linear to clavate, obtuse, stigmatic surfaces variously papillate. Achenes obovoid or oblong with phytomelanin in achene walls, usually 3–5-ribbed, sometimes 10-ribbed, body rarely flattened with 2 ribs or 5 winged ribs, sometimes glandular, glabrous or variously setuliferous; carpopodium often paler than achene body, rarely indistinct or absent, of several layers of variously enlarged, sometimes ornamented cells, usually symmetrical, rarely eccentric, annular, cylindrical, or stopper-shaped; pappus sometimes absent and reduced to an apical callus, rarely a laciniate crown, or vestigial, occasionally coroniform, usually of setae, commonly uniseriate, rarely biseriate or very rarely multiseriate, usually persistent, sometimes fragile, usually numerous, sometimes few, usually equal or subequal, rarely very short, or occasionally of flattened scales or awn-like scales, rarely of two distinct elements, very rarely of broad laciniate setae, or of few clavate apical appendages (*Adenostemma*); setae commonly barbellate or laciniate, rarely plumose, apices acute or obtuse, usually gradually tapering, sometimes dilated, very rarely conspicuously narrowing. The chromosome base number of the tribe has been suggested to be either $x = 10$ or 17, a wide range of haploid numbers are known from 4–51.

Several recent molecular systematic studies clearly position the Eupatorieae nested within the Heliantheae s.l., and more specifically within a clade with the Peritylinae and *Galeana-Villanova* clades (Panero 2007). Jeffrey (2002), whilst discussing the state of systematics within the family has proposed the creation of the new rank, the supersubtribe. None of these proposals are accepted here and the Eupatorieae are treated in their classical sense, separate from the Heliantheae.

A tribe of c. 182 genera in 17 subtribes, containing c. 2200 spp., largely restricted to the New World with two main centres of speciation, Mexico and Brazil. The tribe contains several pantropical and pansubtropical weeds; many species are used medicinally.

Several taxa are cultivated within the Flora region but one, as yet, does not appear to have naturalized: *Bartlettina sordida* (Less.) R.M.King & H.Rob. (grown in Zimbabwe), a native to Mexico and Guatemala, is a 2 m tall shrub with opposite, broad, ovate, 3-veined leaves, dense pyramidal to corymbose inflorescences and blue to purple corollas with densely puberulous abaxial surfaces to the corolla lobes.

---

[241] By D.J.N. Hind

Chromolaena odorata (L.) R.M.King & H.Rob., a widespread plant in the New World tropics, is reported as a relatively recent introduction in Zimbabwe, and is based solely on one herbarium collection (*Baretta* 17a, U). The species is a widespread weed in South Africa and has the potential to spread in the Flora area; it is keyed below but not worthy of a full account other than the following short-description:

Perennial herb to subshrub, 1–3 m tall. Leaves opposite, petiolate, lamina 4–10 × 1.5–5 cm, ovate to triangular, basally 3-veined, margins coarsely crenate or entire. Inflorescences corymbose. Capitula discoid; involucre cylindrical, 10–12 × 4–5 mm; phyllaries 4–6-seriate, imbricate, all deciduous at maturity. Florets 20–25 per capitulum, corollas white or pink, 5–6 mm long; style arms linear, well-exserted. Achenes blackish, c. 4 mm long, 5-ribbed, ribs setuliferous; carpopodium annular; pappus setae uniseriate, c. 6 mm long.

1. Phyllaries and florets 4 . . . . **148. Mikania**
– Phyllaries 5–numerous; florets 4–numerous 2
2. Pappus of 3–5 viscid-tipped knobs . . . . **149. Adenostemma**
– Pappus of bristles, hairs or scales, sometimes absent 3
3. Pappus usually of 5 awned scales, sometimes awnless, reduced or absent . . **150. Ageratum**
– Pappus of fine or coarse hairs 4
4. All phyllaries deciduous but remaining appressed until lost and leaving a flat to convex receptacle . . . . *Chromolaena odorata*
– Phyllaries persistent and usually spreading with age 5
5. Phyllaries 5–7(12); florets 4–5; style base not enlarged, pubescent; pappus setae >20; plants suffrutices (in Africa) . . . . **151. Stomatanthes**
– Phyllaries and florets numerous; style with basal node, glabrous; pappus setae <15; plants shrubs (in Flora area) . . . . **147. Ageratina**

147. **AGERATINA** Spach

Ageratina Spach, Hist. Nat. Vég. Phan. **10**: 286 (1841). —King & Robinson in Phytologia **19**(4): 208–229 (1970); in Phytologia **24**: 79–104. (1972); in Phytologia **69**: 61–86 (1990). —Turner in Phytologia Mem. **11**: i–iv, 1–272 (1997). —Smith in F.T.E.A., Compositae **3**: 831–833 (2005). —Hind & Robinson in Fam. Gen. Vasc. Pl. **8**: 514 (2006)[2007].

Batschia Moench in Methodus: 567 (1794), nom. illeg., non *Batschia* J.F. Gmel. (1791) [BORAGINACEAE], nec *Batschia* Mutis ex Thunb. (1792) [MENISPERMACEAE], nec *Batschia* Vahl (1794) [LEGUMINOSAE].

Ageratiopsis Sch.Bip. ex Benth. in Benth. & Hook.f., Gen. Pl. **2**: 246 (1873), nom. nud. pro syn.

Mallinoa J.M.Coult. in Bot. Gaz. **20**: 47 (1895).

Kyrstenia Neck. ex Greene in Leafl. Bot. Observ. **1**: 8 (1903).

Perennial herbs or shrubs, usually erect, rarely scandent, sparingly to densely branched. Stems terete, striate. Leaves opposite, rarely subopposite or alternate, short- to long-petiolate, laminas narrowly elliptical to deltoid, margins entire to toothed or lobed, serrate or crenate in most species, 3-veined to pinnate venation, rarely glandular-punctate in subgenus *Ageratina*, more common in other subgenera. Inflorescences laxly to densely corymbose; pedicels short to moderately long. Capitula homogamous, discoid; phyllaries c. 30, 2- or 3-seriate, distant to weakly subimbricate, mostly subequal, spreading at maturity, 2-ribbed or more usually indistinctly ribbed; receptacle usually slightly convex, glabrous or with minute scattered hairs. Florets 10–60, often sweetly scented; corollas white or lavender, usually with slender basal tube and campanulate or narrowly funnelform limb; lobes triangular, distinctly longer than wide, inner surface densely papillose, outer surface smooth, glabrous or glanduliferous, usually with hairs; anther collar cylindrical, usually elongate; anther appendage large, ovate-oblong, longer than wide; style base usually enlarged, glabrous; style arms linear, rarely slightly broadened distally, densely papillose with projecting cells on lateral and outer surface, papillae usually long, often with glands along adaxial surface. Achenes prismatic or fusiform, usually 5-ribbed, setuliferous or glanduliferous or both; carpopodium distinct, symmetrical, without prominent upper rim, cylindrical, rounded or shortly stopper-shaped; pappus setae uniseriate, 5–40, barbellate, elongate often easily deciduous, capillary, often enlarged distally, apical cells acute, often with shorter outer series of setulae.

A genus of about 265 spp. from the tropics and subtropics of the New World with two spp. becoming somewhat weedy in the Old World, both of these are present in the Flora area, and *Ageratina vernalis* (Vatke & Kuntz) R.M.King & H.Rob. is cultivated, but does not appear to have escaped and naturalized.

Leaves rhomboid and distinctly petiolate. . . . .**1.** *adenophora*
Leaves lanceolate-elliptic and pseudopetiolate or short-petiolate, base long-attenuate . . . .
. . . .**2.** *riparia*

1. **Ageratina adenophora** (Spreng.) R.M.King & H.Rob. in Phytologia **19**(4): 211 (1970). —Agnew & Agnew, Upland Kenya Wild. Fl., ed. 2: 204 (1994). —Beentje, Kenya Trees Shrubs Lianas: 553, fig. (1994). —Smith in F.T.E.A., Compositae **3**: 831–833 (2005). Type: 'Crescit in alta planitie Mexicana inter Carpio et Gasave, alt. 1200 hex. Floret Majo.' (P-Bonpl – P00320084 holotype).[242]

    Eupatorium glandulosum Kunth in Humb., Bonpl. & Kunth, Nov. Gen. Sp. Pl. **4** (ed. folio): 96 (1818), nom. illeg., non *E. glandulosum* Michx. (1802).

    Eupatorium adenophorum Spreng., Syst. Veg., ed. 16, **3**: 420 (1826), nom. nov. pro *E. glandulosum* Kunth. —McVaugh, Fl. Novo-Galiciana **12**: 351–352 (1984). —Blundell, Wild Fl. E. Afr.: 166, fig. 93 (1987).

Shrub to 2 m. Stems erect, somewhat woody, branched, with short glandular hairs, becoming more densely pubescent towards the apex, with an annular scar at nodes. Leaves opposite, petiole to 2.5 cm long, large leaves sometimes with additional small leaves in pseudowhorl, lamina rhomboid, 1–6.5 × 0.5–4.5 cm, margins serrate, apex acuminate, shortly scabrid-hairy, mainly 3-veined from base. Inflorescence much branched, somewhat leafy, paniculate, with numerous capitula; common stalks long, pubescent. Capitula 4–7 mm diam.; phyllaries bi-seriate, lanceolate, 3.5–5 × 0.5–1 mm, 2-ribbed, margins scarious and with longer hairs, apex acute, glandular-hairy. Florets ± 72; corolla 3.5 mm long, white to cream, tube filiform and prominently 5-veined, limb campanulate, lobes triangular, c. 0.2 mm long, thickened, sparsely hairy externally; anthers c. 1 mm long, basally sagittate, apical anther appendages ovate; style 3.5–4.5 mm long, bifid for 0.5–1.5 mm, style-arms exserted for c. 1.5 mm, papillate. Achenes 1 mm long, 5-ribbed, glabrous; carpopodium distinct; pappus setae ± 10, barbellate, deciduous, c. 3 mm long, basally connate, slightly broader towards apex.

Zambia. C: Lusaka SE, Chisuko River bank, 15°39'E, 28°32'S, 890 m, 14.x.1999, *Bingham & Nefdt* 12041 (K). S: Mazabuku Dist., Mubuyu Farm, 16°00'S, 28°07'E, 1000 m, 9.x.1992, *Bingham* 8562 (K, MO).

Also in Uganda, Kenya, South Africa, Pacific Islands, Australia, S America, native of Mexico. Montane forest and forest margins, lower bamboo zone, riverine and swamp forest; 1950–2600 m. *Bingham* 8562 notes it is a 'weed in fish ponds'.

Conservation Status: Although there are very few collections from the Flora area, this is a widespread weed in the Old World tropics and certainly not threatened; LC (Least Concern).

2. **Ageratina riparia** (Regel) R.M.King & H.Rob. in Phytologia **19**(4): 216 (1970). —Hind in Fl. Masc., Composées **109**: 237–239 (1993). Type: 'Aus dem Garten der Herren Haage und Schmidt in Erfurt erhielt der Petersburger Garten das in Rede stehende Eupatorium, das sehr wahrscheinlich aus den Gebirgen Amerika's stammt.' Herbarium material, if extant, probably in LE. FIGURE 6.5.**40**.

    Eupatorium riparium Regel, Gartenflora **15**: 324, t. 525 (1866).

---

[242] There is a pencilled number, '38', on the label of the holotype specimen.

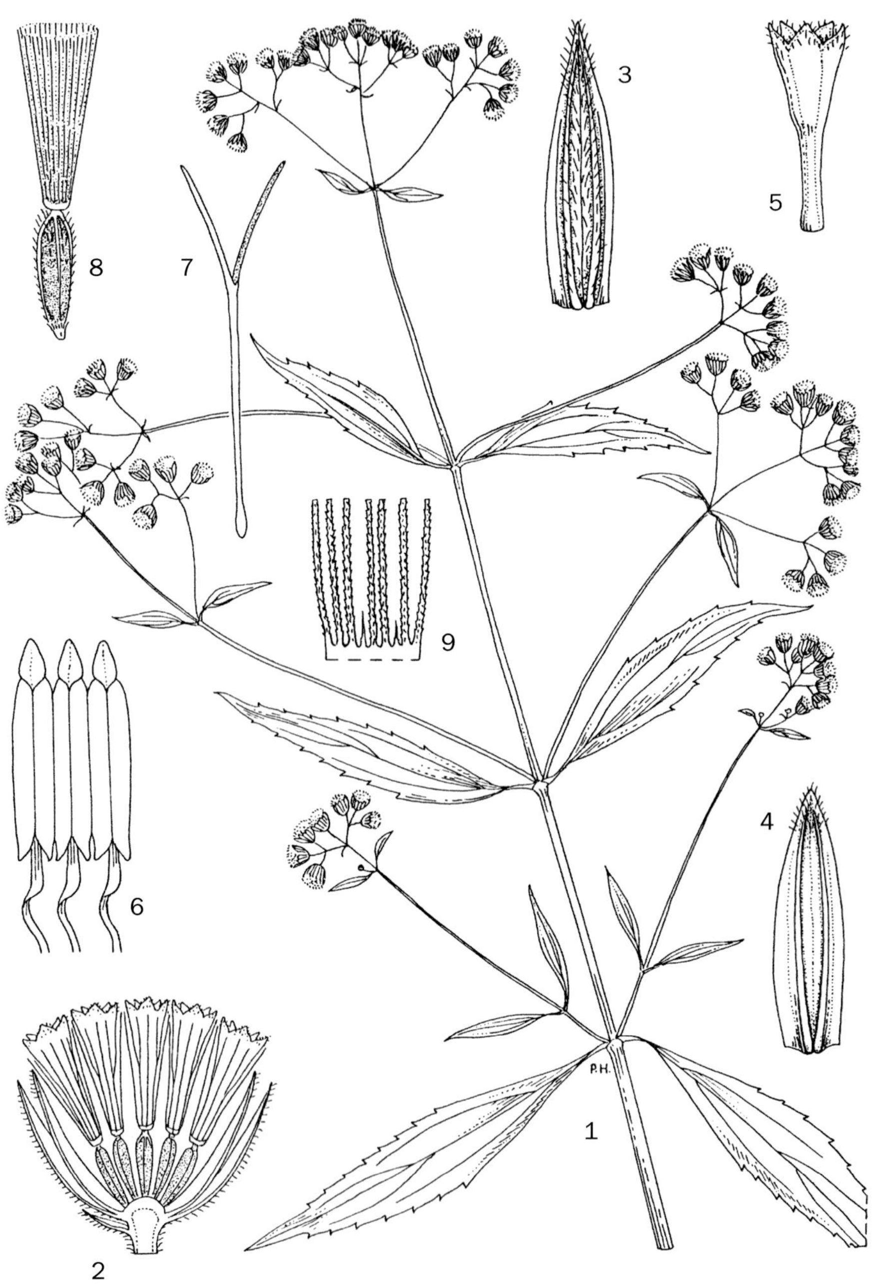

Fig. 6.5.**40**. AGERATINA RIPARIA. 1, flowering shoot (× 2/3); 2, l.s. capitulum (× 6); 3, outer phyllary (× 9); 4, inner phyllary (× 9); 5, corolla (× 9); 6, partial anther cylinder opened out (× 36); 7, style (× 12); 8, achene (× 9); 9, detail of base of pappus (greatly magnified). All from *Cadet* 6276. Drawn by Pat Halliday. From Flore des Mascareignes.

Eupatorium harrisii Urb., Symb. Antill. **1**: 460 (1899). Type: 'Hab. in Jamaica prope Cinchona, [5000 ft, 28.3.1994] m. Mart. flor.: *Harris* in Bot. Dep. Herb. n. 5624.' (UCWI holotype, F145358, F0050126, GH00007454, S-R-2127).

Eupatorium ventillanum Cuatrec. in Ann. Missouri Bot. Gard. **52**(3): 305 (1965). Type: 'Peru, Dept Amazonas, Prov. Chachapoyas: scrub forest along Rio Ventilla, 1–2 km W of Molinopampa, 2350–2400 m alt, locally abundant in colonies on river banks flowers white, 23–22 July 1962, *J. J. Wurdack* 1476.' (US00145773 holotype, F0043801, GH00008049, LIMA, LP002044, NY00169261, TEX00373395, UC1308379).

Ageratina ventillana (Cuatrec.) R.M.King & H.Rob. in Phytologia **19**(4): 217 (1970).

Subshrub or rampant shrub, to c. 1 m tall/long. Stems terete, striate, pubescent, becoming glabrous towards base. Leaves opposite, petiolate to pseudopetiolate, petiole 5–13 mm long, lamina lanceolate-elliptic, 3.5–7(10) × 1–1.5 cm, apices acute, base attenuate, 3-veined from base, subglabrous above, pubescent beneath, margins regularly serrate. Inflorescences laxly corymbose, lateral branches subtended by 2 small foliaceous bracts and numerous capitula grouped in dense corymbs, pedicellate, pedicels ± hirsute, bracteolate, bracteoles 1 or 2, rarely more. Capitula c. 2–3 mm diam.; involucres campanulate, c. 2 mm diam.; phyllaries 12, biseriate, linear to oblong, 4–5 × 0.5 mm, apices obtuse, outer pubescent, inner laciniate towards apices, abaxial surface with 2 or 3 prominent ribs, margins scarious. Florets c. 20, corollas 3–3.5 mm long, white. Achenes prismatic, 5-angled, black, 1.5–2 mm long, body and angles setuliferous; pappus setae biseriate, outer series short, inner c. 15, 2.5–3 mm long, barbellate, white, fragile or caducous.

Zimbabwe. C: Chishawasha Seminary and Mission, near Salisbury, c. 1432 m, 7.ix.1957, *Phipps* 733 (K, SRGH).

A native of Mexico and the Antilles but now a widespread weed in the tropics and invasive in Australia. Apparently rare in the Flora area, absent in the neighbouring F.T.E.A. and Central Africa areas and present only in Natal. Preferring the margins of, to bodies of, water.

Conservation Status: Even though apparently rare in the Flora area, because the species is a widespread weed, it is recorded as LC (Least Concern).

148. **MIKANIA** Willd.

Mikania Willd., Sp. Pl., ed. 4, **3**: 1742 (1803), nom. cons. —Robinson in Contr. Gray Herb. **104**: 55 (1934). —Holmes in Bot. Jahrb. Syst. **103**(2): 211–246 (1982). —Lisowski, (Asterac. Fl. Afr. Cent. 2) Fragm. Flor. Geobot. **36** Suppl. 1: 457–467 (1991). —Bremer, Anderberg, Karis & Lundberg in Bremer, Asterac. Cladist. Classif.: 647 (1994). —Mesfin Tadesse, Fl. Ethiopia & Eritrea **4**(2): 343–344 (2004). —Smith in F.T.E.A., Compositae **3**: 833–837 (2005). —Hind & Robinson in Kubitzki, Fam. Gen.Vasc. Pl. **8**: 516–517 [2006](2007).

Carelia Juss. ex Cav. in Anales Ci. Nat. **6**: 317 (1802)[1803], non *Carelia* Ponted. ex Fabr. (1759) (= *Ageratum* L.), nec Less. (1832) (= *Radlkoferotoma* Kuntze).

Corynanthelium Kunze in Linnaea **20**(1): 19 (1847).

Morrenia Kunze in Linnaea **20**(1): 19 (1847), nom. nud. pro syn.

Kanimia Gardner in London J. Bot. **6**: 446–447 (1847); —Baker in Mart., Fl. Bras. **6**(2): 368–371 (1876).

Willoughbya Neck. ex Kuntze, Revis. Gen. Pl. **1**: 371 (1891).

Usually climbers or vines, or spreading weak shrubs, sometimes erect perennial herbs, subshrubs or shrubs. Stems terete to hexagonal, striate or rarely winged, glabrous or variously pubescent. Leaves opposite or in verticels of 3–4, usually persistent but sometimes easily detached, sessile to long-petiolate, lamina narrowly linear to broadly ovate, climbers with sometimes distinctly dimorphic with two different forms on vegetative and flowering growth, sometimes dissected into broad or narrow segments, base narrowly cuneate to cordate or hastate, margins entire to toothed or lobed, apex rounded to short-acuminate, 3-veined from base to pinnate, lower surface with or without glandular punctate. Inflorescence paniculate with diffuse, thyrsoid, racemose, spicate, corymbose, or subcymose branches. Capitula sessile or with short pedicels, usually immediately subtended by subinvolucral bract, or with subinvolucral bract on pedicel, usually monoecious, rarely dioecious; phyllaries 4, distant, subequal, uniseriate, persistent, often with swollen bases. receptacle flat, glabrous. Florets 4, usually white, sometimes pink; corollas funnelform or with distinct basal tube and campanulate limb, glabrous to pubescent or glandular-punctate on outer surface, inner surface of throat smooth or rarely short-papillose, corolla-lobes mostly triangular and long as wide, sometimes narrowly oblong to 3 times as long as wide, papillate or epapillate on inner surface, often with fringe of hair-like cells along inside of margin; apical anther-appendages large, ovate to oblong, 1–2 × as long as wide; style-base often stout, not or gradually narrowed upwardly on shaft, usually smooth; style branches narrowly linear, without enlarged apices, densely papillose. Achenes 4–10-ribbed, prismatic, glabrous or setuliferous or glandular-punctate; carpopodium short-cylindrical; pappus setae 1–2-seriate, 35–60, capillary, barbellate, persistent, often somewhat broadened distally, very rarely pappus absent.

A genus of about 430 spp., pantropical, although principally neotropical with a few apparent natives in the Old World tropics; three spp. in the Flora area.

Gibbs Russell (in Kirkia **10**(2): 499, 1977) referred to '*M. cordata* (Burm.f.) B.L.Rob.' from 'Zimbabwe: N, W, C, E, S'. This certainly referred to one of two spp. in the account below, although his description is to 'generic' to be able to tell which. The species in question was one of his 'soak zone' plants, most probably *M. chenopodifolia*.

Key modified from Holmes (1982)

1. Inflorescences lax cymose panicles; pedicels 5–15 mm long; corolla tube 2× or more length of throat . . . . **1.** *chevalieri*
– Inflorescences open to dense corymbs or corymbose panicles; pedicels <5 mm; corolla tubes about same length as throat 2
2. Leaves lanceolate, base distinctly sagittate . . . .**2.** *sagittifera*
– Leaves ovate, base cordate to auriculate . . . .**3.** *chenopodifolia*

1. **Mikania chevalieri** (C.D.Adams) W.C.Holmes & McDaniel in Phytologia **31**(3): 274 (1975). Type: 'SOUTHERN NIGERIA. –[In swampy scrub by side of] Jamieson River, Sapoba, Benin, [12-] November 1949, *Keay and Meikle* 524' (K – which consists of 2 sheets 000374816, 000374817 – and spirit material – holotype). FIGURE 6.5.**41**.

   Mikania cordata (Burm.f.) B.L.Rob. var. *chevalieri* C.D.Adams in J. W. African Sci. Assoc. **8**: 136 (1964).

   Mikania laxa A.Chev. in Expl. Bot. Afrique Occ. Franc. **1**: 360 (1920), nom. nud. (based on *Chevalier* 22651), non DC. (1836).

Stems terete to angular, glabrous, rarely pilose, green to purple; internodes 10–15 cm long. Leaves petiolate, petiole 10–25 mm long; lamina triangular to narrow ovate, 3–10 × 3–7 cm, base cordate, 3–5-veined from base, upper surface usually glabrous, lower surface glabrous to puberulent, sometimes glandular-punctate, often purplish, margins entire to crenate or undulate, often denticulate, apices acuminate. Inflorescences lax cymose panicles, branchlets angular, glabrous to appressed-pilose, pedicels angular, 5–15 mm, glabrous. Capitula 8–10 mm long; subinvolucral bract linear to lanceolate, 7–8 mm long (see note below), pilose, purplish; phyllaries linear to narrowly lanceolate, 7–8 mm long, glabrous to appressed-pilose, green to purplish, apices acuminate. Corollas 4.5–5.5 mm, white to greenish, tube 3–4 mm long, slender, throat campanulate, 1–1.3 mm long, lobes triangular-ovate, short. Achenes 3–3.5 mm long, dark brown, glandular; pappus setae c. 60, c. 6 mm long, white to carneous.

Zambia. W: Mwinilunga, 18.v.1969, *Mutinmushi* JMM3202 (K, NDO). Mwinilunga Dist., beside Kampemba River, 16.v.1986, *Philcox et al.* 10336 (K etc.).

Also known from D.R. Congo, Sierra Leone, Liberia, Nigeria, Cameroon and Angola. Florets are slightly scented according to *Keay & Meikle* 524. Equatorial forests, gallery forest, scrub margins, and swampy ground on river margins.

Conservation Status: A widespread species. not threatened, although poorly collected within the Flora area LC (Least Conern). The *Philcox* et al. 10336 collection has slightly denser inflorescences than much of the material from the western part of this species distribution, and has conspicuous spreading subinvolucral bracts, but otherwise agrees.

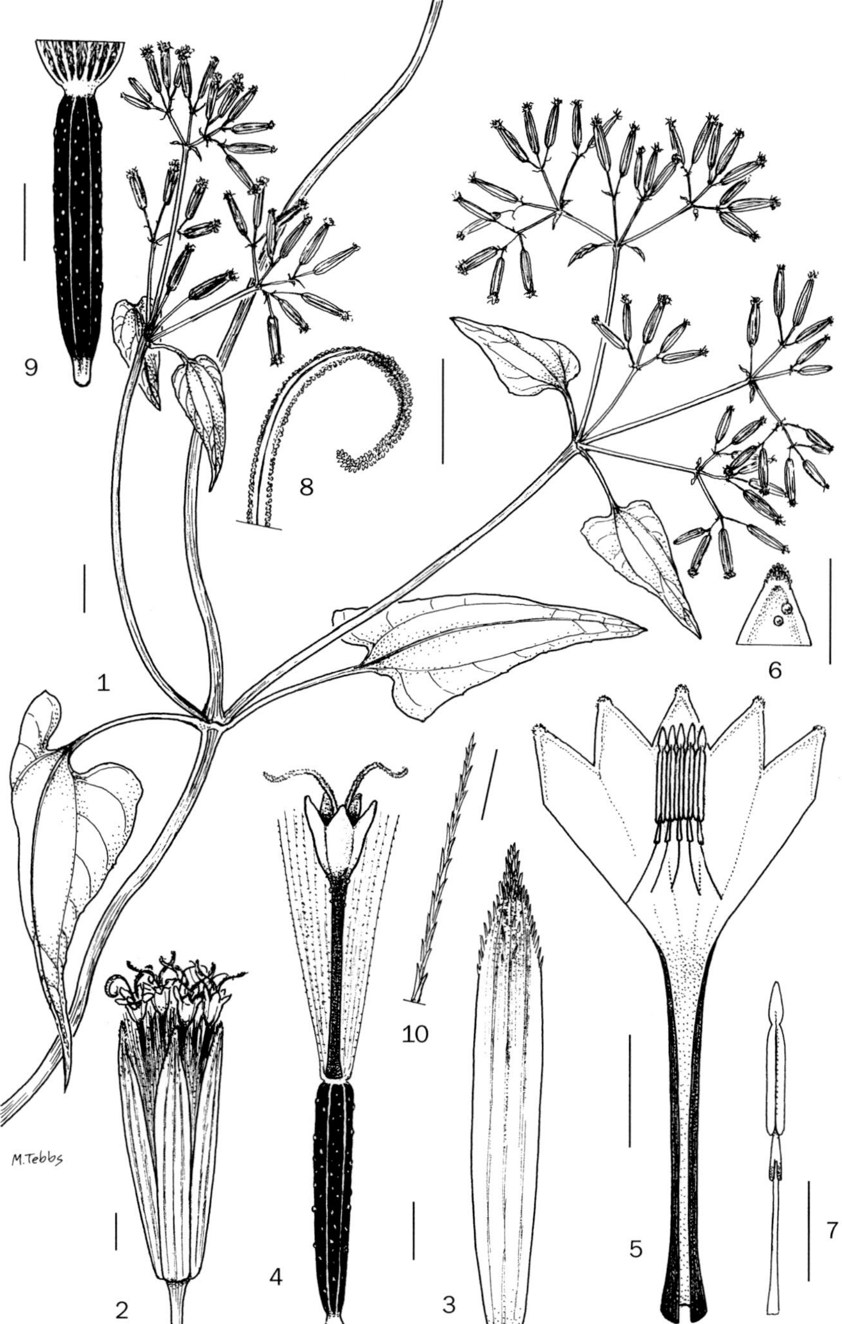

Fig. 6.5.**41**. MIKANIA CHEVALIERI. 1, flowering shoot; 2, capitulum; 3, phyllary; 4, floret; 5, corolla opened out showing attachment of stamen filaments; 6, detail of abaxial view of corolla lobe; 7, stamen; 8, detail of style arm; 9, mature achene (showing base of pappus only); 10, detail of apex of pappus seta. 1–8 from *Milne-Redhead* 4098; 9–10 from *Mutimushi* 3202. Scale bars: 6–8, 10 = 0.5 mm; 2–5, 9 = 1 mm; 1 = 10 mm. Drawn by Margaret Tebbs.

Adams specified phyllaries 7–8 mm long (which is what the holotype material shows) but Holmes made no comment on this (i.e. being much larger) and this is significant because the measurement given by Holmes, of 3–4 mm, would mean that they would be much shorter than the corollas.

2. **Mikania sagittifera** B.L.Rob. in Contr. Gray Herb. **104**: 68 (1934). —Holmes in Bot. Jahrb. Syst. **103**(2): 242, fig. 15 (1982). —Maquet in Fl. Rwanda **3**: 573, fig. 174/4 (1985). —Lisowski, (Asterac. Fl. Afr. Cent. 2) Frag. Flor. Geobot. **36** Suppl. 1: 466 (1991). — Mesfin Tadesse, Fl. Ethiopia & Eritrea **4**(2): 344 (2004). Type: 'Kabulabula, on the Chobe River, Bechuanaland, July 1930, *van Son*, no. 28,729 ... collected by the Vernay-Lang Kalahari Exped.' (GH00010517 holotype, BM, K000273403, PRE – from TRV).

 Mikania scandens sensu Oliv. & Hiern in F.T.A. **3**: 301 (1877) p.p., non Willd.

 Mikania scandens f. *angustifolia* O.Hoffm. in Warburg, Kunene-Sambesi-Exped.: 405 (1903). Type: 'Am Longa oberhalb des Lazingua im Wasser am Uferrand, 1250 m ü. M. ([*Baum*] Nr. 679, blühend am 26. Januar 1900.)' (B† holotype, BR8877451, K00273404).

 Mikania angustifolia (O.Hoffm.) R.E.Fr. in Wiss. Ergebn. Schwed. Rhodesia-Kongo-Exped. 1911-1912 **1**: 328 (1916), nom. illeg. non Kunth (1818).

 Mikania cordata sensu Lind & Tallantire, Comm. Fl. Pl. Uganda, ed. 2: 182 fig. p.p. quoad ic. 121 (1975). —Agnew & Agnew, Upland Kenya Wild Fl., ed. 2: 204 (1994) p.p., non (Burm.f.) B.L.Rob.

Shrub to 2 m. Stem little branched, straw-coloured to reddish brown, terete to sub-angled, glabrous to pubescent, more densely so at nodes, with simple, short, multicellular hairs. Leaves opposite, petiole 0.6–3.3 cm, lamina distinctly sagittate, 1.5–6.3(9.5) × 0.4–3.2 cm, margins minutely toothed to serrate, apex long-acuminate, upper surface glabrous to very sparsely pubescent, lower surface glandular-punctate, glabrous to pubescent usually with 3 distinct and 2 less distinct veins originating from the base. Inflorescence a dense, leafy corymb on an axillary peduncle 3.5–9.3 cm long, branched, clusters of capitula in panicles. capitula 6–8 × 1.5–2 mm; pedicels <6 mm long, becoming pendulous after fruit dispersal. phyllaries equal, light green, sometimes with lilac tinge, chaffy, lanceolate, boat-shaped, 3-veined, 4–7 × 1–1.5 mm, apical margins fimbriate, glabrous to sparsely hairy at base. Corolla 3.5–4.5 mm long, dull white, with scattered glands, lobes acute, <1 mm long, margins thickened; anthers purplish, with white, lanceolate appendages, protruding beyond corolla lobes; style-arms exserted c. 2–3 mm. Achenes brown-black, 1–2 mm long, glandular; carpopodium evident, cream to straw-coloured; pappus 3–5 mm long, barbellate, connate at base, ± spreading.

Botswana. N: Khwai River 1 hour boat trip upstream from Gobegha Lagoon, Okavango Swamp, 23°02'E, 19°04'S, 1000 m, 7.iii.1972, *Gibbs Russell & Biegel* 1541 (K, SRGH). **Zambia**. B: Nangweshi, 26.vii.1962, *Fanshawe* F6963 (K, NDO). N: Bangweula Swamps, Ncheta Island, 3.ix.1969, *Verboom* 2643 (K). C: Mumbwa, 2.vi.1967, *Fanshawe* 6641 (K, NDO). S: Livingstone Dist., Katambora, 884 m, 8.vii.1955, *Gilges* 399 (K, SRGH). **Zimbabwe**. W: Diana's Pool, Matopos, 20.iv.1941, *Hopkins* 8012 (K, SRGH). **Malawi**. C: Namitete R. above bridge on Lilongwe-Fort Jameson Rd, 1150 m, 5.ii.1959, *Robson* 1477 (BM, K).

Also known from Cameroon, D.R. Congo, Rwanda, Burundi, Uganda, Kenya, Tanzania, Angola, Namibia, South Africa. Swampy forest and papyrus or sedge swamps; 900–1850 m.

Conservation Status: A widespread species. LC (Least Concern).

3. **Mikania chenopodiifolia** Willd., Sp. Pl., ed. 4, **3**: 1745 (1803). —Holmes in Bot. Jahrb. Syst. **103**(2): 221, fig. 5 & 6 (map) (1982). —Maquet in Fl. Rwanda **3**: 573 (1985). —Lisowski, (Aster. Afr. Cent. 2) Fragm. Flor. Geobot. **36** Suppl 1: 460 (1991). Type: 'Habitat in Sierra Leona. (v. s.)' [*Thunberg* s.n.] (B-W holotype).

 Mikania capensis DC., Prodr. **5**: 198 (1836). —Harvey in Harvey & Sonder, Fl. Cap. **3**: 59 (1865). —Holmes in Bot. Jahrb. Syst. **130**(2): 214, fig. 1 & 2 (map) (1982). —Maquet in Fl. Rwanda **3**: 573, fig. 174/3 (1985). —Lisowski, (Asterac. Fl. Afr. Cent. 2) Fragm. Flor. Geobot. **36** Suppl. 1:

458 (1991). —Mesfin Tadesse, Fl. Ethiopia & Eritrea 4(2): 344 (2004).[243] Types: 'ad Cap. Bonae Spei (*Drege*!) in sylvis Coloniae distr. orient. *Burch*[*ell*]! cat. geog. n. 3674. in distr. Uitenhage (*Eckl.*!). *Krebs*! pl. exs. n. 199. *Eupatorium scandens* Thunb. Fl. Cap. 627. non Linn. ... (v.s.)' [In the woody ravine by the Spring at Blaauwekrans. Bathhurst Div.] 19.ix.1813, *Burchell* 3674 (G-DC-G00495168 lectotype, GH000273406, K000273406), lectotypified by Holmes (1982: 214).[244]

Mikania oxyota DC., Prodr. **5**: 198 (1836). Type: 'in Africâ australi ad Van Staades Rivier legit cl. *Drege*! ... (v.s.)' (G-DC-G00495615 – '[Drège] Zw. Van Staades - Gamtosrivier. R. I', holotype, probable K000273409).

Mikania dioscoreifolia DC., Prodr. **5**: 198 (1836). Type: 'in ins. Madagascar legit cl. *Goudot*, locis Ambanivoules dictis. ... (v. s. comm. ab. amic. B. Delessert.)'(G-DC-G00495164 – s.n., holotype).

Mikania dioscoreifolia var. β? *bojeri* DC., Prodr. **5**: 198 (1836). Type: [Madagascar:] 'in sylvâ Bé-Fouroun super arbores scandens. ... Trixis? dioscoreaefolia Bojer! in litt. 1835. (v.s. comm. à cl. *Bojer.*)' (G-DC-G00495167 – '2', holotype).

Mikania dioscoreifolia var. γ? *crenata* DC., Prodr. **5**: 198 (1836). Type: [Madagascar:] 'in sylvis cum var. β, sed eâ multò minor. Trixis? crenata Bojer! in litt. 1835. ... (v.s. comm. à cl. *Bojer.*)' (G-DC-G00495166 – '3', holotype).

Mikania thunbergioides Bojer ex DC., Prodr. **5**: 198 (1836). Type: 'in ins. Madagascar prov. Bé-Tani-Mena in sylvae Bé-Fouroun apricis super dumeta scandens. ... (v.s.)' (G-DC-G00495165 – '1', holotype).

Mikania floribunda Bojer ex DC., Prodr. **5**: 198 (1836). Type: 'in sylvis umbrosis prov. Bé-Zonzong ins. Madagascar legit cl. *Bojer*. (v.s.)' (G-DC-G00495163 – '5', holotype).

Mikania scandens sensu Oliv. & Hiern in F.T.A. **3**: 301 (1877). —Brenan in T.T.C.L.: 156 (1949). —Humbert, Fl. Madagasc., Composées **189**: 202, fig. 39/5–9 (1960), non Willd.

Mikania cordata sensu Andrews, Fl. Pl. Sudan **3**: 43 (1956). —Adams in F.W.T.A., ed. 2, **2**: 286 (1963). —Cufodontis, (Enum. Pl. Aethiop.) Bull. Jard. Bot. Natl. Belg. **36** (3, suppl.): 1080 (1966). —Agnew & Agnew, Upland Kenya Wild Fl., ed. 2: 204 (1994) p.p., non (Burm.f.) B.L.Rob.

Shrub to 5–9 m. Stem terete to ± angled, poorly-branched, sometimes ± winged, brown to reddish brown, glabrous to pubescent, nodes sometimes hairy, internodes 10–15 cm. Leaves petiolate, petiole 1.1–7.5 cm, sparsely hairy to pubescent, lamina ovate, cordate, 2.2–10(13) × 1.1–7.0 cm, glabrous or sparsely hairy above, glabrous or sparsely hairy to sometimes pubescent and glandular-punctate beneath, usually 3-veined from base with 2–3 weaker secondary veins, base auriculate, sub-sagittate to hastate, margins dentate to laxly serrate, ± revolute, apex acuminate to long-acuminate. Inflorescence densely corymbose, usually on leafy axillary branches, 2.5–16 × 3–10 cm, branched, clusters of capitula arranged in panicles, branches glabrous to pilose or villous, capitula pedicellate, pedicels 1–5 mm, angular, glabrous to pilous. subinvolucral bract lanceolate, 3–6 mm long, pilose, apices acute to acuminate. Capitula 6–8.5 × 2–3 mm. phyllaries green, sometimes with purplish tinge, equal, narrowly lanceolate, boat-shaped, 4.5–8 mm long, base sometimes appearing swollen, apical margins shortly fimbriate, glabrous to pubescent, sometimes glandular. Corollas 3.5–5 mm long, white, sparsely glandular, especially at lobes, tube c. 2.5 mm long, throat subcylindrical to semi-campanulate, 1.3–1.7 mm long, lobes triangular, recurved in drying; apical anther appendages lanceolate, 1–2½ times longer than wide, protruding beyond corolla lobes; style exserted c. 1–2.6 mm. Achenes 3.5–4 mm long, fawn to brown or black, glandular; carpopodium small, cream to straw-coloured; pappus (3)5–6 mm, minutely barbellate, connate at base, white to carneous.

Zambia. N: Isoka Dist., Mushitu, near Lupita-Kikutu area, 20.viii.1965, *Lawton* 1270 (K). W: Solwezi Dist., R. Mwafwe, 18.vii.1930, *Milne-Redhead* 728 (K). C: Chakwenga headwaters, 100-129 km East of Lusaka, 8.ix.1963, Robinson 5653 (K). **Zimbabwe**. W: Matobo Dist., Yam Chesterfield, (4800 ft), iv.1959, *Miller* 5868 (BR). C: Near Salisbury, 13.xii.1936, *Eyles* 8832 (K). E: Stapleford Forest Reserve, 12.vi.1934, *Gilliland* 288 (BM, K). **Malawi**. N: Rumphi Dist., Nyika-Kafwimba Forest, (6250 ft.), 28.ix.1969, *Pawek* 2774 (K). C: Kota-Kota Dist., Nchisi Mountains, 2.viii.1946, *Brass* 17105 (BR, K, NY). S: Mlanje Dist., Mlanje Mts, 18.vii.1946, *Brass* 16875 (K, MO, NY, UC). **Mozambique**. N: Niassa

---

[243] Mesfin Tadesse's reference to this species in Ethiopia is somewhat dubious.

[244] A set of six duplicates was apparently collected. This is the site of the Weenen Massacre of Voortrekkers (some 230 plus 250 of their porters) by the Zulu on 17th February 1838 in what is now KwaZulu-Natal, South Africa! Blaauwekrans (= Bloukrans) refers to the bluish cliff faces present in the area.

Prov., Rio Messalo, 23 km along the road from Marrupa to Nungo, 500 m, 7.viii.1981, *Jansen et al.* 110 (K). Z: Lusa River, 6 m. W of Gurué, 3.vii.1942, *Hornby* 4557 (K, PRE). MS: Sofala Province, Cheringoma Dist., Cheringoma Plateau, 18°32'32"S, 34°55'25", 185 m, 22.vii.2006, *Goyder* et al. 4085 (K). GI: Inhambane Dist., Rivane, nr. Inhambane town, v.1940, *Gomez e Souza* 2263 (K, PRE). M: Delagoa Bay, 1893, *Junod* 441 (BR, K).

Also in Senegal, Gambia, Guinea Bissau, Sierra Leone, Liberia, Ivory Coast, Ghana, Togo, Nigeria, Cameroon, Equatorial Guinea, Bioko, Principe, São Tomé, Annobón, Gabon, Central African Republic, D.R. Congo, Rwanda, Burundi, Sudan, Ethiopia, Uganda, Kenya, Tanzania, Angola, South Africa, Madagascar, Comoro Is. Very widespread, in forest, swamp forest, on river-banks, and bamboo edges; ± 0–3000 m.

Conservation Status: Widespread species; LC (Least Concern).

Smith (2005) indicated that *M. capensis* effectively only differed from *M. chenopodiifolia* by differences of degree and not of kind. If the nature of the inflorescence, the relative length of the subinvolucral bract, and the capitulum length, as emphasized by Holmes (1982) had proven constant then I would have tended to disagree with Smith. However, the distinctions given by Holmes (1982: 217) do not agree with several of his determinations in K, and several cited specimens of Holmes' two species concepts are from the same locality making any sensible separation into *M. capensis* and *M. chenopodiifolia* impossible. In addition many specimens have combinations of characters of Holmes' two species concepts, so in this treatment I am accepting Smith's view and have provided a fuller synonymy.

149. **ADENOSTEMMA** J.R.Forst. & G.Forst.

Adenostemma J.R.Forst. & J.G.Forst., Char. Gen. Pl., ed. 2: 89 (1776). —Grierson in
Ceylon J. Sci. (Biol. Sci.) **10**: 42–80 (1972). —King & Robinson in Phytologia **29**(1): 1–20
(1974). —Dyer, Gen. S. Afr. Fl. Pl. **1**: 663 (1975). —Panigrahi in Kew Bull. **30**(4): 647–655
(1975). —Gibbs Russell in Kirkia **10**(2): 498–499 (1977). —Lisowski, (Asterac. Fl. Afr. Cent.
2) Fragm. Flor. Geobot. **36** Suppl. 1: 467–472 (1991). —Hind in Fl. Masc., Composées **109**:
229–232 (1993). —Koyama in Mem. Natl. Sci. Mus. (Tokyo) **37**: 159–168 (2001); in Bull.
Natl. Sci. Mus., B (Tokyo) **28**: 49–60 (2002). —Mesfin Tadesse, Fl. Ethiopia & Eritrea **4**(2):
344–345 (2004). —Smith in F.T.E.A., Compositae **3**: 819–824 (2005). —Hind & Robinson in
Fam. Gen. Vasc. Pl. **8**: 518–519 [2006](2007). —Orchard in Telopea **13**(1–2): 341–348 (2011).
Lavenia Sw., Prodr.: 112 (1788).

Annual or short-lived perennial herbs. Leaves opposite, subsessile to petiolate or pseudopetiolate, petiolate portion usually winged (at least in upper part), lamina narrowly elliptical to broadly ovate or hastate, crenate to strongly serrate, acute to slightly acuminate. Inflorescence of solitary capitula or very laxly cymose. Capitula homogamous and discoid; phyllaries 10–30, biseriate, ± overlapping, somewhat fused at base, equal to subequal; receptacle covered with discrete oval deeply concave scars, epaleaceous. Florets all hermaphrodite, 10–60; corolla narrowly funnelform or with narrow basal tube and broadly campanulate limb, usually white, usually with hairs or glands on outer surface, hairs often moniliform; lobes c. 1.5 times longer than wide, non-papillose; anther collar usually strongly expanded below; apical anther appendages distinctly shorter than wide; style shaft with or without long hairs; style arms slightly to strongly clavate, often forming most showy part of head, fleshy, rounded apically, scarcely mamillose below. Achenes slightly curved, usually 3-angled without distinct ribs or 5-angled, body often tuberculate, otherwise smooth, glabrous; carpopodium forming a prominent asymmetrical knob; pappus of usually 3 or 5 terete clavate knobs, knobs with tips and upper outside surface covered with an elongated mass of viscid glands.

A genus of about 26 spp., pantropical. A notoriously difficult genus within which to determine species, as shown by the varying Flora treatments; it is clear that many species are very variable. A discussion of some of the problems is provided by Smith (2005). Orchard (2011) interpreted the taxonomy somewhat differently and included one of our taxa (*A. viscosum*) within the synonymy of *A. lavenia* (L.) Kuntze; this is not accepted here.

1. Mature achenes without tubercles .... **3.** *mauritianum*
– Mature achenes tuberculate 2
2 Leaves petiolate to 3.5 cm, lamina ovate to rhomboid; capitula usually 5 mm diam. or less, few to numerous; corolla < 1.5 mm long with 3–4(5) lobes .... **1.** *viscosum*
– Leaves sessile to short-petiolate (to 1.6 cm), lamina lanceolate to ovate-lanceolate; capitula generally > 5 mm diam., few. corolla 2 mm or longer with 5–6 lobes .... **2.** *afrum*

1. **Adenostemma viscosum** J.R.Forst. & G.Forst., Char. Gen. Pl., ed. 2: 90, t. 45 (1776). — de Candolle, Prodr. **5**: 112 (1836). —Oliver & Hiern in F.T.A. **3**: 299 (1877). —Hoffmann in Pflanzenw. Ost-Afrikas C.: 406 (1895). —Hind in Fl. Masc., Composées **109**: 231 (1993). —Retief & Herman, Pl. N. Prov. S. Africa: 290 (1997). —Smith in F.T.E.A., Compositae **3**: 820–821 (2005). Type: 'Society Is., Tahiti, *Forster* s.n.' (K00009772 lectotype), lectotypified by Panigrahi (1975: 647).[245]

Adenostemma perrottetii DC., Prodr. **5**: 110 (1836). —Brenan in Mem. New York Bot. Gard. **8**(5): 463 (1954). —Andrews, Fl. Pl. Sudan **3**: 8 (1956). —Adams in F.W.T.A., ed. 2, **2**: 286 (1963). — Cufodontis, (Enum. Pl. Aethiop.) Bull. Jard. Bot. Natl. Belg. **36** (3, suppl.): 1079 (1966). —Burkill, Useful Pl. W. Trop. Afr., ed. 2, **1**: 442 (1985). —Maquet in Fl. Rwanda **3**: 568, fig. 173/1 (1985). —Lisowski, (Asterac. Fl. Afr. Cent. 2) Fragm. Flor. Geobot. **36** Suppl. 1: 469 (1991). —Agnew & Agnew, Upland Kenya Wild Fl., ed. 2: 203 (1994). —Mesfin Tadesse, Fl. Ethiopia & Eritrea **4**(2): 345–346 (2004). Type: 'in paludibus turfosis peninsulae Promontorii Viridis Africanae legit cl. *Perrottet* [27]. ... (v.s. comm. à cl. inv.).' (G-DCG00465535 holotype).

Erect to semi-procumbent annual, or short-lived perennial, herb to 1.3 m. Stem fleshy, glabrous or sparsely pubescent, sometimes rooting from lower nodes. Leaves subsessile to petiolate, petiole 0.7–3.5 cm, lamina ovate to rhomboid, 2–15(22) × 1–7(12.5) cm, base attenuate to cuneate, sometimes forming narrow wings along petiole, margins crenulate to serrate, apex obtuse to acute, glabrous or sub-glabrous, 3-veined from base. Inflorescence bracts 0.6–5 cm long at nodes of inflorescence branches. Capitula few to numerous, ± 3–5(7) × 4–5 mm when mature; pedicels 1.2–2.5 cm, glandular-pubescent; phyllaries 2–3 × 0.5–0.75 mm, flat, apex obtuse to acute, margins entire or ciliate, basally pubescent. Florets (20)30–42; corollas white, 1–1.5 mm long, basally wider, then narrowing into tube before expanding again into campanulate limb with 3–4(5) (variable within capitulum) short, broadly triangular lobes; anthers free, 3–4(5), ± 0.2 mm long. style ± 2 mm, bifid for 1/3 of length, arms dilated. Achenes 2.2–3(5) mm long, immature or aborted achenes purplish, generally smooth with small glands, becoming brown, tuberculate and glandular with maturity; pappus of 3 knobs, 0.75–1 mm long.

Zambia. N: Abercorn Dist., Abercorn, Lucheche Mushitu, 1676 m, 22.v.1952, *Siame* 218 (K, LCS). W: Mpongwe, 30.iii.1964, *Fanshawe* F8406 (K, NDO). C: Serenje, iii.1969, *Williamson* 1576 (K, SRGH). S: Livingstone, 30.iv.1972, *Fanshawe* F11432 (K, NDO). **Zimbabwe**. N: Sebungwe, Zambesi Valley, Vulanduli Camp, 610 m, ix.1955, *Davies* 1533 (K, SRGH). W: Wankie, Victoria Falls, 880 m, 13.iv.1979, *Ncube* 53 (K, SRGH). E: Inyanga, Honde Valley, 792 m, 17.iv.1958, *Phipps* 1106 (K, SRGH). S: Belingwe, Mount Buhwa, c. 1200 m, 2.v.1973, *Biegel et al.* 4263 (K, SRGH). **Malawi**. N: Nkhata Bay Dist., 10 mi. S of N. B. Junction, Nkwazi forest, c. 610 m, 24.vi.1977, *Pawek* 12795 (K, MO). C: Ncheu Dist., Kapeni stream near Mtonda Mission, 12.vi.1967, *Salubeni* 749 (K, SRGH). S: Cholo Mt., 1200 m, 21.ix.1946, *Brass* 17724 (K, NY).

Also in Senegal, Guinea Bissau, Sierra Leone, Liberia, Ivory Coast, Ghana, Nigeria, Cameroon, Bioko, São Tomé, Annobón, Gabon, Central African Republic, D.R. Congo, Burundi, Sudan, Ethiopia, Uganda, Kenya, Tanzania, Angola, Mascarenes, South Africa. India, Sri Lanka, Australia, Pacific Islands, Indonesia. Wet ground, often near streams, and in shade, sometimes actually growing in water; 1–2000 m.

Conservation Status: A very widespread species; LC (Least Concern).

---

[245] One other sheet at K (K00009771), and 2 others at BM represent original material according to Panigrahi, although he did not state that they were isolectotypes.

In common with the other spp. present in the Flora area, it is likely that in good conditions (of continuous moisture/water) that this species may be a (short-lived) perennial, something frequently noted on collecting labels.

2. **Adenostemma afrum** DC., Prodr. **5**: 112 (1836) as '*caffrum*'[246]. —Harvey in Harvey & Sonder, Fl. Cap. **3**: 58 (1865). —Andrews, Fl. Pl. Sudan **3**: 8 (1956). —Adams in F.W.T.A., ed. 2, **2**: 286 (1963). —Cufodontis, (Enum. Pl. Aethiop.) Bull. Jard. Bot. Natl. Belg. **36** (3, suppl.): 1078 (1966). —Burkill, Useful Pl. W. Trop. Afr., ed. 2, **1**: 442 (1985). —Maquet in Fl. Rwanda **3**: 568, fig. 173/2 (1985). —Lisowski, (Asterac. Fl. Afr. Cent. 2) Fragm. Flor. Geobot. **36** Suppl. 1: 471 (1991). —Agnew & Agnew, Upland Kenya Wild. Fl., ed. 2: 203, t. 82 (1994). —Mesfin Tadesse, Fl. Ethiopia & Eritrea **4**(2): 346 (2004). Type: [South Africa: 'Kafferland'] 'in Africâ austr. dist. Cafferland legit cl. *Drege* [5081]. ... (v.s.).' (G-DC-G00465494 holotype). FIGURE 6.5.**42**.

Adenostemma schimperi Sch.Bip. ex A.Rich., Tent. Fl. Abyss. **1**: 382 (1848). Type: [Ethiopia:] 'Crescit in ripis rivulorum prope Adoua (*Quartin Dillon* [s.n.] et *Schimper* [112]), mense Septembre florens.' *Quartin-Dillon* s.n. (P syntype); *Schimper* 112 (BM000543877, BM000584316, BR8360533, BR8872388, HBG504081, K00027388, LG90024481 syntypes).

Adenostemma viscosum sensu Oliver & Hiern in F.T.A. **3**: 299 (1877) p.p., non J.R.Forst. & G.Forst.

Adenostemma lavenia (L.) Kuntze var. *longifolium* Chiov. in Racc. Bot. Miss. Consol. Kenya: 62 (1935). Type: [Kenya:] 'South side of Lake Naivasha, British East Africa. altitude 1870 meters, July 12–16, 1909, *E. A. Mearns* 687.' (US00603707 holotype, BM000543880, BR9861664).

Adenostemma afrum var. *asperum* Brenan in Mem. New York Bot. Gard. **8**(5): 462 (1954). Type: [Malawi:] 'Kota-kota District: Nchisi Mountain, marshy ground in Brachystegia woodland, herb, stem reddish, flowers white, 1400 m., July 27, 1946, [*Brass*] 16979 (typus varietatis)' (K000273389, ?NY syntypes).[247]

Adenostemma afrum var. *longifolium* (Chiov.) S.A.L.Sm. in F.T.E.A., Compositae **3**: 822 (2005).

Vigorous, erect or semi-procumbent annual herb to 1.5 m, rooting at nodes on wet ground or in water. Stem ± fleshy, large stems woody and hollow, sometimes ribbed, glabrous or sparsely pubescent, sometimes with glandular hairs. Leaves sessile or shortly petiolate, lamina triangular, broadly ovate, ovate to lanceolate, 4–21 × 0.3–8.5 cm, margins sub-entire to serrate, crenulate or sharply dentate, sometimes minutely revolute, apex obtuse to acuminate, glabrous or pubescent. Inflorescence a loose terminal cyme, bracts ± amplexicaul, to 4.4 cm long. Capitula few to several, 6–10 mm diam. when mature; pedicels 0.4–7.2 cm, pubescent; phyllaries 4–6.5 × 1–1.5 mm, flat, basally pubescent, apex obtuse to acute. Florets ± 50; corollas 3–4 mm long, white, 5–6 lobed. anthers free or weakly fused, 5(6), ± 1 mm long; style conspicuous, 4–6.5 mm long, bifid for ± 2/3 of length, arms swollen and flattened, glabrous or with a few black hairs. Mature achenes brown, 2–3.5 mm long, tuberculate, glands present on tubercules (and sometimes between); pappus of (2)3–4(5) knobs.

Botswana. N: Chobe River, Kasane, 17°40'S, 25°10'E, 1000 m, 21.i.1972, *Gibbs-Russell & Siegel* 1352 (K, SRGH). SE: Palapye, Mosemi Spring Kloof, 975 m, 6.iii.1957, *de Beer* 121 (K, SRGH). **Zambia**. B: Senanga, 27.vii.1962, *Fanshawe* F6980 (K, SRGH). N: Abercorn Dist., Kali Dambo, 1524 m, 21.iii.1952, *Richards* 1137 (K). W: Ndola, 6.iii.1954, *Fanshawe* F933 (K, NDO). C: Lusaka, Mt. Makulu Research Station, 12 mi. S of Lusaka, 25.iv.1957, *Angus* 1561 (K, SRGH). S: Namwala, in Nkala River, 16.ii.1962, *Mitchell* BLM13/5 (K, NDO). **Zimbabwe**. N: Sebungwe Dist., Gokwe, 30.v.1947, *West* 2354 (K, SRGH). W: Matobo, Matopos, ii.1957, *Garley* 125 (K, SRGH). C: Charter Dist., Ngezi River, at Salisbury-Fort Victoria Road 30°50'E, 18°30'S, 1200 m, 30.v.1972, *Gibbs-Russell* 2009 (K, SRGH). E:

---

[246] Treated as orthographical variant, Art. 61.6, ICN (Madrid)

[247] Brenan & collab. were quite explicit in their introduction to the expedition's collections that material existed in both NY and K. The K sheet is oddly marked as holotype – a fact accepted by Smith (2005) – yet no such designation was mentioned in the protologue. If it is felt necessary to accept this as a distinct variety selection of the material in K as lectotype would be appropriate.

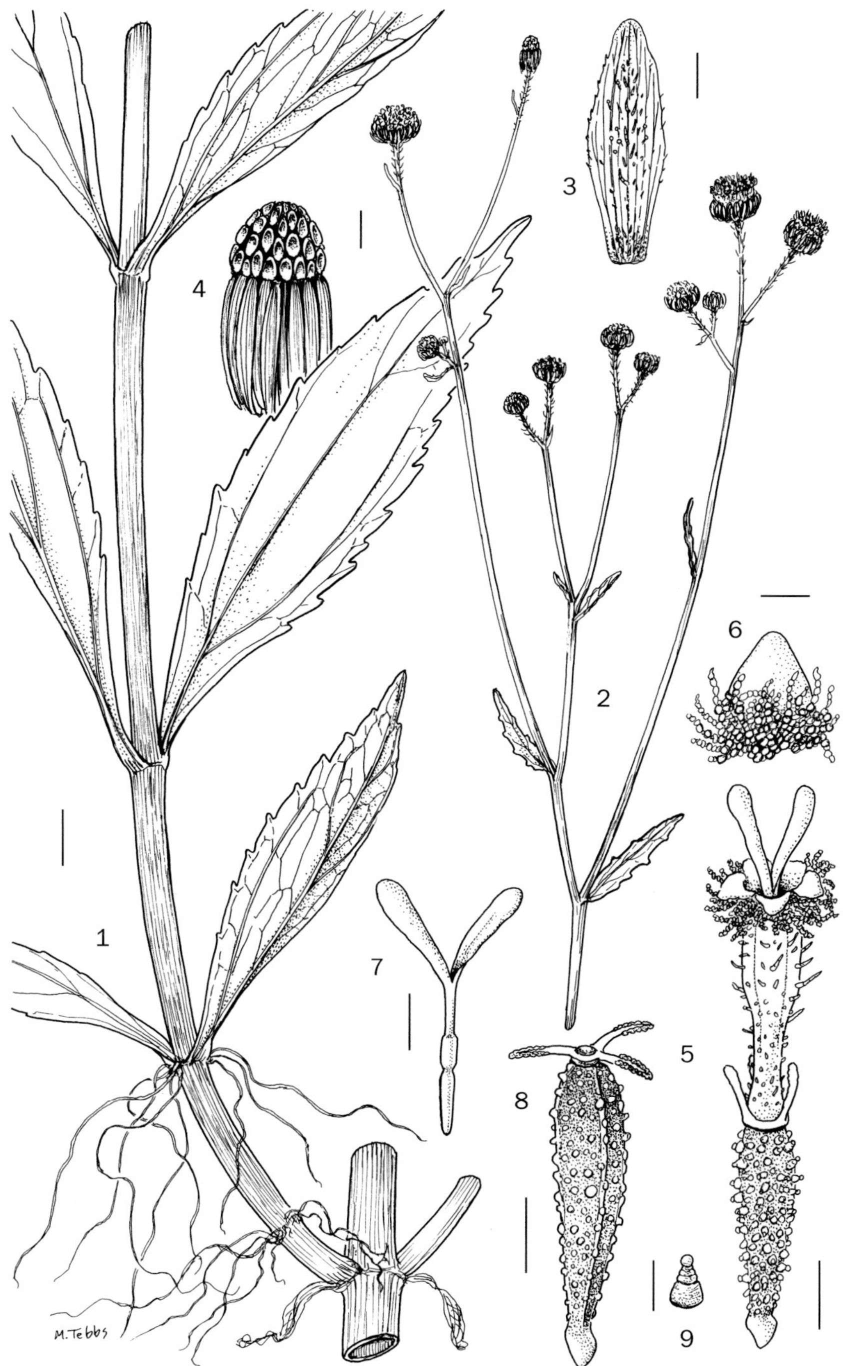

Fig. 6.5.**42**. ADENOSTEMMA AFRUM. 1, lower part of vegetative stem; 2, flowering shoot; 3, phyllary; 4, receptacle post achene loss; 5, floret; 6, detail of corolla lobe; 7, style; 8, mature achene; 9, detail of tubercle on achene surface. 1 from *Smith* 198; 2 from *Russell* 2855; 3–9 from *Biegal & Russell* 3915. Scale bars: 9 = 0.1 mm; 6 = 0.5 mm; 3–5, 7–8 = 1 mm; 1–2 = 10 mm. Drawn by Margaret Tebbs.

Melsetter, Iona Farm, 1432 m, 5.iii.1953, *Chase* 4817 (K, SRGH). S: Bikita Dist., S base Mt Horzi, 1050 m, 9.v.1969, *Biegel* 3072 (K, SRGH). **Malawi**. N: Mzimba, Vip[h]ya Plateau, 35 km SW of Mzuzu on road to Mzimba, 1600 m, 8.iii.1977, *Grosvenor & Renz* 1090 (K, SRGH). C: Dedza, Dedza Mountain slopes, 4.v.1968, *Salubeni* 1087 (K, SRGH).

Also in Sierra Leone, Guinea, Ghana, Nigeria, Cameroon, Central African Republic, D.R. Congo, Rwanda, Burundi, Sudan, Ethiopia, Uganda, Kenya, Angola, South Africa. Lake sides, wet ground and swamps, especially in disturbed sites, sometimes in water up to 2 m deep; 1050–2300 m.

Conservation Status: A widespread species; LC (Least Concern).

Mesfin Tedesse (2004) keyed out this species as a perennial, differing from Lisowski (1991) who considered it an annual. Smith (2005) made no comment. Doubtless in areas with permanent water the species might become a short-lived perennial. The separation of the two varieties is a little artificial, although following Smith's treatment the material found in our Flora area is attributable to var. *asperum* Brenan.

3. **Adenostemma mauritianum** DC., Prodr. **5**: 110 (1836). —Bojer, Hortus Maurit.: 177 (1837). —Andrews, Fl. Pl. Sudan **3**: 8 (1956). —Cufodontis, (Enum. Pl. Aethiop.) Bull. Jard. Bot. Natl. Belg. **36** (3, suppl.): 1079 (1966). –Adams in F.W.T.A., ed. 2, **2**: 286 (1968). —Maquet in Fl. Rwanda **3**: 568, fig. 173/3 (1985). —Lisowski, (Asterac. Fl. Afr. Cent. 2) Fragm. Flor. Geobot. **36** Suppl. 1: 467 (1991). —Hind in Fl. Masc., Composées **109**: 229, t. 80 (1993). —Agnew & Agnew, Upland Kenya Wild Fl., ed. 2: 203 (1994). —Mesfin Tadesse, Fl. Ethiopia & Eritrea **4**(2): 345 (2004). Type: 'in insulae Mauritii sylvis montosis (*Comm.*, *Sieb. Bouton*! etc.) ... *Lavenia erecta* Sieber! fl. maur. exs. 2. n. 119 et 364.' ... (v.s.)' *Bouton* s.n. (K000273390 syntype); *Sieber* 119 (BR9861640, K000273391, M0104526 syntypes); *Sieber* 364 (HAL0114065, P00558717 syntypes); *Commerson* s.n. (P00558715, P00558716 syntypes).

   Adenostemma viscosum sensu Oliver & Hiern in F.T.A. **3**: 299 (1877) p.p., non J.R. Forst. & G. Forst. —Humbert, Fl. Madagasc., Composées **189**: 200, fig. 39/1–4 (1960).

Annual, possibly sometimes a short-lived perennial, herb 0.2–1.2 m. Stem fleshy, ascending, rooting at base, sometimes striate, glabrous or sparsely pubescent, more so towards the apex. Leaves petiolate, petiole 0.5–7.2 cm, lamina broadly triangular to ovate, 3.2–16 × 1.7–9(11) cm, base attenuate to form narrow wings along petiole, margins serrate, dentate or crenulate, apex acute, lamina membranous, darker above than beneath, glabrous or sub-glabrous with few scabrid hairs. Inflorescence branching ± dichotomously, bracts 0.5–5 cm long, sessile, capitula pedicellate, pedicels 1.2–5 cm, densely pubescent. Capitula 4–10 mm diam. when mature; phyllaries green, flat, narrowly elliptic to oblong, 4–5 × 1–1.5 mm, sparsely pubescent, margins entire or ciliate, apex obtuse to acute. Florets 21–28; corollas white, 2–3 mm long, tube sparsely stipitate-glandular, gradually widening into limb, lobes (4)5; anthers free, 5, ± 1 mm long; style 3.5–4 mm long, bifid for more than ½ length, arms swollen and flattened, glabrous. Mature achenes 2.5–4 mm long, smooth, usually indistinctly 5-gonous, glabrous or sometimes with minute stipitate glands, never tuberculate; carpopodium prominent; pappus of 3–4(5) knobs, (0.7)1–1.2 mm long.

Zambia. W: Chingola Dist., Mushishima Botanical Reserve, c. 12 km W of Chingola, 18.v.1986, *Philcox et al.* 10363 (BR, K, LISC). S: Livingstone, N bank of Zambesi, 17°54'S, 25°55'E, 914 m, *Rogers* 7425 (K). **Zimbabwe**. E: Umtali, Vumba Mts, 2 miles W of Childrens' Camp, 19.v.1962, *Noel* 2443 (K, SRGH). **Malawi**. N: Chitipa Dist., Misuku Hills, Mughesse rain forest, 1676 m, 17.ix.1975, *Pawek* 10125 (K, MO, MA). S: Blantyre Dist., Ndirande Mountain, c. 1400 m, 31.v.1970, *Brummitt* 11189 (K). **Mozambique**. Z: Mabu Mts, 16°52'S, 36°22'E, 1371 m, 28.x.2008, *Harris et al.* 682 (K).

Also in Sierra Leone, Ivory Coast, Ghana, Nigeria, Cameroon, Bioko, D.R. Congo, Rwanda, Burundi, Uganda, Kenya, Tanzania, South Africa, Madagascar and Mauritius. Forest, forest edges, swamp forest, stream banks, edges of damp footpaths, bamboo zone; 1300–2750 m.

Conservation Status: A widespread species; LC (Least Concern).

150. **AGERATUM** L.

Ageratum L., Sp. Pl. **2**: 839 (1753). —Johnson in Ann. Missouri Bot. Gard. **58**(1): 6 (1971). —King & Robinson in Phytologia **24**(2): 112–117 (1972). —Dyer, Gen. S. Afr. Fl. Pl. **1**: 663 (1975). —King & Robinson, Genera of the Eupatorieae (Asteraceae). Monogr. Syst. Bot. **22**: 142 (1987). —Sharma in Feddes Repert. **98**: 557 (1987). —Lisowski, (Asterac. Fl. Afr. Cent. 2) Fragm. Flor. Geobot. **36** Suppl. 1: 472–479 (1991). —Bremer, Anderberg, Karis & Lundberg in Bremer, Asterac. Cladist. Classif.: 649 (1994). —Mesfin Tadesse, Fl. Ethiopia & Eritrea **4**(2): 346–347 (2004). —Smith in F.T.E.A., Compositae **3**: 827–829 (2005). —Hind & Robinson in Fam. Gen. Vasc. Pl. **8**: 522–523 [2006](2007).

Annual to perennial herbs or subshrubs, few- to many-branched, base often decumbent with numerous adventitious roots. Stems terete, striate, sparsely puberulous to sparsely hirsute. Leaves opposite or sometimes alternate, short- to long-petiolate, lamina elliptical or lanceolate to deltoid or ovate, margins entire to dentate, apex not or scarcely acuminate, lower surface usually with large sessile or partially sunken glandular-punctate, usually 3-veined from or near base. Inflorescence cymose to subcymose, sometimes subumbellate, pedicels short to moderately long. Capitula homogamous, discoid; phyllaries 30–40, distant, 2–3-seriate, equal or subequal, lanceolate, markedly indurate and often with scarious lateral margins; receptacle conical, glabrous or paleaceous. Florets 20–125; corollas white, blue, or lavender, funnelform or with distinct basal tube; lobes triangular, about as long as wide, papillose on inner surface, partially papillose and sometimes hispidulous on outer surface; anther collar cylindrical; anther appendage large, somewhat longer than wide; style base not enlarged, glabrous; style arms linear, usually strongly and densely papillose. Achenes prismatic, 4–5-ribbed, glabrous or ribs setuliferous; carpopodium distinct, usually large and asymmetrical; pappus lacking, or coroniform, or of 5 or 6 free, flattened, sometimes awn-like scales.

A genus of about 40 spp. of Central and S America. One species (*A. houstonianum* Mill.) is widely cultivated, and another (*A. conyzoides* L.), although sometimes cultivated, is a widespread weed throughout the tropics in both the Old and New Worlds; both these occur in the Flora area.

Leaf base obtuse or broadly cuneate, never cordate or truncate; phyllaries oblong-lanceolate, glabrous to sparsely pilose, eglandular; style-arms exserted for less than 1.6 mm . . . . **1.** *conyzoides*
Leaf base cordate to truncate; phyllaries linear-lanceolate, hirsute and stipitate-glandular; style-arms exserted for more than 1.8 mm . . . . **2.** *houstonianum*

1. **Ageratum conyzoides** L., Sp. Pl. **2**: 839 (1753), nom. cons. —Bojer, Hortus Maurit.: 177 (1837). —Harvey in Harvey & Sonder, Fl. Cap. **3**: 57 (1865). —Oliver & Hiern in F.T.A. **3**: 300 (1877). —Hoffmann in Pflanzenw. Ost-Afrikas C.: 406 (1895). —Williams, Useful Ornam. Pl. Zanzibar & Pemba: 110 (1949). —Andrews, Fl. Pl. Sudan. **3**: 8 (1956). —Humbert, Fl. Madagasc., Composées **189**: 201, fig. 39/10–13 (1960). —Lind & Tallantire, Comm. Fl. Pl. Uganda, ed. 2: 182, fig. 122 (1975). —Adams in F.W.T.A., ed. 2, **2**: 287 (1963). —Cufodontis, (Enum. Pl. Aethiop.) Bull. Jard. Natl. Belg. **36** (3, suppl.): 1079 (1966). —Johnson in Ann. Missouri Bot. Gard. **58**(1): 26 (1971). —Maquet in Fl. Rwanda **3**: 570, fig. 174/1 (1985). —Burkill, Useful Pl. W. Trop. Afr., ed. 2, **1**: 443 (1985). —Banda & Morris, Common Weeds Malawi: 10, fig. 11 (1986). —Lisowski, (Asterac. Fl. Afr. Cent. 2) Fragm. Flor. Geobot. **36** Suppl. 1: 473–477 (1991). —Hind in Fl. Masc., Composées **109**: 234, t. 81/1–7 (1993). —Agnew & Agnew, Upland Kenya Wild Flow., ed. 2: 204, t. 80 (1994). —Mesfin Tadesse, Fl. Ethiopia & Eritrea **4**(2): 347–348 (2004). —Smith in F.T.E.A., Compositae **3**: 827–829 (2005). Type: 'Habitat in America.' '*Eupatorium Senecionis facie folio Lamii*' in Hermann,

Parad. Bat.: ic.161 (1698), typ. cons., iconotype, lectotypified by Reveal in Jarvis *et al.*, Regnum Veg. **127**: 15 (1993).[248] FIGURE 6.5.**43A**.

Ageratum ciliare L., Sp. Pl. **2**: 839 (1753). Type: 'Habitat in Bisnagaria.' '*Centaurium ciliare minus Bisnagaricum, Origani foliis amplioribus, floribus in Umbellis*' in Plukenet, Phytographia: t. 81, f. 4 (1691), iconotype, lectotypified by Robinson in Jarvis & Turland, Taxon **47**(2): 351 (1998). [India:] 'Karnataka, Hassan District, Hassan', 21.ix.1971, *K.N. Gandhi* HFP 2103 (US2792681 epitype), epitypified by Robinson in Jarvis & Turland (1998: 351).

Ageratum hirtum Lam., Encycl. **1**: 54 (1783). Type: 'Cette plant croît au Cap de Bonne-Espérance, & m'a été communiqué par *M. Sonnerat.* (v.s.)' (P-LA 306/12 holotype).

Ageratum humile Salisb., Prodr. Stirp. Chap. Allerton: 188 (1796), nom. illeg. superfl. based on *A. conyzoides* L.

Ageratum latifolium Cav., Icon. **4**: 33, tab. 357 (1797). Type: 'Habitat prope Limam in imperio Peruviano, floretque Iunio. ... Vidi siccam in herbario Domini Ludovici Née. Nominatur vulgo *Teatina*.' (MA221915[249] holotype, LECB0000203 – annotated as 'SYNTYPUS? by V. Byalt 12 XII 2013').

Ageratum coeruleum Desf., Tabl. École Bot.: 98 (1804), nom. nud.

Ageratum hirsutum Lam. in Poir., Encycl. Suppl. **1**: 242 (1810). Type: 'Illustr. tab. 672, fig. 2. (Voyez AGÉRATE, n°. 1.)'

Sparganophorus obtusifolius Lag., Gen. Sp. Pl.: 25 (1816). Type: 'Hab. in N[ova]. H[ispania]. ...' (?MA if extant holotype).

Cacalia mentrasto Vell., Fl. Flumin.: 339 (1825) [7 Sept.–28 Nov. 1829]; Fl. Flumin. Icon. **8**: tab. 69 (1827)[29 Oct. 1831]. Type: 'Habitat maritimis, mediterraneisque. Floret Aug.' Herbarium material unknown.[250]

Ageratum cordifolium Roxb., Fl. Indica **3**: 415 (1832). Type: 'Beng.[al] Oochunti. [sic!]. An annual found in the vicinity of Calcutta during the rainy and cold seasons; flowering time the cold season' (?BM holotype).[251]

Ageratum conyzoides var. β *hirtum* (Lam.) DC., Prodr. **5**: 108 (1836).

Ageratum brachystephanum Regel, Gartenflora **3**: 245, t. 108 (1854). Type: '... Wagener aus Colombia … Caracas.' (?LE holotype).

Ageratum suffruticosum Regel, Gartenflora **3**: 389, t. 108 (1854). Type: 'Die beistehende Abbildung is nach einem Exemplar gemacht, das ins freie Land gepflanzt, im Garten des Herrn Fröbel prächtig blühete.' Herbarium material unknown.

Ageratum nanum hort. ex Sch.Bip. in Koch & Fintelm., Wochenschr. Gart. Pflanzenk. **1**: 26 (1858), nom. nud.

Ageratum muticum Griseb., Fl. Brit. W. I.: 356 (1861). Type: 'HAB. Jamaica!, *Wullschl.*, *March*; [Cuba!, Peru!]'[252] 'Cuba occ. *Wright* 1631=500' (GOET000997, GOET000998, PH00000712, S-R-158 syntypes).[253]

Ageratum odoratum E.Vilm., Fl. Pleine Terre, ed. 2: 42 (1866), nom. illeg.[254]

Carelia conyzoides (L.) Kuntze, Revis. Gen. Pl. **1**: 325 (1891), as '*conyzodes*'.

---

[248] Grierson (in Dassanayake & Fosberg, Rev. Handb. Fl. Ceylon **1**: 141, 1980) selected the lectotype as 'Clifford: 396, *Ageratum* 1', BM000646956. Whilst having no grounds to reject this lectotypification the material chosen by Grierson is actually material of *Eclipta prostrata* (L.) L., and not *Ageratum*! See Barrie *et al.*, Taxon **41**: 508–512 (1992) for an explanation, Reveal & Jarvis, Taxon **58**: 1011 (2009), for the conservation proposal and Applequist, Taxon **62**: 1316 (2013) for the outcome of the vote on the proposal.

[249] The upper of the three labels on the lefthand side of the sheet gives the vernacular name as 'Theatina'.

[250] No herbarium material has yet been found for Vellozo's work, and there is no real need to lectotypify the name.

[251] BR5232642 may also represent type material.

[252] Square brackets in original!

[253] One syntype, *March* 1354, is in K and one annotated sheet, but lacking collector, may represent a Wullschlagel collection. *Wright* 1631 is proposed as a syntype in GOET.

[254] She cited: '*Syn. lat.* – Ageratum conyzoides, *Lin.*, A. album, *Steud.*'

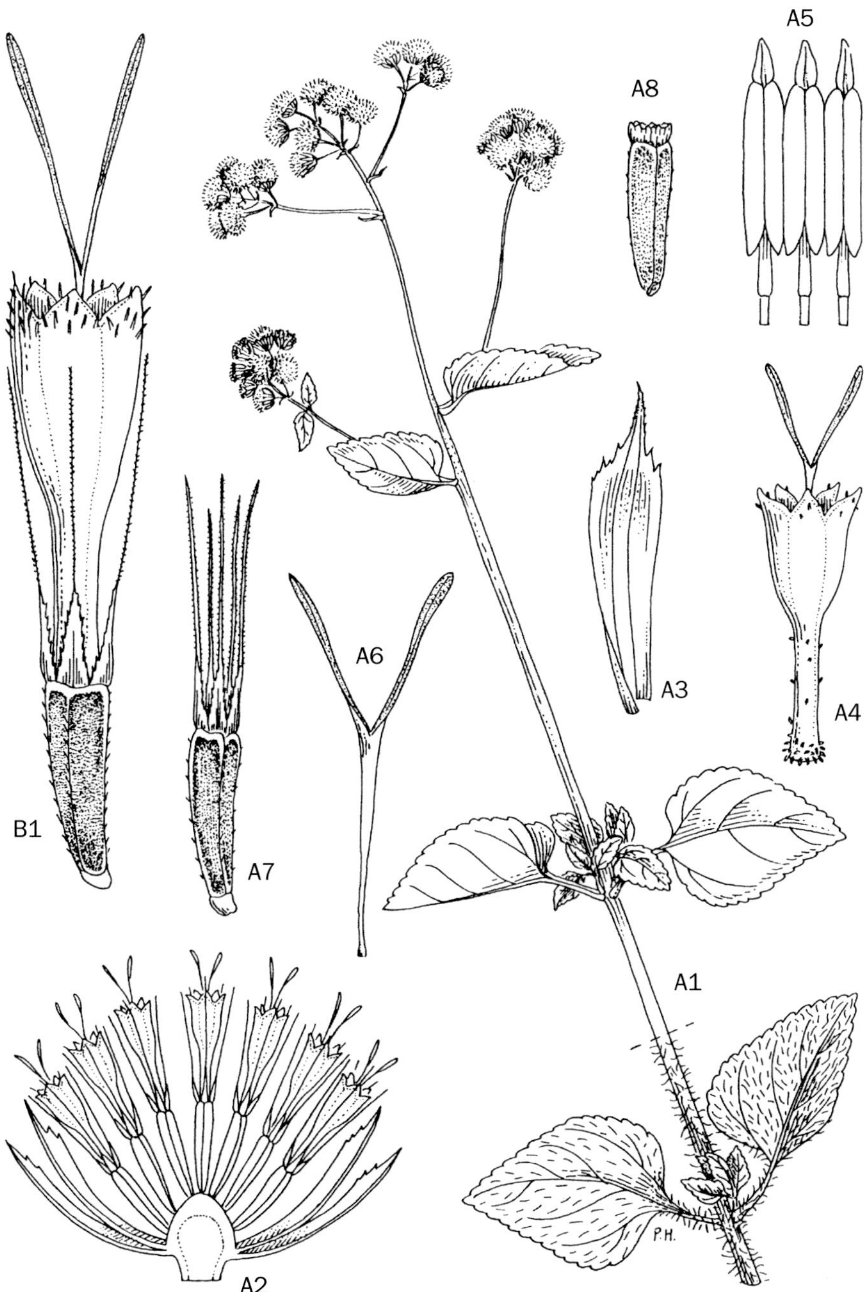

Fig. 6.5.**43**. A. —AGERATUM CONYZOIDES. A1, flowering stem ($\times\,^2/_3$); A2, l.s. capitulum ($\times$ 8); A3, outer phyllary ($\times$ 12); A4, corolla ($\times$ 18); A5, anther cylinder opened out ($\times$ 24); A6, style ($\times$ 18); A7, achene with awned pappus scales ($\times$ 12); A8, achene with truncate pappus scales ($\times$ 12). B. — AGERATUM HOUSTONIANUM. B1, floret ($\times$ 18). A1–7 from *McWhirter* 100; A8 from *McWhirter* 102; B1 from *Rouillard* 344. Drawn by Pat Halliday. From Flore des Mascareignes.

Carelia conyzoides [unranked] α *robusta* Kuntze, Revis. Gen. Pl. **1**: 325 (1891). Types: 'St. Thomas, Portorico.' [PUERTO RICO. Caguas, 8 Mar 1874], *Kuntze* 235 (NY00115550 syntype); [VIRGIN ISLANDS. St. Thomas, Feb 1874], *Kuntze* 116 (NY00115549 syntype).[255]

Carelia conyzoides [α *robusta* Kuntze] var. *alba* Kuntze, Revis. Gen. Pl. **1**: 325 (1891). Type: ?'Honkong, Batavia.'[256]

Carelia conyzoides [unranked] β *umbrosa* Kuntze, Revis. Gen. Pl. **1**: 325 (1891). Type: ?'Hongkong.'[253]

Carelia conyzoides [β *umbrosa* Kuntze] var. *coerulea* Kuntze, Revis. gen. Pl. **1**: 325 (1891). Type: 'Hongkong.' (NY00163096 holotype).[257]

Carelia conyzoides [unranked] γ *pusilla* Kuntze, Revis. Gen. Pl. **1**: 325 (1891). Type: ?'Turong in Anam.'[253] See following name.

Carelia conyzoides [γ *pusilla* Kuntze] var. *alba* Kuntze, Revis. Gen. Pl. **1**: 325 (1891). Type: ?[258]

Carelia brachystephana (Regel) Kuntze, Revis. Gen. Pl. **1**: 325 (1891).

Carelia mutica (Griseb.) Kuntze, Revis. Gen. Pl. **1**: 325 (1891).

Ageratum conyzoides var. *inaequipaleaceum* Hieron. in Bot. Jahrb. Syst. **19**(1): 44 (1895). Type: 'Colombia: crescit prope Popayan, alt. s.m. 1500–2000 m ([*Lehmann*] n. 4666). – Floret mensibus Octobri–Martio.' (B† holotype, F0049608 = F550707, GH00000745 – a fragment of the holotype, including 1 leaf and a small apical portion of a flowering shoot).[259]

Ageratum conyzoides f. *album* (Willd.) B.L.Rob. in Contr. Gray Herb. **42**: 462 (1913).

Ageratum humile Larrañaga, Escritos Damaso Antonio Larrañaga **1**: 406 (1922) [Pub. Inst. Hist. Geog. Urug.], nom. illeg., non Salisbury (1796). Type: not cited.[260]

Ageratum arsenei B.L.Rob. in Contr. Gray Herb. **64**: 3 (1922). Type: 'MEXICO: Cercado near Monterey, *Bro. Arsène*, 12/11 1911' (K000488663 holotype, GH00000642 – 'slight fragment' from the material in K).

Annual herb to 1 m tall, malodorous. Stem erect to occasionally decumbent, terete, sparsely pilose and whitish pubescent, especially at nodes, to subglabrous. Leaves opposite to occasionally sub-opposite above, petiole pubescent 0.3–3 cm, blade ovate to rhomboid, 2.2–8.5 × 1.2–6 cm, base obtuse, upper surface subglabrous to sparsely pilose, lower surface moderately pilose, margins crenate to serrate, apex acute or obtuse. Inflorescences much-branched, open, of 1–several clusters of ± 7–40 capitula in open to dense cymose clusters on pubescent and pilose bracteolate peduncles. Capitula 4–5 mm diam.; pedicels 2–12 mm long, extending in fruit to 37 mm; phyllaries biseriate, green, ovate to oblong-lanceolate, usually 2-ribbed, outer series 3.5–4.5 × 0.9–1 mm, inner slightly narrower, margins basally entire, erose to lacerate at apex, rarely entire, apex abruptly contracted to point, occasionally short-acuminate, often purple, glabrous to subglabrous to very sparsely pilose. Florets 40–57, corolla tubular to sub-urceolate, 2–2.2 mm long, tube white, limb externally glabrous to sub-glabrous, lobes acute, blue, white, mauve, or purple; anthers basally rounded; style-arms blue, white, mauve or purple, exserted for 0.8–1.2 mm, papillate. Achenes 1.2–1.8 mm long, glabrous between ribs; pappus of 5 triangular scales 1.9–2 mm long, usually acuminate into a slender awn (see note), margins scabrid.

Zambia. B: Near Senanga, 1040 m, 29.vii.1952, *Codd* 7639 (K, PRE). N: North Luangwa National Park, Muchinga Escarpment, 950 m, 25.iii.1994, *Smith* (K). W: Ndola, 10.v.1953, *Fanshawe* F2 (K, NDO). C: Chilanga Fish Farm, 1200 m, 5.v.1963, *Lusaka Natural History*

---

[255] Syntypes according to Wetter & Zanoni (1985: 329).

[256] No material was cited by Wetter & Zanoni (1985).

[257] According to Wetter & Zanoni (1985: 329) this specimen is 'CHINA. Hong Kong, [Victoria Peak], 27 Jan 1875, *Kuntze* 3404.'

[258] Wetter & Zanoni (1985: 329) cited the type as 'VIETNAM. Turong [= Da Nang], 24 Feb 1875, *Kuntze* 3598' (NY 00163097 holotype).

[259] Strangely cited as 'Holotype: GH; isotype: NY.' by Johnson (1971).

[260] At the end of the diagnosis Larrañaga noted '– an *A*[*geratum*] *conizoides* [sic!].' The 'Escritos …' is an interesting work published in three volumes. Volume 1, containing 'Diario de la Historia Natural, Diario de la Chacara and Apuntes y observaciones de Historia Natural' (pp. i–xxiii, [1]–439) contains *Ageratum humile* in the 'Apuntes …'. The full title of the 'Apuntes …' is 'Apuntes y observaciones de historia natural que hice en el Janeyro en 1817.', pp. 387–426.

Club 266 (K). E: Nsefu Game Camp, 750 m, 15.x.1958, *Robson* 135 (K). S: Livingstone Dist., Kalambara, 30.vi.1956, *Gilges* 644 (K, SRGH). **Zimbabwe**. N: Darwin Dist., Chimenda Reserve, 6.ix.1958, *Phipps* 1322a (K, SRGH). W: Wankie Dist., Victoria Falls, 880 m, 18.vi.1978, *Mshasha* 74 (K, SRGH). E: Melsetter Dist., Haroni-Lusitu Junction, 10.i.1969, *Mavi* 871 (K, SRGH). **Malawi**. N: Nkhata Bay Dist., Nkhata Bay, 480 m, 10.v.1970, *Brummitt* 10547 (K). C: Lilongwe Dist., Agricultural Research Station, 31.iii.1955, *Jackson* 1532 (K, SRGH). S: Blantyre Dist., Ndirande Mountain Forest, 28.v.1985, *Kwatha & Balaka* 126 (K, MAL). **Mozambique**. N: Mandimba Dist., Mandimba, 760 m, *Leach & Rutherford-Smith* 10854 (K, SRGH). T: Zumbo Dist., Zumbo, 550 m, 22.vi.1900, *Baum* 1003 (K). MS: ± 30 km north of Dombe, ± 150 m, 4.vi.1971, *Biegel & Pope* 3551 (K, SRGH). GI: Chibuto Dist., Baixo Chananga, Propriedade Santos Gil, 26.viii.1963, *Macedo & Macuácua* 1151 (K, LMJ). M: Matutuíne Dist., Salamanga,7.vi.1948, *Gomes e Sousa* 3739 (K).

Pantropical. A weed in cultivated areas and post-cultivation, often a pioneer on cleared land such as logging tracks, also in grassland and *Acacia* woodland; 0–2150 m.

Conservation Status: A widespread pantropical weed; LC (Least Concern).

Like the following species two forms were recognized by Johnson (1971) based on the presence of truncate pappus scales versus pappus scales tapering into setae. The two forms are not formally recognized in this account. Available chromosome counts suggest that *A. conyzoides* is a tetraploid ($n = 20$) and *A. houstonianum* a diploid ($n = 10$).

2. **Ageratum houstonianum** Mill., Gard. Dict., ed. 8, no. 2 (1768). —Robinson in Proc. Amer. Acad. Arts **51**: 532 (1916); in Contr. Gray Herb. **68**: 6 (1923). —Adams in F.W.T.A., ed. 2, **2**: 287 (1968). —Johnson in Ann. Missouri Bot. Gard. **58**(1): 21 (1971). —Burkill, Useful Pl. W. Trop. Afr., ed. 2, **1**: 445 (1985). —Lisowski, (Asterac. Fl. Afr. Cent. 2) Fragm. Flor. Geobot. **36** Suppl. 1: 477–479, fig. 96 (1991). —Hind in Fl. Masc., Composées **109**: 234, t. 81/9 (1993). —Agnew & Agnew, Upland Kenya Wild Flow., ed. 2: 204 (1994). —Mesfin Tadesse, Fl. Ethiopia & Eritrea **4**(2): 348 (2004). —Smith in F.T.E.A., Compositae **3**: 830–831 (2005). Type: 'The second sort was found growing naturally at La Vera Cruz, by the late Dr. William Houstoun, who sent the seeds to Europe, which have so well succeeded in many gardens as to become a weed in the hot-beds. There is a variety of this with white flowers, which arises from the same seeds.' (BM001009250 holotype). FIGURE 6.5.**43B**.

    Ageratum mexicanum Sims, Curtis's Bot. Mag. **52**: t. 2524 (1824). Type: 'Raised by Mr. TATE, of the Sloane Street nursery, from seeds brought from Mexico by Mr. BULLOCK.'[261], [262]

    ?*Cacalia mentrasto* Vell., Fl. Flumin.: 339 (1825)[7 Sept. – 28 Nov. 1829]; —Fl. Flumin. Icon. **8**: t. 69 (1827)[29 Oct. 1831]. Type: 'Habitat maritimis, mediterraneisque, Floret Aug.'[263], [264]

    Ageratum conyzoides L. var. *mexicanum* (Sims) DC., Prodr. **5**: 108 (1-10 Oct. 1836).

    Carelia houstoniana (Mill.) Kuntze, Revis. Gen. Pl. **1**: 325 (1891).

    Ageratum wendlandii hort. ex Vilm., Fl. Pl. Terre, Suppl. 2 (1884).[265]

---

[261] It is unclear if any herbarium material was preserved. Usually cited as 'non Sweet (1825)' but Sweet cited Sims's taxon, and was clearly based on the same type material; it is highly unlikely to have been meant as a new name at all.

[262] *Ageratum mexicanum* 'Sweet', Brit. Fl. Gard. **1**: tab. 89 (1825), nom. illeg., citing *A. mexicanum* Sims and clearly based upon the same material.

[263] Herbarium material of this work is as yet unknown.

[264] This is more commonly placed in synonymy of *Ageratum conyzoides* L.

[265] Whilst distinguishing characters were provided, and the plant cultivated, it is unknown if herbarium material was prepared.

Ageratum mexicanum [unranked, ?var.] a) *majus* Voss, Vilm. Blumengärtn., ed. 3, **1**: 445 (1894).[266]

Ageratum mexicanum [unranked, ?var.] a) *majus* f. *coeruleum* Voss, Vilm. Blumengärtn., ed. 3, **1**: 445 (1894).[267]

Ageratum mexicanum [unranked, ?var.] a) *majus* f. *albiflorum* Voss, Vilm. Blumengärtn., ed. 3, **1**: 445 (1894).[268]

Ageratum mexicanum [unranked, ?var.] a) *majus* f. *lasseauxii* (Carrière) Voss, Vilm. Blumengärtn., ed. 3, **1**: 445 (1894).

Ageratum mexicanum [unranked, ?var.] b) *wendlandii* (Vilm.) Voss, Vilm. Blumengärtn., ed. 3, **1**: 445 (1894).[269]

Ageratum mexicanum [unranked, ?var.] c) *nanum* Voss, Vilm. Blumengärtn., ed. 3, **1**: 445 (1894). Types: 'Sorten find: "Cupid", ... "Cannel's dward" ... "Little Dorrit:, ... "Swanley blue", ... "Johanna Pfitzer" ... "M. Délaux" ... "Elisabeth Kurtz", ... "Kind von Dresden" ... "Perle bleue" ... "Duke of Albany", ... "Malverne Beauty: ... "Louise Bonnet", ... "Snowflake:, ... "Perle blanche" ...' Herbarium material unknown.[270]

Ageratum houstonianum var. *muticescens* B.L.Rob. in Proc. Amer. Acad. Arts **51**: 532 (1916). Type: 'Mexico: Wartenberg, near Tantoyuca, prov. Huasteca, collected in 1858, *L. C. Ervendberg*, no. 100 (Type, in Gray Herb.)' (GH00000667 holotype, P00742244).[271]

Ageratum houstonianum var. α *typicum* f. b. *niveum* B.L.Rob. in Contr. Gray Herb. **68**: 6 (1923). Type: 'Cultivated as "Ageratum Dwarf White" and "Little Dorrit (white) " etc. Founded on material purchased under these names of the Joseph Breck & Sons Corporation, grown for the Department of Agriculture at Glen Echo, Maryland, and collected 13 Sept. 1922, *O. M. Freeman*, nos. 5074, 5080 (U.S., Gr.).' *Freeman* 5074 (GH00000634 syntype); *Freeman* 5080 (GH00000633, GH00000635 syntypes).[272]

Ageratum houstonianum var. α *typicum* f. c. *luteum* B.L.Rob. in Contr. Gray Herb. **68**: 6 (1923). Type: 'Cultivated as "Ageratum nanum luteum." Founded on material supplied by the Joseph Breck & Sons Corporation and cultivated at Glen Echo for the Department of Agriculture, collected 24 Aug. 1922, *O. M. Freeman*, no. 5084 in part. (U.S., Gr.).' (GH00000631 – as '5084', GH00000632 – as '5084', GH00000636 – as '5084a', US syntypes).

Ageratum houstonianum var. β *angustatum* B.L.Rob. in Contr. Gray Herb. **68**: 6 (1923). Type: 'Found naturalized in the Maritime Alps at Menton, France, in the spring of 1878, *Walther* (hb. Univ. Zürich, tracing and small fragm. Gr.)' (Z holotype, GH00000753 – 3 leaves and an inflorescence branch from the holotype, in a capsule, together with a sketch of the holotype sheet).

Ageratum houstonianum var. γ *multicescens* B.L.Rob. f. d. *isochroum* B.L.Rob. in Contr. Gray Herb. **68**: 6 (1923). Type: 'Found in the wild at Wartenberg, near Tantoyuca, Prv. Huasteca, southern Mexico, *Ervendberg*, no. 100 (Gr.); cultivated under the names "Stella Gurney," "Cope's Pet," and "Blue Perfection".'[273]

Ageratum houstonianum var. γ *multicescens* B L.Rob. f. e. *versicolor* B.L.Rob. in Contr. Gray Herb. **68**: 6 (1923). Type: 'Founded on material sold by the Joseph Breck & Sons Corporation under the name "Ageratum nanum luteum" and grown at Glen Echo, Maryland, for study at the Department of Agriculture in Washington, collected 29 July, 1922, by *O. M. Freeman*, no. 5084 in part (U. S., Gr.). (GH00000636 – as '5084a', US syntypes).

---

[266] This variety was merely mentioned with a short description, 'mit folgenden Sorten: f. coeruleum, ...' It is unknown if any herbarium material was prepared.

[267] This forma was mentioned under var. a) *majus* and simply noted as having 'himmelblau bis etwas graublau' corollas. It is unknown if any herbarium material was prepared.

[268] This forma was merely noted as a white flowered form under a) *majus*; it is unknown if any herbarium material was prepared.

[269] The following synonyms were also cited against this name: '(syn. *A. Wendlandii compactum* hort., *A. coeruleum nanum* hort., *Phalacraea coelestina nana* hort.), Wendlands M. L., ...'.

[270] It is unknown if Voss prepared any herbarium material from these numerous cultivars.

[271] Oddly, Smith (1971) cited the holotype as in MO; Robinson was quite specific when he said 'Gray Herb.'

[272] At present it is unclear which of the syntypes are in US.

[273] The type material is the same as the variety, q.v.

Ageratum conyzoides var. *houstonianum* (Mill.) T.R.Sahu in Feddes Repert. **93**(1-2): 64 (1982).
Ageratum conyzoides subsp. *houstonianum* (Mill.) M.Sharma in Geobios, New Rep. **3**(2): 152 (1984).

Usually annual or short-lived perennial, erect to decumbent, simple or sparingly branched herb or sub-shrub to 1 m. Stem terete, whitish pubescent and long, simple hairs. Leaves opposite to occasionally sub-opposite above, petiole pubescent 0.4–3.2 cm, lamina triangular to ovate, 1.3–5.8 × 1–5.3 cm, base cordate to abruptly truncate, dark green, moderately hirsute above, scarcely paler and glabrous to sparsely hirsute beneath, margins crenate to serrate, apex acute or obtuse. Inflorescences terminal, ± 7–60 capitula in tight cymose clusters on pubescent and hirsute bracteolate branches. Capitula 4–7 mm diam.; pedicels 0–6 mm extending to 12 mm in fruit; phyllaries green, lanceolate, usually 2-ribbed, outer series 3.5–4.5 × 0.8–0.9 mm, inner slightly narrower, margins entire, apex acuminate, often purple, hirsute and stipitate-glandular especially basally. Florets 70–103, corollas funnel-shaped, 2.5–3 mm long, tube white, limb blue, mauve, or white, lobes acute, sparsely to moderately short-pubescent; anthers basally subcordate; style-arms blue, mauve or white, clavate, exserted for ± 2.5 mm, papillate. Achenes 1.2–2 mm long, sub-glabrous, with occasional setae between ribs; pappus 0–3 mm long, of 5 free triangular scales, lacerate, acuminate into a slender awn, margins scabrid, or pappus awnless, scales coroniform, truncate, lacerate, or occasionally pappus absent.

Zambia. N: Kasama Dist., 25.ix.1960, *Robinson* 3852 (K). W: Solwezi Dist., Solwezi, 24.ix.1936, *Milne-Redhead* 1188 (K). **Zimbabawe**. E: Umtali Town, 31.x.1967, *Mavi* 333 (K, SRGH). **Malawi**. N: Mzimba Dist., Chikangawa, 1177 m, 9.viii.2004, *Chapama* et al. 158 (FRM, K). S: Zomba Dist., Zomba Plateau, 1830 m, 4.v.1970, *Brummitt & Banda* 10367 (K).

Also in Cameroon, D.R. Congo, Sudan, Ethiopia, Kenya, Tanzania, eSwatini, South Africa, Madagascar, Mascarenes; originally from Central America, and widespread as a garden escape. A weedy escape on stream-banks and moist waste ground; 0–1850 m.

Conservation Status: A widespread garden escape; LC (Least Concern).

Johnson (1971) recognized two forms of this species, one with truncate pappus scales, the other with longer setiferous scales, together with poor distinction of the glandular nature of the phyllaries. The two forms are not formally recognized here, although both were recognized by Smith (2005) at varietal level, together with a key.

151. **STOMATANTHES** R.M.King & H.Rob.

Stomatanthes R.M.King & H.Rob. in Phytologia **19**(7): 430 (1970); in Kew Bull. **30**(3): 463–465 (1975); Genera of the Eupatorieae (Asteraceae). Monogr. Syst. Bot. **22**: 69 (1987). —Bremer et al. in Bremer, Asterac. Cladist. Classif.: 664 (1994). —Mesfin Tadesse, Fl. Ethiopia & Eritrea **4**(2): 348 (2004). —Smith in F.T.E.A., Compositae **3**: 825 (2005). —Hind & Robinson in Kubitzki, Fam. Gen. Vasc. Pl. **8**: 510–574 [2006](2007). —Grossi & Katinas in Syst. Bot. **38**(3): 830–848 (2013). —Hind & Goyder in Kew Bull. **69**(4)–9545: 1–9 (2014).
Stomatanthes R.M.King & H.Rob. subgen. *Verticifolium* R.M.King & H.Rob. in Kew Bull. **30**(3): 463 (1975).
Criscianthus Grossi & J.N.Nakaj. [in Grossi, Katinas & Nakajima] in Phytotaxa **141**(1): 33 (2013).

Perennial herbs or subshrubs/suffrutices. Leaves alternate, opposite or ternate to pseudoverticillate, lamina elliptical or oblanceolate to ovate or orbicular, rarely linear, margins entire to markedly dentate or lobulate. Inflorescences usually pyramidal to thyrsoid-paniculate, sometimes corymbose. Capitula discoid; phyllaries 4–12, distant to weakly subimbricate, (1)2–3-seriate, unequal to subequal; receptacle scarcely convex, epaleaceous. Florets 4–11; corollas white, funnelform or nearly tubular, glabrous or glanduliferous with few to many hairs outside; lobes as long as wide to nearly 1.5 × as long as wide. lower parts of filaments short, thick, straight; anther collars cylindrical; apical anther appendages ovate or slightly shorter than wide; style base lacking basal node, covered with numerous hairs; style arms linear to filiform or with clavate tips, papillose at least below, tips when greatly enlarged with smooth surfaces. Achenes prismatic, 5–8-ribbed, often glandular setuliferous, setulae (at least in some African taxa) with basally branched twin-hairs; carpopodium usually distinct; pappus setae uniseriate, numerous, barbellate, persistent, apical cells with obtuse or acute tips.

A genus of 17 spp., E, central and S Africa, Brazil, Paraguay, Uruguay. Four spp. are currently known from Africa (Hind & Goyder 2014), two occurring in the Flora area.

The original key distinction between the achenes of the two species from the Flora area has now been found to be false, i.e. that the achenes of *S. zambiensis* 'has few or no setae', especially if referring to pappus setae (since they are as equally numerous), and certainly when referring to the achene setulae. However, the species can be easily distinguished on the basis of leaf characters. The recent 'new circumscription' of the genus by Grossi & Katinas (2013) only included three species (*S. africanus*, *S. helenae* and *S. meyeri*), *S. zambiensis* was excluded (but later placed in the newly recognized *Criscianthus*), as were the South American taxa (also left unplaced). This narrow view is not accepted here, especially since their observations are not supported by the material I have available.

Lower leaves alternate, upper opposite, subopposite or alternate . . . . **1.** *africanus*
Lower leaves in whorls or pseudoverticels of three, upper alternate, rarely opposite . . . . .
. . . . **2.** *zambiensis*

1. **Stomatanthes africanus** (Oliv. & Hiern) R.M.King & H.Rob. in Phytologia **19**(7): 430 (1970). —Burkill, Useful Pl. W. Trop. Afr., ed. 2, **1**: 492 (1985). —King & Robinson, Genera of the Eupatorieae: 71, t. 6 (1987). —Lisowski, (Asterac. Fl. Afr. Cent. 2) Fragm. Flor. Geobot. **36** Suppl. 1: 452, fig. 93 (1991). —Agnew & Agnew, Upland Kenya Wild Flow., ed. 2: 204, t. 83 p.p. (1994). —Mesfin Tadesse, Fl. Ethiopia & Eritrea **4**(2): 349 (2004). —Smith in F.T.E.A., Compositae **3**: 825–827 (2005). —Hind & Goyder in Kew Bull. **69**(4)–9545: 5 (2014).

   Eupatorium africanum Oliv. & Hiern in F.T.A. **3**: 301 (1877). —Andrew, Fl. Pl. Sudan **3**: 29 (1956). –Adams in F.W.T.A., ed. 2, **2**: 285 (1963). —Cufodontis, (Enum. Pl. Aethiop.) Bull. Jard. Bot. Natl. Belg. **36** (3, suppl.): 1080 (1966). —Maquet in Fl. Rwanda: 572, fig. 174/2 (1985). Types: 'Nile Land. Niamniam-land, [südlich von Gumba, 5 Febr. 70] *Schweinfurth* [2897]. [Sudan–D.R. Congo border, Niamniam Land, Gumba, *Schweinfurth* 2897]. Mozamb. Dist. Morambella [sic!], alt. 1200 ft,, [18] January [63], *Dr. Kirk.*' *Schweinfurth* 2897 (K000273395 lectotype, US), lectotypified by Smith (2005: 825); *Kirk* s.n. (K000273398 syntype).

   Vernonia humilis C.H.Wright in Bull. Misc. Inform., Kew **1897**(128–129): 269 (1897). Types: [Malawi:] 'German East Africa: Lower plateau, north of Lake Nyasa, *J. Thomson* [7]. British Central Africa [= Tanzania]: Mt. Mlanje, 6000 ft., *McClounie*, 30.' *McClounie* 30 (K000768733 syntype); *Thomson* 7 (K000768732 syntype).

   Vernonia malosana Baker in Bull. Misc. Inform., Kew **1898**(139): 148 (1898). Types: [Malawi:] 'BRITISH CENTRAL AFRICA. Mounts Malosa and Zomba, alt. 4000–6000 ft., *Whyte* [s.n.]' 'Nyassaland. Mt Malosa 4000 to 6000 ft. Nov. & Dec. 1896. *A. Whyte*, s.n.' (K000273397 syntype); 'Nyassaland. Mt Znuba [sic!] 4000 to 6000 ft Dec 1896. *A. Whyte* s.n.' (K000768734 syntype).[274]

   Eupatorium helenae Buscal. & Muschl. in Bot. Jahrb. Syst. **49**(3–4): 505 (1913). Type: 'ZAMBIA. Copperbelt: Steppe zwischen Broken-Hill und Buana-Mukuba, 18. Jan. 1910, *Aosta* 410' (B† holotype).

   Eupatorium africanum var. *vanmeelii* Lawalrée in Bull. Jard. Bot. État Bruxelles **19**(3): 220 (1949). Type: 'CONGO BELGE: HAUT-KATANGA: Tugulu, savane boisée, assez commun, novembre 1946., *Van Meel* 155' (BR6726935 holotype).

   Stomatanthes helenae (Buscal. & Muschl.) Lisowksi, (Asterac. Fl. Afr. Cent. 2) Fragm. Fl. Geobot. **36** Suppl. 1: 456 (1991). —Grossi in Candollea **66**(2): 361–366 (2011).

Perennial subshrub/suffrutex 20–120 cm high, with 1–several stems arising from a woody xylopodium 1–3.5 cm in diam. Stems erect, rigid, woody, somewhat branched, colour from straw-colour through reddish brown to black, short-pubescent, more densely towards apex. Leaves alternate to sub-opposite, sessile to short-petiolate, petiole to 6 mm, blade lanceolate to ovate, sometimes linear, rarely scale-like in lower parts of stem, 0.8–6.2(8) × 0.2–2.2(4.5) cm, margins often slightly thickened, entire to serrate,

---

[274] The latter collection is presumably the source of Baker's citation of the species from Mt Zomba.

sometimes irregularly lobulate, pubescent, especially on veins, sometimes glabrous or pubescence confined to leaf margins, sparsely to conspicuously glandular-punctate beneath or eglandular. Inflorescence of dense corymbs; bracts 3.5–10 mm long, linear to linear-lanceolate and foliaceous; pedicels 2–10 mm, increasing in length in fruit to 18 mm. Capitula 8–14 × 3–6 mm; phyllaries 5–7, 1–2-seriate, outer series (when present) 3–5 × 1–1.5 mm, somewhat shorter than inner series, 4–9 × 1–3.5 mm, green, lanceolate to oblong-ovate, pubescent, ± glandular-punctate towards apices, apex obtuse, sometimes apices reddish or purplish, with scarious, fimbriate margins; receptacle flat. Florets 4–5, corolla 5–7 mm long, whitish, apex narrowly campanulate, 5-lobed, outer surface ± covered with hairs and/or small glands; anthers brown, basally obtuse with ovate apical appendages, anther-collars thickened; style-arms linear, very long, often contorted, short-appendaged. Achenes brown to blackish, 3–4 mm long, ± 6-ribbed, ribs pale, body and ribs sparsely to ± densely setuliferous, setulae of twin-hairs, often divergently forked and with short branches at base, apices acute, body sparsely glandular-punctate or eglandular; carpopodium straw-coloured to whitish; pappus setae uniseriate, 5–8 mm long, straw-coloured, coarsely barbellate.

Zambia. N: Mbala Dist., Chiulungoma Road, 21.viii.1967, *Sanane* 5 (K). W: Mwinilunga Dist., 26.xi.1937, *Milne-Redhead* 3415 (K). Solwezi Dist., Solwezi Dambo, 10.ix.1952, *White* 3206 (K, OXF). E: Nyika Plateau, 2150 m, 21.x.1958, *Robson* 211 (K). S: Mazabuka Dist., 24.viii.1962, Angus 3301 (K, OXF). **Zimbabwe**. N: Miami Dist., Exp. farm in Mahobohobo area, 4.x.1946, *Wild* 1282 (K, SRGH). C: Salisbury Dist., x.1917, *Walters* 2447 (K). E: Inyanga Dist., Rhodes Estate, 7.xi.1944, *Hopkins* in GH 13121 (K, SRGH). **Malawi**. N: Nkhata Bay Dist., 15 m SW Mzuzu, 1540 m, 1.x.1977, *Phillips* 2964 (K, MO). C: Kota-Kota Dist., Chenga Hill, 1600 m, 9.ix.1946, *Brass* 17596 (K, NY). S: Mlanje Dist., Mlanje Mountain, above Fort Lister, 1550 m, 8.vi.1970, *Brummitt* 11364 (K). **Mozambique**. N: Plateau de Lichinga, 13°35′S, 35°20′E, 1300 m, xii.1932, *Gomes Sousa* 1068 (K). Z: Close to Moebede, 50 miles from Nawagra, Quelimane Dist., 610 m, 1.i.1948, *Faulkner* Kew 174 (K). MS: Chimanimani Mountains, between Skeleton Pass and The Plateau, 27.ix.1966, *Grosvenor* 200 (K, SRGH). Border lantine ostt. von Umtali, 10.xii.1913, *Peter* 52879 (K).

Also in Guinea, Sierra Leone, Ivory Coast, Ghana, Nigeria, Cameroon, Gabon, Congo, D.R. Congo, Rwanda, Burundi, Sudan, Ethiopia, Uganda, Kenya, Tanzania, Angola, South Africa. Pyrophyte, often found on regularly burned highland grassland, a weed of cultivation, also in woodland. grows in dense clumps; (350)1050–2460 m.

Conservation Status: A widespread species; LC (Least Concern).

Stomatanthes helenae has been separated from *S. africanus* on the basis of narrower leaves, glabrous phyllaries and pappus colour (Lisowski, 1991: 452). Grossi (2011) clearly still recognized the species and neotypified the basionym, providing an extended English description and a key to African species as recognized by her. However, several collections from Malawi (North Prov.: Nyika Plateau) have both glabrous phyllaries and glandular punctate lower leaf surfaces, but otherwise are identical with many collections of the very variable *S. africanus*. The two species are clearly sympatric in a large part of their supposed distribution and are best treated as synonymous.

2. **Stomatanthes zambiensis** R.M.King & H.Rob. in Kew Bull. **30**(3): 465 (1975). — Hind & Goyder in Kew Bull. **69**(4)–9545: 5 (2014). Type: 'ZAMBIA. Mporokoso District: Mporokoso-Kawimbe, close to Mporokoso, 1200 m, 7 Jan. 1960, *Richards* 12084' (K000273399 holotype, M0104528). FIGURE 6.5.**44**.

Criscianthus zambiensis (R.M.King & H.Rob.) Grossi & J.N.Nakaj. in Phytotaxa **141**(1): 34 (2014).

Perennial subshrub to c. 1 m tall, erect. Rootstock a small woody xylopodium. Stems few, simple or sometimes laxly branched in upper half, moderately to densely tomentose beneath inflorescence, rapidly glabrescent below, hairs eglandular. Leaves usually scale-like in lower quarter, ternate in middle section becoming alternate, or rarely opposite, above just beneath inflorescence, membranaceous or subcoriaceous, broad-elliptic, 3–7 × 5–15 mm, venation prominent and somewhat pale beneath, margins serrate to coarsely so, or even lobed in upper half, apex short-acute. Inflorescence a lax compound cyme. Phyllaries 7–10, 2-seriate, imbricate, ± gradate, narrow-elliptic, outer

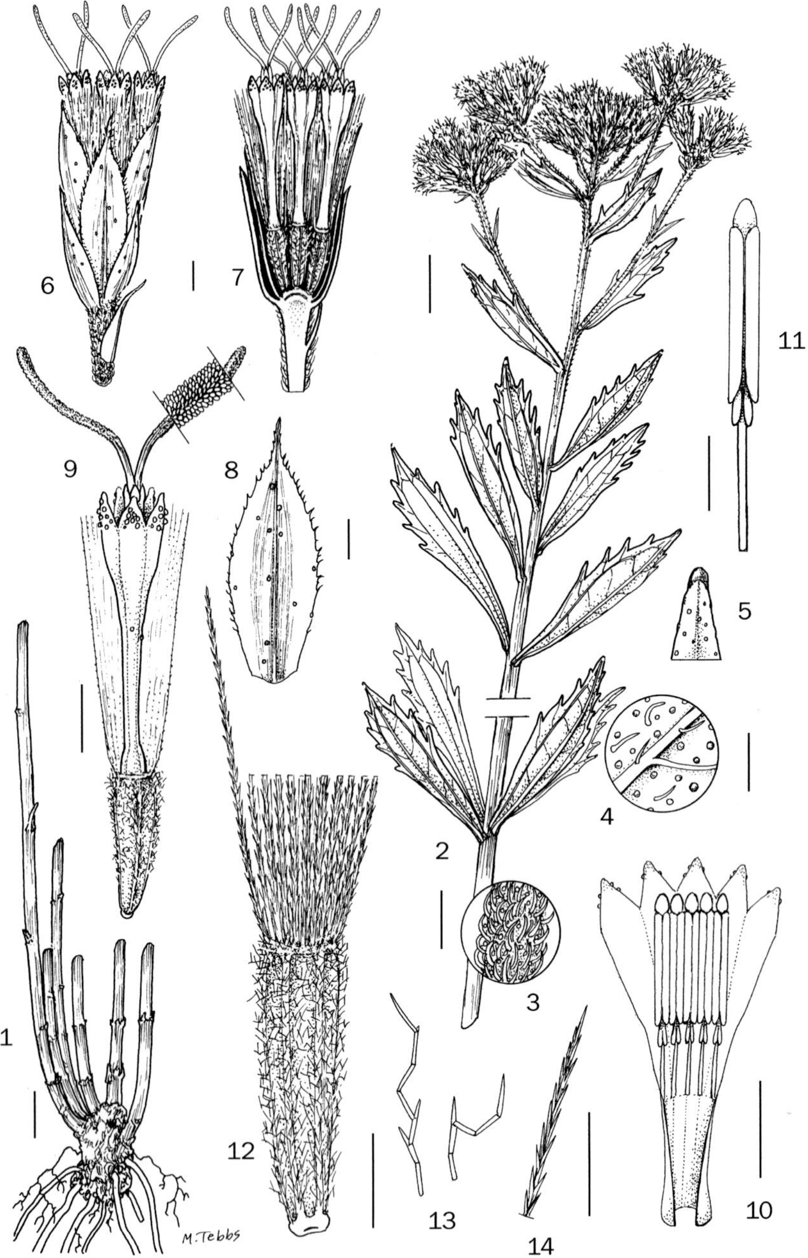

Fig. 6.5.**44**. STOMATANTHES ZAMBIENSIS. 1, Xylopodiaceous rootstock and the bases of several stems showing scale-like leaves; 2, apex of flowering stem showing whorled leaves of lower or mid-stem, and subalternate to alternate leaves closer to inflorescence; 3, detail of stem pubescence; 4, detail of lower surface of leaf; 5, detail of leaf margin tooth; 6, capitulum; 7, l.s. capitulum; 8, phyllary; 9, floret with magnified detail of style arm; 10, corolla opened out to show attachment of filaments; 11, stamen; 12, achene and partial pappus; 13, detail of branched achene setulae; 14, detail of apex of pappus seta. 1, 12–14 from *Mphamba* 950; 2–11 from *Pawek* 3437. Scale bars: 11, 14 = 0.5 mm; 3–4, 6–10 = 1 mm; 1–2 = 10 mm. Drawn by Margaret Tebbs.

3.5–5 × 0.7 mm, canaliculate, apices acute to acuminate, inner 5–6 × 1.2–1.5 mm, flat, glabrous, apices narrow-caudate. Florets 4 or 5; corollas c. 4 mm long, upper part narrowly infundibuliform, c. 1.2 mm diam., corolla tube c. 2.5 × 0.3 mm, white, glabrous or very sparsely glandular-punctate outside, more densely so on abaxial corolla lobe surfaces; apical anther appendages ovate, acute; style arm apices not expanded. Achenes 2.5–3 mm long, 5-ribbed, ribs concolorous with body, body black, glabrous to sparsely or moderately setuliferous, setulae of simple twin-hairs or hairs sympodially branched, body eglandular to densely glandular-punctate; carpopodium conspicuous; pappus setae 4–4.5 mm, coarsely barbellate, fawn, apical cells acute.

Zambia. N: Abercorn Dist., Kambole Escarpment, 1650 m, 19.ii.1957, *Richards* 8251 (K). Katenga Falls, Kambole, 1500 m, 21.ii.1957, *Richards* 8303 (K). E: Nykika National Park, c. 0.5 km SW of Zambian Govt. Rest House, 17.iv.1986, *Philcox* et al. 9962 (K, TEX/LL, US). **Malawi**. N: Rumpi Dist., Nyika, Chowo Rock, [c. 2195 m], 30.iii.1970, *Pawek* 3437 (K). Mafinga Hill Top, Chitipa. 2181 m, 5.viii.2007, *Chapama* et al. 705 (FRIM, K). Nyika National Park. Vitinthiza Peak. 1950 m, 29.vii.2009, *Mphamba* 950 (FRIM, K).

On rocks, in mixed open woodland, or in bush amongst long grass in loamy soil and usually in areas subject to frequent burning; 1200–2195 m.

Conservation Status: A species with a restricted distribution. None of the collections cited give any indication of local abundance. I prefer to note the species as DD (Data Deficient) until further fieldwork is carried out.

Grossi & Katinas (2013) excluded *S. zambiensis* from the genus, on the basis that it has 'verticillate phyllotaxis and lacks ramose hairs'. The phyllotaxis is a red herring and branched setulae are certainly present on the achenes of material outside of the Abercorn District (the achenes from here are glandular-punctate and sometimes possess twin-hairs), suggesting their revised circumscription might have been premature, or inadequate material was examined. Grossi & Nakajima [in Grossi, Katinas & Nakajima] (2013) have recently described the new genus *Criscianthus*, recognizing *C. zambiensis* – which I have immediately placed in synonymy (Hind & Goyder 2014).

INDEX TO BOTANICAL NAMES

Accepted names in roman, synonyms in *italic.* Bold page numbers indicate main entries of accepted names (names with descriptions and in the keys), and illustrations.

Abasoloa La Llave & Lex. 95
  taboada La Llave 97
Acanthambrosia Rydb. 133
ACANTHOSPERMUM Schrank 23, **172**
  australe (Loefl.) Kuntze 172, **174**, **175**
  australe sensu Wild 173, 176
  brasilum Schrank 175
  glabratum (DC.) Wild 172, **173**, **174**, 176
  hirsutum DC. 175
  hispidum DC. 172, **173**, 175
  humile (Sw.) DC. 173
  xanthioides (Kunth) DC.
    [unranked] β *hispidum* (DC.) Kuntze 173
    [var.] α *obtusifolium* DC. 175
    [var.] β *acutifolium* DC. 175
    [var.] γ *glabratum* DC. 173
Acanthotheca DC. 19
Acanthoxanthium (DC.) Fourr. 135
  spinosum (L.) Fourr. 144
    subsp. *catharticum* (Kunth) D.Löve 144
Achaenopodium K.Brandegee 127
Achyropappus Link & Otto 81
Acispermum Neck. 28
ACMELLA Rich. 24, **145**
  biflora (L.) Spreng. 113
  caulirhiza Delile **146**
  lanceolata Link ex Spreng. 97
  oppositifolia (Lam.) R.K.Jansen 145
  radicans (Jacq.) R.K.Jansen **147**
    var. radicans 146, 147, **149**
  repens (Walter) Rich. 145
Actimeris Raf. 127
Actinomeris Nutt. 127
Adenospermum Hook. & Arn. 25
  tuberculatum Hook. & Arn. 26
ADENOSTEMMA J.R.Forst. & G.Forst. 188, **196**
  afrum DC. 197, **198**, **199**
    var. *asperum* Brenan 198, 200
    var. *longifolium* (Chiov.) S.A.L.Sm. 198
  biflorum (L.) Less. 113
  lavenia (L.) Kuntze 196
    var. *longifolium* Chiov. 198
  mauritianum DC. 197, **200**
  perrottetii DC. 197
  schimperi Sch.Bip. ex A.Rich. 198
  viscosum J.R.Forst. & G.Forst. 196, **197**
  viscosum sensu Oliver & Hiern 198, 200

Adventina Raf. 166
  ciliata Raf. 170
  parviflora Raf. 167, 170
AGERATINA Spach **188**
  adenophora (Spreng.) R.M.King & H.Rob. **189**
  riparia (Regel) R.M.King & H.Rob. **189**, **190**
  ventillana (Cuatrec.) R.M.King & H.Rob. 191
  vernalis (Vatke & Kuntz) R.M.King & H.Rob. 189
Ageratiopsis Sch.Bip. ex Benth. 188
AGERATUM L. 188, 191, **201**
  arsenei B.L.Rob. 204
  brachystephanum Regel 202
  ciliare L. 202
  coeruleum Desf. 202
  conyzoides L. **201**, 202, **203**, 205
    f. *album* (Willd.) B.L.Rob. 204
    subsp. *houstonianum* (Mill.) M.Sharma 207
    var. *houstonianum* (Mill.) T.R.Sahu 207
    var. *inaequipaleaceum* Hieron. 204
    var. *mexicanum* (Sims) DC. 205
    var. β *hirtum* (Lam.) DC. 202
  cordifolium Roxb. 202
  hirsutum Lam. 202
  hirtum Lam. 202
  houstonianum Mill. 201, **203**, **205**
    var. *muticescens* B.L.Rob. 206
    var. α *typicum*
      f. b. *niveum* B.L.Rob. 206
      f. c. *luteum* B.L.Rob. 206
    var. β *angustatum* B.L.Rob. 206
    var. γ *multicescens* B.L.Rob.
      f. d. *isochroum* B.L.Rob. 206
      f. e. *versicolor* B.L.Rob. 206
  humile Larrañaga 204
  humile Salisb. 202
  latifolium Cav. 202
  mexicanum Sims 205
    [unranked, ?var.] a) *majus* Voss 206
      f. *albiflorum* Voss 206
      f. *coeruleum* Voss 206
      f. *lasseauxii* (Carrière) Voss 206
    [unranked, ?var.] b) *wendlandii* (Vilm.) Voss 206
    [unranked, ?var.] c) *nanum* Voss 206
  muticum Griseb. 202
  nanum hort. ex Sch.Bip. 202

odoratum E.Vilm. 202
perplexans M.F.Johnson 171
suffruticosum Regel 202
wendlandii hort. ex Vilm. 205
Aizoon L. 179
AMBROSIA L. 23, **133**
  maritima L. **133, 134**
  senegalensis DC. 133
Amellus L.
  abyssinicus (Sch.Bip. ex Walp.) Kuntze 103
  madagascariensis (Baker) Kuntze 108
  pedunculatis Ortega ex Willd. 164
  pungens (Oliv. & Hiern) Kuntze 110
  scandens (Schumach. & Thonn.) Kuntze 107
Amellus? carolinianus Walter 97
Anacis Schrank 28
Anaitis DC. 148
Ancistrophora A.Gray 127
Anisopappus Hook. & Arn. 1, 4
Anthemideae Cass. 3
Anthemiopsis macrophylla Bojer ex DC. 113
Anthemis L.
  cotula Blanco 99
  mysorensis [herb. madr. ex] DC. 180
Apostates Lander 76
Arctotideae Cass. 3
Arctotis glutinosa Sims 20
Argyrochaeta Cav. 130
  bipinnatifida Cav. 131
  parviflora Cav. 131
Arnoldia Cass. 19
Artemisia viridis Blanco 100
Artemisiopsis S.Moore 1, 4
ASPILIA Thouars 25, **115**, 116
  abyssinica (Sch.Bip. ex Walp.) Vatke 120
  africana 116
  angolensis (Klatt) Muschl. 116, **124**
  asperifolia O.Hoffm. 122
  aspilioides (Baker) S.Moore 126
  brachyphylla S.Moore 122
  bracteosa C.D.Adams 120
  chrysops S.Moore 126
  ciliata (Schumach.) Wild 116, **120**
  courtetii O.Hoffm. & Muschl. 118
  culuensis S.Moore 125
  dewevrei O.Hoffm. 120
  engleriana Muschl. 125
  eylesii S.Moore 110
  fischeri O.Hoffm. 120
  fontinalis Hiern 123
  gillettii Wild 126
  gondensis O.Hoffm. 122
  helianthoides (Schumach. & Thonn.) Oliv. &
      Hiern
    subsp. *ciliata* (Schumach.) Adams 94, 120
    subsp. *papposa* (O.Hoffm. & Muschl.)
        C.D.Adams 120

    subsp. *prieuriana* (DC.) C.D.Adams 94, 120
    var. *papposa* O.Hoffm. & Muschl. 120
  holstii O.Hoffm. 125
  huillensis (Hiern) S.Moore 124
  involucrata O.Hoffm. 122
  kakondensis S.Moore 124
  kotschyi (Sch.Bip.) B.D.Jacks. 118
  kotschyi (Sch.Bip.) Oliv. **116**, 119
    var. alba Berhaut 118, **119**
    var. kotschyi **117, 118**, 119
  mendoncae Wild 116, **121**
  monocephala Baker 123
  mossambicensis (Oliv.) Wild 116, **125**
  natalensis (Sond.) Wild 116, **123**
  pluriseta Schweinf. 116, **122**
    subsp. *gondensis* (O.Hoffm.) Wild 122
  polycephala S.Moore 119
  schimperi (Sch.Bip. ex A.Rich.) Oliv. & Hiern
      120
  tanganyikensis Lawalrée 126
  thouarsii DC. 116
  vernayi Brenan 126
  vulgaris N.E.Br. 122
  wedeliiformis Vatke 125
  welwitschii O.Hoffm. 123
    var. *serrata* Hiern 123
  zombensis Baker 110
    var. *longifolia* S.Moore 110
Astereae Cass. 3, 23
Athroisma DC. 1, 2, 3
Athroismeae 1, 2, 3, 4, 23
Athronia Neck. 145
Balbisia Cav. 163
Balbisia Willd. 163
  canescens Pers. 164
  divaricata Cass. 164
  elongata Willd. 164
  pedunculata Ortega ex Hoffmanns. 164
Bartlettina sordida (Less.) R.M.King & H.Rob.
    187
Bartolina Adans. 163
Batschia J.F.Gmel. 188
Batschia Moench 188
Batschia Mutis ex Thunb. 188
Baziasa Steud. 166
  urticifolia (Kunth) Steud. 170
Bellis ramosa Jacq. 97
Berkheyopsis diffusa O.Hoffm. 58
BIDENS L. 23, 36
  abyssinica Sch.Bip. ex Walp. 59
    var. *glabrata* Vatke 59
    var. *incisifolia* Hochst. ex Chiov. 60
  acuticaulis Sherff 37, **50**, 53
    var. acuticaulis **51, 52**
    var. filirostris (P.Taylor) T.G.J.Rayner **51, 52**
  acutiloba Sherff 58
  adhaerescens Vell. 62

ambacensis (Hiern) Sherff 48
ambigua S.Moore 44
artemisiifolia (Jacq.) Kuntze 35
bampsiana Lisowski 43
baumii (O.Hoffm.) Sherff 36, 38, **41**, **42**
bequaertii De Wild. 44
    var. *amplior* Sherff 44
bipinnata Baill. 34
bipinnata L. 37, **61**
    var. *minor* Memm. 61
bipinnata sensu auct. mult. non L. 60
biternata (Lour.) Merr. & Sherff 37, **59**
    var. *glabrata*
        f. *abyssinica* (Sch.Bip.) Sherff 60
        f. *lasiocarpa* (O.E.Schulz) Sherff 60
californica DC. 62
cicutifolia Tausch 61
chinensis (L.) Willd.
    var. *abyssinicus* (Sch.Bip. ex Walp.)
      O.E.Schulz 60
ciliata De Wild. 50
cochlearis Merxm. 53
crataegifolia (O.Hoffm.) Sherff 39
    var. *burttii* Sherff 40
crocea Welw. ex O.Hoffm. 38, **44**
    var. *ornata* Sherff 45
    var. *verrucifera* S.Moore 44
cuspidata Sherff 40
cylindrica Sherff 60
decomposita Wall. ex DC. 61
decussata Pav. ex DC. 62
dielsii Sherff
    var. *intermedia* Sherff 41
diversa Sherff 37, **53**
    subsp. *filiformis* (Sherff) T.G.J.Rayner 53
drummondii Wild 49
elongata Tausch 61
exilis (Sherff) Lisowski 41
fervida hort. ex Colla 61
filiformis Sherff 53
formosa (Bonato) Sch.Bip 34
gardneri Baker 53
gossweileri Sherff 44
grantii (Oliv.) Sherff.
    var. *stapfioides* Sherff 58
hirsuta Nutt. 63
hispida Kunth in Humb. 62
insecta (S. Moore) Sherff 46
insignis Sherff 40
kasaiensis Lisowski 45
kilimandscharica (O.Hoffm.) Sherff 37, 38, **39**
    var. *oxymera* Sherff 40
    var. *retrorsa* Sherff 40
kirkii (Oliv. & Hiern) Sherff 38, **46**
kotschyi Sch.Bip. ex Walp. 59
lasiocarpa O.E.Schulz 60
leptoglossa (Sherff) Lisowski 40

leptolepis Sherff 55
    f. *pallida* Sherff 55
leucantha (L.) Willd. 63
    f. *discoidea* (Sch.Bip.) Krauss 63
    var. *pilosa* (L.) Griseb. 63
lindleii Sch.Bip. 34
lindleyii Sch.Bip. 34
malawiensis Mesfin 37, **50**
meruensis Sherff 40
microcarpa Sherff 58
modesta Sherff 43
moorei Sherff 37, **43**
    [var.] *verrucosa* Sherff 43
myrrhidifolia Tausch 61
negriana (Sherff) Cufod. 50
nyikensis Sherff 41
oblonga (Sherff) Wild 38, **41**
occidentalis (Hutch. & Dalziel) Mesfin 50
ochracea (O.Hoffm.) Sherff 38, **45**
oligoflora (Klatt) Wild 37, **49**
onisciformis Sherff 49
palustris Sherff 44
    var. *cubangona* Sherff 45
paupercula Sherff 50
    var. *filirostris* P.Taylor 51
phalangiphylla Sherff 44
pilosa L. 37, **62**, 63
    f. *discoidea* Sch.Bip. 63
    f. *radiata* Sch.Bip. 63
    subvar. *discoidea* (Sch.Bip.) Pit. 63
    subvar. *radiata* (Sch.Bip.) Pit. 63
    var. *abyssinica* (Sch.Bip. ex Walp.) Fiori 60
    var. *bipinnata* (L.) Hook.f. 61
    var. *decomposita* (Wall. ex DC.) Hook.f. 61
    var. *discoidea* Sch.Bip. 63
        [forma] *pinnata* Kuntze 63
        [forma] *subbiternata* Kuntze 63
    var. *glabrata* (Vatke) Engl. 60
    var. *leucantha* (L.) Kuntze 63
        [forma] *subsimplicifolia* Kuntze 63
    var. *leucantha* Sch.Bip. 63
        [forma] *subbiternata* Kuntze 63
        [forma] *ternata* Kuntze 63
    var. *minor* (Blume) Sherff 63
    var. *quadriseta* (Hochst. ex Oliv. & Hiern)
      Engl. 60
    var. *radiata* (Sch.Bip.) J.A.Schmidt 63
pinnatipartita **38**
prolixa S.Moore 56
punctata Sherff 56
quadriseta Hochst. ex Oliv. & Hiern 60
ramtilla Wall. 180
reflexa Link 62
rhodesiana Sherff 41
richardsiae Sherff 41
robustior S.Moore 39
ruandensis Sherff 43

rubicundula Sherff 36, **55**
    f. *alba* T.G.J.Rayner 55
rubra De Wild. 53
rufovenosa Sherff 46
schimperi Sch.Bip. ex Walp. 37, 48, **56**, **57**
    var. *brachycera* Sherff 58
    var. *greenwayi* Sherff 58
    var. *leiocera* Sherff 58
    var. *leptocera* Sherff 56, 58
    var. *punctata* (Sherff) Sherff 58
schlechteri Sherff 46
    var. *wildii* Sherff 46
setigera (Sch. Bip.) Sherff
    subsp. setigera 50
somaliensis Sherff
    var. *bukobensis* Sherff 43
steppia (Steetz) Sherff 37, **46**, **47**
    var. *ambacensis* (Hiern) Sherff 48
    var. *elskensii* Sherff 48
    var. *garusonis* Sherff 48
    var. *inarmata* Sherff 48
    var. *kalamboensis* Sherff 48
    var. *leptocarpa* Sherff 48
sulphureus (Cav.) Sch.Bip. 35
sundaica Blume 62
    var. *minor* Blume 62
triplinervia Kunth 35
uhligii Sherff 48
ukambensis S.Moore 39
urceolata De Wild. 36, **53**, **54**
    var. *leptolepis* (Sherff) T.G.J.Rayner 55
volkensii O.Hoffm. 39
BLAINVILLEA Cass. 25, **91**
  acmella (L.) Philipson 94
  dichotoma (Murray) Cass. ex Hemsl. 92, 93, 94
  gayana Cass. 92, 94, 116
    var. *lanceolata* Chiov. 92
  hispida Edgew. 92
  latifolia (L.f.) DC. 92, 94
  polycephala Gardner 92
  prieureana DC. 94, 116, 120
  racemosa Gardner 92
  rhomboidea Cass. 92, 94
    var. *lanceolata* (Poir.) DC. 92
    var. *polycephala* (Gardner) Baker 92
    var. *racemosa* (Gardner) Baker 92
  rhomboidea sensu Oliv. & Hiern 92
Blaxium Cass. 19
Blepharispermum DC. 1, 2, 3
Bolophyta Nutt. 130
Brachypoda prostrata Raf. 99
Brachystegia manga De Wild. 108
Brotera Spreng. 67
  contrayerva Spreng. 70
  sprengelii Cass. 70
  trinervata (Willd.) Pers. 70

Buphthalmum L.
  diffusum Vahl ex DC. 99
  scandens Schumach. & Thonn. 107
Cacalia mentrasto Vell. 202, 205
Caesulia radicans Willd. 66
Calendula L. 5
  cuneata Thunb. 19
  viscosa Andrews 19
Calenduleae Cass. **4**
Callilepis DC. 4
Calliopsis Rchb. 28
Calyptocarpus burchellii (Hook.) Sch.Bip. 92
Cardueae Cass. 3
Carelia Juss. ex Cav. 191
  brachystephana (Regel) Kuntze 204
  conyzoides (L.) Kuntze 202
    [α *robusta* Kuntze] var. *alba* Kuntze 204
    [β *umbrosa* Kuntze] var. *coerulea* Kuntze 204
    [γ *pusilla* Kuntze] var. *alba* Kuntze 204
  conyzoides [unranked]
    α *robusta* Kuntze 204
    β *umbrosa* Kuntze 204
    γ *pusilla* Kuntze 204
  houstoniana (Mill.) Kuntze 205
  mutica (Griseb.) Kuntze 204
Carelia Ponted. ex Fabr. 191
Carphostephium Cass. 163
Castalis Cass. 19
Centrospermum Kunth 172
  xanthioides Kunth 175
Cephalobembix Rydb. 81
Ceratocephalus Burm. ex Kuntze 145
  exasperatus (Jacq.) Kuntze 147
Chaenactidinae B.G.Baldwin 76
Chaenocephalus Griseb. 127
Chamaestephanum Willd. 81
Chromolaena odorata (L.) R.M.King & H.Rob. 188
Chrysanthellina Cass. 25
CHRYSANTHELLUM Rich. 24, **25**
  americanum (L.) Vatke 28
  argentinum Ariza & Cerana 26
  boliviense Sch.Bip. 26
  indicum DC. **26**
    subsp. afroamericanum B.L.Turner **26**, **27**
    var. *afroamericanum* B.L.Turner 26
  procumbens Pers. 28
  tuberculatum (Hook. & Arn.) Cabrera 26
  weberbaueri Chung 26
CHRYSANTHEMOIDES Fabr. 5, **15**
  monilifera (L.) Norl. **16**, **17**
    subsp. rotundata (DC.) Norl. **18**
    subsp. septentrionalis Norl. **16**, 19
Chrysogonum peruvianum L. 150
Chrysomelea Tausch 28
  lanceolata (L.) Tausch 31
Chrysostemma Less. 28

Clipteria Raf. 95
  dichotoma Raf. 99
Collaea Spreng. 25
Colobogyne Gagnep. 145
Conopsis Nutt. 28
COREOPSIS L. 23, 25, **28**, 36
  sect. *Calliopsis* (Rchb.) Nutt. 28
  sect. *Eublepharis* Nutt. 28
  sect. *Gyrophyllum* Nutt. 28
  sect. *Leptosyne* (DC.) O.Hoffm. 29
  sect. *Pugiopappus* (A. Gray) S.F.Blake 29
  sect. *Tuckermannia* (Nutt.) S.F.Blake 29
  ambacensis Hiern 48
  artemisiifolia Jacq. 35
  aspilioides Baker 126
  atkinsoniana Douglas ex Lindl. 29
  baumii O.Hoffm. 43
  biternata Lour. 59
  corymbifolia Buch.-Ham. 61
  corymbifolia Hamilt. 61
  cosmophylla Sherff 45
  crassifolia Dryand. ex Aiton 29
  crataegifolia O.Hoffm. 39
  curtisiae Sherff 46
  exaristata O.Hoffm. 56, 58
  exilis Sherff 41
  formosa Bonato 34
  galericulata Sherff 182
  goffardi Sherff 41
  grandiflora Hogg ex Sw. 29
  heterogyna Fernald 31
  injucunda Sherff 48
  insecta S.Moore 46
  isokoensis Sherff 49
  kilimandscharica O.Hoffm. 39
  kirkii Oliv. & Hiern 46
  lanceolata L. 23, **29**, **30**
    var. *pumila* Moldenke 31
    var. *villosa* Michx. 29
    var. α *glabella* Michx. 29
    var. α *succisifolia* DC. 31
    var. β *angustifolia* Torr. & A.Gray 31
    var. γ *crassifolia* (Dryand. ex Aiton) Heynh. 31
  leptoglossa Sherff 40
  leucantha L. 62, 63
  leucanthema L. 62
  lupulina O.Hoffm. 38
  mattfeldii Sherff 49
  multiflora Sherff 48
  oblonga Sherff 41
  oblongifolia Nutt. 31
  ochracea O.Hoffm. 45
  ochraceoides Sherff 45
  oligantha Klatt 49
  oligoflora Klatt
    var. *robusta* Sherff 49

pinnatipartita O.Hoffm. 38
  scabrifolia Sherff 43
  schlechteri (Sherff) Burtt Davy 46
  steppia Steetz 48
  tinctoria Nutt. 29
  vulgaris Sherff 48
  whytei S.Moore 38
Coreopsoides Moench 28
  lanceolata (L.) Moench 29
Coronocarpus Schumach. & Thonn. 115
  kotschyi (Sch.Bip.) Benth. 118
Corynanthelium Kunze 191
Cosmea Willd. 31
  bipinnata (Cav.) Willd. 34
  sulphurea (Cav.) Willd. 35
  tenuifolia (Lindl.) Heynh. 34
  tenuifolia (Lindl.) J.W.Loudon 34
COSMOS Cav. 23, **31**, 36
  ×*spectabilis* Carrière 34
    var. *alba* Carrière 34
    var. *rosea* Carrière 34
  artemisiifolius (Jacq.) M.R.Almeida 35
  aurantiacus Klatt 35
  bipinnatus Cav. **32**, **33**
    var. β *exaristatus* DC. 34
  caudatus Kunth 32
  formosa Bonato 34
  hybridus Goldring 34
  sulphureus Cav. 32, **33**, **35**
    var. *exaristatus* Sherff 35
  tenuifolius Lindl. 34
Cosmus Pers. 31
Cotula L.
  alba (L.) L. 97
  oederi Murray 97
Crassina Scepin 148
  intermedia (Engelm.) Kuntze 152
  leptopoda (DC.) Kuntze 152
  multiflora (L.) Kuntze 152
  peruviana (L.) Kuntze 152
    var. *flava* Kuntze 152
  tenuiflora (Jacq.) Kuntze 152
  verticillata (Andrews) Kuntze 152
Criscianthus Grossi & J.N.Nakaj. 208, 207, 211
  zambiensis (R.M.King & H.Rob.) Grossi & J.N.Nakaj. 209, 210, 211
Cryphiospermum P.Beauv. 64
  abyssinicum (Sch.Bip. ex Walp.) Schweinf. 178
  repens P.Beauv. 66
Dahlia Cav.
  coccinea Cav. 23
  imperialis Roezl ex Ortgies 23
  sp. 23
Diglossus Cass. 71
Dilepis Suess. & Merxm. 67
  dichotoma Suess. & Merxm. 70

DIMORPHOTHECA Moench 5, 11, **19**
  sect. *Acanthotheca* (DC.) J.C.Manning &
    Goldblatt 19
  sect. *Arnoldia* (Cass.) DC. 19
  sect. *Blaxium* (Cass.) DC. 19
  sect. *Castalis* (Cass.) DC. 19
  sect. *Meteorina* (Cass.) DC. 19
  barberae Harv. 19
  cuneata (Thunb.) DC. **19**
  ecklonis DC. 19
  fruticosa (L.) Less. 19
  zeyheri Sond. 19, **20**, **21**
Diplosastera Tausch 28
Diplothrix DC. 148
Dipterotheca Sch.Bip. 115
  kotschyi Sch.Bip. 118
Ditrichum Cass. 127
Dyssodia tenuiloba (DC.) B.L.Rob. 23
Echetrosis Phil. 130
  pentasperma Phil. 131
Echinodium Poit. ex Cass. 172
  prostratum Poit. ex Cass. 175
ECLIPTA L. 24, **95**
  adpressa Moench 97
  alba (L.) Hassk. 100
    forma *prostrata* (L.) Hassk. 100
    forma α *erecta* (L.) Hassk. 100
    forma β *longifolia* (Schrad. ex DC.) Hassk.
      100
    forma γ *zippeliana* (Blume) Hassk. 100
    forma β *longifolia* (Schrad. ex DC.) Hassk.
      100
    var. *longifolia* Bettfr. 100
    var. α *erecta* (L.) Miq. 100
    var. β *zippeliana* (Blume) Miq. 100
    var. γ *prostrata* (L.) Miq. 100
    var. δ *parviflora* (Wall. ex DC.) Miq. 100
  angustata Umemoto & H.Koyama 101
  angustifolia C.Presl. 100
  arabica Steud. 99
  brachypoda Michx. 97
  dentata B.Heyne ex Wall. 98
  dichotoma Raf. 99
  dubia Raf. 99
  erecta L. 97, 99
    var. *brachypoda* (Michx.) Torr. & A.Gray 99
    var. *prostrata* (L.) Baker 100
    [var.] β *diffusa* DC. 98
    [var.] γ *latifolia* Willd. ex Walp. 100
  filicaulis Schumach. & Thonn. 146
  flexuosa Raf. 99
  hirsuta Bartl. 99
  linearis Otto ex Sweet 98
  longifolia Schrad. 98
    var. *linearis* Fisch. & E.Mey. 99
  longifolia Schrad. ex DC. 98
  marginata Boiss. 100

  marginata Steud. 99
  nutans Raf. 99
  oederi (Murray) Weigel 97
  palustris DC. 99
  palustris G.Forst. ex Spreng. 98
  parviflora Wall. ex DC. 98
  patula Schrad. 98
  patula Schrad. ex DC. 99
  patula (Schrad. ex DC.) Endl. 99
  philippinensis Gand. 100
  procumbens Michx. 97, 99
    var. *brachypoda* (Michx.) A.Gray 100
    [var.] β *patula* DC. 99
  prostrata (L.) L. **95**, **96**, 100
    var. *aureoreticulata* Y.T.Chang 101
    var. *dixitii* An.Kumar & K.K.Khanna 101
    var. *zippeliana* (Blume) J.Kost. 100
    [var.] β *undulata* (Willd.) DC. 98
  pumila Raf. 99
  punctata L. 97
  simplex Raf. 99
  strumosa Salisb. 97
  sulcata Raf. 99
  thermalis Bunge 98
  tinctoria Raf. 99
  undulata Willd. 97
  zippeliana Blume 98
Ecliptica Rumph. ex Kuntze 95
  alba (L.) Kuntze 100
  alba β *zippeliana* (Blume) Kuntze 100
  alba γ *prostrata* (L.) Kuntze 100
  alba δ *parviflora* (Wall. ex DC.) Kuntze 100
Eisenmannia Sch.Bip. 91
  clandestina Sch.Bip. 91, 92
Electra DC. 28
Eleutheranthera prostrata (L.) Sch.Bip. 100
Enalcida Cass. 71
Encelia (Geraea) albescens A.Gray 128
Enhydra DC. 64
ENYDRA Lour. 24, **64**
  anagallis Gardner 66
  fluctuans Lour. **64**, **65**
  [sub *Enhydra*] *heloncha* DC. 66
  [sub *Enhydra*] *longifolia* (Reinw.) DC. 66
  [sub. *Enhydra*] *paludosa* (Reinw.) DC. 66
  woollsii F.Muell. 66
Epilepis Benth. 28
Eriocarpha Cass. 85
Eriocoma Kunth 85
  elegans (K.Koch) Kuntze 88
  hibiscifolia (Benth.) Kuntze 87
  pyramidata (Sch.Bip.) Kuntze 88
Erlangea calycina S.Moore 58
Ethulia bidentis L. 67
EUPATORIEAE Cass. 2, **187**
Eupatoriophalacron Mill. 95
  album (L.) Hitchc. 100

Eupatorium
  adenophorum Spreng. 189
  africanum Oliv. & Hiern 208
    var. *vanmeelii* Lawalrée 208
  chilense Molina 67
  glandulosum Kunth 189
  glandulosum Michx. 189
  harrisii Urb. 191
  helenae Buscal. & Muschl. 208
  riparium Regel 189
  scandens Thunb. 195
  ventillanum Cuatrec. 191
FLAVERIA Juss. 24, **67**
  australasica sensu Wild 70
  bidentis (L.) Kuntze **67**, 70
    [forma] β *angustifolia* Kuntze 68
  bonariensis DC. 68
  capitata Juss. ex Sm. 68
  chilensis J.F.Gmel. 67
  chiloensis Juss. 67
  contrayerba (Cav.) Pers. 68, 70
    var. *latifolia* Phil. 68
  repanda Lag. 70
  trinervata (Willd.) Baillon 70
  trinervia (Spreng.) C.Mohr 67, **68**, **69**
Franseria Cav. 133
Gaertneria Lam. 133
Gaertneria Medik. 133
Gaertneria Retz. 133
Gaertneria Schreb. 133
Galinsogaea Himpel 166
Galinsoga? *oblongifolia* (Hook.) DC. 98
GALINSOGA Ruiz & Pav. 25, **166**
  aristulata E.P.Bicknell 171
  bicolorata H.St.John & D.White 171
  brachystephana Otto 170
  brachystephana Otto ex Heer & Regel 170
  caracasana (DC.) Sch.Bip. 170
  ciliata (Raf.) S.F.Blake 171
  eligulata Cuatrec. 171
  hirsuta Baker 167
  hispida Benth. 170
    var. *albiflora* Fenzl 170
    var. *purpurascens* Fenzl 170
  hispida (DC.) Hieron. 171
  humboldtii Hieron. 171
  laciniata Retx. 167
  parviflora Cav. **167**, **168**, 171
    var. *caracasana* (DC.) A.Gray 170
    var. *genuina*
      f. *parceglandulosa* Thell. 169
    var. *hispida* DC. 170, 171
    var. *quadriradiata* (Ruiz & Pav.) Poir. 170
    var. *semicalva* A.Gray 167
    var. β *adenophora* Thell. 169
  quadriradiata Ruiz & Pav. 167, **168**, **169**, 171
    [unranked] *hispida* (DC.) Thell. 171

    [unranked] *quadriradiata*
      f. *albiflora* (Fenzl) Thell. 171
      f. *purpurascens* (Fenzl) Thell. 171
      f. *vargasiana* Thell. 171
  quadriradiata sensu Merxm. 169
  quinqueradiata Ruiz & Pav. 167
  semicalva (A.Gray) H.St.John & D.White 169
    var. *percalva* S.F.Blake 169
  sphaerocephala Jones ex S.F.Blake 169
  urticifolia (Kunth) Benth. 170
Galinsogea Kunth 163
Galinsogea Willd. 166
Galinsoja Roth 166
Gallinsoga J.St.Hil. 166
Galophthalmum Nees & Mart. 91
  brasiliense Nees & Mart. 91
Gattenhoffia Neck. 19
Gazania rigens var. leucolaena (DC.) Roessler 23
Geigeria Griess. 4
Georgia bipinnata (Cav.) Spreng. 34
Gnaphalieae (Cass.) Lecoq & Juillet 1
Grangea Adans. 23
Grangeinae Benth. 23
GUIZOTIA Cass. 24, **178**
  abyssinica (L.f.) Cass. **179**, 183
    var. *angustior* (DC.) Oliv. & Hiern 180
    var. *sativa* (Roxb. ex Sims) Oliv. & Hiern 180
  bidentoides Oliv. & Hiern 38
  collina S.Moore 182
  eylesii S.Moore 182
  kassneri De Wild. 182
  nyikensis Baker 182
  oblonga (Hutch.) Hutch. & Bullock 182
  oleifera (DC.) DC. 180
    var. *sativa* (Roxb. ex Sims) DC. 180
  ringoetii De Wild. 182
  scabra (Vis.) Chiov. 179, **180**, **181**, 183
    var. *sotikensis* (S.Moore) Robyns 182
  schultzii Hochst. ex A.Rich. 180
    var. *sotikensis* S.Moore 182
Hamulium Cass. 127
Helenieae Lindl. 1, 23
HELIANTHEAE Cass. 1, 2, 3, **22**, 23
  supersubtribe *Eupatoriodinae* (Cass.) C.Jeffrey 187
HELIANTHUS L. 24, **160**
  angustifolius L. 23, 160
  annuus L. 160, 161
    var. *debilis* (Nutt.) Anashch. 161
  argophyllus Torr. & A.Gray 160, **161**
  cucumerifolius Torr. & A.Gray 161
  debilis Nutt. **161**
    f. *cucumerifolius* (Torr. & A.Gray) Voss 161
    subsp. cucumerifolius (Torr. & A.Gray) Heiser 160, **161**, **162**

var. *cucumerifolius* (Torr. & A.Gray)
        A.Gray 161
    oleifer Wall. 179
    speciosus Hook. 157
Heliopsis Pers. 145
    buphthalmoides (Jacq.) Dunal 145
    platyglossa Cass. 179
Hemiambrosia Delpino 133
Hemixanthidium Delpino 133
Hingstonia Raf. 127
Hingtsha Roxb. 64
    repens Roxb. 66
Hinterhubera Sch.Bip. 25
    kotschyi Sch.Bip. 26
Hopkirkia DC. 81
    anthemoidea DC. 82
HYPERICOPHYLLUM Steetz 24, **75**, 76
    angolense (O.Hoffm.) N.E.Br. **76**, **77**
    brevipapposum Gilli 79
    compositarum Steetz 76, **78**
    elatum (O.Hoffm.) N.E.Br. 76, **80**, 81
    nyassicum Gilli 80
    scabridum N.E.Br. 78
    speciosum (Lawalrée) Gilli 80
    speciosum (Lawalrée) Lawalrée 80
    tessmannii (Mattf.) Pope 76, **81**
Hysterophorus Adans. 130
Inuleae Cass. 1
Inuleae sensu lato 4
Jaegeria Kunth
    abyssinica (L.f.) Spreng. 179
    urticifolia (Kunth) Spreng. 170
Jaumea Pers. 75, 76
    angolensis O.Hoffm. 78
    compositarum (Steetz) Oliv. & Hiern 79
    elata O.Hoffm. 80
    helenae Buscal. & Muschl. 80
    johnstoni Baker 79
    scabrida (N.E.Br.) Lawalrée 78
    speciosa Lawalrée 80
    tessmannii Mattf. 81
Kanimia Gardner 191
Kerneria Moench
    bipinnata (L.) Gren. & Godr. 61
    pilosa (L.) Lowe 63
        var. α *radiata* (Sch.Bip.) Lowe 63
        var. β *discoidea* (Sch.Bip.) Lowe 63
Kleinia Mill.
    compositarum (Steetz) Kuntze 79
    oliveri Kuntze 79
Kyrstenia Neck. ex Greene 188
Lactuceae Cass. 2
Lavenia Sw. 196
Leachia Cass. 28
    lanceolata (L.) Cass. 29
Leighia speciosa (Hook.) DC. 157
Lejica Hill ex DC. 148

Lepisiphon dentatus Turcz. 16
Leptosyne DC. 28
Limnogenneton Sch.Bip. ex Walp. 176
    abyssinicum Sch.Bip. ex Walp. 176
Lipochaeta DC. 113
    ovata R.C.Gardner 113
LIPOTRICHE R.Br. 25, **102**, 113
    abyssinica (Sch.Bip. ex Walp.) Orchard **103**
    albinervia 111
    brownii DC. 107, 109
    marlothiana (Klatt) D.J.N.Hind 102, **104**, **105**
    pungens (Oliv. & Hiern) Orchard 103, **109**, 112
        subsp. caudata (Wild) D.J.N.Hind 110, **111**,
        112
        subsp. pungens **110**
    richardsiae (Wild) D.J.N.Hind 102, **104**
    robinsonii (Wild) D.J.N.Hind 103, **112**
    scandens (Schumach. & Thonn.) Orchard
        103, **107**
        subsp. dregei (DC.) Orchard 108, **109**
        subsp. madagascariensis (Baker) D.J.N.
        Hind **108**, 109
        subsp. *subsimplicifolia* (Wild) Orchard 108,
        109
    triternata (Klatt) Orchard 106
Mallinoa J.M.Coult. 188
Mandonia Wedd. 163
Melampodium australe Loefl. 175
Melanthera Rohr 102, 113
    abyssinica (Sch.Bip. ex Walp.) Vatke 103
        var. *angustifolia* Chiov. 103
    acuminata S.Moore 110
    albinervia O.Hoffm. 110, 111
        subsp. *acuminata* (S.Moore) Wild 111
        subsp. *caudata* Wild 111, 112
    angustifolia Gilli 108
    baumii O.Hoffm. 106
    biflora (L.) Wild 113, 115
        var. *ryujyuensis* (H.Koyama) K.Ohashi &
        H.Ohashi 113
    brownii (DC.) Sch.Bip. 107
    cuanzensis Hiern 108
        var. *altior* Hiern 108
    djalonensis A.Chev. 103
    gossweileri S.Moore 110
    letestui Philipson 111
    madagascariensis Baker 108
    marlothiana O.Hoffm. 104
    monochaeta Hiern 106
    pungens Oliv. & Hiern 110, 111, 112
        var. *albinervia* (O.Hoffm.) Beentje 111
    richardsiae Wild 104
    rivicola Muschl. 106
    robinsonii Wild 112
    scaberrima Hiern
        var. *angustifolia* S.Moore 108
    scandens (Schumach. & Thonn.) Brenan 107

subsp. *madagascariensis* (Baker) Wild 108
subsp. *subsimplicifolia* Wild 108
scandens (Schumach. & Thonn.) Roberty 107
subsp. *dregei* (DC.) Wild 109
schinziana S.Moore 106
seineri Muschl. 106
sokodensis Muschl. ex Hutch. & Dalziel 103
swynnertonii S.Moore 108
triternata (Klatt) Wild 106
ugandensis S.Moore 110
varians Hiern 106
Mendezia DC. 148
Menotriche Steetze 115
strigosa Steetz 125
Meteorina Cass. 19
Meyera Schreb. 64
fluctuans (Lour.) Spreng. 66
guineensis Spreng. 66
MICRACTIS DC. 24, **176**
abyssinica (Sch.Bip. ex Walp.) Chiov. 178
bojeri DC. **176**, **177**, 183
Micrelium Forssk. 95
asteroides Forssk. 97
tolak Forssk. 97
Mieria La Llave 81
MIKANIA Willd. 188, **191**
angustifolia (O.Hoffm.) R.E.Fr. 194
capensis DC. 194, 196
chenopodiifolia Willd. 192, **194**, 196
chevalieri (C.D.Adams) W.C.Holmes &
McDaniel **192**, **193**
cordata (Burm.f.) B.L.Rob.
var. *chevalieri* C.D.Adams 192
cordata sensu Andrews 195
cordata sensu Lind & Tallantire 194
dioscoreifolia DC. 195
var. β? *bojeri* DC. 195
var. γ? *crenata* DC. 195
floribunda Bojer ex DC. 195
laxa A.Chev. 192
oxyota DC. 195
sagittifera B.L.Rob. 192, **194**
scandens f. *angustifolia* O.Hoffm. 194
scandens sensu Oliv. & Hiern 194, 195
thunbergioides Bojer ex DC. 195
Milleria Houst. ex L.
chiloensis Juss. 67
contrayerba Cav. 67
Minyranthes Turcz. 183
heterophylla Turcz. 184
Mirasolia (Sch.Bip.) Benth. & Hook.f. 156
diversifolia Hemsl. 158
Montagnaea DC. 85
MONTANOA Cerv. 24, 85
bipinnatifida (Kunth) K.Koch 85, 87, **88**
elegans K.Koch 88
heracleifolia Brongn. ex Groenl. 88

hibiscifolia Benth. **85**, **86**
hibiscifolia (Benth.) D'Arcy 87
hibiscifolia (Benth.) Sch.Bip. ex K.Koch 87
pittieri B.L.Rob. & Greenm. 87
pyramidata Sch.Bip. 88
samalensis J.M.Coult. 87
wercklei Berger 87
Morrenia Kunze 191
Mutisieae Cass. 2, 3, 4
Nauenburgia Willd. 67
trinervata Willd. 70
Neojeffreya Cabrera 4
Niebuhria biflora (L.) Britten 113
Nothoschkuhria 82
Ochronelis Raf. 127
Oedera trinervia Spreng. 70
Oligocarpus Less. 11
Oligogyne burchellii Hook.f. 92
Orcya Vell. 172
adhaerescens Vell. 175
OSTEOSPERMUM L. 4, 5, **11**
sect. *Acanthotheca* (DC.) Norl. 19
sect. *Blaxium* (Cass.) Norl. 19
sect. Xenismia 11
barberae (Harv.) Norl. 19
afromontanum Norl. 8, 9, 11
amplexicaule Steud. 12
corymbosum sensu DC. 12
var. β *rotundifolium* DC. 12
var. γ *lasiocaulon* DC. 12
var. γ *parvifolium* Harv. 12
dichotomum E.Mey ex DC. 12
ecklonis (DC.) Norl. 19
fruticosum (L.) Norl. 19
glaberrimum O.Hoffm. 13
glandulosum A.Spreng. 12
hamiltonii S.Moore 15
helichrysoides DC. 12
imbricatum L. **12**
subsp. nervatum (DC.) Norl. 11, **12**
var. *helichrysoides* (DC.) Norl. 13
lanatum Spreng. 12
macrocarpum DC. 18
moniliferum L. 16, 18
subsp. *septentrionalis* (Norl.) J.C.Manning &
Goldblatt 16
var. *rotundatum* (DC.) Harv. 18
monocephalum (Oliv. & Hiern) Norl. 6
muricatum E.Mey. ex DC. 11, **13**, **14**
subsp. *longiradiatum* Norl. 15
var. *asperum* Harv. 13
var. *glabratum* Harv. 13
nervatum DC. 12
nyikense Norl. 10
picridioides DC. 9
retirugum DC. 12
rotundatum DC. 18

scariosum DC. 9
    subsp. *integrifolium* (Harv.) Norl. 10
    subsp. *setiferum* (DC.) Norl. 10
    vaillantii (Decne.) Norl. 8, 9
Paleista? brachypoda (Michx.) Raf. 99
Paleista Raf. 95
    procumbens [(Michx.)] Raf. 99
PARTHENIUM L. 23, **130**
    argentatum A.Gray 131
    hysterophorus L. **131, 132**
    lobatum Buckley 131
    luteum Spreng. 179
    pinnatifidum Stokes 131
Pectis pinnata Lam. 82
Phaethusa Gaertn. 127
Phyllimena Blume ex DC. 64
Plagiocheilus erectus Rusby 26
Platycarpheae V.A. Funk & H.Rob. 1, 3
Platycarphella V.A.Funk & H.Rob. 1
Platypteris Kunth 127
Plucheeae (Benth.) A.Anderb. 1
Polygyne Phil. 95
    inconspicua Phil. 100
Polymnia Kalm
    abyssinica L.f. 179
    frondosa Bruce 179
    grandis hort. ex Kunth 88
Priestleya Sessé & Moc. ex DC. 85
Psathurochaeta DC. 102
    dregei DC. 109
Ptilostephium Kunth 163
Pugiopappus A.Gray 28
Quadribractea moluccana (Blume) Orchard
    92
Radlkoferotoma Kuntze 191
Ramtilla DC. 179
    oleifera DC. 180
      var. *angustior* DC. 180
Ridan Adans. 127
Rosenia Thunb. 4
Rothia Schreb.
    pinnata (Lam.) Kuntze 84
      α *pallida* Kuntze 84
      β *purpuascens* Kuntze 84
Rudbeckia L.
    hirta L. 23
    laciniata L. 23
Sabazia Cass.
    microglossa DC. 167
      var. β *puberula* DC. 167
    urticifolia (Kunth) DC. 170
      var. *venezuelensis* Steyerm. 171
Sanvitalia longepedunculata M.E.Jones 148
Sanvitaliopsis Sch.Bip. ex Greenm. 148
Saubinetia J.Rémy 127
Schismus P.Beauv. 28
Schkuhria Moench 183

SCHKUHRIA Roth 24, **81**
    abrotanoides Roth 82
      var. *isopappa* (Benth.) Hieron. 84
      var. *pomasquiensis* Hieron. 84
    advena Thell. 84
    bonariensis Hook. & Arn. 82
    coquimbana Phil. 84
    isopappa Benth. 84
    pinnata (Lam.) Kuntze 84
    pinnata (Lam.) Kuntze ex Thell. **82, 83**
      var. *abrotanoides* (Roth) Cabrera 84
SCLEROCARPUS J.Jacq. ex Murr. 24, **154**
    africanus Jacq. **154, 155**
Sebastiania Bertol. 25
Selleophytum Urb. 29
Senecioneae Cass. 2
SIGESBECKIA L. 24, **183**
    abyssinica (Sch.Bip. ex Walp.) Oliv. & Hiern
      178
    abyssinica (Sch.Bip.) Oliv. & Hiern 183
    brachiata Roxb. 184
    caspia Fisch. & C.A.Mey. ex Hohen. 184
    emirnensis Baker 178
    esquirolii H.Lév. & Vaniot 186
    filarszkyi Pit. 184
    formosana Kitam. 184
    glabrescens (Makino) Makino 184
    gracilis DC. 184
    humilis Koidz. 184
    iberica Willd. 184
    microcephala DC. 184
    orientalis L. **183, 185**
      f. *angustifolia* Makino 184
      f. *glabrescens* Makino 184
      f. *pubescens* Makino 184
      var. *angustifolia* Makino ex Makino 184
      var. *glabrescens* Makino ex Makino 184
      var. *pubescens* Makino ex Makino 184
      var. *tenggerensis* Hochr. 184
      subsp. *glabrescens* (Makino) H.Koyama 186
      subsp. *glabrescens* (Makino) Kitam. 184
      subsp. *pubescens* (Makino) H.Koyama 186
      subsp. *pubescens* (Makino) Kitam. 186
    pubescens (Makino) Makino 184
    triangularis Cav. 184
Sobreya Pers. 64
Sobreyra Ruiz & Pav. 64
Sogalgina Cass. 163
Solenotheca Nutt. 71
Sparganophorus obtusifolius Lag. 202
Sphagneticola trilobata (L.) Pruski 23
Spilanthes Jacq.
    abyssinica Sch.Bip. ex A.Rich. 146
    africana DC. 146
    botterii S.Watson 147
    caulirhiza (Delile) DC. 146
      β *madagascariensis* DC. 146

exasperata Jacq. 147
   [var.] β *cayennensis* DC. 147
filicaulis (Schumach. & Thonn.) C.D.Adams
   146
  leucophaea Sch.Bip. ex Klatt 147
  mauritiana (Rich. ex Pers.) DC. 94
   f. *madagascariensis* (DC.) A.H.Moore 146
  ocymifolia
   var. *acutiserrata* A.H.Moore 147
  pseudoacmella (L.) L. 97
  radicans Jacq. 147
Stemmatella Wedd. ex Benth. 166
Stemmatella Wedd. ex Sch.Bip. 166
  lehmannii Hieron. 170
  sodiroi Hieron. 170
  urticifolia (Kunth) O.Hoffm. ex Hieron. 171
   var. *eglandulosa* Hieron. 171
Stemmodontia biflora (L.) W.Wight 113
Stenocarpha S.F.Blake 166
STOMATANTHES R.M.King & H.Rob. 188,
  207
  subgen. *Verticifolium* R.M.King & H.Rob. 207
  africanus (Oliv. & Hiern) R.M.King & H.Rob.
   208, 209
  helenae (Buscal. & Muschl.) Lisowksi 208, 209
  meyeri R.M.King & H.Rob. 208
  zambiensis R.M.King & H.Rob. 208, **209**,
   211
SYNEDRELLA Gaertn. 25, **89**
  nodiflora (L.) Gaertn. **89**, **90**
TAGETES L. 23, **71**
  arvensis Rojas Acosta 72
  bonariensis Pers. 72
  corymbosa Sw. 74
  diversifolia sensu Martineau 157
  erecta L. 23, 75
  glandulifera Schrank 72
  glandulosa Schrank ex Link 72
  macroglossa Pol. 74
  minuta L. **71**, **73**
  montana hort. ex DC. 72
  patula L. 71, **74**, 75
  porophyllum Vell. 72
  remotiflora Kunze 74
  riojana M.Ferraro 72
  rotundifolia Mill. 156
  tenuifolia Kunth 74
  tinctoria Hornsch. 72
Tatraotis paludosa Reinw. 66
Tetracarpum Moench 81
Tetragonotheca abyssinica (L.f.) Ledeb. 179
Tetraotis Reinw. 64
  longifolia Reinw. 66
Tilesia baccata (L.) Pruski 107
TITHONIA Desf. ex Juss. 24, **156**
  subgen. *Mirasolia* Sch.Bip. **156**
  aristata Oersted 157

diversifolia (Hemsl.) A.Gray 156, **158**, **159**
  heterophylla Griseb. 157
  macrophylla S.Watson 157
  rotundifolia (Mill.) S.F.Blake **156**
  speciosa (Hook.) Griseb. 157
  speciosa (Hook.) Klatt 157
  tagetiflora Desf. 157
  tagetiflora Lam. 157
  uniflora Desf. ex J.F.Gmel. 156
  vilmoriniana Pamp. 157
Tragoceros Kunth 148
TRIDAX L. 25, **163**
  procumbens L. 163, **164**, **165**
   var. *canescens* (Pers.) DC. 164
   var. *ovatifolia* B.L.Rob. & Greenm. 164
Trifenestrata Norl. 5
Trigonotheca natalensis Sch.Bip. 109
Tripterachaenium Kuntze. 5
  humile (Turcz.) Kuntze 10
  scariosum (DC.) Kuntze 10
TRIPTERIS Less. **5**, 11
  afromontana (Norl.) B.Nord. 8
  aghillana DC. 5, **9**
   var. *integrifolia* Harv. 9
  angustissima S.Moore 8
  cheiranthifolia Sch.Bip. 8
  cuneifolia Schweinf. & Asch. 8
  flexuosa Harv. 9
  glandulosa var. dentata Harv. 9
  goetzei O.Hoffm. 6
  gossweileri Mattf. 6
  gweloensis Mattf. 6
  humilis 9
  lordii Oliv. & Hiern 8
   var. *racemosa* Balf.f. 8, 9
  monocephala Oliv. & Hiern 5, **6**, **7**
  natalensis Harv. 9
  nyikensis 5, 9, **10**, 11
  racemosa Wagner sensu Vierh. 8
  rhodesica R.E.Fr. 6
  rueppellii Schweinf. & Asch. 8
  setifera DC. 9
  vaillantii Decne. 5, **8**, 9
   var. racemosa 8
Tuckermannia Nutt. 28
Ucacou Adans. 89
  nodiflorum (L.) Hitchc. 89
Uhdea bipinnatifida Kunth 88
Uhdea Kunth 85
Urbanisol Kuntze 156
  heterophyllus (Griseb.) Kuntze 157
  aristatus (Oerst.) Kuntze 157
  tagetifolius (Desf.) Kuntze 157
   [unranked] α *normalis* Kuntze 157
   [unranked] β *speciosus* (Hook.) Kuntze 157
   var. *diversifolius* Kuntze
    f. *grandiflorus* Kuntze 157

Vargasia DC. 166
  caracasana DC. 170, 171
VERBESINA L. 25, **127**
  alba L. 97
  argentea Gaudich. 113
  australis Baker 130
  biflora L. 113
  canescens Gaudich. 113
  ciliata Schumach. 120
  conyzoides Trew 97
  dichotoma Murray 92
  encelioides (Cav.) A.Gray **128**, **129**
    var. *cana* (DC.) B.L.Rob. & Greenm. 130
  lanceolata Poir. 92
  microptera (DC.) Herter 130
  nodiflora L. 89
  prostrata L. 97
  pseudoacmella L. 97
  pusilla Poir. 97
  sativa Roxb. ex Sims 179, 180
  scabra Phil. 128
  strigulosa Gaudich. 113
Vermifuga Ruiz & Pav. 67
  corymbosa Ruiz & Pav. 67
Vernonia Schreb.
  humilis C.H.Wright 208
  malosana Baker 208
Vernonieae Cass. 2
Veslingia Vis. 179
  scabra Vis. 180
Viborgia Moench 166
Vigolina Poir. 166
  acmella (Roth) Poir. 167
Villanova Ortega 130
Wahlenbergia Schumach. 64
Wedelia 102, 111, 116
  abyssinica sensu Eyles 126
  affinis De Wild. 123
  africana sensu Eyles 122
  albiflora Hiern 125
  angolensis Klatt 124
  argentea (Gaudich.) Merr. 113
  aristata Less. 113
  biflora (L.) DC. 113
    var. *canescens* (Gaudich.) Fosberg 113
    var. *ryukyuensis* H.Koyama 113
    var. *scabriuscula* (DC. ex Decne.) Hochr. 113
  canescens (Gaudich.) Merr. 113
  chamissonis Less. 113
  ciliata (Schumach.) Soldano 121
  cryptocephala Peter 89
  diversipapposa S.Moore 126
  gossweileri S.Moore 92
  helianthoides (Schumach. & Thonn.) Isawumi
    subsp. *ciliata* (Schumach.) Isawumi 120
    subsp. *papposa* (O.Hoffm. & Muschl.)
      Isawumi 121

    subsp. *prieureana* (DC.) Isawumi 121
  huillensis Hiern 124
  instar S.Moore 126
  katangensis De Wild. 123
  kotschyi (Sch.Bip.) Isawumi 118
    var. *alba* (Berhault) Isawumi 119
  kotschyi (Sch.Bip.) Soldano 118
  menotriche Oliv. & Hiern 125
  mossambicensis Oliv. 125
  natalensis Sond. 123
  oblonga Hutch. 182
  psammophila Poepp. 99
  ringoetii De Wild. 120
  scabriuscula (DC. ex Decne.) Engl. ex K.Schum. 113
  strigulosa (Gaudich.) K.Schum. 113
  triseta Peter 92
  triternata Klatt 104
  wallichii Less.
    var. *megalantha* H.Chuang 113
Welwitschiella O.Hoffm. 22
Wiborgia oblongifolia Hook. 98
Wiborgia Roth 166
  acmella Roth 167
  brachystephana (Otto) Heynh. 170
  parviflora (Cav.) Kunth 167
  urticifolia Kunth 170
Wiborgia Thunb. 166
Willoughbya Neck. 191
Wirtgenia Sch.Bip. 115
  abyssinica Sch.Bip. ex Walp. 120
  kotschyi (Sch.Bip.) A.Rich. 118
  schimperi Sch.Bip. ex A.Rich. 120
WOLLASTONIA DC. ex Decne. 24, 102, **112**, 115
  biflora (L.) DC. **113**, **114**
  dentata (H.Lév. & Vaniot) Orchard 98
  scabriuscula DC. ex Decne. 113
  strigulosa (Gaudich.) DC. ex Decne 113
  zanzibarensis DC. 113
Wuerschmittia Sch.Bip. 102
  abyssinica Sch.Bip. 103
Wuerschmittia Sch.Bip. ex Walp. 102
  abyssinica Sch.Bip. ex Walp. 103
Xanthidium Delpino 133
XANTHIUM L. 23, **135**
  sect. *Acanthoxanthium* DC. 135
  sect. II. *Akanthoplium* Wallr. 135
  abyssinicum Wallr. 138
  acerosum Greene 140
  acutilobum Millsp. & Sherff 141
  acutum Greene 140
  affine Greene 139
  albinum (Widder) Scholz & Sukopp 142
    subsp. *ripicolum* (Holub) Dostál 143
  ambrosioides Hook. 144
  americanum Walter 136, 138

antiquorum Wallr. 138, 139
arcuatum Millsp. & Sherff 141
arenarium Lasch 139
aridum H.St.John 142
armatum Humb. ex Wallr. 144
australe Millsp. & Sherff 141
barcinonense Sennen 141
brasilicum Vell. 138
brevirostre Wallr. 138
bubalocarpon Bush 140
californicum Greene 140
    var. *oligacanthum* (Piper) Widder 142
calvum Millsp. & Sherff 141
campestre Greene 140
canadense Hook. 139
canadense Mill. 136, 138, 140
canescens (Costa) Widder 144
catharticum Kunth 144
cavanillesii Schouw 139
cenchroides Millsp. & Sherff. 141
chasei Fernald 142
chinense Mill. 136
    var. *globuliforme* C.Shull 142
cloessplateanum D.Z.Ma 145
commune Britton 140
    forma *wootonii* Cockerell 140
cordifolium Stokes 136
crassifolium Millsp. & Sherff 141
cuneatum Moench 136
curvescens Millsp. & Sherff 141
cylindricum Millsp. & Sherff 141
decalvatum Widder 140, 142
dioscoridis Gundelsh. 139
discolor Wallr. 138
echinatum
    subsp. *italicum* (Moretti) O.Bolòs & Vigo 143
    var. *cavanillesii* (Schouw) O.Bolòs & Vigo 143
    var. *italicum* (Moretti) O.Bolòs & Vigo 143
eriocarpon Wallr. 144
fuscescens Jord. & Fourr. 139
glabratum (DC.) Britton 140
glanduliferum Greene 140
globosum C.Shull 141
homothalamum Spreng. 136
inaequilaterum DC. 138
indicum DC. 138
inflexum Mack. & Bush 140
italicum Moretti 138
    var. *albinum* Widder 142
japonicum Widder 141
laevigatum Muhl. ex Wallr. 138
leptocarpum Millsp. & Sherff 141
longirostre Wallr. 139
macounii Britton 140
macrocarpum DC. 136, 139
    var. *laciniatum* Pouzolz 139
    [var.] β *glabratum* DC. 138

maculatum Raf. 136
medium Nossotovsky 144
mongolicum Kitag. 142
monoicum Gilib. 136
multifidum Larrañaga 144
natalense Widder 142
nigri Ces. 139
occidentale Bertol. 138
occidentale Poepp. 139
oligacanthum Piper 140
orientale L. 136
    f. *laciniatum* (Pouzolz) Thell. 142
    var. *albinum* (Widder) Adema & M.T.Jansen 142
    var. *riparium* (Itzigs. & Hertsch) Adema & M.T.Jansen 142
oviforme Wallr. 139
palustre Greene 140
parvifolium DC. 144
pensylvanicum Gand. 141
pensylvanicum Wallr. 139
    var. *laciniatum* C.Shull & Sherff 142
priscorum Wallr. 138
pungens Wallr. 138, 142
    var. *cylindricum* (Millsp. & Sherff) Widder 142
    var. *denudatum* Widder 142
    var. *globosum* (C.Shull) Widder 142
riparium Itzigs. & Hertsch 139, 142
riparium Lasch 139, 142
ripicolum Holub 142
roxburghii Wallr. 138
saccharatum Wallr. 139
    subsp. *aciculare* Widder 142
    subsp. *commune* (Britton) Widder 142
sibiricum Patrin 141
    var. *jingyuanense* H.G.Hou & Y.T.Lu 143
    var. *subinerme* (C.Winkl.) Widder 141
silphiifolium Greene 139
speciosum Kearney 139
sphaerocephalum Salzm. 139
spinosum Beyrich 144
spinosum L. 136, **137, 143**
    f. *laciniatum* Scheuerm. & Thell. 144
    f. *praecocius* Bitter ex Widder 144
    var. *canescens* Costa 144
    var. *inerme* Bel 144
    var. *laciniatum* (Scheuerm. & Thell. ex Widder) Widder 144
    var. *pseudinerme* Widder ex Parodi 144
    var. *synacanthum* Widder 144
    [var.] β *brachyacanthum* DC. 144
strumarium Beyrich 139
strumarium L. **136**, 145
    f. *purpurascens* Priszter 143
    subsp. *brasilicum* (Vell.) O.Bolòs & Vigo 143
    subsp. *cavanillesii* (Schouw) D.Löve & Dans. 142

subsp. *sibiricum* (Patrin ex Widder) Greuter 143
  var. *arenarium* (Lasch) R.Uechtr. 139
  var. *canadense* (Mill.) Torr. & A.Gray 138
  var. *glabratum* (DC.) Cronquist 142
  var. *hausmanni* Widder 141
  var. *japonicum* (Widder) H.Hara 142
  var. *wootonii* (Cockerell) W.C.Martin & C.R.Hutchins 142
strumarium Schimp. 138
varians Greene 139
wootoni Cockerell 140
xanthocarpon Wallr. 144
Ximenesia Cav. 11, 127
  australis Hook. & Arn. ex DC. 128
  encelioides Cav. 128
    var. α *hortensis* DC. 128
    var. β *pachyptera* DC. 128
    var. γ *oblongifolia* DC. 128
    var. δ? *cana* DC. 128
  microptera DC. 128
Zandera D.L.Schulz 183

ZINNIA L. 24, **148**
  sect. *Diplothrix* (DC.) A.Gray 148
  subgen. *Diplothrix* (DC.) Torres 148
  australis F.M.Bailey 153
  elegans Jacq. 23, 153
    [unranked] α *violacea* (Cav.) DC. 153
    [unranked] β *alba* DC. 153
    [unranked] γ *purpurascens* DC. 153
    [unranked] δ *coccinea* (Lindl.) DC. 153
  hybrida Ruiz & Pav. ex Sims 150
  intermedia Engelm. 152
  leptopoda DC. 152
  mendocina Phil. 152
  multiflora L. 150
  pauciflora L. 148, 150
  peruviana L. 148, **150, 151**
  revoluta Cav. 150
  tenuiflora Jacq. 150
  verticillata Andrews 150
  violacea Cav. 150, **153**
    var. *coccinea* Lindl. 153

FAMILIES OF VASCULAR PLANTS REPRESENTED IN THE FLORA ZAMBESIACA AREA

PTERIDOPHYTA

(Flora Zambesiaca families and family number. Published 1970)

Actiniopteridaceae		Gleicheniaceae	9	Parkeriaceae	
see Adiantaceae	18	Grammitidaceae	20	see Adiantaceae	18
Adiantaceae	18	Hymenophyllaceae	15	Polypodiaceae	21
Aspidiaceae	27	Isoetaceae	4	Psilotaceae	1
Aspleniaceae	23	Lindsaeaceae	19	Pteridaceae	
Athyriaceae	25	Lomariopsidaceae	26	see Adiantaceae	18
Azollaceae	13	Lycopodiaceae	2	Salviniaceae	12
Blechnaceae	28	Marattiaceae	7	Schizaeaceae	10
Cyatheaceae	14	Marsileaceae	11	Selaginellaceae	3
Davalliaceae	22	Oleandraceae		Thelypteridaceae	24
Dennstaedtiaceae	16	see Davalliaceae	22	Vittariaceae	17
Dryopteridaceae		Ophioglossaceae	6	Woodsiaceae	
see Aspidiaceae	27	Osmundaceae	8	see Athyriaceae	25
Equisetaceae	5				

GYMNOSPERMAE

(Flora Zambesiaca families and family number. Volume 1(1) 1960)

Cupressaceae	3	Cycadaceae	1	Podocarpaceae	2

ANGIOSPERMAE

(Flora Zambesiaca families, volume and part number and year of publication)

Acanthaceae			Aristolochiaceae	9(2)	1997
tribes 1–5	8(5)	2013	Asclepiadaceae		
tribes 6–7	8(6)	2015	see Apocynaceae part 2	7(3)	2020
Agapanthaceae	13(1)	2008	Asparagaceae	13(1)	2008
Agavaceae	13(1)	2008	Asphodelaceae	12(3)	2001
Aizoaceae	4	1978	Avicenniaceae	8(7)	2005
Alangiaceae	4	1978	Balanitaceae	2(1)	1963
Alismataceae	12(2)	2009	Balanophoraceae	9(3)	2006
Alliaceae	13(1)	2008	Balsaminaceae	2(1)	1963
Aloaceae	12(3)	2001	Barringtoniaceae	4	1978
Amaranthaceae	9(1)	1988	Basellaceae	9(1)	1988
Amaryllidaceae	13(1)	2008	Begoniaceae	4	1978
Anacardiaceae	2(2)	1966	Behniaceae	13(1)	2008
Anisophylleaceae			Berberidaceae	1(1)	1960
see Rhizophoraceae	4	1978	Bignoniaceae	8(3)	1988
Annonaceae	1(1)	1960	Bixaceae	1(1)	1960
Anthericaceae	13(1)	2008	Bombacaceae	1(2)	1961
Apocynaceae	7(2)	1985	Boraginaceae	7(4)	1990
subfam. Apocynoideae	7(2)	1985	Brexiaceae	4	1978
subfam. Asclepiadoideae	7(3)	2020	Bromeliaceae	13(2)	2010
subfam. Periplocoideae	7(3)	2020	Buddlejaceae		
subfam. Rauvolfioideae	7(2)	1985	see Loganiaceae	7(1)	1983
subfam. Secamonoideae	7(3)	2020	Burmanniaceae	12(2)	2009
Aponogetonaceae	12(2)	2009	Burseraceae	2(1)	1963
Aquifoliaceae	2(2)	1966	Buxaceae	9(3)	2006
Araceae	12(1)	2011	Cabombaceae	1(1)	1960
Araliaceae	4	1978	Cactaceae	4	1978

Caesalpinioideae
  see Leguminosae   3(2)   2006
Campanulaceae   7(1)   1983
Canellaceae   7(4)   1990
Cannabaceae   9(6)   1991
Cannaceae   13(4)   2010
Capparaceae   1(1)   1960
Caricaceae   4   1978
Caryophyllaceae   1(2)   1961
Casuarinaceae   9(6)   1991
Cecropiaceae   9(6)   1991
Celastraceae   2(2)   1966
Ceratophyllaceae   9(6)   1991
Chenopodiaceae   9(1)   1988
Chrysobalanaceae   4   1978
Colchicaceae   12(2)   2009
Combretaceae   4   1978
Commelinaceae   –   –
Compositae
  tribes 1–5   6(1)   1992
  tribes 12–14   6(5)   2025
Connaraceae   2(2)   1966
Convolvulaceae   8(1)   1987
Cornaceae   4   1978
Costaceae   13(4)   2010
Crassulaceae   7(1)   1983
Cruciferae   1(1)   1960
Cucurbitaceae   4   1978
Cuscutaceae   8(1)   1987
Cymodoceaceae   12(2)   2009
Cyperaceae   14   2020
Dichapetalaceae   2(1)   1963
Dilleniaceae   1(1)   1960
Dioscoreaceae   12(2)   2009
Dipsacaceae   7(1)   1983
Dipterocarpaceae   1(2)   1961
Dracaenaceae   13(2)   2010
Droseraceae   4   1978
Ebenaceae   7(1)   1983
Elatinaceae   1(2)   1961
Ericaceae   7(1)   1983
Eriocaulaceae   13(4)   2010
Eriospermaceae   13(2)   2010
Erythroxylaceae   2(1)   1963
Escalloniaceae   7(1)   1983
Euphorbiaceae   9(4)   1996
Euphorbiaceae   9(5)   2001
Flacourtiaceae   1(1)   1960
Flagellariaceae   13(4)   2010
Fumariaceae   1(1)   1960
Gentianaceae   7(4)   1990
Geraniaceae   2(1)   1963
Gesneriaceae   8(3)   1988
Gisekiaceae
  see Molluginaceae   4   1978
Goodeniaceae   7(1)   1983
Gramineae
  tribes 1–18   10(1)   1971
  tribes 19–22   10(2)   1999
  tribes 24–26   10(3)   1989
  tribe 27   10(4)   2002
Guttiferae   1(2)   1961
Haloragaceae   4   1978
Hamamelidaceae   4   1978
Hemerocallidaceae   12(3)   2001
Hernandiaceae   9(2)   1997
Heteropyxidaceae   4   1978
Hyacinthaceae   13(3)   2023
Hydnoraceae   9(2)   1997
Hydrocharitaceae   12(2)   2009
Hydrophyllaceae   7(4)   1990
Hydrostachyaceae   9(2)   1997
Hypericaceae
  see Guttiferae   1(2)   1961
Hypoxidaceae   12(3)   2001
Icacinaceae   2(1)   1963
Illecebraceae   1(2)   1961
Iridaceae   12(4)   1993
Irvingiaceae   2(1)   1963
Ixonanthaceae   2(1)   1963
Juncaceae   13(4)   2010
Juncaginaceae   12(2)   2009
Labiatae
  see Lamiaceae, Verbenacaeae
Lamiaceae
  Viticoideae, Pingoideae   8(7)   2005
Lamiaceae
  Scutellaroideae-
    Nepetoideae   8(8)   2013
Lauraceae   9(2)   1997
Lecythidaceae
  see Barringtoniaceae   4   1978
Leeaceae   2(2)   1966
Leguminosae,
  Caesalpinioideae   3(2)   2007
  Mimosoideae   3(1)   1970
  Papilionoideae   3(3)   2007
  Papilionoideae   3(4)   2012
  Papilionoideae   3(5)   2001
  Papilionoideae   3(6)   2000
  Papilionoideae   3(7)   2002
Lemnaceae
  see Araceae   12(1)   2011
Lentibulariaceae   8(3)   1988
Liliaceae sensu stricto   12(2)   2009
Limnocharitaceae   12(2)   2009
Linaceae   2(1)   1963
Lobeliaceae   7(1)   1983
Loganiaceae   7(1)   1983
Loranthaceae   9(3)   2006
Lythraceae   4   1978
Malpighiaceae   2(1)   1963
Malvaceae   1(2)   1961
Marantaceae   13(4)   2010
Mayacaceae   13(2)   2010
Melastomataceae   4   1978

Meliaceae	2(1)	1963
Melianthaceae	2(2)	1966
Menispermaceae	1(1)	1960
Menyanthaceae	7(4)	1990
Mesembryanthemaceae	4	1978
Mimosoideae		
see Leguminosae	3(1)	1970
Molluginaceae	4	1978
Monimiaceae	9(2)	1997
Montiniaceae	4	1978
Moraceae	9(6)	1991
Musaceae	13(4)	2010
Myristicaceae	9(2)	1997
Myricaceae	9(3)	2006
Myrothamnaceae	4	1978
Myrsinaceae	7(1)	1983
Myrtaceae	4	1978
Najadaceae	12(2)	2009
Nesogenaceae	8(7)	2005
Nyctaginaceae	9(1)	1988
Nymphaeaceae	1(1)	1960
Ochnaceae	2(1)	1963
Olacaceae	2(1)	1963
Oleaceae	7(1)	1983
Oliniaceae	4	1978
Onagraceae	4	1978
Opiliaceae	2(1)	1963
Orchidaceae	11(1)	1995
Orchidaceae	11(2)	1998
Orobanchaceae		
see Scrophulariaceae	8(2)	1990
Oxalidaceae	2(1)	1963
Palmae	13(2)	2010
Pandanaceae	12(2)	2009
Papaveraceae	1(1)	1960
Papilionoideae		
see Leguminosae	–	–
Passifloraceae	4	1978
Pedaliaceae	8(3)	1988
Periplocaceae		
see Apocynaceae part 2	7(3)	2020
Philesiaceae		
see Behniaceae	13(1)	2008
Phormiaceae		
see Hemerocallidaceae	12(3)	2001
Phytolaccaceae	9(1)	1988
Piperaceae	9(2)	1997
Pittosporaceae	1(1)	1960
Plantaginaceae	9(1)	1988
Plumbaginaceae	7(1)	1983
Podostemaceae	9(2)	1997
Polygalaceae	1(1)	1960
Polygonaceae	9(3)	2006
Pontederiaceae	13(2)	2010
Portulacaceae	1(2)	1961
Potamogetonaceae	12(2)	2009
Primulaceae	7(1)	1983
Proteaceae	9(3)	2006
Ptaeroxylaceae	2(2)	1966
Rafflesiaceae	9(2)	1997
Ranunculaceae	1(1)	1960
Resedaceae	1(1)	1960
Restionaceae	13(4)	2010
Rhamnaceae	2(2)	1966
Rhizophoraceae	4	1978
Rosaceae	4	1978
Rubiaceae		
subfam. Rubioideae	5(1)	1989
tribe Vanguerieae	5(2)	1998
subfam.Cinchonoideae	5(3)	2003
Rutaceae	2(1)	1963
Salicaceae	9(6)	1991
Salvadoraceae	7(1)	1983
Santalaceae	9(3)	2006
Sapindaceae	2(2)	1966
Sapotaceae	7(1)	1983
Scrophulariaceae	8(2)	1990
Selaginaceae		
see Scrophulariaceae	8(2)	1990
Simaroubaceae	2(1)	1963
Smilacaceae	12(2)	2009
Solanaceae	8(4)	2005
Sonneratiaceae	4	1978
Sphenocleaceac	7(1)	1983
Sterculiaceae	1(2)	1961
Strelitziaceae	13(4)	2010
Taccaceae		
see Dioscoreaceae	12(2)	2009
Tecophilaeaceae	12(3)	2001
Tetragoniaceae	4	1978
Theaceae	1(2)	1961
Thymelaeaceae	9(3)	2006
Tiliaceae	2(1)	1963
Trapaceae	4	1978
Turneraceae	4	1978
Typhaceae	13(4)	2010
Ulmaceae	9(6)	1991
Umbelliferae	4	1978
Urticaceae	9(6)	1991
Vacciniaceae		
see Ericaceae	7(1)	1983
Vahliaceae	4	1978
Valerianaceae	7(1)	1983
Velloziaceae	12(2)	2009
Verbenaceae	8(7)	2005
Violaceae	1(1)	1960
Viscaceae	9(3)	2006
Vitaceae	2(2)	1966
Xyridaceae	13(4)	2010
Zannichelliaceae	12(2)	2009
Zingiberaceae	13(4)	2010
Zosteraceae	12(2)	2009
Zygophyllaceae	2(1)	1963